OPTIMALITY IN BIOLOGICAL AND ARTIFICIAL NETWORKS?

THE INTERNATIONAL
NEURAL NETWORKS SOCIETY SERIES

Harold Szu, Editor

OPTIMALITY IN BIOLOGICAL AND ARTIFICIAL NETWORKS?

Edited by

DANIEL S. LEVINE
University of Texas, Arlington

WESLEY R. ELSBERRY
Texas A&M University

Routledge
Taylor & Francis Group

LONDON AND NEW YORK

First published 1997 by Lawrence Erlbaum Associates

Published 2018 by Routledge
2 Park Square, Milton Park, Abingdon, Oxon OX14 4RN
52 Vanderbilt Avenue, New York, NY 10017

Routledge is an imprint of the Taylor & Francis Group, an informa business

Copyright © 1997 by Taylor & Francis

Library of Congress Cataloging-in-Publication Data

Levine, Daniel S.
    Optimality in biological and artificial networks? / Daniel S.
Levine, Wesley R. Elsberry.
        p.   cm.
    Includes bibliographical references and index.
    ISBN 0-8058-1561-9(alk. paper)
    1. Neural networks (computer science) 2. Artificial intelligence.
I. Elsberry, Wesley R. II. Title
QA76.87.L477   1996
006.3--DC20   96-42260
                                                    CIP

Cover design by Debra Karrel

ISBN 13: 978-0-8058-1561-0 (hbk)
ISBN 13: 978-1-138-87648-4 (pbk)

Contents

SECTION III: OPTIMALITY IN LEARNING, COGNITION, AND PERCEPTION

SECTION IV: OPTIMALITY IN DECISION, COMMUNICATION, AND CONTROL

List of Contributors

Hervé Abdi, University of Texas at Dallas, School of Human Development GR41, Box 830688, Richardson, TX 75083, or CCUB-CRIC Faculte des Sciences Mirandes, Universite de Bourgogne, 21004 Dijon Cédex, France (herve@utdallas.edu or herve@u-bourgogne.fr)

Samy Bengio, INRS —Télécommunications, 16, Place du Commerce, Verdun, Québec, Canada H3E 1H6 (bengio@inrs-telecom.uquebec.ca)

Yoshua Bengio, Université de Montréal, Dept IRO, C.P. 6128, Succ. Centre-Ville, Montréal, Québec, Canada, H3C 3J7 (bengioy@iro.umontreal.ca)

Basabi Bhaumik, Department of Electrical Engineering, Indian Institute of Technology (Delhi), Hauz Khas, New Delhi, 110016, India (bhaumik@ee.iitd.ernet.in)

Raymond T. Bradley, Institute for Whole Social Science, 35400 Telarana Way, Carmel, CA 93923

Sylvia Candelaria de Ram, 2080 Boise Drive, Las Cruces, NM 88003 (cogandcomm@delphi.com)

Gail A. Carpenter, Center for Adaptive Systems, Boston University, 677 Beacon Street, Boston, MA 02215 (gail@cns.bu.edu)

David C. Chance, HC-1 BOX 139, Eagle Rock, MO 65641 (dchance660@aol.com)

John Y. Cheung, School of Electrical Engineering, University of Oklahoma, Norman, OK 73019

Jocelyn Cloutier, AT&T Bell Laboratories, 101 Crawfords Corner Road, Holmdel, NJ 07733 (jocelyn@big.att.com)

Mark R. DeYong, Intelligent Reasoning Systems, Inc., 7801 North Lamar (Suite E216), Austin, TX 78752 (mark@irsinc.com)

Robert E. Dorsey, School of Business Administration, University of Mississippi, 218 Conner Hall, University, MS 38677

Wesley R. Elsberry, 6070 Sea Isle, Galveston, TX 77554 (welsberr@orca.tamu.edu)

Thomas Eskridge, Intelligent Reasoning Systems, Inc., 7801 North Larar (Suite E216), Austin, TX 78752 (tom@irsinc.com)

Jan Gecsei, Université de Montréal, Dept IRO, C.P. 6128, Succ. Centre-Ville, Montréal, Québec, Canada H3C 3J7 (gecsei@iro.umontreal.ca)

Richard Golden, University of Texas at Dallas, School of Human Development GR41, Box 830688, Richardson, TX 75083 (golden@utdallas.edu)

Bernie Jackson, Department of Computer Science, Stanford University, Jordan Hall, Stanford, CA 94305

Arun Jagota, Department of Computer Science, 225 Applied Science, Baskin Center, University of California at Santa Cruz,Santa Cruz, CA (jagota@icsi.berkeley.edu)

Jayadeva, Department of Electrical Engineering, Indian Institute of Technology (Delhi), Hauz Khas, New Delhi, 110016, India (jayadeva@ee.iitd.ernet.in)

John D. Johnson, School of Business Administration, University of Mississippi, 218 Conner Hall, University, MS 38677 (johnson@bus.olemiss.edu)

Asa W. Lawton, School of Computer Science, University of Oklahoma, Norman, OK 73019

Sam Leven, For a New Social Science, 4681 Leitner Drive West, Coral Springs, FL 33067 (samnets@aol.com)

Daniel S. Levine, Department of Psychology, University of Texas at Arlington, Arlington, TX 76019, or Department of Mathematics, University of Texas at Arlington, 411, S. Nedderman Drive, Arlington, TX 76019 (b344dsl@utarlg.uta.edu)

Sue Lykins, Departmentof Psychology, Oklahoma State University, Stillwater, OK 74078

Haluk Öğmen, Department of Electrical Engineering, University of Houston, Houston, TX 77204 (ogmen@uh.edu)

Alice J. O'Toole, University of Texas at Dallas, School of Human Development GR41, Box 830688, Richardson, TX 75083 (otoole@udallas.edu)

Ian Parberry, Department of Computer Sciences, University of North Texas, P.O. Box 13886, Denton, TX 76203 (ian@ponder.csci.unt.edu)

Karl H. Pribram, Director, Center for Brain Research and Informational Sciences, Radford University, Radford, VA 24142 (kpribram@ruacas.ac.runet.edu)

Ramkrishna V. Prakash, Department of Electrical Engineering, University of Houston, Houston, TX 77204 (prakash@uh.edu)

Paul Prueitt, Highland Technologies, Inc., 4303 Forbes Boulevard, Lanham, MD 20706 (paul@htech.com)

Gershom-Zvi Rosenstein, Department of Physiology, Hadassah Medical School, Hebrew University, 91010 Jerusalem, Israel (rosen@md2.huji.ac.il)

David G. Stork, Ricoh California Research Center, 2882 Sand Hill Road, Suite 115, Menlo Park, CA 94025 (stork@crc.ricoh.com)

Graham D. Tattersall, New House, Mill Road, Friston, Suffolk IP17 IPH, United Kingdom (gdt@sys.uea.ac.uk)

Dominique Valentin, University of Texas at Dallas, School of Human Development GR41, Box 830688, Richardson, TX 75083 (valentin@utdallas.edu)

Scott Walker, Department of Computer Science, Stanford University, Jordan Hall, Stanford, CA 94305

Paul J. Werbos, 8411 48th Avenue, College Park, MD 20740 (pwerbos@note.nsf.gov)

Preface

This book is the third of a series of books based on conferences sponsored by the Metroplex Institute for Neural Dynamics (M.I.N.D), an interdisciplinary organization of Dallas-Fort Worth area neural network professionals in both academia and industry. M.I.N.D. sponsors a conference every year or two on some topic within neural networks. The topics are chosen (a) to be of broad interest both to those interested in designing machines to perform intelligent functions and those interested in studying how these functions are actually performed by living organisms, and (b) to generate discussion of basic and controversial issues in the study of mind. The subjects are chosen for depth and fascination of the problems covered, rather than for the availability or airtight conclusions; hence, well-thought-out speculation is encouraged at these conferences. Thus far, the topics have been as follows:

> May 1988 — Motivation, Emotion, and Goal Direction in Neural Networks
> June 1989 — Neural Networks for Adaptive Sensory-Motor Control
> October 1990 — Neural Networks for Knowledge Representation and Inference
> **February 1992 — Optimality in Biological and Artificial Networks?**
> May 1994 — Oscillations in Neural Systems
> May 1995 — Neural Networks for Novel High-Order Rule Formation

A book based on the May 1988 conference was published by Lawrence Erlbaum Associates, Inc. (LEA), in 1992. A book based on the October 1990 conference was published by LEA in 1994. The current book is based on the February 1992 conference, and one based on the May 1994 conference is in its early stages.

The topic of optimality was chosen because it has provoked considerable discussion and controversy in many different academic fields (see, in particular, Schoemaker, 1991). There are several aspects to the issue of optimality. First is it true that actual behavior and cognitive function of living animals, including humans, can be considered optimal in some sense? Is there a measurable *utility function*, to use the economists' term, or at least a utility function deducible on theoretical grounds, that all actions ultimately serve to maximize? Or is most actual human or animal behavior better described by what the economist and cognitive scientist Herbert Simon (1979) called *satisficing* — in colloquial terms, "muddling through" or "making do" with solutions that may not be the best possible, but are in some measurable sense good enough? The answer to this question is still unknown, which is one reason for the question mark in this book's title.

Second, what *is* the utility function for biological organisms, if any, and can it be described mathematically? Even if all behavior does not fit the maximization paradigm, as Schoemaker (1991) has suggested, optimality might provide a *normative* criterion for which behaviors are desirable or should be encouraged. This kind of normative criterion can also guide the design of artificial neural networks to perform engineering tasks, whether in robotics, pattern recognition, business applications such as scheduling, or a variety of other situations. If not all biological behavior is in fact optimal, this also suggests that although designers of intelligent machines should understand the biological functions of the brain as well as possible, they should not adhere slavishly to "biological realism" in the architectures for their machines.

So the questions posed by the participants in this conference tended to fall into the categories of (a) *how* to optimize particular functions, in both biological and artificial networks and (b) *whether* particular functions are in fact performed optimally by particular biological or artificial networks. Rather than organize the chapters by what stance they took on optimality, it seemed more natural to organize them either by what level of questions they posed or by what intelligent functions they dealt with. This led to four major sections, including the following authors:

What Is the Role of Optimality?
Daniel Levine
Paul Werbos
Sam Leven
David Stork, Bernie Jackson, and Scott Walker
Wesley Elsberry
Mark DeYong and Thomas Eskridge

Quantitative Foundations of Neural Optimality
Paul Prueitt
Ian Parberry
Richard Golden
Graham Tattersall
Robert Dorsey and John Johnson
Arun Jagota

Optimality in Learning, Cognition, and Perception
David Chance, John Cheung, Sue Lykins, and Asa Lawton
Samy Bengio, Yoshua Bengio, Jocelyn Cloutier, and Jan Gescei
Gail Carpenter
Hervé Abdi, Dominique Valentin, and Alice O'Toole
Jayadeva and Basabi Bhaumik

Optimality in Decision, Communication, and Control
Haluk Öğmen and Ramkrishna Prakash
Gershom-Zvi Rosenstein
Sylvia Candelaria de Ram
Raymond Bradley and Karl Pribram

The chapters in the first section set some general frameworks for discussing optimality, or the lack of it, in biological artificial systems. The second section deals with some general mathematical and computational theories that help to clarify what the notion of optimality might entail in specific classes of networks. The chapters in the third section begin with optimizing rules for changing connection weights to facilitate associative learning, then move on to optimizing various processes in visual pattern perception. The chapters in the final section deal with optimality in the context of many different high-level issues, including exploring one's environment, understanding mental illness, linguistic communication, and finally, social organization.

The diversity of topics covered in this book is designed to stimulate interdisciplinary thinking and speculation about deep problems in intelligent system organization. This can have the unfortunate side effect of creating confusion for the reader by leading the reader to believe that many of the chapters are unrelated. At the suggestion of one of the book's anonymous reviewers, we have attempted to mitigate this possible confusion by writing prefaces at the start of each chapter, preceding the abstract. These prefaces are designed not only to frame the problem posed by the chapter's authors but also to show salient relationships between the chapter and others in this book.

In addition to the chapter authors, we acknowledge contributions made to this volume by several other individuals and organizations. The conference on which this book is based was made possible by generous financial support from two other organizations in addition to M.I.N.D. One was For a New Social Science (NSS), a nonprofit research foundation based in Coral Springs, Florida, that also cosponsored the M.I.N.D. 1995 conferences and supported the 1990 conference. The purpose of NSS, as stated by its founder, Dr. Sam Leven, is "turning the findings and techniques of science to the benefit of social science." It seeks to develop more predictive methodological bases for areas ranging from economics to management theory to social psychology — in some cases, to replace foundational assumptions dating from the time of David Hume and Adam Smith, based on a static and unrealistic model of human behavior, with new foundational assumptions that draw on modern knowledge of neuroscience, cognitive science, and neural network theory. The other organization that supported the conference as the International Neural Network Society (INNS), through its Texas Area Special Interest Group (SIG), and administered by the then Executive Director of INNS, Morgan Downey. INNS, founded in 1987, has become the flagship interdisciplinary organization for neural network researchers and practitioners, through several World Congresses on Neural Networks that draw around 500 attendees and the society's official Journal, Neural Networks (published by Elsevier). INNS is now involved with this book in another respect, having joined forces with Lawrence Erlbaum Associates, Inc., to promote an INNS Book Series of which this book is a part.

The speakers and poster presenters at the conference included one author for each chapter in this book except for Chapters 2 and 17, and several other distinguished neural network researchers: Stephen Grossberg of Boston University (whose talk is mentioned in the editors' preface to Gail Carpenter's chapter); Harold Szu of the Naval Surface Warfare Center (whose ideas are alluded to in Robert Dorsey and John Johnson's chapter); Steven Hampson of the University of California at Irvine; and Subhash Kak of Louisiana State University. These speakers made strong contributions to the dialogue. Some could not contribute chapters to the book because of other time commitments, and others were not asked to contribute because the anonymous reviewers expressed the need to focus the dialogue more sharply, but their influence is felt in the points raised by the chapter authors.

The other members of the Metroplex Institute for Neural Dynamics lent us considerable organizational support, especially Alice O'Toole and Raju Bapi, who were with us on the M.I.N.D. executive committee at the time of the conference. The University of Texas at Dallas provided the excellent Conference Center with state-of-the-art facilities at which the meeting took place.

We owe a debt of thanks to the staff of Lawrence Erlbaum Associates, Inc., particularly to Judi Amsel and Ray O'Connell, our editors at different stages; our unknown copyeditor; and Arthur Lizza and Sondra Guideman, our production editors. Ray, in particular, promoted this book as a natural sequel to the book on knowledge representation and inference.

Finally, we thank our wives, Lorraine Levine and Diane Blackwood, for their patience and support. Their intuitive understanding of and proximity to our editorial efforts made them in effect cocreators with us.

Daniel S. Levine
Wesley R. Elsberry

References

Schoemaker, P. J. H. (1991).The quest for optimality: A positive heuristic of science? *Behavioral and Brain Sciences,* **14,** 205-245.
Simon, H. A. (1979). *Models of Thought.* New Haven, CT: Yale University Press.

I

WHAT IS THE ROLE OF OPTIMALITY?

1

Don't Just Stand There, Optimize Something!

Daniel S. Levine
University of Texas at Arlington

*Daniel Levine's chapter, **Don't Just Stand There, Optimize Something!**, attempts to give a general theory for how much influence optimization has on human decision making. Levine considers the roles of optimization both at the descriptive level (how **do** we make decisions in reality?) and the normative level (how **should** we make decisions?).*

*Levine compares actual human decision making with **self-actualization**, Abraham Maslow's description of optimal human mental functioning. He proposes a tentative neural network theory for self-actualization that posits an explanation for why it doesn't always happen. A submodule of his network calculates a utility function of its present state, and another node (analogous to a function of the frontal lobes) imagines alternative states and calculates their utility functions. If an alternate state is seen as "better" than the current state, this generates "negative affect" which drives the network to seek a new, and presumably closer to optimal, state. But the strength of the network's approach to a new state is regulated by a complex chemical transmitter system. If this strength is insufficient, the network can get "stuck in local minima" in the familiar fashion of back propagation networks.*

*Being stuck in a local minimum is not necessarily bad. It may be analogous to **satisficing**, the term coined by Herbert Simon for reaching the first decision that is good enough to satisfy current needs, even if it is known not to be optimal (a concept also discussed in the chapters by Golden and by Werbos). Also, Levine points out, as does Leven's chapter, that rational optimization of **all** decisions may lead to spending too much time and effort on decisions whose consequences don't merit this effort. Based on previous work of Pribram, he suggests that the frontal lobes, hippocampus, and amygdala combine into a system that decides which goals are worth how much effort to optimize. He makes a distinction, also made in Werbos' chapter, between optimizing at "macro" and "micro" levels.*

ABSTRACT

This chapter deals at a philosophical level with two questions about human cognitive functioning: (1) *Do* we always optimize some variable that provides an advantage to us? and (2) *Should* we always optimize some variable?   The first question is answered with a resounding "No." Some of the influence of irrational constructs, such as metaphors, on cognition is explored. Then a tentative neural network theory is proposed for self-actualization, an optimal state, partly rational and partly intuitive, that is achieved only a small portion of the time by most people and more consistently by a minority of people. The second question is left open: some situations are given whereby detailed rational strategies are counterproductive, but there still may exist a broadly normative utility function that combines reason and intuition.

1. INTELLECTUAL ISSUES

The conference on which this book is based has roots going back to the early 1970s. At that time many neural network theorists sought to explain human behavior broadly as maximizing positive reinforcement and/or minimizing negative reinforcement. The most important of these optimality theorists were Harry Klopf, Paul Werbos, and Gregory (now Gershom-Zvi) Rosenstein (the last two being contributors to this volume). Let us look at where these scholars derived their inspiration. Part of it came from analogy with economics, particularly microeconomics: just as consumers and producers are assumed to maximize profit, minimize cost, and so on, organisms maximize biological reinforcement, which is treated as a sort of "net income" (cf. Rosenstein, Chapter 19). But the inspiration for optimality also came from evolutionary theory. Ever since the age of Darwin, there has been a strong teleological itch among biologists, a tendency to see prevailing behavior as somehow justified from an evolutionary standpoint, as *serving a purpose*.

Yet in all disciplines (less in economics than anywhere else, cf. Leven, Chapter 3) there has been a countervailing tendency to see rationality, particularly human rationality, as flawed, to see Edgar Allan Poe's imp of the perverse (Stedman & Woodberry, 1894) in some of the actions of biological organisms. How, using optimality principles alone, can we explain addictive gambling, neurotic self-punishment, sexual attraction to toxic personalities, election of obvious scoundrels to political office? The list goes on and on. The title of my chapter is actually a variant of one used in the commentary (on the article of Schoemaker, 1991) by Paelinck (1991), who in turn took it from a cartoon in which it is an exhortation from an American economics professor to his students. Sometimes, mathematical theories needed to justify behavior within a rubric of "optimizing something" lead to absurdly tortuous utility functions.

The debate over how much human behavior is rational goes on in every scientific and social scientific discipline, with major effects on the philosophical foundations of these fields (Cohen, 1981; Kyburg, 1983; Schoemaker, 1991). Sometimes (Jungermann, 1983), this debate has been couched in terms of optimism versus pessimism, with the believers in pervasive rationality being counted as optimists. But look at the optimality question from another view-point, that of the social reformer. If you are interested in ridding the world of unjust war, poverty, or environmental pollution, each of which is at least partly caused by human actions, you hope that these actions do *not* represent optimal human behavior. That is, people are capable of "better" than war or poverty or pollution. Hence, from the social reform viewpoint, Junger-mann's "optimists" become "pessimists," and his "pessimists" become "optimists"!

Even from an evolutionary viewpoint, Stephen Gould has shown that evolution does not necessarily imply "progress." Gould (1980, p. 50) reviewed two principles Darwin had propounded about nonadaptive biological change. One is that "organisms are integrated systems and adaptive change in one part can lead to nonadaptive modifications of other features." The other is that "an organism built under the influence of selection for a specific role may be able, as a consequence of its structure, to perform many unselected functions as well." Darwin disagreed with other biologists of his day who were stricter believers in optimality, such as Alfred Russel Wallace. He believed, as I do, that whereas evolution may lead to optimization of some functions, this process could have accidental by-products that are not always optimal (see Stork, Jackson, & Walker, Chapter 4, for a specific biological example). What Gould said about biological functions in general is particularly true of neuropsychological functions.

In another sense, evolutionary theory does not tell us the whole story about human choice. Natural selection only means that traits will be selected that promote *survival* (of the individual or of his or her genes). It does not mean that traits will be selected that enhance the *quality* of life in senses that most of us would agree on, the best use of human potential.[1] In later sections, I develop a tentative neural theory of *self-actualization*, defined by Abraham Maslow (1968, 1972) as the state of optimal human potential. Maslow noted that self-actualization is achieved consistently by about 1% of the population and on rare occasions by most other people. This is far less often than would be predicted if evolution selected for a self-actualizing tendency.

Schoemaker (1991) asked what is the level at which the concept of optimality is meaningful. He asked whether optimality is "(1) an organizing principle of nature, (2) a set of relatively unconnected techniques of science, (3) a normative principle for rational choice and social organization, (4) a metaphysical way of looking at the world, or (5) something else still" (p. 205). The chapters in this volume vary widely in their viewpoints, but the largest segment seems to have arrived at a general consensus. The majority of authors herein, and of scientists in general, believe that optimality contains elements of both (1) and (3) of Schoemaker's choices. It is *an* organizing principle of nature but not *the* organizing principle, that is, it does not point to a universal rule. The chapters in this volume by DeYong and Eskridge, Elsberry, Leven, and Werbos particularly point to optimization as a useful tool for understanding consciousness or intelligence, in spite of having major limitations. A system for vision, or cognition, or motor control may be optimal in its overall organization but suboptimal in parts, or vice versa.

In dynamical systems in general, and neural network systems in particular, the crucial distinction is between competing attracting states of the system. This includes the distinction, now already a cliché after less than 10 years in wide usage, between global and local minima of an energy (or cost, or error) function. Ironically, the bugbear of nonoptimal local minima now particularly haunts back propagation networks, which originated with Werbos' (1974/1993) effort to link brain theory to the optimization rubrics of economics!

---

[1] It can be convincingly argued that under the current threats of nuclear war and environmental catastrophe, enhancing human potential is necessary for our survival as a *species*. If so, natural selection does not even ensure survival.

2. FRONTAL LOBE DAMAGE AS A PROTOTYPE OF NONOPTIMAL COGNITION

The hint that nonoptimality is pervasive in human cognition is important to those developing machines to perform higher-level cognitive functions, because those functions involve a mixture of reason and intuition. It suggests that although designers of such machines should study neuroscience and neuropsychology, they should not adhere slavishly to the buzzword of biological realism. This is because someone might devise an intelligent machine that is less vulnerable than our brains to, say, cognitive dissonance (Festinger, 1957) or conflict between reason and emotion. In fact, such a hypothetical machine might even be based on the same types of components as our brains are but with those components combined in a novel architecture. If so, as Lorenz (1966) and Werbos (Chapter 2) suggest, the long-awaited missing link between animals and a truly humane being might be ourselves!

Since the frontal lobes integrate sensory, semantic, affective, and motor systems among others (Pribram, 1991), theories of their function would seem to bear on the issue of optimality. There have been several recent neural network simulations of cognitive effects of frontal lobe damage (Bapi & Levine, 1994; Cohen, Dunbar, & McClelland, 1990; Dehaene & Changeux, 1989, 1991; Leven & Levine, 1987; Levine & Parks, 1992; Levine & Prueitt, 1989). These networks model behavioral circuits combining cognition, motivation, and reinforcement, in which frontal connections (with the limbic system, hypothalamus, thalamus, caudate nucleus, and perhaps midbrain) play a controlling role. I suggest that these frontal damage effects can be treated as prototypical examples of nonoptimal human cognitive functioning.

David Stork (personal communication) has objected that any lesioned system's functioning is *of course* suboptimal and does not indicate the system's normal processes. However, our models treat frontal damage as *weakening*, not breaking, a connection. This is because the frontal lobes provide the most direct link, but not the only link, between sensory areas of the cerebral cortex and affective areas of the limbic system and hypothalamus (Nauta, 1971). Hence, optimal cognitive function (which Levine, Leven, & Prueitt, 1992, compared to self-actualization) requires balance of activities and connection weights in many brain areas. This balance, I conjecture, is disrupted not only by focal brain damage but by many other contingencies, including bad education or maladaptive social customs (society's "phobias" or "obsessive-compulsive neuroses"). Figure 1.1 shows the continuum of human cognitive function from least to most integrated.

The networks of Levine and Prueitt (1989) incorporated two generic frontal lesion effects: perseveration in formerly reinforcing behavior, and excessive attraction to novelty for its own sake. Many familiar human and social phenomena are analogs of these two types of effects. For example, one form of perseverative behavior is prejudice against a group of people because of an early bad experience. Sometimes, in fact, the prejudiced individual will base a habit of prejudice not on direct experience with Blacks, Jews, women, laborers, mathematicians, and so forth, but on what he or she *has heard* about the group. If that kind of conditioning is obtained from an entire social circle, or from influential individuals such as parents, it can override later, more favorable, direct experience with the group in question.[2] One form of excessive novelty preference is following fads, whether in political beliefs, scientific outlook, or drug usage.

3. THE PROMISE AND SPECTER OF OUR METAPHORS

Lakoff and Johnson (1980) noted how much human thought is structured by metaphors. The metaphors we use are unconscious, frequently culturally based, and create whole systems of analogies that become embedded in our common language without our being aware of their source. One of their key examples was the metaphor "ARGUMENT IS WAR." They gave the following examples of common American English phrases informed by that metaphor:

> Your claims are *indefensible.*
> He *attacked every weak point* in my argument.
> His criticisms were *right on target.*
> I *demolished* his argument.
> I've never *won* an argument with him.
> You disagree? Okay, *shoot.*
> If you use that *strategy,* he'll *wipe you out.*
> He *shot down* all of my arguments. (p. 4; authors' italics)

Lakoff and Johnson emphasized that the war metaphor is not the only possible way to view arguments. By contrast, they asked us to "Imagine a culture where an argument is viewed as a dance, the participants are seen as performers, and the goal is to perform in a balanced and aesthetically pleasing way.   In such a culture, people would view arguments differently, experience them differently, carry them out differently, and talk about them differently."

In much the manner that frontal patients on a card sorting test develop an unbreakable positive feedback loop between their habits and their decision criteria (Milner, 1964), people frequently develop hard-to-break positive feedback loops between their metaphors and their belief systems. For example, when in graduate school I had an argument with a roommate about equality between men and women. My roommate, whose cultural background was more sexist than mine, said at one point in the discussion, "But I should be the *man* in the house." What was happening, I believe, is that he had among his mental constructs the metaphor "POWER IS MASCULINITY," and thereby used the term *man* metaphorically to mean boss. But since he used the word *man* instead of the word *boss,* his wording made it sound absurd that a woman should play the role, thus ridiculing the idea that a woman should have power equal to a man's.

Hence, our conditioning (either from experience or from teaching) determines our metaphors, which in turn limit our further conditioning. How this takes place is at the heart of analogical reasoning, which is one of the major current boundary areas between connectionist theory and traditional artificial intelligence (Barnden, 1994; Jani, 1991). I am by no means

---

[2] Some preliminary neural network theories of how prejudice arises and how it might be overcome are discussed in Chapters 4 and 5 of Levine (1996).

arguing that the cure for all social problems involves overcoming our metaphors and being "rational." Metaphors, in addition to adding poetry to life, enable creative leaps. Analogies lead to hunches supported by evidence insufficient for a formal proof but sufficient to suggest that a formal proof may be along the way. Analogies, some of them far-fetched, are a source of creative ideas even in the most proof-oriented of fields; the late mathematician Lipman Bers of Columbia University (personal communication) once said humorously that most mathematical proofs are derived from either "cheap tricks or bad jokes."

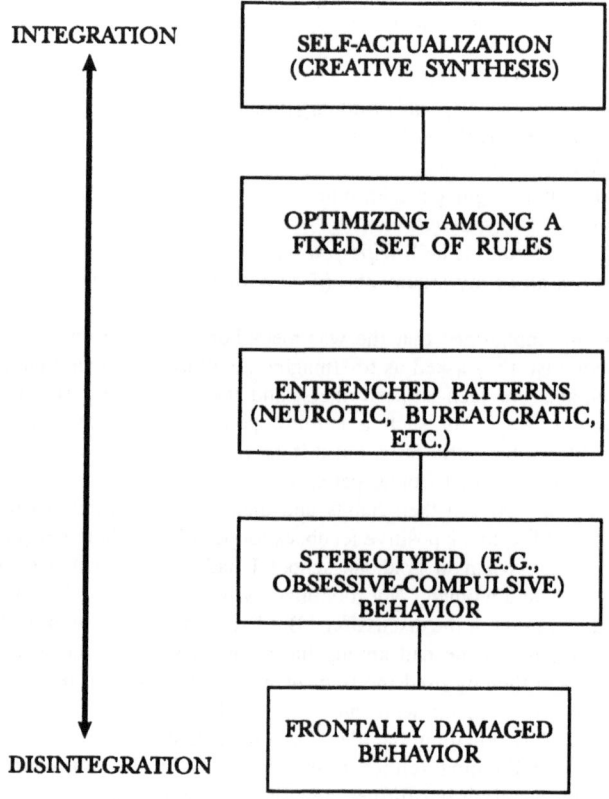

Fig. 1.1. Continuum of behavioral patterns from frontally damaged to self-actualized, with stereotyped or entrenched behavior in between. (From Levine & Leven, 1995, adapted with permission.)

I am merely arguing that we leave the way open for even our most beloved metaphors to be challenged if they seem inappropriate for a changed context. A suggestion for how this might be done arises from a role ascribed to the hippocampus in Pribram (1991). Pribram reviewed a large body of literature implicating two regions of the limbic system in the processing of familiarity versus novelty: the amygdala functions to create a sense of familiarity, whereas the hippocampus is involved in innovation. If part of a scene is currently being positively or negatively reinforced, hippocampectomized animals are unable to respond to changes in the part of the scene that is *not* being currently reinforced. It is precisely that type of change to which the innovator has to be able to respond, using the "peripheral vision" of his or her consciousness. A sexist man, for example, will never change his view that women are naturally subordinate unless he can be influenced by continual encounters with women acting effectively and pleasantly in powerful roles. Moreover, such encounters are most effective when they occur at times when his mind is focused on something unrelated to gender roles (such as needing to get a job done).

4. A SPECULATIVE FRONTOLIMBIC SYSTEM

In Pribram's view, the frontal lobes serve to mediate the balance between hippocampal and amygdalar functioning. He reviewed work that shows the division of the frontal lobes into three main functional subunits: the ventral part that is extensively connected with the amygdala, the dorsal part that is connected with the hippocampus, and the medial part that is connected with the somatosensory cortex. All three of these sets of connections are bidirectional. Based in part on event-related potential data (Deecke et al., 1984), Pribram suggested that these three areas of the frontal cortex are respectively concerned with appropriateness of actions, with setting of priorities, and with practicality of subgoals. A very simplified scheme for the significance of interactions among the frontal lobes, hippocampus, and amygdala is shown in Fig. 1.2.

This frontocortical-limbic scheme might have some more direct implications for optimality theory, as well. Pribram made the distinction between what he calls *efficient* processing, connected with functions of the hippocampus, and *effective* processing, connected with functions of the frontal cortex. What he meant by these terms was that efficient processing occurs when optimal ("least effort") choices are made in a fairly known environment, whereas effective processing occurs when choices are made in an environment that may be unknown (such as a new city after a move), choices that may not be optimal but "do the job." Hence, the frontal lobes, in their executive function (Pribram, 1973, 1991) are apparently deciding *when* it is appropriate to make optimal choices and when it is more appropriate to satisfice (cf. Golden, this volume; Simon, 1979; Werbos, this volume).

4.1. Self-Actualization: Why Doesn't it Happen All the Time?

More suggestions about functional interactions between the prefrontal cortex and subcortical areas can be obtained from network analysis of the idea of self-actualization due to Maslow (e.g., 1968, 1972). This term was intended to mean human functioning at the highest possible level. One of the major characteristics Maslow found in self-actualizing people — and in average people during temporary episodes of self-actualization known as *peak experiences* — is that such people tend to resolve ambiguities in a way that synthesizes conflicting interests within the mind rather than deciding between them. Hence, these people bridge typical dichotomies such as serious versus playful, masculine versus feminine, strong versus generous, rational versus emotional, by innovative solutions to problems.

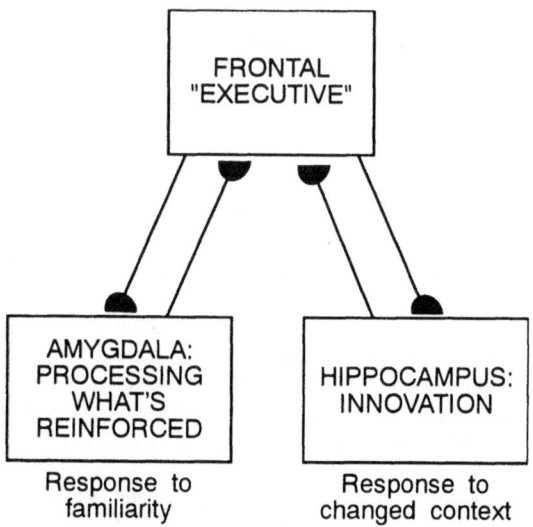

Fig. 1.2. Schematic of three interconnected brain regions that perform different functions in rule formation. (From Levine, 1995, reprinted with permission.)

To understand self-actualization, or its absence, in neural networks, let us start with the notion of simulated annealing (Ackley, Hinton, & Sejnowski, 1985; Kirkpatrick, Gelatt, & Vecchi, 1983). Simulated annealing is a widely used probabilistic method to move a system out of a suboptimal local minimum of an energy function, and closer to an optimal global minimum. In Fig. 1.3, I propose an alternative to simulative annealing, one that seems related to human introspection. The basic needs of the organism are encoded by a competitive (on-center off-surround) module as in Cohen and Grossberg (1983). The Cohen-Grossberg theorem tells us that such a competitive network has a Lyapunov function, called V, and always approaches a steady state that is at least a local minimum of that function. My proposal is to supervise this competitive module by a "world modeler" module, possibly analogous to part of the prefrontal cortex (cf. Ingvar, 1985). The world modeler makes "copies" of various possible states of the need subsystem and calculates the Lyapunov function for each. If V of the current state is larger than V of some other projected state, this sends a signal to a "negative affect" module that in turn sends random noise back to the need subnetwork, which can move it out of an unsatisfying local minimum in much the same manner as in a Boltzmann machine (Ackley et al., 1985).

The network of Fig. 1.3 is a step toward modeling the overarching function of the prefrontal cortex, as described by Damasio (1991): "*To select the responses most advantageous for the organism in a complex social environment*" (p. 404, author's italics). In more detail:

> The *primary value* used for the selection is the *state of the soma,* understood as a combination of the state of viscera, internal milieu, and skeletal musculature. The *primary signal* used for the process of response selection is a somatic response, which we call a *somatic marker.* (Damasio, 1991, p. 404)

Is Damasio's somatic marker described by a Lyapunov function such as the one in Fig. 1.3?

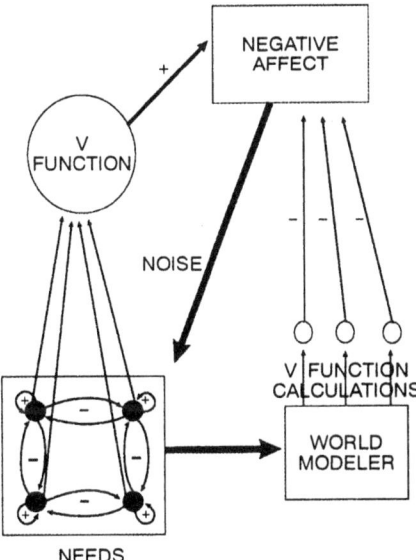

Fig. 1.3. Alternative scheme for simulated annealing. If the current state of the module has a larger energy function than some alternative state imaged by the world model module, this activates the negative affect module, which sends noise to perturb the needs module. (From Levine, 1994, reprinted with permission.)

The scheme of Fig. 1.3 is a first approximation: it must be expanded to include context-dependent biases within the needs module. Maslow (1968) discussed the *hierarchy of needs*: Survival needs like safety or food tend to be satisfied first, then needs for love or belonging, and finally needs for achieving one's potential. A homeless person, for example, tends to accept a job that stifles creativity more easily than does an affluent person. This suggests a scheme whereby a few overwhelming needs suppress perception of mismatch from a global energy minimum in the state of other needs. Maslow (1968, p. 26) said, and Hofstede (1980) confirmed, that this hierarchy is not a strict all-or-none progression: some personalities and cultures can more easily than others accept frustration of a lower-level need in order to try to resolve the "whole picture." Leven (1987) posited three major styles of problem solvers: "Dantzig" or direct solvers, who try simply to achieve an available solution; "Bayesian" solvers, who play the percentages and try to maximize a measurable criterion; and "Godelians," who combine intuition and reason into innovative solutions. Hence, any neural model of self-actualization and the needs hierarchy includes wide parameter variations based on personality differences.

Leven's three types of solvers may differ in the extent to which they accept "satisficing" solutions that satisfy only some of their needs. The idea of a suboptimal *local minimum* may at times need to be reconfigured as a minimum for an energy function over part, but not all, of the needs module. If a subset B of the need set N suppresses the other needs excessively, the contribution of nodes in N-B to the calculation of the energy function V is also weakened. This is because, in Cohen and Grossberg (1983), V has the form

$$V(\bar{x}) = -\sum_{i=1}^{n} \int_{0}^{x_i} b_i(y) d_i'(y) dy + \frac{1}{2} \sum_{j,k=1}^{n} c_{jk} d_j(x_j) d_k(x_k),$$

where the c_{jk} are positive constants and the d_j are monotone nondecreasing differentiable functions. This equation shows that if the system is in a state that primarily differs from an optimal state in those node activity variables v_i for which i is a member of N-B, the affective error signal from this mismatch will be weakened.

The system of Fig. 1.3 can be regulated at many levels. The two main processes to be regulated are (a) the competitive needs module itself, whereby tonic signals can move the module's behavior toward either "winner-take-all" or stable coexistence, and (b) the gain from the negative affect error signal to production of "simulated annealing" noise. As to possible brain loci for controllers and modulators, either the needs module or the error signal may be identifiable with part of the amygdala, which has been implicated in calculations of emotional valuation (LeDoux, 1991; Pribram, 1991). Effects of fronto-amygdalar connections, in addition to those arising from the "world modeler" module of Fig. 1.3, could include control of the gain of the "noise" signal from the "negative affect" module. This suggestion comes from the clinical observation (e.g., Milner, 1964) that frontal patients sometimes express frustration at their own ineffective actions, but this frustration does not make them change their actions.

The amygdala (especially its central and basolateral nuclei) is also heavily innervated by noradrenergic projections from the locus ceruleus (Foote, Bloom, & Aston-Jones, 1983). In addition to enhancing novel or significant inputs (Hestenes, 1992), noradrenaline (NA) influences cognitive attributions and beliefs. Individuals with low NA levels tend toward learned helplessness and lowered confidence in their ability to affect events (Leven, 1992; Samson, Mirin, Hauser, Fenton, & Schildkraut, 1992). A milder form of learned helplessness, with an intermediate NA level, might lead to passivity about satisfying the "higher" needs in the set N-B (discussed earlier) if the "lower" needs in B are already met. In other words, the person may feel *confident* only about satisfying a small set of needs. In the network of Fig. 1.3, NA signals affect the competitive needs module, making its dynamics more winner-take-all with a low NA level, and more coexistent with a high NA level. Grossberg (1973) showed that tonic excitatory signals tend to uniformize activities in a competitive network. The NA signal could similarly tonically arouse the needs module, as shown in Fig. 1.4. Larger NA moves this module toward equilibria that satisfy a greater number of needs.[3]

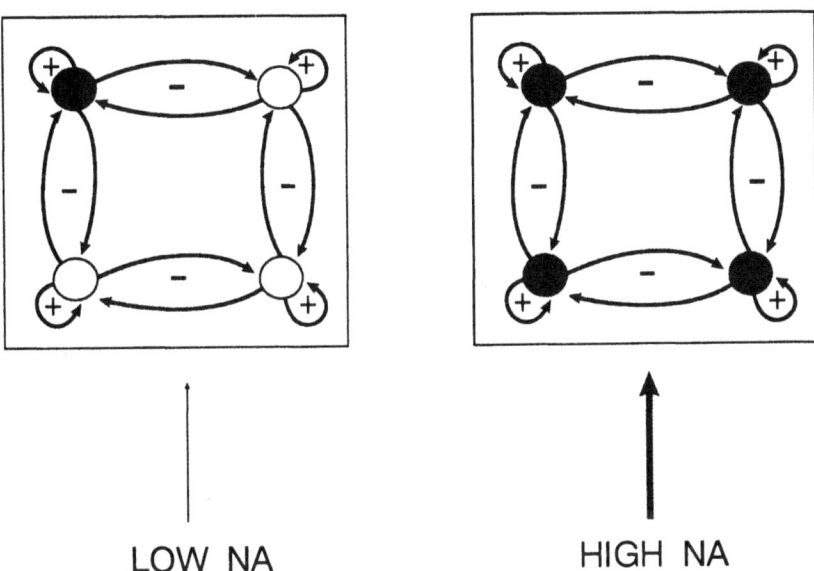

LOW NA   HIGH NA

Fig. 1.4. Effect of noradrenaline (NA) level on an on-center off-surround module, such as the needs module of Fig. 1.3. Dark circles indicate nodes with positive asymptotic activity. (From Levine, 1994, reprinted with permission.)

4.2. Self-Actualization and Information Processing

Now that a tentative theory of interactions between drives (Maslow's "hierarchy of needs") has been outlined, let us look further at what may constitute satisfaction of a "self-actualization drive." The discussion in this part continues and extends that in Levine et al.

---

[3] This model is a simplistic first approximation, because the role of noradrenaline is far too complex to be captured by a single parameter. More detailed network hypotheses about the interplay of NA and other transmitters such as dopamine and serotonin, and the brain regions involved, appeared in Leven (1992) and Luciano (1995).

(1992). We return to the point of Maslow (1968, 1972) that self-actualization involves creative synthesis of previously conflicting concepts or beliefs. Recall that Fig. 1.1 depicts a continuum of human behavior from the most "disintegrated" to the most "integrated." Decisions based on winner-take-all choices to act strong or generous, playful or serious, and so forth, are more effective than decisions made by people with frontal lobe damage. Also, choices are based on rational judgment are more effective than choices based on entrenched habits. But if the claims of both conflicting entities (e.g., "strength" and "generosity") are powerful enough, still more effective choices (even if they are riskier) are available from syntheses of the two alternatives. For example, one might combine generosity and strength into being powerful so as to empower others, or combine playfulness and considerateness into "if it harm none, do as you will." Because these high-level syntheses involve a blend of rational, intuitive, and instinctive processes, that is, all of MacLean's (1980) "three brains," they should, as Fig. 1.1 suggests, engage the prefrontal cortex, which is the chief communicator between the three brains.

Different degrees of self-actualization lead to different methods of resolving ambiguity. Wegner and Vallacher (1977, p. 124 ff.), for example, reported studies of general impressions people formed about women who were depicted as both kind (leading to a positive evaluation) and sexually promiscuous (leading to a negative evaluation). Most people adopted either the univalent strategy of resolution (she's "good" or "bad," not both), the aggregative strategy (she's "a bit of each"), or the integrative strategy (e.g., she's "happy go lucky" — an answer that ties the paradox together). The integrative strategy involves an ability, and decision, to transfer to a higher level in conceptual space if no decision made at a lower level is satisfactory. Levine (1989) described one possible way to implement ambiguity-dependent interlevel switching in an adaptive resonance theory (ART) network. Previous models (Dehaene & Changeux, 1991; Levine & Prueitt, 1989) posited rule-coding neurons in the prefrontal cortex. The choices between rules in those networks, however, were within one level (e.g., sorting cards on the basis of color versus shape of design). The prefrontal cortex also seems to make choices *between levels*, that is among *types* of rules. Examples of rules that need intact frontal lobes to be learned effectively are (a) choose whichever object is the most novel (Pribram, 1961); (b) alternate moving to the left and right (Goldman & Rosvold, 1970); and (c) press each of several panels once, regardless of order (Brody & Pribram, 1978). The greater the degree of self-actualization, the higher the level of rules that will tend to be chosen. This depends, in an ART network, on a parameter called *vigilance* that connotes probable, but not certain, match. This vigilance level in turn, I believe, depends on interactions between the neurotransmitters noradrenaline and serotonin; these are beyond the scope of this chapter, but related experimental results are found in Hestenes (1992).

5. BUT IS NONOPTIMALITY SO TERRIBLE?

Self-actualization (SA) is not an example of rational optimization in the classical sense used in economics (e.g., Lancaster, 1966). Rather, SA is related to the function of deciding *when* it is appropriate to optimize a rational utility function. It seems plausible that SA actually is equivalent to optimizing some other utility function that incorporates both reason and intuition, but at this stage of knowledge it is premature to try to quantify this function. I agree with

Werbos (Chapter 2) that a truly intelligent system (i.e., one that optimizes utility *over time*) needs capacities other than classical rationality. Such a system clearly, Werbos emphasizes, needs emotion, in order to have a criterion for determining what is reinforcing. It also needs attraction to novelty, in order to encourage exploration (see also Öğmen & Prakash, Chapter 18).

Meanwhile, I am indebted to Sam Leven for the following example (discussed in Leven, 1987) of a situation whereby "it is not optimal to optimize":

> You are in a supermarket, buying weekly groceries. You arrive at the aisle containing breakfast cereals. There are fifty different brands. Each comes in three different sizes. Each has different quantities of ten vitamins than the rest, a different price, and, of course, a different flavor. Buy a box of cereal.

The optimizing — rational — man[4] faces a combinatorial nightmare. Following the broadly accepted model introduced by Lancaster (1966), our economic man identifies the categories on which he should judge the cereals ("characteristics") and creates a matrix of rankings: a very difficult multidimensional optimization problem.

The frontal lobes say: hey wait a minute, devoting all that time to the cereal choice which doesn't make a lot of difference shows (in the terminology of Pribram, 1991) trouble setting priorities, and even some impropriety. Hence, we only try to optimize overall decision schemes, not every single decision (see also the discussion of economics and business in Werbos, this volume). An analogous "macro-optimality" without "micro-optimality" occurred in Bullock and Grossberg's (1988) model of arm motor control. These authors argue that their model, which sets control goals but does not establish an optimal trajectory, is more flexible than a competing model, the "minimum jerk model," which interprets every intermediate position of a muscle as the minimum of some utility function. This might be the ambiguity that we have to live with at all levels, sensory, motor, and associative. Perhaps optimality itself is just a metaphor!

REFERENCES

Ackley, D. H., Hinton, G. E., & Sejnowski, T. J. (1985). A learning algorithm for Boltzmann machines. *Cognitive Science*, **9**, 147-169.

Bapi, R. S., & Levine, D. S. (1994). Modeling the role of the frontal lobes in performing sequential tasks. I. Basic structure and primacy effects. *Neural Networks*, **7**, 1167-1180.

Barnden, J. (1994). On using analogy to reconcile connections and symbols. In D. S. Levine and M. Aparicio, IV (Eds.), *Neural Networks for Knowledge Representation and Inference* (pp. 27-64). Hillsdale, NJ: Lawrence Erlbaum Associates.

Brody, B. A., & Pribram, K. H. (1978). The role of frontal and parietal cortex in cognitive processing: Tests of spatial and sequence functions. *Brain*, **101**, 607-633.

---

[4] I am using the sexist word *man* because of the common use of the term "Economic Man" in traditional economic models from the early to middle part of this century.

Bullock, D., & Grossberg, S. (1988). Neural dynamics of planned arm movements: emergent invariants and speed-accuracy properties during trajectory formation. *Psychological Review*, **95**, 49-90.

Carpenter, G. A., & Grossberg, S. (1987). A massively parallel architecture for a self-organizing pattern recognition machine. *Computer Vision, Graphics, and Image Processing*, **37**, 54-115.

Cohen, J. D., Dunbar, K., & McClelland, J. L. (1990). A parallel distributed processing model of the Stroop effect. *Psychological Review*, **97**, 332-361.

Cohen, L. J. (1981). Can human irrationality be experimentally demonstrated? *Behavioral and Brain Sciences*, **4**, 317-370.

Cohen, M. A., & Grossberg, S. (1983). Absolute stability of global pattern formation and parallel memory storage by competitive neural networks. *IEEE Transactions on Systems, Man, and Cybernetics*, **13**, 815-826.

Damasio, A. R. (1991). Concluding comments. In H. S. Levin, H. M. Eisenberg, & A. L. Benton (Eds.), *Frontal Lobe Function and Dysfunction* (pp. 401-407). New York: Oxford University Press.

Deecke, L., Heise, B., Kornhuber, H. H., Lang, M., & Lang, W. (1984). Brain potentials associated with voluntary manual tracking: Bereitschaftspotential, conditioned premotion positivity, directed attention potential, and relaxation potential. In R. Karrer, J. Cohen, & P. Tueting (Eds.), *Brain and Information: Event-Related Potentials. Annals of the New York Academy of Sciences*, **425**, 450-464.

Dehaene, S., & Changeux, J.-P. (1989). A simple model of prefrontal cortex function in delayed-response tasks. *Journal of Cognitive Neuroscience* **1**, 244-261.

Dehaene, S., & Changeux, J.-P. (1991). The Wisconsin Card Sorting Test: Theoretical analysis and modeling in a neuronal network. *Cerebral Cortex* **1**, 62-79.

Festinger, L. (1957). *A Theory of Cognitive Dissonance*. Stanford, CA: Stanford University Press, 1957.

Foote, S. L., Bloom, F. E., & Aston-Jones, G. (1983). Nucleus Locus Ceruleus: New evidence of anatomical and physiological specificity. *Physiological Reviews*, **63**, 844-914.

Goldman, P. S., & Rosvold, H. E. (1970). Localization of function within the dorsolateral prefrontal cortex of the rhesus monkey. *Experimental Neurology*, **27**, 291-304.

Gould, S. J. (1980). *The Panda's Thumb*. New York: Norton.

Grossberg, S. (1973). Contour enhancement, short term memory, and constancies in reverberating neural networks. *Studies in Applied Mathematics*, **52**, 213-257.

Hestenes, D. O. (1992). A neural network theory of manic-depressive illness. In D. S. Levine & S. J. Leven (Eds.), *Motivation, Emotion, and Goal Direction in Neural Networks* (pp. 213-257). Hillsdale, NJ: Lawrence Erlbaum Associates.

Hofstede, G. (1980). *Culture's Consequences: International Differences in Work-related Values*. Beverly Hills, CA: Sage.

Ingvar, D. (1985). Memory of the future: An essay on the temporal organization of conscious awareness. *Human Neurobiology*, **4**, 124-136.

Jani, N. G. (1991). Through the eyes of metaphor. Unpublished master's thesis, University of Texas at Arlington.

Jungermann, H. (1983). The two camps on rationality. In R. Scholz, R. (Ed.), *Decision Making Under Uncertainty* (pp. 63-86). Amsterdam: Elsevier.

Kirkpatrick, S., Gelatt, C. D., Jr., & Vecchi, M. P. (1983). Optimization by simulated annealing. *Science*, **220**, 671-680.

Kyburg, H. E., Jr. (1983). Rational belief. *Behavioral and Brain Sciences*, **6**, 231-274.

Lakoff, G., & Johnson, M. (1980). *Metaphors We Live By.* Chicago: University of Chicago Press.

Lancaster, K. (1966). A new approach to consumer theory. *Journal of Political Economy*, **74**, 132-157.

LeDoux, J. E. (1991). Information flow from sensation to emotion: Plasticity in the neural computation of stimulus value. In M. Gabriel & J. Moore (Eds.), *Learning and Computational Neuroscience: Foundations of Adaptive Networks* (pp. 3-51). Cambridge, MA: MIT Press.

Leven, S. J. (1987). Choice and neural process. Unpublished doctoral dissertation, University of Texas at Arlington.

Leven, S. J. (1992). Learned helplessness, memory, and the dynamics of hope. In D. S. Levine & S. J. Leven (Eds.), *Motivation, Emotion, and Goal Direction in Neural Networks* (pp. 259-299). Hillsdale, NJ: Lawrence Erlbaum Associates.

Leven, S. J., & Levine, D. S. (1987). Effects of reinforcement on knowledge retrieval and evaluation. *First International Conference on Neural Networks* (Vol. II, pp. 269-279). San Diego: IEEE/ICNN.

Levine, D. S. (1989). Selective vigilance and ambiguity detection in adaptive resonance networks. In W. Webster (Ed.), *Simulation and AI 1989* (pp. 1-7). San Diego: Society for Computer Simulation.

Levine, D. S. (1994). Steps toward a neural theory of self-actualization. *World Congress on Neural Networks, San Diego, June, 1994* (Vol. 1, pp. 215-220). Hillsdale, NJ: Lawrence Erlbaum Associates.

Levine, D. S. (1995). Learning and encoding higher order rules in neural networks. *Behavior Research Methods, Instruments, and Computers*, **27**, 178-182.

Levine, D. S. (1996). *Common Sense and Common Nonsense.* New York: Oxford University Press, to appear.

Levine, D. S., & Leven, S. J. (1995). Of mice and networks: Connectionist dynamics of intention versus action. In F. Abraham & A. Gilgen (Eds.), *Chaos Theory in Psychology* (pp. 205-219). Westport, CT: Greenwood.

Levine, D. S., Leven, S. J., & Prueitt, P. S. (1992). Integration, disintegration, and the frontal lobes. In D. S. Levine & S. J. Leven (Eds.), *Motivation, Emotion, and Goal Direction in Neural Networks* (pp. 301-335). Hillsdale, NJ: Lawrence Erlbaum Associates.

Levine, D. S., & Parks, R. W. (1992). Frontal lesion effects on verbal fluency in a network model. *International Joint Conference on Neural Networks, Baltimore, Maryland, June 7-11, 1992* (Vol. II, pp. 39-44). Piscataway, NJ: IEEE.

Levine, D. S., & Prueitt, P. S. (1989). Modeling some effects of frontal lobe damage: novelty and perseveration. *Neural Networks* **2**, 103-116.

Lorenz, K. (1966). *On Aggression.* New York: Harcourt, Brace, and World.

Luciano, J. S. (1995). Quantitative analyses of temporal evolution of unipolar depression as measured by psychiatric and neurochemical indices. Unpublished doctoral dissertation, Boston University.

MacLean, P. D. (1980). Sensory and perceptive functions in emotional functions of the triune brain. In A. Rorty (Ed.), *Explaining emotions* (pp. 9-36). Berkeley, CA: University of California Press.

Maslow, A. H. (1968). *Toward a Psychology of Being.* New York: Van Nostrand.

Maslow, A. H. (1972). *The Farther Reaches of Human Nature.* New York: Viking.

Milner, B. (1964). Some effects of frontal lobectomy in man. In J. Warren & K. Akert (Eds.), *The Frontal Granular Cortex and Behavior* (pp. 313-334). New York: McGraw-Hill.

Nauta, W. J. H. (1971). The problem of the frontal lobe: A reinterpretation. *Journal of Psychiatric Research*, **8**, 167-187.

Paelinck, J. H. P. (1991). Don't just sit there, optimise something. *Behavioral and Brain Sciences*, **14**, 230.

Pribram, K. H. (1961). A further experimental analysis of the behavioral deficit that follows injury to the primate frontal cortex. *Experimental Neurology*, **3**, 432-466.

Pribram, K. H. (1973). The primate frontal cortex — executive of the brain. In K. H. Pribram & A. R. Luria (Eds.), *Psychophysiology of the Frontal Lobes* (pp. 293-314). New York: Academic Press.

Pribram, K. H. (1991). *Brain and Perception: Holonomy and Structure in Figural Processing.* Hillsdale, NJ: Lawrence Erlbaum Associates.

Samson, J. A., Mirin, S. M., Hauser, S. T., Fenton, B. T., & Schildkraut, J. J. (1992). Learned helplessness and urinary MHPG levels in unipolar depression. *American Journal of Psychiatry*, **146**, 806-809.

Schoemaker, P. J. H. (1991). The quest for optimality: A positive heuristic of science? *Behavioral and Brain Sciences*, **14**, 205-245.

Simon, H. A. (1979). *Models of Thought.* New Haven, CT: Yale University Press.

Stedman, E. C. and Woodberry, G. E. (Eds.) (1894). *Works of Edgar Allan Poe.* Freeport, NY: Books for Libraries Press (Vol. II, pp. 37-47).

Wegner, D. M., & Vallacher, R. R. (1977). *Implicit Psychology: An Introduction to Social Cognition.* New York: Oxford University Press.

Werbos, P. J. (1974). Beyond regression: new tools for prediction and analysis in the behavioral sciences. Doctoral dissertation, Harvard University; reprinted as Chapters 1-6 of P. J. Werbos (1993), *The Roots of Backpropagation: From Ordered Derivatives to Political Forecasting.* New York: Wiley.

2

Optimization: A Foundation for Understanding Consciousness

Paul J. Werbos[1]
College Park, Maryland

Paul Werbos' chapter, **Optimization: A Foundation for Understanding Consciousness**, *provides a general picture of the role of optimization in deepening insights about brain
organization and human nature. Although Werbos does not believe that optimization explains
every detail of behavior, he argues that optimization, and the related idea of control, provide a
foundation for helping to understand the structure of many types of intelligence in both brains
and electronic computers. He argues further that the insights obtained from neural network
analysis deepen, rather than change, centuries-old ideas from some philosophies and religions.*
*In his preliminary section, Werbos discusses classic deviations from optimality in actual
neural systems. One has to do with the problem of getting "stuck" in local minima. Although
no intelligent system can entirely avoid local minima, Werbos mentions recent work in adaptive
control (some of it his own) on strategies to alleviate the problem. One of these is "shaping":
to train a network on easy versions of a problem, and then gradually readapt the weights to
increase the chances of solving harder versions. Another has to do with attraction to novelty and
exploratory behavior, which is also discussed in the chapters of this book by Prueitt and by
Öğmen and Prakash. Like those other authors, Werbos believes that exploration, although it
deviates from static utility maximization, is actually a necessary component of intelligent
systems. Finally, there is the difference between "micro-" and "macro"-optimality, also discussed
in Levine's chapter. Werbos adduces evidence from business organizations that certain types of
"micromanagement" are actually suboptimal for overall performance of companies.*
*Werbos defines intelligence in terms of macro-optimality, that is, maximizing the value
of a utility function **over time**. (In this light, the schizophrenics discussed in Rosenstein's
chapter might be interpreted as people who are focused strongly on **immediate** maximization of
a variable, called Income, and are ineffective at planning its long-term maximization.) But
Werbos' interpretation of this utility function deviates somewhat from the orthodoxy that*

---

[1]The views herein are purely my personal views, oversimplified in places to make a point. They certainly do
not in any way represent the views of any of my employers past and present, one of whom remains a close friend
and supporter even though he is totally aghast at Section 7 and the Appendix.

*developed in economics, behaviorist psychology, and analytical philosophy. He stresses that not only are reason and emotion not opposites, but an intelligent organism **requires** emotion to make good decisions — a point also made in the chapters by Leven, Levine, and Rosenstein.*

Also, Werbos believes that study of intelligent control requires consideration of things outside the pale of reductionist science, such as consciousness and, possibly, the soul. This may be somewhat related to his view of how humans fail to be optimal: in our ability to articulate our desires, that is, to get our rationality in harmony with our emotions. In this regard he believes, with Konrad Lorenz and others, that humans might be transitional forms to a higher level of evolution. In his appendix, he speculates that something akin to a soul, a part of mental life not reducible to atoms and neurons, could be important for bridging to the higher level.

ABSTRACT

This chapter describes how the concept of optimization — whatever its limitations — can be a useful tool in efforts to understand consciousness and the mind. Such efforts must draw on what has been learned in many disciplines, many cultures, and many centuries. Neural net designs based on optimization offer us a more complete understanding of the phenomenon of intelligence and mind, precise enough to be replicated on electronic computers, yet fully consistent with what we see in the brain and in experiments on overt behavior. A deeper understanding of intelligence and the mind has immediate implications for the problem of consciousness, and for the foundations of psychology and philosophy.

This chapter provides a global summary of these implications, *as seen from* the viewpoint of existentialism, Confucianism, and linguistic analysis — established philosophical traditions which should not be ignored here. Among the six issues discussed are the subjective sense of existence, the levels of intelligence, the foundations of ethics, alternative states of consciousness, concepts of the soul, and the role of quantum theory. In all cases, the chapter presents candid personal views which may be regarded as heresies by a significant fraction of the community. The chapter argues that neural network research can indeed yield important insights into all of these questions, but that it does not provide a basis for overthrowing earlier views in philosophy or for resolving the debate about the existence of the soul; instead, it may help us to understand, unify, sharpen and deepen some very ancient insights. It suggests how one might understand and reinterpret some ancient four-letter words — hope, fear and soul — which have permeated human cultures for millennia, long before the advent of formal philosophy or theology.

1. PRELIMINARIES: IS IT INTELLIGENT TO DO ONE'S BEST?

The title of this section is partly a pun, and partly an appeal to common sense. The word "intelligent" by definition has something to do with the ability to do one's best, to *optimize*. There is a huge literature out there — both in economics and in other social sciences — on humans' ability to foul up, to be irrational, to make mistakes, and to become totally insane; however, it is important that this literature mainly focuses on deviations from the default, reference assumption of perfect optimality. The behavior it describes may be viewed as examples of stupidity (i.e., failures of intelligence) rather than examples of intelligence. We as humans do not really intend to foul up (vis-à-vis our real values) or to waste energy fighting ourselves. Optimality still provides a very powerful intellectual tool, which we can use to create very powerful designs and models, even if these models must be modified later to account for second-order phenomena.

Years ago, for purposes of mathematical research, I proposed that we should actually define an intelligent system as a "generalized system which takes action to maximize some measure of success over the long-term in an uncertain environment which it must learn how to adapt to in an open-minded way." I went on (Werbos, 1986) to define the terms within this definition. This chapter tries to describe the relation between this more precise concept of intelligence and the fuzzier concepts which have emerged simply by observing human beings. The key concept here is that an intelligent system does not *start out* with an optimal strategy of action; instead, it tries to *learn* an optimal strategy, bit by bit, over time.

If you are an intelligent human being, and you can think of ways that other people around you could be a little bit smarter in achieving their goals, then the *deviations* from optimality may seem very important to you.   You are comparing one intelligent system (yourself) against another. But if you are an engineer, trying to build systems which work as well as possible, you will find it truly amazing that human brains of any description perform as well as they do on such a wide variety of very difficult tasks. If you do the very best you can, as an engineer, to develop an optimizing learning system, you will still find imperfections or limitations in what you develop. In fact, it is fascinating how the imperfections of the best possible engineering designs do seem to match the most obvious imperfections of organic brains.

As an example, consider the *local minimum* problem. In realistic terms, it is not possible to build a powerful learning system which can never get caught in a rut, in a vicious cycle or local minimum. Therefore, it should be no surprise at all that people and animals do get caught in ruts, even though their brains do have well-tuned mechanisms to try to minimize the problem. Present-generation artificial neural networks (ANNs) do not get caught in local minima nearly as much as some people feared ten years ago; however, when ANNs are used to solve very complex control problems, it is crucial to use a strategy called "shaping" to avoid terrible local minima. In "shaping" (White & Sofge, 1992), one first trains the ANN to learn a very simple version of the problem; one then uses the resulting weights as initial values for an ANN trained to solve a harder version of the problem; and so on. This parallels the human need to learn "one step at a time."

Please note that learning one step at a time is not the same thing as performing a defined task one step at a time. A single step or stage in the learning process often represents an entire new strategy or concept of how to perform a complex task, a task which may not even be divisible into a sequence of subtasks. For example, in engineering, consider the problem of training a system to balance three connected poles, one on top of the other, like a family of acrobats trying to stand on top of each other without falling over. The first step may be to learn how to balance a simple pole. The second step might be to balance a large pole with a smaller pole on top. The third stage might be to balance two poles of the same size. Four stages of learning may or may not be good enough to solve this training problem. This step-by-step approach can work only if each individual step is easy enough to be learned, but hard enough to force the development of the new concepts (or "hidden variables" or "representation") needed to solve the next more difficult problem. In formal terms, one is always guaranteed to overcome the local minimum problem *if* one somehow can learn to develop the concepts necessary to the task at hand.

Thanks to the learned use of symbolic reasoning, human beings do not get caught in a rut nearly as much as other species. (A human being who lived exactly like a chimpanzee would generally be considered as being caught in a rut.) We can understand this situation better by viewing it in a more positive way: humans often use symbolic reasoning to help them visualize *new, creative* opportunities to enhance their lives, in ways that would not be obvious if they always just followed the path of least resistance. Symbolic reasoning can help us learn and develop new concepts in a more systematic way, based on learned strategies of thought. Even so, the human use of symbolic reasoning has some serious limitations, to be described in section 5. Even the most sane among us are still caught in local minima, in lives that fall short of our ultimate potential, to some degree (Campbell, 1971; Levine, 1994).

Another example of alleged imperfection is the tendency of humans to seek novelty or new information, even at some cost in terms of reinforcement. In actuality, novelty-seeking or exploratory behavior turns out to be an *essential component* of optimizing neural network control systems. It is essential both for stability and for avoiding local minima (Miller, Sutton, & Werbos, 1990; White & Sofge, 1992). In other words, it is essential to our ability to find ever more intelligent and more creative ways of coping with reality. Dreams, heresies, humor, and new challenges are all crucial aspects of human exploratory behavior.

Certainly, the concept of optimization has been abused very often. Many of us have gone through phases of excessive self-control during adolescence, in alternation with periods of excessive exploratory behavior. In management research, it is now well known that "reinforcement" strategies which are based on demeaning, distorted assumptions about human values can reduce productivity substantially. (There is an old adage that productivity is lowest in organizations where people are motivated by fear, mediocre in organizations ruled by greed, and highest in organizations driven by pride or self-respect.) In large organizations of all kinds, managers who try to micro-optimize — assuming that they know everything, and assuming that there is no need for exploratory behavior — often degrade productivity. All of these behaviors are motivated by an honest desire to optimize, but they are in fact grossly suboptimal; they are transitional stages, the kinds of stages or ruts which learning systems get stuck in for a while as they gradually learn better. They may learn better either through creative thinking or through bankruptcy. (Admittedly, however, the phenomenon of intelligence is far less obvious in social systems than in individual human minds.)

In the field of psychology, Stephen Grossberg has argued very often that models based solely on *reinforcement learning* or optimization can only explain about half of the experiments out there. To explain the other half, one must account for "classical conditioning," which requires a subsystem to generate *expectations* about the environment. In fact, the more advanced ANN designs which I have developed (Santiago & Werbos, 1994; White & Sofge, 1992) do contain such an expectations system, because that is crucial to effective optimization in complex engineering applications. Prior to late 1993, there had been no serious, published tests of these particular designs on realistic control challenges, in part because there were simpler versions which were easier to implement; however, by late 1994, five groups of researchers had implemented these designs, and had shown that they do lead to better results across a variety of applications — difficult benchmark problems in bioreactor control, robot arm control, and automatic aircraft landing; simulated missile interception, compared against current state-of-the-art methods used on that problem; and control of a physical prototype of a hypersonic aircraft (Werbos, 1995a). In this chapter, however, I do not review the mathematics of these models in detail, because they are moderately complex and have appeared elsewhere.

A complete review of the literature on rationality and learning would require far more detail than I have provided here. For example, it should consider Raiffa (1968), Von Neumann and Morgenstern (1953); Werbos (1968, 1992b), and the work of Herbert Simon and others. The goal of this chapter, however, is not to *evaluate* the concept of optimality, but rather to *use* the concept in addressing larger questions about consciousness and the mind, and so on.

2. INTRODUCTION: THE ISSUE OF CONSCIOUSNESS

In 1992, a prominent speaker at the conference of the European Neural Network Society declared "an open season on the problem of consciousness." The "problem of consciousness" is a very old problem, and one may legitimately ask why we would suddenly spend so much energy in revisiting it at this time. There are at least two legitimate answers: (a) that fundamentally new insights, developed by interdisciplinary neural network research, let us address the problem of consciousness at a higher level; and (b) that a relaxation of certain academic taboos — restricting analysis to overt behavior *only* (as in classical behaviorism) or to linguistic analysis *only* (as in some university philosophy departments in the United States and United Kingdom) — may now permit us to face up to issues which were hard to address 10 or 20 years ago. These answers lead, however, to further questions:

1. If insights from neural network research are useful, why are so many of the new manifestoes on consciousness written by people with limited knowledge of the real frontiers of the field (i.e., of those aspects which are most relevant to higher intelligence?)
2. Where is there serious philosophical depth in this discussion, above and beyond the classical Anglo-American approach?
3. Just what *is* the problem of consciousness anyway?

This chapter draws heavily on current neural network research, as one might expect, but it also draws on traditions like existentialism and Confucianism, which have critical contributions to make. I do not have enough space here to explain all the vicissitudes and varieties of existentialism or Confucianism; however, these traditions are very important as an antidote to some of the more extreme and parochial approaches to philosophy which have existed in the past in some American universities. Twenty years ago, the leading theory of ethics in the Anglo-American philosophy departments was a theory attributed to Rawls which proceeded entirely by performing a semantic analysis of the word "justice" and of what it should mean (based on assorted assumptions about what good definitions for a word should be), building up to strong recommendations for what policymakers should do all across the board. (Bear in mind that the problem of ethics refers to the problem of purpose and goals in human life; it requires a lot more than just coming up with a formula to keep lawyers happy.) This episode reminds me of a meeting I once attended at the Census Bureau, where famous world-class statisticians proposed to develop a measure of value or utility, for use in allocating federal funds, by simply doing a factor analysis of a complete set of available data series collected by the Bureau. This situation would have been very amusing, except that billions of dollars of federal funds have actually been allocated on the basis of formulas derived in such ways. See Werbos (1990) for a discussion of assorted ways that value measurements have been developed in the government.

Nevertheless, I would agree with the Anglo-American school on at least two basic points: (a) that it is foolish to invest too much energy in worrying about words like "consciousness" until we develop some sort of clear idea of what it is that we are worrying about, an idea of what the word is supposed to mean, and (b) that language, in general, does play a deep and central role in philosophy (Werbos, 1992c).

So what, then, is the "problem of consciousness"? This chapter does not start out by picking out one particular definition of the word "consciousness"; this would be a misleading exercise, because the word really does have many different meanings. Instead, it focuses on six more specific questions that people appear to be asking under this general rubric:

1. How is it possible — objectively — that human beings could ever meet the dictionary definition of "consciousness" — a basic sense of awareness, which allows them to respond to what they are aware of?
2. How is it possible that human beings have a subjective feeling that we do in fact exist, given that we have the various capabilities discussed under questions 1 and 3?
3. How is it possible that human beings show additional capabilities, such as intelligence or emotions or creativity, which we commonly tend to associate with our consciousness?
4. What is it in the brain that distinguishes between states of "consciousness" versus states of "unconsciousness" like sleep?
5. Can the human mind — in its widest scope — be explained entirely in terms of atoms and neurons, or do we need to invoke some sort of "soul" to explain the full range of our experience?
6. Can the human mind or the "soul" be fully explained in terms of algorithms or Turing-machine concepts (generalized to include continuous variables), or must we invoke other concepts like quantum computing (Penrose, 1989)?

This chapter presents my personal opinions on these questions. The reader should be reassured that I am aware of the idiosyncratic nature of my views, and that my strategic goals in the neural network field (Werbos, 1993a, 1994a, 1994b, 1994c) are sufficiently explicit that they leave no room at all for me to entertain any kind of bias against anyone who can advance those goals, regardless of their views on these questions. Because of page limits, this chapter simply explains what my views are, and cites other papers which explain the critical details.

3. THE OBJECTIVE QUESTION OF AWARENESS

Question (1) is hardly a problem at all, objectively — even though it is probably the most semantically correct interpretation of the "problem of consciousness." Not only human beings, but all animals on earth show some degree of awareness of their environment. Awareness — in a literal, objective interpretation of the word — simply refers to the ability of organisms to input and respond to data from the environment. There is no great mystery in explaining why that phenomenon should evolve (i.e., can confer an advantage in survival), and no great mystery in seeing that there are neural circuits capable of providing that simple capability.

Many of the neuroscientists working on "consciousness" would say that they are studying consciousness in the sense of awareness. They study how people become "conscious of a stimulus." (For example, members of that community speaking at the 1994 World Congress on Neural Networks, whose works are cited in Alavi and Taylor, 1994, and Taylor, 1992, made this statement.) That research does not try to explain how awareness exists, in a general sense; rather, it attempts to uncover the specific mechanisms by which information attracts attention and is registered at various levels of the sensory system of the brain. To fully understand these mechanisms — or to understand any other subsystem of the brain — it is crucial to understand how the subsystems contribute to the functioning of the *whole system*; thus, consciousness (defined as awareness) is very much a subset of consciousness as intelligence, to be discussed in Section 5.

It is unfortunate, in my view, that work on sensory input pathways — however important — has been mixed up with discussions of the existence of the soul, based solely on confusions between different definitions of the word "consciousness." Leaping from sensory physiology directly to assertions about the soul is analogous to jumping from the physics of silicon to assertions about computer design, without bothering to learn about chip design or transistors (let alone applications) along the way. In fact, the latter extrapolation makes more sense than the former, because silicon is at least a dominant aspect of chips, whereas sensory input is only one aspect of human intelligence.

Another common fallacy in the neuroscience of consciousness is the search for the site of "consciousness" within the cerebral cortex. This is analogous to the famous "search for the engram," back in the days before we understood that human memory is more distributed — even "holographic" — in nature. Sensory inputs typically get registered at many sites, at many levels in the brain. Each of these sites represents a certain level of "awareness" — a level of responsiveness to stimuli. Some biologists have been very excited to learn that human subjects *state* that they are aware of stimuli which reach certain sites, and *state* that they are unaware of certain others; however, from an objective point of view, this does not imply that one site is magically "conscious," whereas others are not. It only tells us that information in one site is available as an input (direct or indirect) to those areas of the cortex which control the verbal behavior of asserting "I am aware of that stimulus."

It should be emphasized that neither of these fallacies is universal within the neuroscience of consciousness. However, there are many cases where these fallacies have received greater publicity than the valid, underlying science.

4. THE SUBJECTIVE SENSE OF EXISTENCE

From a very strict existentialist point of view, it is nonsense to try to "explain" our own subjective sense of existence. Our subjective sense of existence or awareness is *our starting point*, the foundation on which we build everything else. This question is analogous to a question which novices ask of physicists: "Dr. Einstein, can you explain *why R=T* in general relativity? What underlying phenomena give rise to that equation? What kind of ether do electromagnetic waves travel in?" The point is that Einstein was looking for the lowest level of physical description, that level which inherently cannot be explained as the working out of something more fundamental.   Both Einstein and the existentialists were very active in questioning and revising their views of what exists at the most fundamental level, but they still maintained an effort to build everything else up from that level.

From an objective point of view, we may twist the question around, and ask how it is that organisms could *evolve* a sense of their own existence as such.   Marvin Minsky answered this years ago, by simply pointing out that there are evolutionary advantages in organisms developing models of the self and insights to describe their own thinking.   Once again, there is no real problem here from an objective point of view. When we ask whether *other* human beings have a sense of their own existence, we are essentially just asking the objective question; the answer is obviously "yes." (It would still be "yes" even if other humans were actually just programs in a vast virtual reality game, as long as those programs demonstrated the pertinent objective capabilities.) From an objective point of view, one may go further and argue that sane, self-aware organisms will naturally tend to accept the existentialist view of taking their own existence and awareness as a starting point, because this is an honest reflection of how their natural thought-processes work. (See section 5.)

From a strict Anglo-American point of view, neither of these answers is entirely satisfactory, because they seem to assume that there really do exist organisms on earth, that there is such a thing as biological evolution, and so forth. If we limit our thinking to nothing but the manipulation of words, without ever grounding ourselves in any sort of direct perception of reality, then we can in principle permit any fantastic combination of words to emerge from our mouths. From such a viewpoint, we could just as well worry deeply about issues like why the sun appears to rise every day; after all, can we be *really* sure that the earth revolves about the sun? Even if we accept that there is always some distant degree of uncertainty here (as is appropriate, from an existentialist point of view), it would seem silly to invest a huge amount of emotional energy on quirky little hypothetical contingencies which are poorly integrated into the rest of our concerns and which we have no way to account for in any case.

I do not believe that all American philosophers adhere to the extreme viewpoint I am arguing against here; in fact, I do not spend any further time on that particular species of philosophy here. Also, I do not mean to downplay the issue of how we *know* that the sun is likely to rise tomorrow; studying that issue is quite different from actually worrying about what to do (or how to answer intellectual questions) in case the sun actually does not rise to tomorrow. See section 10.4.6.4 of White and Sofge (1992) for a discussion of how old questions, like the question of the sun rising tomorrow, do in fact get assimilated into more far-ranging theory in the neural network field. They do have a serious link to the hard-core scientific work to be summarized very briefly in the following section.

5. INTELLIGENCE, EMOTIONS, CREATIVITY AND ETHICS

In most of my research, I have found it preferable to address the issue of "intelligence," rather than the issue of "consciousness," because it expresses more exactly where the hard-core scientific issues really lie. My view of intelligence is itself somewhat controversial, and some psychologists would argue that it is far too narrow; however, even my view requires us to include both emotions and creativity as attributes of intelligence. This is one case where neural net theory does indeed have something to say about conventional views of the mind: *contrary* to popular wisdom, as expressed in Star Trek and such, intelligent androids and the like cannot be devoid of emotional systems, because emotional systems are a necessary component of intelligent systems (Werbos, 1992a, 1992c). There are excellent reasons to expect this conclusion to apply even with fuzzier, less specialized views of "intelligence."

In my own research, I have defined an *intelligent system* as a system capable of maximizing some kind of measurement of utility or reinforcement or performance or goal-satisfaction (with or without prior knowledge of how that measure is defined as a function of other variables) *over time*, in an environment whose dynamics are not known in advance, so that the system must learn both the dynamics and a strategy of action in real time through experience. It must be a *generalized* system, capable of adapting to "any" noisy, nonlinear environment, if given enough time to adapt. (See White & Sofge, 1992, Chapter 10, for more precise concepts to replace the word "any.") This definition implicitly includes the ability to solve complex problems which, in turn, implies some degree of creativity. Neural net designs now exist, on paper, which appear fully capable of meeting this definition (Werbos, 1992c; White & Sofge, 1992), although there are a few points where the approach is clear but the details have yet to be worked out (Werbos, 1993a, 1994c). Some psychologists would complain that human beings are not totally rational or optimal; however, realistic neural net designs have imperfections which are similar in many ways to those of humans.

Why are "emotions" necessary as part of such an intelligent system? The technical arguments are given in more detail in the sources already cited. Crudely speaking, any "intelligent" system — by my definition or any other — should at least have some ability to learn how to take actions at the present time which lead to better outcomes (by some criterion) in the future. It should have some degree of *foresight*. Foresight also turns out to be essential even to stability in conventional control systems like chemical plants controllers trying to maintain operation at a fixed set-point (Werbos, 1996). There are really only two ways to achieve "foresight" in the general case, where we can't cheat by exploiting linearity or the like: (a) by building explicit plans for what we will do and what will happen, extending all the way into the distant future as far as we care about; (b) by developing an *evaluation* system, or "Critic," which can be used to predict the *long-term benefit* of the various *near-term* alternative outcomes of alternative actions. (One can, of course, combine both planning and a Critic.) Whenever it is not possible to plan the future exactly — because of uncertainties or variables beyond one's control — then an adaptive Critic becomes essential.

When there are many, many variables to be considered (as in human decision-making), it is not enough to have one large evaluation system which produces a *global* evaluation of the entire state of one's environment. It is important to have *individual* evaluations, analogous to prices, for each of the important variables or objects in one's environment. This idea — the idea of calculating a positive or negative evaluation for each object — corresponds exactly to Freud's notion of "emotional charge." (It also relates to the ancient idea of "hopes" or "fears" attached to individual objects or variables. Hope and fear refer specifically to the "emotional" reactions — positive or negative weights — placed on different variables, based on their implications for the future success of the organism. The words "good" and "bad" also express such assessments by the organism.) Backpropagation itself originated in 1974 as a surprisingly direct translation of Freud's concept of "emotional energy" or "psychic energy" into mathematics; those concepts are also the basis of the most powerful neurocontrol systems in engineering applications today (Werbos, 1994b, 1995a). Grossberg (1982) argued that an emotional system is needed even to replicate the simplest kinds of memory capabilities found in the human brain. Levine and Leven (1992) also discussed the importance of emotional systems at some length.

Classical views of intelligence have often assumed that intelligence is either a binary variable (either you have it or you don't) or a continuous variable (everything from microbes to superhumans has a certain degree of it). A careful examination of the real-time optimization designs now available (White & Sofge, 1992) suggests, instead, that intelligence is more like a *quantized* or *discrete* variable. (Continuous variables like brain size and metabolic level also have some significance, contrary to what is politically correct; if they were irrelevant, evolution would have settled on a zero-cost zero-weight brain.) For example, even with simple supervised learning networks — which probably exist as local circuits in the brain (Werbos, 1994a) — there are fundamental, qualitative differences between different types of design: local designs based on fixed preprocessors, feedforward designs with adaptable hidden units, and simultaneous-recurrent networks adapted by simultaneous backpropagation. These different types of design yield distinct quantum levels of capability in approximating functions (Werbos, 1993a).

At a more global level, Bitterman (1965a, 1965b) demonstrated years ago that there are basic, qualitative differences between intelligence in different classes of vertebrates, as seen in experiments on behavior. He also showed that these differences have definite links to the qualitative differences in the gross cellular architecture between brains from different classes of vertebrates. These differences, in turn, can be related to clear-cut differences which exist between different levels of design in artificial neural networks; for example the "error critic" design in White and Sofge (1992, Chapter 13) requires something like a merger of limbic (critic) cortex and general (neuroidentification) cortex, which does in fact underlie the historical evolution of neocortex in the mammal, whose removal (according to Bitterman) generates the removal of processing capabilities which happen to be related to error critics. To an engineer, it is astonishing that anyone would have simply assumed qualitatively equivalent behavior from well-designed systems with radically different components and structures; however, behaviorist dogma historically made it very difficult to study these basic realities. (A cynic might argue that the behaviorists were trying to defend themselves against the charge that experiments with animals might not tell us directly about humans. Another explanation is that behaviorists were trying to save the world from the dangers of racism — including racism against snails and microbes.) The requirement for an emotional system applies, however, even to the *simplest* level of intelligence within vertebrates; *all* vertebrate brains do possess a limbic system.

What would it take to achieve a quantum level of intelligence which can truly adapt to "any" environment, up to the full potential of the universal Turing machine? In Werbos (1992b) and White and Sofge (1992, Chapter 13), I argued that full Turing machine capabilities require the use of explicit symbolic reasoning. The naive next step is to conclude that human beings — who seem capable of symbolic reasoning by use of words or mathematics — represent a quantum step in the evolution of intelligence, above other mammals. From the viewpoint of everyday experience, this would seem highly probable, at first.

On the other hand, formal symbolic reasoning is a surprisingly recent phenomenon. It is easy enough for humans to *utter* words, but the conscious *manipulation* of words or equations by use of formal symbolic logic and related techniques is relatively new. In fact, the articulation of experience into formal logical propositions or equations is also new. Without such articulation, symbolic reasoning as such has little value. Of equal importance are those forms of "visualization" which translate back from formal symbols into presymbolic "images." The general development of symbolic reasoning over the past few millennia has been charted in some detail by Sapir (in comparative linguistics) and by Max Weber (in comparative sociology). For ideological reasons, Max Weber has become quite popular in recent years and Sapir has not, but the history they summarize remains quite serious.

In the neural network field, Jim Anderson (e.g., Anderson, Spoehr, & Bennett, 1994) has done extensive modeling of how humans learn arithmetic. Based on his empirical findings, he has argued that humans possess "two" learning mechanisms: (a) a highly developed and fine-tuned "sensory" system, shared with other mammals, and (b) a "buggy alpha test version" of formal symbolic reasoning. After all, if symbolic reasoning is the foundation of human technology and civilization, how do we explain the fact that human technology and civilization is only a few thousand years old? The obvious answer (elaborated on in Werbos, 1992c) is that humans represent a *recent, unstable transitional life-form*, which has only recently evolved just enough capability for symbolic reasoning to let it muddle through a few technological design problems, on a one-in-a-million basis (which is still enough to start a technological civilization, when there is a culture available to disseminate new ideas, as has been observed even in chimpanzees). We ourselves are the "missing link" between the mammalian and the symbolic levels of intelligence. Perhaps there will never be such a thing as a fully perfected symbolic reasoner, but it is clear that humans have not exhausted whatever potential does exist. These ideas may be seen as an explanation for related observations by Lorenz, as discussed by Levine in Chapter 1.

One might then pose the problem of consciousness as follows: Are human beings really "conscious" or "intelligent"? Perhaps not, in the larger scheme of things. In Werbos (1992c), I explained how simple wiring changes, related to the balance between the waking state and the dreaming state, might be central to human abilities in symbolic reasoning. (These, in turn, might be related to the unique wiring of the human thalamic reticular nucleus as discussed by Alavi and

Taylor, 1994, and Taylor, 1992). If so, there is little doubt that such capabilities could be wired into a computer as well. Computers could be made "conscious" or "intelligent" at a level beyond that of human brains today, if we were crazy and suicidal enough to want to do this.

In my view, the biggest symptom of our lack of evolution is our inability to master the most fundamental aspects of symbolic reasoning: the ability to accurately articulate our true goals and values, in a way which is totally in harmony with the presymbolic aspects of our thought and allows us to master symbols instead of being mastered by them. In crude language, the problem is that we lie to ourselves. (In psychiatrists' terms, we overuse denial as a defense mechanism.) We lack the ability to simply articulate — in a direct, honest way — the information coming to us from all of our feelings and our everyday experience of life. My examples of Anglo-American philosophers and statisticians, in Section 2, are not isolated examples. To perform reasoning effectively, humans must learn even the most basic things the hard way, like dogs learning to walk on two feet. It is natural for humans to learn symbolic reasoning, when they have enough time and help and intelligence, but the process can be very difficult. The basic foundation of Confucian ethics — to learn to know oneself, and to be "true" to oneself — may be viewed as a clear expression of (and aid to) that learning process. In this view, the mark of a sane human being is an attitude towards life which includes a kind of total openness to the empirical data which comes to us from our senses and from our emotionally charged feelings, and an easy two-way communication and harmony between the symbolic and nonsymbolic aspects of our intelligence. This is close to the Freudian ideal of "sanity."

From a more formalistic viewpoint, Confucian ethics may be justified as follows. As Bertrand Russell pointed out, there can be no logical, operational answer to questions like "What *should* we do with our lives?" because the word "should" has no operational, objective content. However, there *can* be an operational answer to the question: "What *would* I do if I were wise? What "answers" to the problems of ethics would *satisfy me* — put me in a state of stable mental equilibrium with respect to my acceptance of these "answers" — if I fully understood myself, my feelings, and my environment?" These questions are inherently meaningful and operational because they address the *I*, the self, which *can* be understood — in part because of neural network research (Werbos, 1992c). Using these questions as the foundations of ethics leads one to the pursuit of *integrity*, as defined by Confucius. As a practical matter, one can never expect to achieve a complete and perfect understanding of one's environment and oneself, any more than one can expect to play a perfect game of chess; however, this does not invalidate the effort.

This section should not be interpreted as an endorsement of all the secondary ideas which have evolved in Confucianism over the years. Confucianism — like Christianity, Marxism, Islam, Buddhism, and Western science — has accumulated its share of obnoxious barnacles, due to the universal existence of power seekers, opportunists masquerading as zealots, gullible followers, and groupthink.

6. STATES OF "CONSCIOUSNESS" VERSUS "UNCONSCIOUSNESS"

There is a radical difference between the concept of consciousness as "wakefulness" and the concept of consciousness as "intelligence." Neural network theory already provides some insight into the reasons why intelligent organisms must have multiple states of consciousness.

For example, in Werbos (1987) and White and Sofge (1992), I argued that some form of "dreaming" or "simulation" is essential to the efficient adaptation (or effective foresight) of advanced reinforcement learning systems. After Sutton and I had long discussions of that paper (cited by Sutton) at GTE in 1987, Sutton performed simulations (described in Miller et al., 1990) demonstrating this point empirically. This interpretation of dreaming is basically equivalent to the theory developed independently by LaBerge (see LaBerge & Rheingold, 1990), who is perhaps the world's leading dream researcher.

As noted in the previous section, I have also suggested how an intermediate stage of consciousness, linked to hypnosis (Werbos, 1992c), may be important to human abilities with language. Deep sleep (and its substates?) remains a mystery, but there are new possibilities for linking that phenomenon to neural network research (Werbos, 1993a). More research is needed here, especially to pin down the link between neural net models and brain circuits, but there is good reason to expect success in this work, if sufficient effort is applied.

7. WHAT ABOUT THE SOUL?

Up to this point, I might hope that any truly rational scientist, reviewing the evidence carefully, would at least respect the views I have expressed. From this point on, I have no such illusions.

Sections 5 and 6 argued that everything people associate most passionately with human consciousness — intelligence, emotions, creativity, dreams, and so on — can be fully understood in terms of classical neural network models, consistent with the Turing theory of computation. Werbos (1994a) gave an overview of how these new models fit with specific circuits in the brain as well. By Occam's Razor, this suggests that the hypothesis of a "soul" is totally unnecessary and should be abandoned. This is clearly a highly rational conclusion to draw, and I remember believing in this conclusion very intensely back at ages 8 through 19. However, on a purely personal basis, I have come around to the view that something like a "soul" — a part of the mind and the self which cannot be reduced to atoms and neurons — is in fact necessary in order to explain the full range of human experience. Like Shaw (1965), I am concerned with dimensions of experience more subtle than those which are usually cited in these discussions, and my use of the word *soul* is not intended in any way as a reference to theology (as discussed later).

Based on past experience, I predict that most readers will feel surprise at seeing the last two sentences in print. Many readers — including some creative and prominent people — will quietly voice agreement, but wonder where we go from here. A few canny old psychiatrists may even snigger, "So someone else has discovered that you need Jung as well as Freud to come to terms with the full spectrum of human experience. So what else is new?" A few psychologists will immediately leave the room, for fear that the physicists will denounce them as practitioners of voodoo and steal all their federal funding if they are seen consorting with people who express such views. (These fears are not entirely based on fantasy, either.) A very few readers will actually feel honest, subjective uncertainty about the issue, and really seek evidence for and against. (That was my stance in 1969-1971, the period when I really first developed backpropagation, ADAC, and other backpropagation-based critic designs, though I only published

Werbos, 1968, then.)   A fair number of articulate readers — including many powerful administrators — will instantly think about two questions: (a) Has an eccentric lunatic just walked into the room? Is this another Eccles (1993)?; (b) If we make room to discuss the soul hypothesis on an equal footing with the "standard" alternative, do we risk losing the insights of neural net research and unleashing forces of sheer craziness and illogical thinking which could overwhelm us?

A chapter this brief cannot resolve the concerns of all these groups. However, I would like to comment on (a) and (b). In 1964, when I first read Hebb's ideas about these issues, I was in complete agreement with his views. Hebb was trying to explain the idea of Occam's Razor, which we now understand more precisely (White & Sofge, 1992, Chapter 10). He described how prior expectations — which encourage us not to invoke "expensive" assumptions which complicate our understanding of the universe — are important in science, above and beyond empirical data as such. For example, he pointed toward laboratory work in parapsychology. He argued that most scientists would probably agree with the conclusions of that work, *if* they judged the statistics as they do with most scientific papers they read.   However, because those conclusions have a huge improbability "cost" a priori, we still tend to disbelieve them, if we look at prior empirical information. Based on Section 5, I would take this a step further: I would argue, even now, that *all* laboratory data we now have regarding human abilities, from problem solving through to parapsychology, is *still* not convincing enough to justify the soul hypothesis.

In fairness to the parapsychologists, I should confess that I do not know their literature well enough to draw strong conclusions. There is an analogy here between parapsychology and the study of ancient history: it requires reliance on a huge body of secondary sources, many of them quite willing to stretch the truth in favor of diverse biases (some in favor and some against), so that it would take a huge effort to make a truly judicious analysis. Even if one did all that work, one should recall the example of Aristotle, who produced a wonderfully judicious resolution of the scientific issues of the time; judicious or not, it was dead wrong. Thus even if the results from parapsychology were very clear-cut, the average scientist could not afford to know enough to find a compelling reason to believe them.

Given this situation, how could I — or any other scientist, thinking for himself or herself — give any credence at all to the soul hypothesis? In my own case, the answer lies in direct, personal observation of what I see around me. *I do not expect* all rational scientists to agree with me, because they do not share the same base of experience. But I do not accept the idea that I myself, in formulating my own views, must discard any personal experience which has not been socialized through the laboratory. I like to believe that my interest in the human mind, and my acceptance of the existentialist/Confucian viewpoint back in 1964, was the real cause of my making these observations — which I did not allow myself to accept for several years.

Just how eccentric is it to be open to the soul hypothesis based on personal experience? Years ago, the National Science Foundation commissioned a study of the underlying values of Americans, through the National Opinion Research Center (NORC) at the University of Chicago, a leading center for surveys and sociology and the like. Among the difficult issues they addressed was the nature of beliefs and experience related to the soul hypothesis. They discovered that personal experiences played a far greater role than they had expected. Even more surprising, they found that the percentage of people claiming such experience increased monotonically with education and other measures of success. The investigators have reported (Greeley & McCready, 1975) the great surprise they encountered when they presented this finding to their review board. A skeptic on the board pointed out that their statistical results would predict that 70% of that very board (composed of PhDs) would have answered "yes" to a highly inflammatory-looking question. After this, 70% of the board did in fact come forward, reluctantly, and validate the prediction — to the great surprise of everyone in the room. My own views of the soul hypothesis and the relevant experience are considerably more complex and idiosyncratic than what was reported in Greeley and McCready (1975), but whether I am a lunatic or not, I am certainly not a very eccentric one (except perhaps in my willingness to articulate taboo ideas, when my session chair asks me to address a controversial issue). There are many serious, technical people who take the soul hypothesis seriously, and they merit equal time.

Would these statistics be different for people who — in addition to being well-trained — are highly independent, creative thinkers, the kind of people who have demonstrated more than anyone else their ability to ignore conventional wisdom and arrive at their own viewpoint? Let us consider the four greatest physicists of this century, the pioneers who rebuilt the foundations of modern physics — Einstein, Schrodinger, Heisenberg and DeBroglie. Einstein often used the word *God*, and is alleged to have been a mystic; however, what I have seen of his writings suggests that this was not anything more than the erudite but firmly "secular" theology I have seen often, expressed in similar ways, at the local Unitarian church. On the other hand, records of the conversations between Schrodinger and Einstein make it clear that Schrodinger was deeply interested in things like Sufi mysticism — something which is far more than mere allegory. Heisenberg consistently described his physics in Vedantic terms, and invited well-known yogis to expound their views at the Copenhagen Institute. DeBroglie is said to have been a follower of Bergson's vision of collective intelligence, which seems like a close relative of Teilhard de Chardin's views. All in all, the 70% figure would seem to be in the ballpark here.

Would the soul hypothesis per se undermine the effort to understand the mind in a scientific way? On the contrary, one might argue that efforts to repress this idea (or to hand it over to television preachers) would be as conducive to sanity as any other kind of gross repression of thought.

The greatest abuse of the soul hypothesis has come from power seekers who try to use it as an excuse for making other people follow their orders in a blind, unthinking manner, without opening themselves to personal experience or to mathematical or scientific efforts to understand that experience. The formulation I propose here still starts from the Confucian/existentialist view; that view clearly argues that we should try to be true to our *entire* self — including both the brain *and* the soul. If neural network mathematics is useful in understanding the *general* phenomenon of intelligence — regardless of the hardware that implements this intelligence — it should, in principle, be useful even in explaining other forms of intelligence. The Appendix to this chapter describe some of my personal thoughts on this point, for those who take the hypothesis seriously. Section 8 will explain why I use this ancient four-letter word "soul," despite the unfortunate associations it conjures up in the minds of some readers.

8. QUANTUM COMPUTING, MIND AND SOUL

Quantum computing is a serious and exciting new area for research. However, like the neuroscience of consciousness, it has spawned massive confusion, both in the public and in the scientific community, in part because it combines two complex research areas — quantum field theory (QFT) and advanced computing. Even within the scientific community, there are relatively few people who truly understand the basics of *both* of these areas.

In this section I argue that there is a realistic possibility that quantum computing might produce generic, useful computational capabilities, and that related capabilities might even exist in the "soul" (if the soul exists) but probably not in the brain. However, I suggest that these capabilities could only become intelligible after we reorient this research in new directions. Before explaining these points, I must first review some basic facts which are well understood already by the relevant specialists.

Some people imagine that a valid understanding of computation in the brain must make reference to quantum theory because, after all, electrons and protons and so on are governed by quantum theory. But one could apply the same logic to computer chips as well; they too are made of electrons, protons and neutrons. In actuality, quantum theory *is* used routinely by designers of fundamental electronic devices like transistors and gates; the literature on electronics is already quite full of concepts like quantum wells, tunneling junctions, band gaps, Bohm-Aharanov rings, and so on. But all of this is at the *device level*. One uses quantum theory, for example, to design a device which performs a task like the logical "AND" operation. Then, when combining low-level devices together to make a useful computer *system*, one relies mainly on classical, digital logic or (as in artificial neural networks) on classical, simple analog concepts. Penrose (1989) did a reasonably accurate job of describing the kind of logic that we use when we build up systems from devices. Our new designs in the neural network field have many advantages in terms of cost and throughput, but they still fit into this general framework.

In formal terms, all of the computer systems in use today — from personal computers through to biologically inspired holographic systems — can be understood as "Turing machines." They fit into a universal theory of computing systems developed decades ago by Alan Turing.

Quantum computing is a novel effort to design computer systems which exploit fundamental effects in QFT which cannot be reduced to Turing machines. Early work in this field was inspired by suggestions from Richard Feynman, one of the co-inventors of QFT. An excellent survey was published by David Deutsch (1992), one of the leading researchers in this area. Deutsch developed a new universal theory of computing, analogous to Turing's, but expanded to incorporate quantum effects. Deutsch and other workers in this field demonstrated that quantum effects can be used to perform tasks which cannot be performed nearly as well by Turing machines. Nevertheless, the tasks described so far appear more like curiosities, rather than the basis of any generic technology. Deutsch expressed doubt as to whether any of this will ever have practical significance to any form of generic computing technology; however, he hoped that it was too early to tell. This literature provides no basis at present for believing that quantum effects are important in any way to the phenomenon of intelligence.

Within the fields of psychology and neural networks, many researchers have suggested that field effects or even three-dimensional Schrodinger equations could be important to intelligent systems (Pribram, 1991; Werbos, 1993b). But the computational mechanisms proposed in that literature are *not* examples of quantum computing as just defined. They are fully within the range of what can be simulated (albeit inefficiently) on conventional digital computers. They are fully within the range of what can be implemented efficiently in the kind of hardware used for artificial neural networks.

Hameroff and his collaborators (see Pribram, 1994) recently proposed that coherence effects like those used in lasers might produce true quantum computing effects within cell microtubules. There are excellent computational reasons to predict that microtubules do play a crucial role in "intelligence" in the brain (Werbos, 1992a, 1994a); however, this does not require quantum computing effects. For Hameroff's coherence effects to work, Penrose has calculated that they would somehow have to involve correlations across 10,000 neurons or more. There is no indication of what new computational capabilities such a correlation would lead to, and no indication that such effects would have anything to do with what we see happening at that level in the brain. It is not obvious that laser-like activity could be possible in assemblies of neurons.

All of these negative conclusions and loose ends appear very discouraging at first. However, they are really quite typical of any research field in its early stages. The neural network field went through a similar period of discouragement, between the publication of Minsky's book on perceptrons and the work which led to the popularization of backpropagation (Werbos, 1994b). Fifteen years ago, the most serious, well-informed analysis of fuel cells in transportation appeared quite negative; however, new approaches and breakthroughs have made this the lead candidate for the automobile of the future, and the subject of a major joint initiative between the President of the United States and the automotive industry. There is a legitimate basis for hoping that new approaches might work as well in the field of quantum computing.

Conventional approaches to quantum computing are inspired mainly by the Copenhagen or the many-worlds interpretations of QFT, and by conventional digital, sequential computing. But there are other interpretations of QFT in existence. Regardless of which interpretation is actually true, in an objective sense, they are all close enough that they give some valid intuition about the phenomena themselves. One interpretation which I have developed (Werbos, 1994d) is the idea that quantum effects can be explained by assuming that causality runs forward *and* backward, symmetrically, in quantum experiments. Thus, when people use special crystals to demonstrate basic quantum effects, there is a kind of settling down through a resonance between past and future — like a Hopfield net or a simultaneous-recurrent net (Werbos, 1992a), but *without the need to wait* for convergence through iteration in forward time. Even if the human brain has no such capabilities, I can imagine a possibility (with 20% probability?) that this could be used to increase the power of optical neural networks. It is questionable that humanity would benefit much from such technology, but the intellectual issue is worth resolving.

Because Penrose has generated some strong visceral reactions among physicists, I need to make a few side comments here, for the physicist. In my alternative interpretation of quantum theory, I am not hypothesizing that "quantum causality" (as Schwinger would define it) is violated; rather, I am merely highlighting the well-known fact that ordinary time-forward causality — causality as defined in the original Bell-Shimony work — is violated by standard quantum electrodynamics (QED). (In my papers, for example, I cite well-known work by Von

Neumann and DeBeauregard on this point.) I am not assuming any deviations from QED in this argument. My alternative interpretation is relevant here only as a way of getting intuition *about* QED. Likewise, I am not talking about a kind of computing which would require astronomical energies; ordinary Bell's Theorem experiments have been conducted at very ordinary levels of energy, using the same kinds of photorefractive crystals that people use in optical implementations of ANNs. As this book goes to press, both Elizabeth Behrman of Wichita State University and John Caulfield of Alabama A&M University have claimed serious progress in developing ideas and designs of this sort, involving realistic optical computing hardware.

One reviewer — a non-physicist — has asked for a simple example of backwards causality in quantum physics. The simplest example I know was discussed in my 1974 paper on quantum foundations (cited in Werbos, 1993c, 1993d), based on the account of nuclear exchange reactions in Segre's book *Nuclei and Particles.* Suppose you could design a cannon which, without any automatic control system, could, whenever an enemy rocket is about to come up over the horizon, automatically swivel into exactly the right angle, and fire at the exact time, so that it will hit the target exactly when the target first appears over the horizon, *even if the target is fired after the cannon must fire to meet it.* If anyone ever built such a cannon, one might attribute it to magic or precognition, or suspect over-the-horizon radar and cheating. But neutrons, shooting pi mesons out to oncoming protons, have displayed exactly such a "precognition." The conversion of the oncoming proton to a neutron proves that charged mesons are exchanged. More relevant, but complicated, examples (involving optics and Bell's Theorem) are cited in Werbos (1993c, 1993d). Behavior like this may sound mysterious, but it is fully consistent with the model of a universe governed by partial differential equations.

Taking this further, some of my friends have suggested that quantum effects and holographic processing could possibly explain the aspects of experience which I attribute to "soul." As an example, one of these friends has cited the work on remote viewing of H. E. Puthoff and Russell Targ at SRI International in the 1980s, funded by the Department of Defense. Unfortunately, I do not have easy access to that work, and I do not have strong feelings about its validity. However, the concept of remote viewing does exemplify the kind of phenomenon which — if true — would present an interesting challenge to physics and psychology. It is easier to discuss than the more complex phenomena which I find more interesting.

Quantum effects and holographic effects by themselves could not begin to explain something like remote viewing. The kinds of mechanisms which we observe in the brain — the mechanisms which drive the creation of chemical bonds, the flux of electromagnetic fields, and the movement of currents — are based entirely on quantum electrodynamics (QED), an aspect of QFT which is well understood in phenomenological terms. QED fully incorporates quantum effects, and it underlies all forms of holography now known to the human species. It is not a deep, dark mystery. If quantum and holographic effects were enough to give us a capability to see a picture of a remote location far away, based on a receiving device as small as a human brain on the surface of the earth, then the scientists in the military — who are very familiar with QED — would have built such a device long ago. The military have spent billions of dollars, across many research labs and universities, trying to improve the resolution of their imaging of distant objects, using devices much larger than a brain, exploiting all kinds of interference effects at all kinds of frequencies in the electromagnetic spectrum. On occasion, highly creative physicists like Schwinger and Hagelstein have demonstrated that coherence effects can accomplish things which more pedestrian experimentalists had thought impossible; however, these things fall far short of remote viewing à la Puthoff and Targ.

Based on this work, we may be reasonably sure that "remote viewing" would require one or more of (a) a complex signal processing system and "antenna," (b) some kind of explicit cabling system or network to connect remote sites, and (c) additional physical fields beyond those covered by QED. Even in biological signal processing systems, such as the bat's sonar processing, it is clear that a large and visible chunk of the brain is needed to perform signal processing for something much less complex than remote viewing. All this suggests that we need to face up to a stark, binary decision here: either to reject the proposed class of phenomena altogether, or to consider the possibility of information processing structure (like invisible networks or invisible signal processing or "intelligence" in the universe itself) beyond what we can see in the atoms of the brain. It is rational to feel uncertain (i.e., to assign probabilities) between these two alternatives, but it is not rational to imagine that one can avoid the choice itself through some kind of fuzzy logic. As noted in Section 7, the "soul" alternative has a high a priori improbability cost; however, it need not be a lot worse than the assumption of unseen "dark matter" among astronomers, if one considers the amazing variety of biological systems on earth adapted to exploit diverse sources of energy. Still, as discussed in Section 7, there are good reasons to respect those scientists who consider the improbability cost too high to consider.

The foregoing argument does not suggest that quantum effects, holography or complex vibrational states in large molecules are unimportant to biological intelligence. It merely suggests that they would not be enough by themselves to explain phenomena like remote viewing. It reinforces the conclusion from earlier paragraphs that there is little if any indication of true quantum computing in the brain itself even if we should postulate effects like remote viewing. However, once we postulate such effects, we can begin to imagine the possibility of yet another level of intelligence, beyond the level of single-stream symbolic reasoning, based on effects such as time-symmetric causality or the processing of multiple streams of symbols in parallel. Such possibilities are extremely speculative, of course, at the present time.

APPENDIX: A FEW PERSONAL THOUGHTS ABOUT THE SOUL

The editor of this book has asked me to say something more specific about my views on the nature of the soul, and its relation to other themes in this book. This request is reasonable; however, my thoughts on this point should not be considered part of the chapter proper, because they are inextricably linked to idiosyncratic aspects of personal observations and experience. In the absence of shared experience and lengthier, more complete explanations, I would not expect a rational reader to agree with the details of my views. I would ask the classical materialist simply to skip this appendix; it is, at best, a "what if" piece, asking what we might conclude *after* we agree that the soul does exist.

My own experience is perhaps closer to the kind described by Jung (see Campbell, 1971) than to the kind described by Greeley and McCready (1975), although I can relate somewhat to both. Greeley and McCready state that the experiences they refer to *are not limited* to any religious or ethnic group, but that most educated people tend to become much more involved in their own religious heritage and more committed to its beliefs after undergoing such experience. I find this disappointing, and perhaps further evidence that we are still a transitional species. To the extent that there is common experience out there, logic suggests that it should push us toward more common conclusions, rather than toward greater provincialism and sectarianism. It is one thing to appreciate the living culture and past experience of one's provincial heritage; it is a totally different thing to endorse florid theories of bureaucratic rather than empirical origin (like the Government Printing Office Style Manual or lists of prison sentences in purgatory), without paying full attention to the global heritage of humanity as a whole.

After one accepts that the soul exists, one's prior probabilities (per Hebb's argument) change substantially. One naturally tries to learn from the experience of others, as well as oneself. In anthropology, the example of penicillin is famous: penicillin (in bread mold) was used in healing for many, many years by African witch doctors, but totally ignored by scientists because they did not like the *explanations* used by the witch doctors; knowing about that example, we may try to learn what we can from the experience of many cultures, without letting ourselves be put off by our disrespect for their explanations of their experience. Of course, we must be careful about the ways in which rumor and wishful thinking tend to distort experience in predictable ways (especially when they tend to deify people in power).

After having explored more cultures and people than can be summarized here, I feel confident that no one on earth has a legitimate basis for describing the nature of the soul in any detail. The exploration has been worthwhile for other reasons, and important insights are to be found in obscure cultures, but none of these people begins to approach the level of qualitative understanding we would want to demand, as scientists. In understanding the soul, we are like tenth century people interested in astronomy; some important information is available, but if we *demand* full understanding in our lifetimes, we will only set ourselves up to become victims of other people's fantasies. A rational, honest, intelligent human being would have to take the approach described by Raiffa (1968) in decision analysis: to accept uncertainty as an unavoidable fact, and to live with it as best we can. We may choose to work hard to grow in understanding, but to do this effectively we must admit the limitations we face. We need to play these issues by ear, to maintain some balance and detachment, to rely heavily on direct observation (which we constantly try to enhance), and to maintain several alternative working hypotheses.

In examining historical ideas about the soul, I am amazed at the florid details of religious mythologies which contradict each other and are rather easy to explain away in psychoanalytic terms as creations of the mind (Campbell, 1971). On the other hand, it is hard at times to avoid some degree of respect for the extreme Buddhist viewpoint that *everything* we see can be explained away as a creation of the mind, including the walls and the floor; however, such feelings can be explained away as a consequence of our present ignorance, and are comparable to the pessimism of certain neuroscientists regarding our understanding of the brain (Werbos, 1994b, p. 2). Still, the existence of an alternative explanation does not disprove the concept.

As a humorous aside, I can imagine someone arguing that everything we see is a product of Mind, and that Mind in turn is governed by backpropagation — ergo that backpropagation is the foundation of everything. Even as the inventor of backpropagation, I would find that idea a bit too much.

If we find that florid mythologies are unsatisfying (and are too large a set to select from, anyway), our best hope for avoiding chaotic, pure phenomenology is to use some of the same ideas we use in science, including Occam's Razor. In fact, even mystics use expressions like "As above, so below," and expound the idea of *monism* — the idea that the soul and the body are governed by the same set of natural laws, laws which are no less precise and universal for being as yet unknown. Even the New Testament is full of references to things that can only be "revealed" or understood in a future age when humanity is ready, as a result of learning over time. Is it not possible that *mathematics* is a crucial part of what is necessary for such understanding, and part of what we have really been learning in the past two millennia?

From this perspective, then, can we imagine how a universe governed by some kinds of mathematical laws that we can conceive of — either from differential equation theory or information processing theory — could generate such a phenomenon as "soul"? Despite knowing more about information processing than about differential equations, I still find it hard to imagine information processing as a foundation to explain everything. The problem is that all forms of Mind that we are familiar with (and can conceive of) inherently require something outside themselves to relate to (Jung in Campbell, 1971). Finkelstein (1985) and others have looked for reformulations of quantum theory in terms of Mind based on quantum neural networks; however, it is my understanding that such efforts have not gotten very far. If we cannot yet conceive of a universe governed by information processing concepts, then we are left with the alternative of partial differential equations, an approach studied at length by physicists such as Einstein.

Any differential-equation-based Cosmos would presumably be governed by thermodynamic principles, like those we experience here which generate Darwinian selection, or a generalization to account for causality forward and backward in time. (See Werbos, 1994d, for a discussion of relations between these concepts). In a Darwinian Cosmos, the soul might be a kind of living organism, based on fields and forces not yet understood, living in symbiosis with the other part of us. (I am reminded of the Star Trek episode where Dr. Crusher points towards a "ghost" and says something like: "You ... you are not really a spirit ... I now know what you really are, you dirty cheater ... you are nothing but a *life form*." But this "ghost" was not the only one of us guilty of being a life form.) The traditional *alchemical marriage* (Campbell, 1971) can be seen simply as an effort to get both parts working in harmony, in a unified way, in recognition of the fact that this is the only way to get a Pareto optimal result for both parts. When storing information, however, one would normally prefer to store it in more permanent hardware. (Some mystical traditions argue that *all* humans routinely exercise capabilities beyond what they consciously believe in — but that people have difficulties in putting enough learning or experience into their souls to permit easy memory or control of such faculties. Hebb [1949] commented that more brain space and learning time are needed when learning to cope with larger volumes of sensory input.) The quality of symbiosis might depend *both* on actions initiated on the soul side *and* on the normal genetically determined capabilities of the nervous system.

Based on these ideas, one might imagine two kinds of symbiosis — a one-to-one symbiosis, or a many-to-one symbiosis. The latter would match a range of traditional mystical beliefs, from Jung's collective unconscious to Teilhard de Chardin (1972) or the Gaia hypothesis (Lovelock, 1992). If we postulated such a collective intelligence or soul, I would predict that our experience of the soul would be analogous to the experience of a single neuron (or cell assembly) inside a higher order neural network; for example, we may be whipsawed by backpropagation effects at times, or we may find ourselves acting as powerful channels of psychic energy (backpropagation), especially when we crystallize concepts which can help the entire global system to escape from local minima, and to grow in maturity. In *either* model of symbiosis — one-to-one or many-to-one — I would expect that issues related to psychological growth and ego formation, as described by Freud and clarified by neural network models, would apply in a similar way both to the soul and to the brain.

More recently, I find myself influenced by images (Werbos, 1994d), which come closer to older ideas of a much larger web of life, in which people may vary in their degree of immersion in the more local collective intelligence. There has been much interest lately in the Gaia hypothesis (Lovelock, 1992), which has been used, for example, as a rationale for *environmentalism of the spirit* (Gore, 1992). (There have been many treatments of this idea in science fiction as well, including some of the works of Orson Scott Card, Silverberg, and Chalker.) All of this fits well with my own thoughts, but lately I feel there is something fundamentally incomplete in that image. Recently, I find myself more attracted to the old Chinese image, which pictures humanity more as a middle kingdom, poised between earth and sky — demanding a balance between these two strong spiritual connections or parts of our lives.

Some readers may feel that I have left out some crucial things in this very brief account. I agree, and a few of the holes are filled in (albeit still very briefly) in Werbos (1986, 1992c, 1993b, 1994d). As a practical matter, I do not spend a lot of time thinking about these concepts, however great their putative importance, because I recognize how great our ignorance really is; however, there is no doubt that they substantially color my perception of human events, and I like to believe that they do at least represent some improvement over the traditional extremes of florid, fearful ethnocentric mythologies and cold, grey, blind materialism, both of which substantially inhibit the natural human tendency toward spiritual growth.

REFERENCES

Alavi, F., & Taylor, J. G. (1994). Computer simulation of conscious sensory experiences. *World Congress on Neural Networks, San Diego* (Vol. 1, pp. 209-214). Hillsdale, NJ: Lawrence Erlbaum Associates.

Anderson, J. A., Spoehr, K. T., & Bennett, D. J. (1994). A study in numerical perversity: Teaching arithmetic to a neural network. In D. S. Levine & M. Aparicio, IV (Eds.), *Neural Networks for Knowledge Representation and Inference* (pp. 311-335). Hillsdale, NJ: Lawrence Erlbaum Associates.

Bitterman, M. E. (1965a). Comparative analysis of learning, *Science*, **188**, 699-709, 1975.

Bitterman, M. E. (1965b). The evolution of intelligence. *Scientific American*, January 1965.

Campbell, J. (Ed.) (1971). *The Portable Jung.* New York: Viking.

Deutsch, D. (1992). Quantum computing. *Physics World*, June 1992.

Eccles, J. C. (1993). Evolution of complexity of the brain with the emergence of consciousness. In K. H. Pribram (Ed.), *Rethinking Neural Networks: Quantum Fields and Biological Data* (pp. 1-28). Hillsdale, NJ: Lawrence Erlbaum Associates.

Finkelstein, D. (1985). Superconducting causal nets. *International Journal of Theoretical Physics*, **27**, 1985.

Gore, A. (1992). *Earth in the Balance: Ecology and the Human Spirit.* Boston: Houghton-Mifflin.

Greeley, A. M., & McCready, W. C. (1975). Are we a nation of mystics? *New York Times Magazine*, January 26.

Grossberg, S. (1982). A psychophysiological theory of reinforcement, drive, motivation, and attention. *Journal of Theoretical Neurobiology*, **1**, 286-369, 1982.

Hebb, D. O. (1949). *The Organization of Behavior.* New York: Wiley.

LaBerge, S., & Rheingold (1990). *Exploring the World of Lucid Dreaming.* New York: Ballantine.

Levine, D. S. (1994). Steps toward a neural theory of self-actualization. *World Congress on Neural Networks, San Diego* (Vol. I, pp. 215-220). Hillsdale, NJ: Lawrence Erlbaum Associates.

Levine, D. S., & Leven, S. J. (Eds.) (1992). *Motivation, Emotion, and Goal Direction in Neural Networks.* Hillsdale, NJ: Lawrence Erlbaum Associates.

Lovelock, J. E. (1992). The Gaia hypothesis. In L. Margulis & L. Olendzenski (Eds.), *Environmental Evolution.* Cambridge, MA: MIT Press.

Miller, W., Sutton, R., & Werbos, P. J. (Eds.) (1990). *Neural Networks for Control.* Cambridge, MA: MIT Press.

Penrose, R. (1989). *The Emperor's New Mind: Concerning Computers, Minds and the Laws of Physics.* Oxford, UK: Oxford University Press.

Pribram, K. H. (1991). *Brain and Perception.* Hillsdale NJ: Lawrence Erlbaum Associates.

Pribram, K. H. (Ed.) (1994). *Origins: Brain and Self Organization.* Hillsdale, NJ: Lawrence Erlbaum Associates.

Raiffa, H. (1968). *Decision Analysis: Introductory Lectures on Making Choices Under Uncertainty.* Reading, MA: Addison-Wesley.

Santiago, R., & Werbos, P. J. (1994). New progress toward truly brain-like intelligent control. *World Congress on Neural Networks, San Diego* (Vol. 1, pp. 27-33). Hillsdale, NJ: Lawrence Erlbaum Associates.

Shaw, B. (1965). *Back to Methusaleh.* In *The Complete Plays of Bernard Shaw.* London: Paul Hamlyn.

Taylor, J. G. (1992). From single neuron to cognition. In I. Aleksander & J. Taylor (Eds.), *Artificial Neural Networks 2* (ICANN92 Proceedings, pp. 11-16). Amsterdam: North Holland.

Teilhard de Chardin, P. (1972). *Activation of Energy.* New York: Harcourt-Brace.

Von Neumann, J., & Morgenstern, O. (1953). *The Theory of Games and Economic Behavior.* Princeton, NJ: Princeton University Press.

Werbos, P. J. (1968). Elements of intelligence. *Cybernetics* (Namur), No. 3.

Werbos, P. J. (1986). Generalized information requirements of intelligent decision-making systems. *SUGI 11 Proceedings*, SAS Institute, Cary, NC. A version updated later in 1986 is available from the author.

Werbos, P. J. (1987).   Building and understanding adaptive systems: a statistical/numerical approach to factory automation and brain research. *IEEE Transactions on Systems, Man, and Cybernetics*, January/February, 1987.

Werbos, P. J. (1990).   Rational approaches to identifying policy objectives. *Energy*, **15**, 171-186.

Werbos, P. J. (1992a).   The cytoskeleton: Why it may be crucial to human learning and to neurocontrol. *Nanobiology*, **1**, 75-95.

Werbos, P. J. (1992b).   Neurocontrol: where it is going and why it is crucial, in I. Aleksander and J. Taylor (Eds.), *Artificial Neural Networks II* , New York: North Holland. Updated versions are reprinted in Werbos (1994b) and in M. Gupta (Ed.), *Intelligent Control Systems*, IEEE, New York, 1996.

Werbos, P. J. (1992c).   Neural networks and the human mind: new mathematics fits ancient insights. *IJCNN92-Beijing Proceedings*, IEEE, New York, 1992. An updated version appears in Werbos (1994b).

Werbos, P. J. (1993a).   Supervised learning: Can it escape from its local minimum. *World Congress on Neural Networks, Portland, Oregon* (Vol. III, pp. 358-363). Hillsdale, NJ: Lawrence Erlbaum Associates.

Werbos, P. J. (1993b).   Quantum theory and neural systems: alternative approaches and a new design.   In K. H. Pribram (Ed.), *Rethinking Neural Networks: Quantum Fields and Biological Data* (pp. 299-314). Hillsdale, NJ: Lawrence Erlbaum Associates.

Werbos, P. J. (1993c).   Quantum theory, computing and chaotic solitons. *IEICE Transactions on Fundamentals*, **E76-A** (5), May.

Werbos, P. J. (1993d).   Chaotic solitons and the foundations of physics: a potential revolution. *Applied Mathematics and Computation*, **56**, 2/3.

Werbos, P. J. (1994a).   The brain as a neurocontroller: New hypotheses and new experimental possibilities.   In K. H. Pribram (Ed.), *Origins: Brain and Self Organization* (pp. 680-706). Hillsdale, NJ: Lawrence Erlbaum Associates.

Werbos, P. J. (1994b).   *The Roots of Backpropagation: From Ordered Derivatives to Neural Networks and Political Forecasting.* New York: John Wiley and Sons.

Werbos, P. J. (1994c).   How we cut prediction errors in half by using a different training method. *World Congress on Neural Networks, San Diego* (Vol. II, pp. 225-230). Hillsdale, NJ: Lawrence Erlbaum Associates.

Werbos, P. J. (1994d).   Self-organization: reexamining the basics, and an alternative to the big bang.   In K. H. Pribram (Ed.), *Origins: Brain and Self Organization* (pp. 16-52). Hillsdale, NJ: Lawrence Erlbaum Associates. For more mathematical treatments focusing solely on the issue of quantum foundations as such, see Werbos (1993c).

Werbos. P. J. (1995a).   Neural networks for flight control: a strategic and scientific assessment. In M. Padgett (Ed.), *1994 Workshop on Neural Networks, Fuzzy Systems, Evolutionary Systems and Virtual Reality*. Bellingham, WA: Society of Photo-Optical Instrumentation Engineers (SPIE).

Werbos, P. J. (1996).   Control. In E. Fiesler & R. Beale (Eds.), *Handbook of Neural Computation* (Chapter F1.9). New York: Oxford University Press.

White, D., & Sofge, D. (1992).   *Handbook of Intelligent Control: Neural, Fuzzy and Adaptive Approaches.* New York: Van Nostrand.

3

Negotiating Inside the Brain — and Out: The Microfoundations Project

Sam Leven
Radford University and For a New Social Science

Sam Leven's chapter, **Negotiating Inside the Brain — and Out: The Microfounda-** *tions Project,* *sets out to challenge the notion that human decision making is based on the rational, systematic search for optimal solutions. He starts by tracing the history of this concept: the Frenchman de la Mettrie in the eighteenth century gained the enmity of the church through his "man is a machine" concept, but in this century such mechanistic ideas are the orthodoxy. Recently, an underground against this orthodoxy has gained momentum in many fields, as Leven reviews. In computer science and artificial intelligence, Terry Winograd has argued the need for context and Marvin Minsky has sought a "society of mind." In experimental psychology, management theory, and (the ultimate goal of Leven's theorizing) economics, an increasing number of scholars have suggested the importance of emotional and automatic as well as rational processes. In logic and semantics, revolutionary scholars have suggested ways to partially systematize types of processing that follow general rules other than the strictly rational ones.*

Leven reviews some of his own earlier work whereby the theories of many social scientists are joined into a general idea of "triunity" of reason, affect, and habits. This triunity also draws on older ideas from such eminent neurobiologists as Paul MacLean, Karl Pribram, and A. R. Luria. Like some of the other authors in this book (particularly Levine, Öğmen and Prakash, Rosenstein, and Werbos), Leven believes that affect, habit, and novelty perform useful functions in human information processing. In a series of neural network models, Leven and Wesley Elsberry have incorporated these capabilities into a model of negotiation. Each of their two "negotiators" contains a "habitual" module that is similar to a Hopfield net, a "rational" module similar to a back propagation net, and an "affective" module similar to an adaptive resonance theory net. Each negotiator also has a frontal lobe-like module that serves the function of integrating all three, along lines that Pribram has suggested. Also, Leven and Daniel Levine have combined affective and semantic capabilities in a different manner to model consumer preference in an actual product buying situation (Coca-Cola). Although all this work is far from complete, it suggests that modern neuropsychological modeling can provide economists with a plausible alternative to the rational optimization models that still dominate their discipline.

ABSTRACT

In neural networks (Anderson, 1991) and economics (Arrow & Hahn, 1971), the dominance of optimizing models produces inevitable distortions (Simon, 1986, 1991). Both fields consider cooperative and distributed domains (Hayek, 1952; Pribram, 1991) — and both suffer from assumed local global optimization.

A heterodox view for modelling both fields may be maintained (Leven, 1987b, 1992). Automaticity, reason, and emotion all play a role in memory, analytic, and creative processes. Their neural substrates and interactions are modelled; an application to decision making is described.

Men are not narrow in their intellectual interest by nature; it takes special and rigorous training to accomplish that end. (Viner, 1958, p. 380).

The progress so made is immensely impressive. It is made by sleepwalking, is it wise to "wake up"? I am not sure it is. So I speak now in a very low voice. (Bell, 1988, p. 170).

1. INTRODUCTION

Many fields have assumed rationality — the systematic search for an optimal solution. We argue, from history, science, and common sense, that this assumption is wrong-headed. The failure of the Rationality Principle (Anderson, 1991) imposes heavy burdens on social and natural scientists — and changes the questions neural network modelers of decision making should ask, the techniques they should employ, and the results they can reach.

Following Bruner (1990, p. 19), we maintain that models should be concerned with "situated action" — action situated in a cultural setting, and in the mutually interacting intentional states of the participants. Only such discursive networks, aware of, and affected by, their environments, can capture the sense of thought and the feel of action (Leven, 1992a, 1992b).

In 1747, the physician J. O. de la Mettrie announced that "man is a machine." Behavior and internal function could be analogized to systems subject to numerical control, composed according to an internal logic, and understood to be as systematic as much as any other natural phenomenon (hence, not different as a language-producing creature). Language and thought, like other biological functions, were structured and systematic (not God-given). His systematic medicine and psychology provided the basis for a morality based in the perfection of the human machine, not in religion or Cartesian sentimentality.

De la Mettrie was, in fact, advocating a mathematizable cognitive psychology. Speech and ideas, rather than suggesting human superiority, demonstrated the sophistication of the natural control system involved. Each individual mind is "like a self-performing piano listening to (and amused by) its own playing" (da Fonseca, 1991, p. 31).

Others were not so sanguine about de la Mettrie's system. When his Dutch publishers were blackmailed into revealing his identity, the Calvinist, Lutheran and Catholic Churches joined in a rare combination to ban and burn his books. He was expelled from "tolerant" Holland and forced to flee to Prussia — where his next book was banned and where he died under suspicious circumstances.

Now, a specter looms over both human and natural sciences. It is the legacy of the triumph of de la Mettrie over the forces of "superstition." The modern human behavior orthodoxy that dominates such fields as psychology, computer science, linguistics, economics, and neural networks securely asserts the centrality of optimization and rationality — like de la Mettrie's self-playing piano, all behavior is cognitively controlled. The majority of textbooks in these fields has embraced the Cognitive Revolution, which began with Miller, Galanter, and Pribram (1960) — who had not, of course, read the manuscripts the holy men burned.

The irony of de la Mettrie's triumph is that his success has been total. Discipline upon discipline maintain that distributed systems (especially those dominated by human behavior) follow an optimizing, rational structure. J. R. Anderson (1991, p. 3) could posit — standing as both psychologist and neural network modeler — a "Principle of Rationality. The cognitive system optimizes the adaptation of the behavior of the organism."

The implication of de la Mettrie and Anderson for models of thinking and behavior, which has been embraced across many fields, is that there is a structure of inference involved in human thought process. Economists maintain that all markets, ceteris paribus, are composed of consumers and producers who plan and act rationally — optimizing the value of their rewards (e.g., Arrow, 1990). Computer scientists employ optimizing compilers and rationalizing data structures, Cognitive psychologists find depression to be suboptimal explanation of events (Seligman, 1991). Epistemologists suggest that "semantically valid sets of syntactic operations are 'preferred' (by Nature, that is)" (Matthen, 1989, p. 564).

An intellectual underground in each of the sciences has grown to undertake de la Mettrie's subversive role. Traditional artificial intelligence (AI) has found its Maquis among the elite: Terry Winograd, who first employed schemas in computing, has decried the lack of context and sensibility (Winograd & Flores, 1987) — and Marvin Minsky (1986), who first connected schemas to AI, has sought a "society of mind." Experimental psychologists have betrayed the fallibility of cognition (Neale & Bazerman, 1991) and its susceptibility both to affective influences and to "habits of mind" (Gilovich, 1991). Economists have faced "anomalies" (Thaler, 1991) and the unnerving prospect of moral economic men (Koford & Miller, 1991). Even in linguistics, subversion is carried on by situational linguists (Devlin, 1991) and radicals like Eco (1986) who suggest that language must convey meaning under such revolutionary influences as emotion.

Do these somewhat fuzzy-headed rebels have some basis for the attack on cognition-as-rationality-as-optimization? Is there a physiological substrate that might justify the notion that control of human behavior is not solely rational? Could one, for example, be depressed without having a thinking disorder?

We return to the source to pose such questions. He may lead us past perfection to a more accurate view of decision making (and error-making) behavior. Then, we add modestly, Sigmund Freud's work shall suggest the accuracy of our own work.

2. STRUCTURES OF MIND AND BEHAVIOR

But in what do logical faults consist? ... in the non-observance of the biological rules for the passage of thought. These rules lay down where it is that the cathexis of attention is to be directed each time and when the thought-process is to come to a stop. They are protected by threats of unpleasure, they are derived from experience, and they can be transposed directly into the rules of logic ... The existence of biological rules of this kind can in fact be proved from the feeling of unpleasure at logical faults. (Freud, in Pribram & Gill, 1976, p. 119)

Sigmund Freud, in his Project for a Scientific Psychology (Pribram & Gill, 1976), pointed to three classes of general errors: "threats of unpleasure" (affect inhibits), experience-derived (habits of mind), and those based in logic (inappropriate rule use). These types of error, we should not be surprised to learn, have been rediscovered in modern psychological study.

Recently, Reason (1990, p. 201 ff.) presented three basic classes of mistakes made in industrial accidents. Slips, Reason determined, were "attentional failures": omission, misordering, and mistiming, among others. Lapses were "memory failures" forgetting intentions, omitting planned acts. Mistakes constituted misapplication of a good rule or application of a "bad" one.

That Freud's 1896 analysis stands scrutiny one hundred years later would be a curiosity, were it not for the explanations he offers for the sources of these errors. The rationale he employed for error diagnosis is highly suggestive of a general approach to decision making. Recall that Freud distinguished errors on the bases of affect (emotion), habit (automaticity), and logical operation (semantics). These classes make considerable sense in light of findings by Pribram (1986, 1991) and others that suggest a biological basis for decision processes, including wrong decisions of course.

This approach fits with the analysis of Levine (1986) and Leven (1987b); they match the three classes of decision processes with three broad regions of mind, following approaches by Pribram (1991), Luria (1981), and MacLean (1991). The separate (though not mutually exclusive) areas are tied to three divisions of the frontal lobes, the "executive of the brain" (Levine, 1986; Pribram, 1986).[1]

Pribram (1991) stressed the contributions of three mental frames. He saw practicalities as the embodiment of routinized performances, involving the control of patterned, hierarchical behaviors and producing images of achievement. The tennis player strikes a forehand similarly thousands of times; it is a structured, systematic performance that allows the ability to adjust (or satisfice, as Simon, 1986, would say). Automatic control is regulated by feedback structures tied to amygdalar, hypothalamic, and striatal "command structures."

---

[1] This discussion follows Leven (1988, 1992a).

Pribram's proprieties constitute rule-based behaviors. Here, order is imposed by inference, rather than rote learning. An internal Socratic dialogue establishes fitness of rules and logic. Neocortical and temporal structures collaborate in brain processing of "correct reasoning."

Priorities allow the maintenance and reconciliation of "maps" — spatial, musical, emotional. These are context-sensitive capabilities that create coherence from many "strands" of information. This fronto-limbic capability is based on the making sense of complicated settings.

Pribram (1986) noted that each of these capabilities is mediated within its own third of the frontal lobes. The ability of this executive region to integrate and direct behavior allows sophisticated and "conscious" performance. The categories Pribram employs are somewhat comparable to other, well-known models of regional brain function. MacLean (1991) suggested that the three sets of capabilities constitute "Three Ages of the Brain."[2] He suggested that the motor- and drive-inducing regions of the brain (e.g., basal ganglia and hypothalamus) evolved earliest and allowed primitive creatures to acquire and perform automatic (habitual) tasks. Thus, he called the region Reptilian. The orienting, affective, context-sensitive region he termed Mammalian; this "higher" area is composed mainly of limbic structures. Lastly, MacLean's Neomammalian structure is the "latest" and "most sophisticated"; located in neocortex, it is responsible for logic and analysis.

Similar to the Pribram (and MacLean) model is research in classes of memory function. Tulving (1983 et al.) suggested a threefold typology. Tulving labeled automatic memory, memory for routinized performance, Procedural Memory. Recall of context and affect, "gestalt memory," he termed Episodic. Hierarchical and categorizing (logical recall) is Tulving's Semantic Memory. The three flavors may be recalled separately, even independently: a need (drive) or physical performance may invoke one set of recollections, a mood might evoke another, and a set of inferences could justify a third. The recalled information from the three different qualities of memory can be so different as to seem incomparable — different "keys" and different modalities are being utilized.

We should not be too surprised that these categories, matching physical structures and mental capabilities, match modern classifications of information theory. Nicolis (1991) isolated syntactic (regular, easily analyzed) patterns, semantic (logical, hierarchical) categories, and pragmatic (gestalt-making, context-sensitive) structures. He claimed the three are orthogonal and can only be emulated with very different algorithms. Chandrasekharan (1990, p. 40) found that "mental architectures" must be related to three classes of logic: hierarchical classification, concept matching, and abductive assembly — "interaction among malfunctions [becomes] a composite hypothesis."

In fact, Peng and Reggia (1990, p. 3 ff.) formalized a highly similar three "classes of logic":

---

[2] A brief critique of this model is offered in Leven (1992b).

Deductive:
Given Rule — All balls in the box are black.
+ Case — These balls are from the box.
Conclude — These balls are black.
Inductive:
Given Case — These balls are from the box.
+ Result — These balls are black.
Hypothesis — All balls in the box are black.
Abductive:
Given Rule — All balls in the box are black.
+ Result — These balls are black.
Hypothesize Case — These balls are from the box.

Deduction is automatized, routine. Induction seeks a general rule, based on an array of like rules. Abduction responds to nonspecific cues from the environment to construe a single case into a sensible order,

There is, in fact, a relationship between all the foregoing categories and Peirce's (1965-1966) semiotics. The notion that information and processing structures seem to map one-to-one and into each other suggests higher order relationships as well (Leven, 1992a).

3. OPTIMIZATION, EMPATHY, AND STRUCTURES OF BELIEF

These components of individual neural function, decision making, and logic can be employed to model personal and group action (Leven, 1987b, 1992b). Standard models of organizations support the notion of three types of organization style, strategy, and rationality (Goold & Campbell, 1989).

One, the Classical model, is based on centralized control, rigid routines, and wage-based rewards. Garbage collection is a Classical task: no one finds fulfillment doing the job, regular and careful performance is essential for safety (requiring strict supervision and the assumption that workers shirk), and pay is considered the sole likely reward. Strong leaders and quiet employees are the most effective match. Worker participation in the planning process is absent; the tasks are so routine, there is no distant horizon, and managers assume employees to be barely competent.

Another approach, the Human Relations view, is effective in complex, short-horizon planning regimes. Group solidarity is emphasized: managing a convention facility, for example, involves mediating among competing claims and taking abuse from dissatisfied customers. Employees must cope with internal and external tensions — supervisors must be systematic and understanding. The organization must ration limited resources effectively. That requires tolerance — and a logical approach.

Lastly, the Social Psychological school involves long-term problems and emphasizes the creativity of employees. Work, in this model, is complicated and demanding; workers are valuable and must be solicited for advice. The development of NASA is comparable to the tasks appropriate to this approach: talented people, encountering an ill-defined problem, must create and share unlikely results — and learn to treasure frustration.

What makes the Organization Theory models revealing is their comparability to the three approaches of economic theory. Morishima (1990) has noted that the three approaches (classical, neo-classical, and Keynesian) are not, as frequently maintained, competitors. The three models correspond to different economic and cultural circumstances.

Under conditions of low technology and resource-based economies, wages tend to regress toward a minimum, and work is highly routinized and heavily supervised. The structure of the economy seems absolutely stable; planning horizons are extremely short, even seasonal. Under these circumstances, Morishima explained, the traditional wage-fund, Ricardian approach (control by exploiters) is accurate: many work, few rule.

As basic industries (iron foundries and steel mills) grow, the neo-classical model becomes plausible. Each new product requires massive quantities of other materials and equipment. The dictum of Say's Law, "Supply creates its own demand," is fully appropriate — after all, a rapidly industrializing society has virtually infinite needs. As more output becomes available, more employees are needed to use or transform it. Iron leads to railroads, lead to steel mills, lead to port construction — and, as scale economies take place, lead to export. Here, any employee is valued, because he or she is needed badly. Horizons are, as in the Human Relations model, long enough to plan new upgrades in technology; the complexity and constant newness of the tasks lead to high esprit de corps that accompanies rising wages.

Finally, as an economy attains modernity, large profitable firms require individual risk-takers as employees and indulgent, whimsical customers (changes in fashion often drive the market). Here, Keynes' demand-management approach is appropriate. Workers have high skills and face volatile markets (e.g., Silicon Valley computer firms). Managers are frequently dependent on their employees — they may not even understand what their "charges" do! Morishima suggests that the world looks very different to these folks: change is frequent, technology advances as long as demand is strong, consumers "self-actualize."

Just as large multinational firms may employ elements of all three organization models (Goold & Campbell, 1989), so economic circumstances coexist — even within the same firm. Eliasson (1991) suggested that firms migrate among the three mind-sets as their environments and problems shift; the three different "rationalities" (exploitative, managerial, visionary) must compete and cooperate in changing markets.

Eliasson has sought to model the functions of the firm, within a competitive economic environment. He employs standard computational technique appropriate to the three models, involving different time horizons, different goals, and different abilities to adapt (acquire previously unknown existing technologies). This ambitious project allows the construction of complex economic games that demonstrate the competing goals of the existing firms and the problems involved in absorbing innovations.

As remarkable as the elaborate design is, it lacks certain qualities neural networks could bring to such a simulation. Nets can respond to ambiguous patterns (even with ambiguous responses!). A few models (discussed later) have learning capabilities that traditional dynamic programming models lack (Werbos, 1974, 1990). And, if neural nets bear any relationship to human thought processes, they may provide an opportunity to employ "human-like" qualities of thought.

Leven (1987b) and colleagues (Blackwood, Elsberry, & Leven, 1988; Leven & Elsberry, 1990) developed a model of "discursive networks" (Leven, 1992a). The group simulated dyadic bargaining between two complexes of neural networks (Leven, 1992b). Recent work involved the extension of this work to multilayer economic games (Leven, 1995). The work of this group, they asserted, allows emulation of habits, affect, and logical analysis (Leven, 1987a). The dyadic model allows gestures and expressions to "frame" the logic of negotiation — and to change its nature and outcomes.

The logics employed parallel the ones we have discussed. The task of going "beyond optimization" they have set themselves is what we consider next.

4. IMPLEMENTING A DISCURSIVE MODEL OF MIND

We begin by introducing the original model employed by L&E.[3] Leven (1987a) noted that only a complex environment could emulate human decision making (or even sense making). He recognized that models would have to emulate regional neural processes — and their interaction. He followed a set of analogies: between the Hopfield (Hopfield & Tank, 1986) model and fixed action patterns in instinctive and habitual behaviors; between back-propagation (BP) and semantic processing (as Rumelhart, McClelland, and the PDP Research Group, 1986, noted!); and, between adaptive resonance theory (ART) (Grossberg, 1980) and affective/contextual processing.

As he would elaborate, Leven (1987b) recognized that the automaticity and stability represented by Hopfield would provide a valuable emulation model for rote and motor learning. He recognized, further, its relationship to the classical organization theory model (massively parallel, strictly constrained, optimizing behaviors). Finally, he saw an analogy to traditional models of "economic man." Hopfield required that its fixed learning and easy acceptance of minima be appropriate to the problem (it cannot change). An optimizing economic agent presents great stability of cognitive processing and search style (Arrow & Hahn, 1971). Hopfield's failure to be minimally self-training was the only caveat.

Similarly, he saw the match between back-propagation and logical processes the PDP Group asserted. Linear separability, flexible system design, and the possibility of affecting learning rate and error-passing rate were highly attractive. Only the acknowledged failure to accommodate "dynamic schemas" (Rumelhart et al., 1986, Chapter 14), flexible and complex meanings that changed in time, was a source of disturbance.

Finally, he was aware of the close match between hippocampal CA1-CA3 interaction and adaptive resonance. The sensitivities of ART to arousal (presumably, activation) and gain control

---

[3] Dan Levine (discussed later), Diane Blackwood, and several others have participated in what I call "L&E," for succinctness' sake. Beyond the actual modeling, I am deeply indebted to Bill Hudspeth, Alianna Maren, and (of course), Karl Pribram of the Center for Brain Research. I am grateful, as well, for insights and inspiration from Paul Werbos (National Science Foundation), Joe King (Radford University), and the Neural Networks and Decision Theory classes at Radford University.

(sensor-fused external distractions?) provided the same sensitivity Thompson (1990) called emotional resonance, awareness of interacting (interpersonal) stimuli. ART did not fully specify sources for its parameter values, though, and lacked computational stability (Leven & Elsberry, 1990).

Leven (1987a) recognized the necessity for mutual cooperation of hierarchies of networks — the need to modify both within a set of models (e.g., ARTs) and between them (ART--> Hopfield--> BP). It would take nearly three years (Leven & Elsberry, 1990) to begin providing solutions.

The first implementation allowed each network to be trained independently and to function almost independently. Each network passed output to its closest neighbor, as well as to a "frontal lobe simulator" (which performed a simple averaging and rounding function). Two sets of networks were presented; each had a Hopfield ("Dantzig"), back-propagation ("Bayes"), and adaptive resonance ("Godel") net — and the "Lobes" device. Inputs derived (except at t) from the colleague's opposite — "George" would send "Mikhail" a composite bid of Dantzig (received by Mikhail's Dantzig), Bayes (sent to Mikhail's Bayes), Godel (to Mikhail's Godel), and Lobes (sent to Mikhail's Lobes).

The two sets of networks were trained separately. Each was biased to recognize different inputs and to respond differently. Four sets of parameters were passed, ultimately: Dantzig's "motoric" output of physical gestures (closeness to colleague, eyes opening or narrowing, etc.), Bayes' "semantic" output (quality of argument), Godel's "affective" output (tone of voice, etc.), and Lobes' money bid (averaged from proposed bids of the trio).

Despite the minimal flexibility of the network (all parameters had to be set by the tester), George and Mikhail proved to be valuable negotiators. The two proved highly sensitive to initial conditions — and to each other's bids. It turned out to be fairly common for one bargainer to win for a period of time, with ever increasing return, using a strategy of buying for a low price, selling for a high price, and performing with intimidation, but eventually to suffer a reversal of fortunes. Under that condition, the bargainer would demonstrate withdrawal — to an analogy for helplessness — and dominance. Their expectations could be undermined; their poor responses would evidence a sort of "shock."

Still, the "communication" emulated with such convincing style was based in "canned", almost fully controlled networks. Only Godel (ART) showed the capacity to learn — and the inability to dynamically alter arousal and gain control in response to changes in the (internal and external) environment. Bayes and Dantzig (our analogs for semantic and motoric processes) were immune to change, learning, or even system failure.

The stability-plasticity dilemma (e.g., Grossberg, 1980) loomed: networks that were reliable and predictable lacked the capacity to adapt, whereas adaptive nets lacked sufficient reliability and predictability. We may, however, have found a basis for "making peace" with the dilemma, if not fully overcoming it. As Leven (1995) demonstrated, a self-training and self-repairing back-propagation network can be employed to replace the unsatisfactory Bayes: George and Mikhail can learn new rules, reject old ones, and analyze confusing information with a set of different techniques, within a BP environment.

We shall, below, detail further changes in progress, as our discursive networks attain the capability to interact, react, and learn new responses as well as patterns.

5. BEYOND OPTIMIZATION

We begin by describing our most successful progress, in adapting Bayes to perform self-training and self-repair. Our model in progress, BP-SAM, is constructed from a few components: three (or more) higher-order nets trained in standard logics; a set of lower-order nets, either initially-trained or "empty" (unexposed to data); and "$N2$", a reset and information-control mechanism which determines when it is appropriate to retrain ("repair") a lower-order net and then gate input data from failing nets.

The model presents "superior nets," which are, in current versions, the three underlying logic bases most commonly employed: classical (predicate), modal, and nonmonotonic logics. Classical logic includes propositional logic.

As we wrote (p. 2), modal logic, which is frequently associated with temporal logic, is distinguished from classical logic by inclusion of operators. $L(\grave{I})$ can be interpreted as "I now know that $\grave{I}$," "I now believe that $\grave{I}$," and "$\grave{I}$ is true at all times." Its dual, $M(\grave{I})$, can represent "I do not know that $\grave{I}$ is false," "I do not doubt $\grave{I}$," and "There is some time when $\grave{I}$ is true" (Shoham, 1988). Temporal logic is, largely, a direct extension: "at % o'clock, $\grave{I}$ is true," "I believe that $\grave{I}$ at % o'clock," "between % and D, $\grave{I}$ is true," and their duals.

Nonmonotonic logic (NML) represents inherently contradictory information that remains useful and structured. NML can be applied to everyday behavior and probability. For example, supposing that we have found rules for assigning subjective probabilities to events, how are we to determine whether these rules are reasonable? Since each event is a unique occurrence and involves a different prior knowledge state, it is not possible to ask whether the prediction corresponds to the actual frequency? Many studies have shown that human beings judge likelihoods in bizarre ways, which violate the most fundamental laws of probability theory. People must be doing something right.

NML, then, accommodates predictable contradiction. When Coca-Cola lovers were exposed to New Coke in taste tests, they responded to the new flavor enthusiastically. When the new flavor appeared in stores, they responded with violence against the flavor they had approved. The apparent contradiction may not be so confusing (Leven, 1987b): during the taste test, consumers were highly aroused and sensitive to the physiological taste experience, whereas in the market they related with relaxed affect — and were horrified to find the product with warm associations missing. The rules posited may be inadequate to explain an outcome; NML provides tools with which to cope with results that are common, yet lack simple explanations.

The three "superior" nets perform training upon "empty" nets — these are previously untrained "memory capacity." They also accept input that $N2$ has redirected from lower level nets which have failed to produce satisfactory results.

How could such an environment get established? Where do the "superior nets" get their training? Pribram (1991, Lecture 7) asserts that the structures of thought, in frontal and temporal areas, are prepared by interacting with patterned (already structured) motor and somatosensory processes.

Karmiloff-Smith (1991, p. 179) argued that children learn in three ways: by innate specification, by interaction with the outside world, and by "an endogenous process whereby the mind exploits the knowledge that it has already stored (both innate and acquired), by representing recursively its own representations." This third form of representation Karmiloff-Smith termed "endogenous exploitation via representational redescription and restructuring." This re-representational model consists of three level, she asserted. Her I-level is implicit, effectively stored, and quickly accessed in response to external stimuli. This level is comparable to input to Bayes (the semantic system) from Dantzig (the motoric system). Such an internally sensitive process can self-generated. In such a way, Karmiloff-Smith argued, retarded children can develop automatized language fluency.

The normally fluent child can redescribe language learning at a higher level of abstraction, she maintained — as a "data structure" or a schema. It is ready to become an E-level structure, accessible for conscious consideration. This next level of evaluation ("E-2") is a process of logical derivations.

Lastly, the highest level of abstraction ("E-3") provides the capacity to integrate schematic and socially and contextually cued information, This ability to draw on automaticity and context in building explicit representations is representative of the capacity the "superior nodes" might attain. These patterns, of course, are based on essential notions of order — and its ability to transform sense and other data into a coherent if/then framework.

Thus, the three levels are conformant to Dantzig, Bayes, and Godel. And the "endogenous exploitation via representational redescription and restructuring" Karmiloff-Smith described is the interprocess resonance to which we have ascribed the "superior node" learning algorithm.

As Pribram (1991) and others noted, the state of one system inevitably effects others' effectiveness — and, even, actual functions. Karmiloff-Smith's work demonstrates that the model of "superior nodes" is valid — but, further, it shows that the three processes we have specified (and modeled) are precise matches to experimental evidence.

Although the model we have discussed grossly oversimplifies decision making and learning processes, it does offer an opportunity to begin to test notions of thought process, learning and memory, and affect- and state-dependent behaviors. It is, even now, allowing us to build multiperson negotiation environments, in which bargainers come from different cultural and physical environments. It allows us to plumb the depths of individual affective behavior (Leven, 1992a).

There is abundant evidence that, in all situations, our perceptions, memories, and decision processes are affected by the environment (Gilovich, 1991). This is, of course, no less true for economic actors — even ones who are certain they are optimizing!

We have begun to model everyday (Leven & Levine, 1987) and strategic economic situations (Leven & Elsberry, 1990). The complex problems are beginning to yield novel formulations of the process by which preferences are formed (Leven & Levine, 1996) and by which unintentional exploitation begins (Leven & Elsberry, 1990; Leven, 1995) — no mean problems to consider.

At every turn, we discover the complex ties between motivations (or drives or automaticity), affect (or emotion or context-sensing), and cognition (or semantic or logic processing). And the most impressive part of human behavior to us is its coherence (Levine, Leven, & Prueitt, 1992).

We have just begun to pose the difficult questions. Heaven help us if we ever get good at it.

REFERENCES

Anderson, J. R. (1991). The place of cognitive architectures in a rational analysis. In K. Van Lehn (Ed.), *Architectures for Intelligence* (pp. 1-24). Hillsdale, NJ: Lawrence Erlbaum Associates.

Arrow, K., & Hahn, E. (1971). *General Competitive Analysis.* San Francisco: Holden-Day.

Bell, J. (1988). *Speakable and Unspeakable in Quantum Mechanics.* New York: Cambridge University Press.

Blackwood, D., Elsberry, W. R., & Leven, S. J. (1988). Competing network models and problem solving. *Neural Networks,* 1 (Supplement 1), 10.

Chandrasekharan, B. (1990). What kind of information processing is intelligence. In D. Partridge & Y. Wilks (Eds.), *The Foundations of Artificial Intelligence* (pp. 14-46). New York: Cambridge University Press.

da Fonseca, E. (1991). *Beliefs in Action.* New York: Cambridge University Press.

Eco, U. (1986). *Semiotics and the Philosophy of Language.* Bloomington, IN: Indiana University Press.

Eliasson, G. (1991). Modeling economic change and restructuring. The micro foundations of economic expansion. In P. de Wolf (Ed.), *Competition in Europe* (pp. 33-59). New York: Kluwer.

Gilovich, T. (1991). *How We Know What Isn't So.* New York: Free Press.

Goold, M., & Campbell, A. (1989). *Strategies and Styles.* Cambridge, MA: Blackwell.

Grossberg, S. (1980). How does a brain build a cognitive code? *Psychological Review* 87, 1-51.

Hopfield, J. J., & Tank, D. (1986). Computing with neural circuits: A model. *Science,* 233, 625-633.

Karmiloff-Smith, A. (1991). Beyond modularity: Innate constraints and developmental change. In S. Carey & R. Gelman (Eds.), *The Epigenesis of Mind* (pp. 171-198). Hillsdale, NJ: Lawrence Erlbaum Associates.

Koford, K., & Miller, J. (1991). *Social Norms and Economic Institutions.* Ann Arbor: University of Michigan Press.

Leven, S. J. (1987a, October). *S.A.M.: a triune extension to the A.R.T. model.* Paper presented at the Conference on Networks in Brain and Computer Architecture, Denton, TX.

Leven, S. J. (1987b). Choice and neural process. Unpublished doctoral dissertation, University of Texas at Arlington.

Leven, S. J. (1995). Intelligent, creative, and neurotic agents: Tools for effective neural net user support. In *World Congress on Neural Networks, Washington, DC* (Vol. II, pp. 293-300). Mahwah, NJ: Lawrence Erlbaum Associates.

Leven, S. J., & Elsberry, W. R. (1990). Interactions among embedded networks under uncertainty. *IJCNN International Joint Conference on Neural Networks* (Vol. III, pp. 739-742). San Diego: IEEE.

Leven, S. J., & Levine, D. S. (1987). Effects of reinforcement on knowledge retrieval and evaluation. In M. Caudill & C. Butler (Eds.), *Proceedings of the IEEE First International Conference on Neural Networks* (Vol. II, pp. 269-277). San Diego: IEEE/ICNN.

Leven, S. J., & Levine, D. S. (1996). Multiattribute decision making in context: A dynamical neural network methodology. *Cognitive Science, 20*, 271-299. (This was also the subject of a column in *Science*, September 3, 1993.)

Levine, D. S. (1986). A neural network theory of frontal lobe function. In *Proceedings of the Eighth Annual Conference of the Cognitive Science Society* (pp. 716-727). Hillsdale, NJ: Lawrence Erlbaum Associates.

Levine, D. S., Leven, S. J., & Prueitt, P. S. (1992). Integration, disintegration, and the frontal lobes. In D. S. Levine & S. J. Leven (Eds.), *Motivation, Emotion, and Goal Direction in Neural Networks* (pp. 301-335). Hillsdale, NJ: Lawrence Erlbaum Associates.

MacLean, P. D. (1991). *The Triune Brain Revisited.* Hillsdale, NJ: Lawrence Erlbaum Associates.

Miller, G. A., Galanter, E. H., & Pribram, K. H. (1960). *Plans and the Structure of Behavior.* New York: Holt, Rinehart, and Winston.

Morishima, M. (1990). Economic theory and industrial evolution. In Baranzini, M. & Scazzeri, R. (Eds.), *The Economic Theory of Structure and Change* (pp. 175-197). New York: Cambridge University Press.

Neale, M., & Bazerman, M. (1991). *Cognition and Rationality in Negotiation.* New York: Free Press.

Nicolis, J. (1991). *Chaos and Information Processing.* New York: World Scientific.

Peng, Y., & Reggia, J. (1990). *Abductive Inference and Diagnostic Problem-Solving.* New York: Springer-Verlag.

Pribram, K. H. (1986). The subdivisions of the frontal cortex revisited. In E. Perecman (Ed.), *The Frontal Lobes Revisited* (pp. 11-40). New York: IRBN.

Pribram, K. H. (1991). *Brain and Perception.* Hillsdale, NJ: Lawrence Erlbaum Associates.

Pribram, K. H., & Gill, M. (1976). *Freud's "Project" Reassessed.* New York: Basic Books.

Reason, J. (1990). *Human Error.* New York: Cambridge University Press.

Rumelhart, D. E., McClelland, J. J., & the PDP Research Group (1986). *Parallel Distributed Processing*, Vols. 1 and 2. Cambridge, MA: MIT Press.

Seligman, M. (1991). *Learned Optimism.* New York: Pocket Books.

Shoham, Y. (1988). *Reasoning About Change.* Cambridge, MA: MIT Press.

Simon, H. A. (1986). *Reason in Human Affairs.* Palo Alto, CA: Stanford University Press.

Simon, H. A. (1991). Cognitive architectures and rational analysis: A comment. In K. Van Lehn (Ed.), *Architectures for Intelligence* (pp. 25-39). Hillsdale, NJ: Lawrence Erlbaum Associates.

Thaler, R. (1991). *The Winner's Curse.* New York: Free Press.

Thompson, R. (1990). Empathy and emotional understanding. In N. Eisenberg & J. Strayer, *Empathy and Its Development* (pp. 119-145). New York: Cambridge University Press.

Tulving, E. (1983). *Elements of Episodic Memory.* Oxford, UK: Clarendon Press.

Viner, J. (1958). *The Long View and the Short.* Glencoe, IL: Free Press.

Werbos, P. J. (1990). A menu of designs for reinforcement learning over time. In W. Miller, R. Sutton, & P. J. Werbos (Eds.), *Neural Networks for Control* (pp. 67-96). Cambridge, MA: MIT Press.

4

Nonoptimality in a Neurobiological System

David G. Stork
Ricoh California Research Center and Stanford University

Bernie Jackson and Scott Walker
Stanford University

(Reprinted with permission from Artificial Life II. SFI Studies in the Sciences of Complexity, vol. X, edited by C. G. Langton, C. Taylor, J. D. Farmer, & S. Rasmussen, Addison-Wesley, 1991.)

*David Stork's chapter, **Nonoptimality in a Neurobiological System**, gives a specific biological example to show that evolution does not automatically lead to optimal neural structures. On the contrary, the process of evolution in complex, dynamic environments means that at any given time in the history of a species, structures are present that arose to solve a set of tasks that were relevant to an earlier environment. These same structures often do not lead to optimal performance of tasks that are relevant to the organisms' current environment. This outlook is in agreement with principles propounded in several other chapters (notably those of DeYong, Elsberry, Leven, and Levine), but Stork's chapter is the only one in this book that provides a specific example.*

*Stork's neurobiological example is the circuit mediating a tail-flipping response in the crayfish. In this circuit, there is a ganglion (LG) that excites a neuron called FF, leading to activation of flexor muscles in a particular segment. However, the tailflip response involves **inhibition** of flexion in those muscles. The same sensory stimuli that excite LG also inhibit FF directly by another pathway, and the flexor muscles of that segment indirectly by yet a third pathway. Hence, the excitatory LG-to-FF synapse is "useless" because it is always overridden. The nonoptimality, in terms of tail flipping, of this synapse's existence is supported by computer simulations (in another article by the author, referenced herein). Stork goes on to show that the "useless" synapse serves an evolutionary purpose for a different response, **swimming** without tail flipping. Moreover, his simulations show that if the crayfish ever went back to a situation* *where swimming was again the response of choice, the neural circuit would again be well suited for that response.*

The crayfish results suggest that nonoptimal structures are not only prevalent, but in some cases may serve the purpose of helping the organism adapt to a changed environment of the future. Stork notes that as neural systems get more complex, as in mammals, the amount of potentially useful nonoptimality is likely to be even greater than in the crayfish. This is related to the notion of Prueitt's chapter in this book, that the principle of optimality needs to be (and is) balanced by a complementary principle of optionality, or diversity generation.

ABSTRACT

We simulate the evolution of the neural circuitry subserving the tailflip escape maneuver in the crayfish in order to help explain a paradoxical ("nonoptimal") feature of that circuit. Specifically, a "useless" synapse in the current tailflip circuit can be understood as being a vestige from a previous evolutionary epoch in which the circuit was used for swimming instead of flipping. Such preadaptation effects may underlie a broad range of neural structures throughout the animal world, and illustrate fundamental principles important for artificial life, most notably the locally greedy nature of evolutionary change and that "elegance of design counts for little."

1. INTRODUCTION

The structure and function of every organism — both biological and the vast majority posited for artificial life — depend crucially upon its evolutionary precursors (Bonner, 1988). The form of the human eye and the neural system subserving peripheral visual processing, for example, depend upon the evolutionary history of hominids and pre-hominids (Spinelli, 1987); likewise, the structure of systems subserving hearing (and thus speech recognition), motor control, and so on derived from those of earlier evolutionary epochs Indeed, evolutionary change is so fundamental to our understanding of biological life that Dawkins (1976, 1989) claimed that life without the notion of evolution was virtually unthinkable.

Neural systems of all animals possess structure at birth — there are no *tabulae rasae* anywhere in the animal kingdom. Such structure is absolutely fundamental to the performance of the organism, of course, and even determines what can, and what cannot, be learned from the environment. Moreover, it is increasingly clear that the initial promise of artificial neural networks toward achieving adequate performance on speaker-independent speech recognition, three-dimensional visual object recognition, scene analysis, language understanding, and a host of higher cognitive functions cannot be met without continued progress in understanding constraints, as manifest in network structure (Stork, 1988, 1990). Whereas nearly all researchers in neural networks *design* their networks (or "reverse engineer" what exists in biology), we believe that a deeper understanding of the *sources* of biological structure will also help us create artificial neural systems duplicating or mimicking complex behavior. Such understanding will also support efforts to produce artificial life.

Because biological structure evolved through selection in extremely complex environments, we should not expect that biological solutions will always conform to "good" design principles. The research related here is directed to understanding how "inelegant" — indeed, counterintuitive, or "nonoptimal" — structures might arise through evolution, even in quite simple neural systems. We argue, moreover, that "nonoptimality" should be expected to be even more prevalent in complex neural structures, for instance, the human brain.

Although its roots extend back to the time of Darwin (1866/1968), the concept of *preadaptation* has been recently elaborated by S. J. Gould, E. Mayr and others (Gould, 1982; Gould & Vrba, 1982; Mayr, 1976). Preadaptation is used to describe the process by which an organ, behavior, neural structure, etc., which evolved to solve one set of tasks is later utilized to solve a *different* set of tasks. It illustrates the dichotomy between designed, planned, and "optimal" forms in biology on the one hand, and "nonoptimal" ones on the other.

An example of preadaptation of an organ is the bird wing. The proto-bird wing was too short to be used for flight, and hence must have been used for some other task; the Darwinian fitness at that time did not depend upon flight. Theories of the use of the proto-wing include thermoregulation (the proto-bird spreads or retracts its wings to cool or warm itself), insect catching (the proto-wings are used to knock down insects to be eaten), and reorientation during jumps for insects (the proto-bird can then catch insects from a larger volume of air), and others. Whatever the reason, the proto-wing was indisputably not used for flight. Later in evolution, as the proto-wing became longer, a behavioral threshold was reached in which the limb could be used for flight. Then, a different set of evolutionary pressures were placed on the wing, yielding a lighter and more aerodynamic wing. The later wing, though, had to be built upon the structures that evolved for the previous task. Thus there could be structures in the current bird wing — holdovers from the earlier evolutionary epoch — that are "nonoptimal" for flight (Stork, 1989).

If such structures do not present an excessive biological "cost" (say, in energy or resources), then those structures may remain in the later system. Even if the structure does pose a cost to the organism, that structure might nevertheless remain in the later organisms, since intermediate states in its elimination may prove very detrimental to the organism. In such a case, the structure is "frozen into" the organism, a relic of the earlier evolutionary epoch.

Figure 4.1 illustrates, metaphorically, the process of preadaptation, and can be discussed in terms of neural networks (our primary system of interest). At an early epoch, the network solved Task 1, and might even have been optimal for it. (Optimal is, of course, dependent upon one's measure. We need not be specific here, but state roughly that a circuit which uses the minimum number of components, biological energy, and structure to solve the problem without compromising the organism's ability to solve other problems can be regarded as more optimal than a circuit that doesn't.) At a later evolutionary epoch, a *different* task becomes more relevant. This switch in task might be due to a changing environment, or to the network evolving such that new niches become available (as in the bird wing), and so on. The network is then under *different* evolutionary pressures, and the "energy landscape" is deformed. The network, however, must build upon structures selected based on Task 1 — structures that might not be appropriate for the second task. The result is that the network may be "nonoptimal" for Task 2.

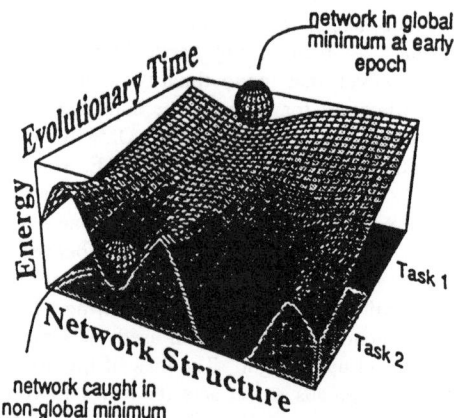

Fig. 4.1. Preadaptation. Metaphorical energy landscape describing performance of a network throughout evolution. Evolutionary time runs from the back of the figure to the front; the "energy" (e.g., a measure of fitness) is vertical, and some index of network structure runs left to right. At an early epoch, the network may have been optimal for solving the task at that time — Task 1 — but later, the appearance of Task 2 deforms the energy landscape. The network might, therefore, be in a nonlocal minimum, and hence "nonoptimal" for Task 2. In our typical crayfish simulations, Task 1 is swimming and Task 2 flipping.

Investigations of preadaptation are important in neurobiology, artificial neural networks, and artificial life. Such studies elucidate the nature of evolutionary change and the function of biological networks (especially since such information cannot be preserved in the fossil record). Preadaptation sheds light on the study of artificial neural networks in at least two ways: it can help guide the "reverse engineering" of biological systems, showing which structures might or might not be relevant to the cognitive task at hand; it can suggest general hybrid evolution-learning neural networks based on biological processes (Keesing & Stork, 1991; Miller, Todd, & Hughes, 1989; Stork & Keesing, 1990, 1991). Since the vast majority of attempts at artificial life incorporate evolution in some form, preadaptation can aid these efforts by clarifying the difference between elegant and simple design principles and the "inelegant" implementations that might be required in living organisms. Likewise, studies such as this one can help to illuminate the processes in evolution.

We have chosen the crayfish tailflip circuit for our simulation studies for several reasons. First, the neural circuitry has been extensively mapped by neurophysiologists (Wine, 1971). Second, the circuit is small enough that realistic simulations can be made using the computer resources available to us. Third, an apparently "nonoptimal" structure is evident in the circuit. Fourth, the circuit is responsible for a behavior that is of the utmost survival value for the crayfish (flipping away from danger), and thus Darwinian selection pressures on the circuit are great. Fifth (and closely related to the previous reason), a highly plausible evolutionary scenario can be made for the circuits. Finally, the crayfish has a phylogenetically close relative, *Anaspides tasmaniae*, which can serve as a sort of "control" organism, since its homologous circuits differ in ways easily linked to its different behavior.

2. CRAYFISH TAILFLIP CIRCUIT

The crayfish tail consists of six segments, each with its own small neural circuit linking pressure-sensitive cells to flexor muscles governing the tail segment. The tailflip escape maneuver is effected by *flexion* of the *anterior* segments (segments 1-3) with no flexion in the *posterior* segments (segments 4-6). Figure 4.2 shows the basic structure of the actual crayfish circuits responsible for this behavior and possible evolution.

Consider carefully the circuit in segment 6, which leads to inhibition of the flexor muscles whenever the sensory interneurons are excited. A neural volley passing from the LG to the FF neuron would lead to excitation of the FF. However, this excitation is counteracted by the direct *inhibitory* connection from the sensory interneuron to the FF itself. There is, moreover, inhibition of the flexor muscle vis the F1 neuron. The synapse between the LG and FF is thereby overridden; it seems to have no purpose. So far as is known, then, the circuit is "nonoptimal."

The question naturally arises: Why does the crayfish have this apparently useless synapse? What can account for such "nonoptimality" in design?

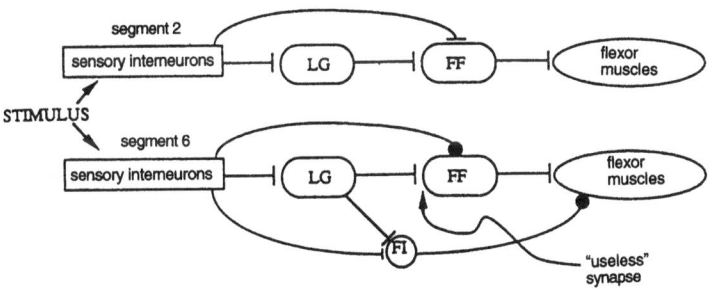

Fig. 4.2. The neural circuitry subserving the tailflip in crayfish. Excitatory synapses are represented by a T and inhibitory synapses by a ●. In the event of a rapid rise in ambient water pressure (from a predator), pressure transducers yield excitatory activation in the sensory interneurons. To effect the tailflip maneuver, each anterior segment (e.g., segment 2) must flex (i.e., the flexor muscles must be excited) and each posterior segment (e.g., segment 6) must not flex (i.e., the flexor muscle must be inhibited). Note especially that one of the excitatory synapses in segment 6 is "useless": any time an excitatory volley passes from neuron LG to FF, the FF neuron is also inhibited (via a direct connection from the sensory interneuron), thereby rendering the excitation ineffective. Furthermore, the only projection of the FF (which is to the flexor muscles) is also overridden by inhibition from the F1 neuron.

2.1. Preadaptation Hypothesis

Dumont and Robertson (1986) hypothesized that the excitatory LG $\Rightarrow$ FF synapse is a vestige from an earlier evolutionary epoch, one in which the proto-crayfish did not flip, but instead merely *swam*. (Simultaneous flexion in all segments leads to swimming, as in the *Anaspides tasmaniae*, which has in each of its six tail segments a circuit homologous to those in the anterior segments of the crayfish.) The hypothesis is that the circuits in the posterior segments originally had the form at the top of Fig. 4.2 (appropriate for swimming), but under a change in task — from swimming to flipping — the circuit evolved by building upon the previous ones. The LG $\Rightarrow$ FF synapse was useful for swimming, but not for flipping, and the circuit evolved other connections to override that synapse. Because that synapse is no longer expressed behaviorally, it is "frozen into" the circuit — a vestige of the earlier epoch, and nonoptimal in the context of the circuit's current use, in much the same way that the appendix has been "frozen" into our digestive system. We provide here computer simulations and further analysis in support of this hypothesis.

3. SIMULATION APPROACH

The overall approach follows a classical Darwinian evolution scenario, shown in Fig. 4.3; a more complete explanation and description of the relationship to actual biology is given in a recent paper (Stork & Keesing, 1992). Each network has a haploid gene, which is expressed to yield the full network, including connectivities and neural response characteristics. Networks then respond to the environment — a simulated pressure wave from a predator — and are selected based on their response. The selected networks then reproduce to give the genes of the next generation, and the cycle continues.

Genotype. The genetic representation and development used in our model system together avoid some of the artificial assumptions made by other modelers of genetic systems. The most important question centers on that of genetic representation of neural connection strengths: is this representation *localized* (each initial connection strength determined by one or a small number of genes) or is it *distributed* (the many connection strengths determined by several genes)?

There is abundant evidence for pleiotropy and a *distributed* genetic representation in biology (Dawkins, 1976; Griffiths & McPherson, 1989; Hall, Greenspan, & Harris, 1982; Wilkins, 1988). It is clear that the information in the entire human genome is insufficient to specify every brain synapse, not to mention those elsewhere in the nervous system. Nor does there seem to be much evidence for "one gene-one synapse." Instead, genetic representation can act in several ways: setting affinities for connections, development rates, and so forth (Purves & Lichtman, 1985; Purves, 1989). Furthermore, there are many cases in which mutations in a single gene or a small number of genes can have distributed consequences, as in many systemic neural disorders such as multiple sclerosis. On the computational and systems levels, a distributed representation has several useful properties. Perhaps most importantly, it permits mutations to make large changes in network structure, thereby leaving small refinements to be accomplished through learning (Keesing & Stork, 1991; Plotkin, 1988; Stork & Keesing, 1990).

Our simulations employ a distributed representation, based on properties of *control genes* and *structural genes* (Hawkins, 1986). The structural genes code for fundamental aspects of the phenotype, here the cell type, neurotransmitters, type of synaptic receptors, and so on; the control genes guide the expression of the structural genes (Fig. 4.4). Thus, for instance, if a particular enhancer from the control genes is activated, it will lead to a *distribution* of the structural genes to be expressed. This captures the fact that certain phenotypic features are expressed in concert. For example, a human photoreceptor contains both photopigment and platelets, as well as other structures unique to photoreceptors; these are all expressed together. (One typically does not find cells with photopigment but no platelets, for instance.) In our model, then, several of these features are represented by a single structural gene; if that gene is activated, *all* of the component phenotypic features are candidates for expression.

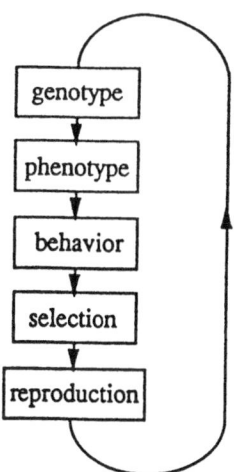

Fig. 4.3. Evolutionary processes. The genes lead via development to a structured network, including interconnections (excitatory and inhibitory), neural-channel properties, and so on. The network then responds to the environment and is selected based on the resulting fitness score. Fitness depends upon the posited task, here either swimming or flipping. The most fit individuals then reproduce to yield the genotypes in the next generation, and the evolutionary processes continue.

Consider just one of the phenotypic traits: cell adhesion molecules (CAMs), implicated in developmental programs for connectivity (Edelman, 1987, 1988). In our model, there are four types of CAMs; during development the initial connectivity between two neurons is specified by the similarity in their CAMs, just as biological CAMs, large cell surface glycoproteins, are homophilic. Suppose that promoter 1 (also sometimes called an enhancer) would lead to the expression of CAM1 and CAM2 (Fig. 4.5). If no other promoters are activated, the final neuron would have those two CAMs expressed. But suppose, moreover, that promoter 2 would lead to CAM2 and CAM4, but *not* CAM1 and CAM3, and analogously for promoter 3, as shown in the figure. (In our simulations, a promoter table describes the relationship between the promoters and the CAM structural genes.) If all three promoters are activated, each would express its corresponding set of CAMs, but prevent other CAMs from being expressed. The final distribution of CAMs expressed in a neuron are then the result of a majority vote for each CAM, as if the promoters competed among themselves to express the individual CAMs. Similarity in the cell surface markers expressed in any two neurons determines the initial interconnectivity — the greater the similarity, the stronger the initial synaptic connection, in accord with homophilic properties of CAMs (Edelman, 1987).

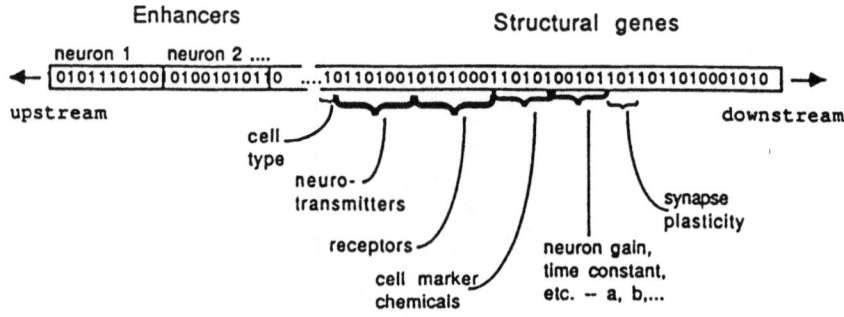

Fig. 4.4. Haploid genome used in simulations. Structural genes (shown downstream, grouped for convenience) govern the phenotypic structures in the network. Enhancers (upstream, grouped by neuron for convenience) govern the expression of the structural genes.

A similar computation occurs for the neurotransmitter to be produced in a neuron; we use twelve candidate neurotransmitters (e.g., GABA, acetylcholine, ... , whereas above we used just four CAMs. In the simulations described here, only one transmitter is expressed (as described by Dale's Law, which is not universally obeyed). Genes coding for acetylcholine and cholineacetyltransferase have been found on two separate chromosome segments in *Drosophila melanogaster* (Greenspan, 1980; Hall et al., 1982) and this suggests that a similar arrangement could exist in the crayfish.

Grouped phenotypic features that lead to a neuron being either a sensory, or an interneuron or a motor neuron are expressed by an analogous mechanism, though with only three (exclusive) attributes rather than twelve. Neural channel properties are computed as the *average* of those from each structural gene activated. Thus if one structural gene would lead to a large number of *Na* channels, while another would lead to a small number, then if both are activated, that actual number expressed will be intermediate. Such features of the model are motivated by recent results on mutations in three different alleles in the Shaker locus, which led to postsynaptic potentials in muscles longer and larger than in the wild type (Jan, Jan, & Dennis, 1977), implying a genetic representation of potassium channels. (See a current paper (Stork & Keesing, 1992) for more detailed discussion of the biological motivation of the model.)

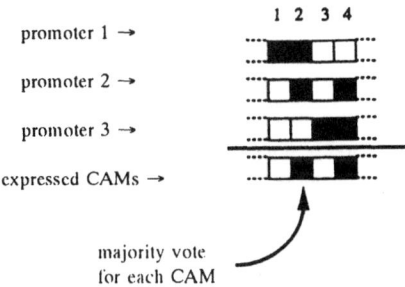

Fig. 4.5. Model for the expression af cell adhesion molecules in a neuron. Suppose that for a given neuron three promoter genes. are activated. In the example shown here, the first leads to activation of the structural genes 1 and 2, which would lead to CAM1 and CAM2; promoter 2 would likewise lead to CAM3 and CAM4, etc. (This relationship between promoters and these structural genes is stored in a look-up table in the simulations, and derives from physiological data on gene expression af CAMs.) The final CAMs expressed in the neuron are the result of a majority vote for each CAM; in the case shown, CAM3 and CAM4 are expressed. (Tie votes are decided by an unbiased random choice.)

What is important here is that the relationship between genetic representation and ultimate phenotype is distributed and indirect.

3.1. Phenotype

Each neuron is thus described by its global type (sensory, interneuron, or motor neuron), its decay rate constant, neural channel concentrations (which determine the excitatory and inhibitory saturation levels), its neurotransmitter type, its synaptic receptor type, and complement of cell adhesion molecules.

The network as a whole is specified by the neural interconnectivities, determined by the similarities of the CAMs (computed as a Hamming distance) on each candidate pair of neurons. We also include a distance-dependent term, making neurons that are physically more separated have lower connectivity for any given CAM similarity. Expressed networks have the form shown in Fig. 4.9 below.

3.2. Behavior

The behavior of each neuron in the network is governed by Hodgkin-Huxley equations of the following form (Grossberg, 1982; Hodgkin, 1964):

$$\frac{dx_i}{dt} = -ax_i + (b - cx_i)\left\{\sum_{j \in G_{ex}} z_{ij}f(x_j) + I_i\right\} + (d - ex_i)\left\{\sum_{j \in G_{in}} z_{ij}f(x_j)\right\} \tag{1}$$

where
- x_i = activity in neuron (depolarization);
- $f(x_j)$ = output spike rate — a compressively nonlinear transfer function of the activity;
- a, b, c, d, e = constants describing ion concentrations, channel densities, and so forth. In particular, a describes the time constant for neural recovery, b and c together with a specify the excitatory saturation level, and likewise d, e, and a specify the inhibitory saturation level;
- z_{ij} = strength of synapse between neurons i and j;
- I_i = external input for neuron i (not due to other neurons);
- G_{ex} = the set of neurons connected to neuron i by synapses leading to excitation;
- G_{in} = the set of neurons connected to neuron i by synapses leading to inhibition.

The right-hand side of the equation consists of three terms. The first denotes a relaxation decay, the second an excitation term (involving the sum over all the inputs that lead to excitation), and the third term, analogously, inhibition. For our task, the input I_i is nonzero only for the sensory neuron, and in that case consists solely of a brief delta-function impulse at $t = 0$.

3.3. Selection

Our selection procedures are based on fitness-proportional reproduction (Goldberg, 1989); the fitness score depends upon the task. For *swimming*, this score is equal to the maximum instantaneous excitation in the network's motor neuron (normalized over the population), corresponding (roughly) to the strength of flexion in the posterior tail segments. For *flipping*, the score is the maximum magnitude of *inhibition* in the motor neuron, corresponding (roughly) to the lack of such flexion. Although other measures of fitness are possible (motor-neuron activity integrated over time, maximum value of the derivative of the activation, etc.), the one we used captures the behaviorally relevant features of flexion. This fitness function is biologically plausible, since the crayfish locomotion is fundamental to its survival. Of course, other traits confer fitness: we are concentrating solely on one of the most important.

The algorithm for selection can be visualized as taking the fitness scores of each of the networks in a population and lining them up in a bar whose length is proportional to the individual scores. Then, points are chosen randomly and independently along the entire length (Fig. 4.6). The networks selected in this way are then reproduced, regardless of the value of their fitness score (see subsection on reproduction). The number of points chosen is equal to the number of individuals.

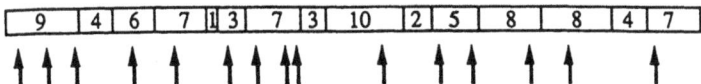

Fig. 4.6. Selection for fitness-proportional reproduction. Each network is represented by a rectangle having a width equal to its fitness score. Selection is achieved by randomly choosing points along the entire population (arrows), which determine which networks survive. Thus, the probability a network survives is proportional to its fitness score, and a network can be selected multiple times. It is possible — though somewhat rare — for a network with a very low fitness score to be selected over a network with high fitness score. (Here the scores have been arbitrarily normalized to maximum = 10.)

Such fitness-proportional reproduction is biologically motivated and generally preferable to schemes in which merely the *most* fit individuals are selected by truncation selection. In general, fitness-proportional reproduction helps to preserve diversity in the genome by permitting some low-fitness networks to pass on their genes.

3.4. Reproduction

Those networks selected in this manner are reproduced using the familiar processes of replication, mutation (p_{bit} flip $= 10^{-2}$/bit/generation), bit insertion (p_{bit}insert $= 10^{-3}$/network/generation), and single-position crossover (75% of pairs), as put forth by Holland (1975) and illustrated in Fig. 4.7. The genetic algorithm parameters — in particular the somewhat high mutation rate — were chosen in order to probe the phenomena as thoroughly as possible using our computer. Based on several runs with different parameters and random number seeds, we found that our fundamental findings did not depend significantly on the choice of parameters over a wide range.

4. RESULTS AND ANALYSIS

All simulations were done on Connection Machine CM-28, either at RIACS (Moffett Field, CA) or Thinking Machines Corporation (Cambridge, MA). Our program consisted of roughly 12,000 lines of C* code, the parallelized version of C; typical simulations required two to three hours. On SIMD (single-instruction multiple data) computers, there is always the question of the level at which parallelization of the problem should be made. The Connection Machine operating system and C* language permit construction of *domains*, which are processed in parallel. Candidate domains for our system were:

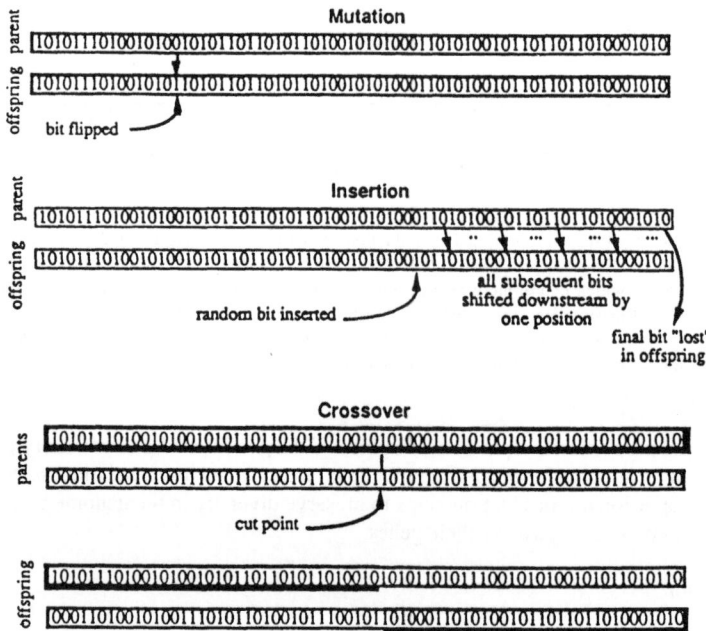

Fig. 4.7. The processes of random mutation, bit insertion, and crossover (shown) as well as replication (i.e., duplication without mutation, not shown) are used between generations.

· individual networks,
· individual neurons, and
· individual synapses.

(The temporal dynamics of the neurons are inherently serial — the integration of Eq. (1) — and thus could not be parallelized. Indeed, this serial integration alone accounted for over 1/4 of the total processing time.)

Thus, for instance, if the code were parallelized at the level of individual networks, then the neurons and synapses would be serially processed. If, on the other hand, individual *neurons* were parallelized, then just the synapses and any finer grain structures would be processed serially, and so on. While paralleling to the finest grain (here synapses) would lead to most rapid calculations, the overhead in inter-processor communication would increase, since each neuron interacts with several other neurons. For our small number of neurons (7), parallelizing at the level of individuals was most efficient. Only if the number of neurons per circuit were larger (roughly 20-30) would the speedup in computation by parallelizing at the *neuron* level outweigh the drawbacks in communication overhead.

The parallel aspect of the our program is that all members of the population are calculated simultaneously on this SIMD machine. Individual neurons and synapses within a network are computed in series. We created the parallel data structure "*domain* individual," a C* domain that allocates one processor (each with 8 kbytes of memory) per crayfish circuit. All the code was on the host VAX, while the data (synaptic strengths, neural activities, etc.) were stored on each physical processor. We ran some simulations with larger populations and found that the fitness curves (see Fig. 4.8(a)-(b)) did not differ significantly from our results for populations of 200 individuals. Whereas the statistics for these larger sets is only slightly more reliable, we found that analyzing individual networks for "nonoptimal" structures — which had to be done laboriously, by hand — became prohibitively time-consuming.

4.1. Preadaptation

Figure 4.8, from a typical simulation, illustrates the basic phenomenon of preadaptation (Stork, Walker, Burns, & Jackson, 1990). The graph on the left shows the population average fitness as a function of generation. At generation 75, the task was changed from *swimming* to *flipping* — the fitness score magnitude of the positive activity in motor neuron is then negative (inhibitory). The population average fitness drops precipitously as the circuits previously selected for swimming are then tested and selected for flipping. Later the fitness levels off (by generation 150) to a mean score of 0.13 (in arbitrary units). The right hand graph shows evolution in the case of rewarding flipping alone — no preadaptation. After 75 generations, the mean score, 0.29 (in the same arbitrary units), is significantly above that of the preadapted networks in the left figure, given the same number of generations rewarding flipping. In short, evolving flipping networks from those previously selected for swimming leads to poorer performance than evolving them from the random networks present at the beginning of each of our simulations. Although, of course, there is a small chance the preadapted networks (Fig. 4.8, left graph) could spontaneously increase in fitness through a fortuitous combination of mutations or crossovers, the networks seemed to be caught in a local minimum (cf. Fig. 4.1).

The structure of preadapted networks differed from those not preadapted (Fig. 4.9). In particular (based on a preliminary analysis of several dozen networks), roughly three times as many "nonoptimal" structures were found in preadapted circuits as in non-preadapted circuits (other variables held constant). The structures we termed "nonoptimal" included neurons unconnected to the rest of the network and synapses whose polarities (e.g., excitatory) were counterbalanced by another projection of the opposite polarity (i.e., inhibitory).

Because non-optimal structures arose more frequently in preadapted circuits in our simulations, and because several simulated circuits had nonoptimal forms very closely homologous to those in the biological crayfish (compare Figs. 4.2 and 4.9), our simulations provide support for an understanding of the LG $\Rightarrow$ FF synapse in the crayfish in terms of preadaptation.

A possible objection arises: how can we be sure that the LG ⟹ FF synapse is, indeed, never used by the crayfish for some other purpose? Perhaps we simply have not been clever enough to guess a use. By analogy, very recent work on potassium channels in *Aplysia* on first analysis seemed to show that certain channels were non-functional, and, hence, perhaps nonoptimal (Treistman & Grant, 1993). It was only after the ambient water temperature was raised from the (natural) 10°C to the warmer 1520°C that these channels became active. (This suggested that the channels might help prevent convulsions in the *Aplysia*. As F. H. Crick has remarked, evolution can be more creative than humans!

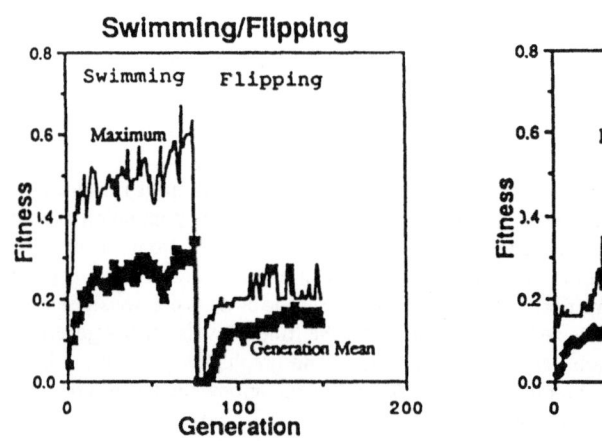
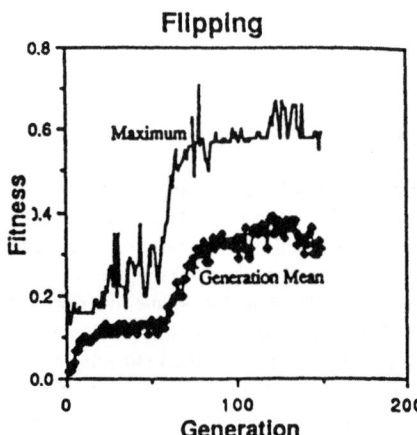

Fig. 4.8. Preadaptation. (Left) The maximum individual fitness and the generation mean fitness for a population selected first for swimming and then (after generation 75) for flipping. (Right) Population selected solely for flipping. The minimum fitnesses were zero at virtually every generation, and hence have not been plotted. The same normalization convention was used for the graphs.

To such objections we respond that the manifest simplicity of the crayfish network and the restricted behavioral repertoire exhibited by the crayfish (at least evident in laboratory studies) seems to limit such hypothetical uses. Of course, a use might be found in the future. It might be possible that the "nonoptimal" synapse and attendant projections give an architectural constraint of some sort, and cannot be removed without great behavioral and fitness cost. (One hypothetical use for the "nonoptimal" circuit is for the inhibitory sensory-FF projection to limit the duration of an excitatory volley — perhaps to make a short "burst" in activity in the motor neuron. Alas, this does not appear to be the case in either the crayfish or our model networks: the inhibition of the FF neuron invariably precedes the excitatory volley through the "useless" synapse.) Given the simplicity and plausibility of the preadaptation scenario provided by Dumont and Robertson and by our simulations, this explanation seems far more acceptable than any current alternative.

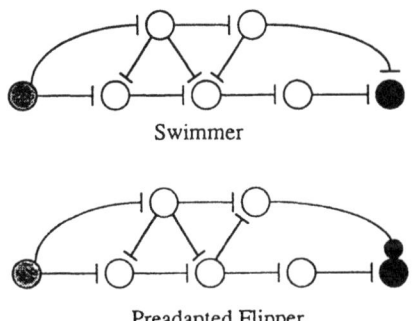

Swimmer

Preadapted Flipper

Fig. 4.9. (Top) Network resulting from evolution by selection for swimming alone. (Bottom) Network after preadaptation scenario. Note in particular the nonoptimal connections in the lower circuit. In both circuits, the sensory neuron is shown at the left and the motor neuron at the right. (As in Fig. 4.2, Ts represent excitatory connections and ●s inhibitory ones.)

4.2. Evolutionary Memory

How can we understand in a deeper way the preservation of genetic information coding for functionally unless structures? Perhaps we can consider genetic information to be "junk." But note: junk is fundamentally different from trash. The junk around our house was at one time useful, and is often stored in an attic in the possibility of being used later. Trash, however, might never have been useful, and is not useful at present. We discard trash; we save junk, even if there is but a small chance that it might be used again. Perhaps the distributed genetic information responsible for the "useless" synapse is "junk" in just this way.

In order to explore this possibility, we performed another set of simulations. We selected first for swimming, and then for flipping (as before), thereby creating a population of networks which possessed a significant fraction of structures "nonoptimal" for the flipping. We then changed the task back again to swimming, in order to see how rapidly and how well the population then evolved for swimming.

Figure 4.10 shows typical results. After selection for swimming then flipping, the population fitness rose very *rapidly* for the subsequent swimming task. The population did this more rapidly than when it had evolved under the first swimming epoch, presumably in part because the later evolution could appropriate structures remaining from the first swimming epoch. The "junk" in the genome permits the crayfish to *rapidly* relearn how to swim, should the environment require it.

Keeping genes that were useful at previous epochs may help explain how evolution can be faster at later epochs, since the structures need only be recalled or reselected, not rebuilt *ex nihilo* (Dawkins, 1976; Wills, 1989).

5. CONCLUSIONS AND FUTURE WORK

Our simulations support an explanation that an apparently "useless" feature of the contemporary crayfish tailflip circuit arose from preadaptation, specifically, that the crayfish circuit was historically selected based on the circuit's ability to have the crayfish swim, and later selection was based on the crayfish's ability to *flip*. As such, there are features "left over" from the earlier (*swimming*) epoch, not selected out, and hence perhaps "nonoptimal" in the current (flipping) circuit. Nevertheless, genes that code for structures that are at one epoch "useless" may be expressed under different environmental circumstances and, thus, permit the system to respond rapidly to changing environments.

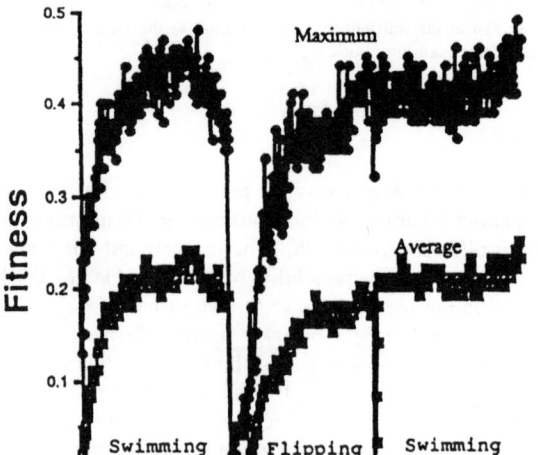

Fig. 4.10. Evolutionary memory. The population was first selected for 200 generations for swimming, then for another 200 generations for flipping. At generation 400, the task was changed back to swimming. Note especially that the recovery of fitness is extremely rapid in this last epoch (after generation 400).

These results, and the theories underlying them, have great import for biological systems and posited artificial life organisms. As Dumont and Robertson (1986) wrote of the evolution of biological networks: "As long as both the end result and all the intervening stages work, elegance of design counts for little." The same phenomena are even more likely to occur in *complex* neural systems (which have more degrees of freedom) because there are more intervening stages between the genes and the behavior they influence. Hence, nonoptimality may permeate neural systems in the animal world. We thus provide an alternate — but not necessarily competing — explanation to that of Edelman (1987) for the large number of silent and perhaps unused synapses throughout the mammalian brain.

It has been argued persuasively that human language has a strong innate, and hence genetic, component (Chomsky, 1957). However, speech seems to have arisen fairly late in hominid evolution, roughly 100,000 years ago (Lieberman, 1984). This epoch is very brief (on an evolutionary time scale), and surely too brief for complex language circuits to arise ex nihilo. Thus, it appears likely that our current language circuits appropriated and built upon structures selected for tasks *other* than language. Perhaps the most plausible use for the circuits before language was orofacial motor control (Lieberman, 1984; Stork, 1989). Generalizing and extrapolating from our crayfish analysis, we can perhaps understand why language may not be "optimal," that is, why grammar contains quirky forms or rules, due to preadaptation.

ACKNOWLEDGMENTS

We gratefully acknowledge DARPA grant DACA-88-C-0012 for Connection Machine time, administered through NASA-Ames Research Center, without which these simulations would have been nearly impossible.

REFERENCES

Bonner, T. (1988). *The Evolution of Complexity.* Princeton, NJ: Princeton University Press, 1988.

Chomsky, N. (1957). *Syntactic Structures.* The Hague: Mouton.

Darwin, C. R. (1968). *The Origin of Species.* Harmondsworth, Middlesex: Penguin. (Original work published 1866.)

Dawkins, R. (1976). *The Selfish Gene.* Oxford: Oxford University Press.

Dawkins, R. (1989). *The Extended Phenotype.* Oxford: Oxford University Press.

Dumont, J. P. C., & Robertson, R. M. (1986). Neuronal circuits: An evolutionary perspective. *Science, 233,* 849-853.

Edelman, G. M. (1987). *Neural Darwinism.* New York: Basic Books.

Edelman, G. M. (1988). *Topobiology.* New York: Basic Books.

Goldberg, D. (1989). *Genetic Algorithms in Search, Optimization and Machine Learning.* Redwood City, CA: Addison-Wesley.

Gould, S. J. (1982). Darwinism and the expansion of evolutionary theory. *Science, 216,* 380-387.

Gould, S. J., & Vrba, E. S. (1982).   Exaptation — A missing term in the science of form. *Paleobiology*, **8**, 4-15.

Greenspan, R. J. (1980).   Mutations of choline acetyltransferase and associated neural defects in *Drosophila melanogaster*. *Journal of Comparative Physiology*, **137**, 83-92.

Griffiths, A. J. F., & McPherson, J. (1989).   *100+ Principles of Genetics*. New York: Freeman.

Grossberg, S. (1982).   *Studies in Mind and Brain*. Boston: Reidel.

Hall, J. C., Greenspan, R. J., & Harris, W. A. (1982).   *Genetic Neurobiology*. Cambridge, MA: MIT Press.

Hawkins, J. D. (1986).   *Gene Structure and Expression*. Cambridge, UK: Cambridge University Press.

Hodgkin, A. L. (1964).   *The Conduction of the Nervous Impulse*. Springfield, IL: C. C. Thomas.

Holland, J. (1975).   *Adaptation in Natural and Artificial Systems*.   Ann Arbor: University of Michigan Press.

Jan, Y. N., Jan, J. Y., & Dennis, M. J. (1977).   Two mutations of synaptic transmission in *Drosophila*. *Proceedings of the Royal Society B*, **198**, 87-108.

Keesing, R., & Stork, D. G. (1991).   Evolution and learning in neural networks: The number and distribution of learning trials affect the rate of evolution. *Neural Information Processing Systems*, held in Denver, 1990. Palo Alto, CA: Morgan-Kauffman.

Lieberman, P. (1984).   *The Biology and Evolution of Language*. Cambridge, MA: Harvard University Press.

Mayr, E. (1976).   *Evolution and the Diversity of Life*. Cambridge, MA: Belknap Press, Harvard University Press.

Miller, C., Todd, P., & Hegde, S. (1989).   Designing neural networks using genetic algorithms. In *Proceedings of the Third International Conference on Genetic Algorithms*. Palo Alto, CA: Morgan-Kaufmann.

Plotkin, H. C. (1988).   Learning and evolution. In H. C. Plotkin (Ed.), *The Role of Behavior in Evolution* (pp. 133-164). Cambridge, MA: MIT Press.

Purves, D., & Lichtman, J. W. (1985).   *Principles of Neural Development*. Sunderland, MA: Sinauer.

Purves, D. (1989).   *Body and Brain*. Cambridge, MA: Harvard University Press.

Spinelli, D. N. (1987).   A trace of memory: An evolutionary perspective on the visual system. In M. A. Arbib & A. R. Hanson (Eds.), *Vision, Brain and Cooperative Computation*. Cambridge, MA: MIT Press.

Stork, D. G. (1988).   Review of *Parallel Distributed Processing: Explorations in the Microstructure of Cognition*, Vols. 1 and 2, edited by D. E. Rumelhart & J. L. McClelland and the PDP Research Group. *Bulletin of Mathematical Biology*, **50**, 202-207.

Stork, D. G. (1989).   Preadaptation and evolutionary considerations in neurobiology. In K. H. Zhao, C. F. Zhang, & Z. X. Zhu (Eds.), *Learning and Recognition —A Modern Approach* (pp. 51-58). Singapore: World Scientific.

Stork, D. G. (1990).   Sources of structure in neural networks for speech and language. In J. Elman & D. E. Rumelhart (Eds.), *Progress in Connectionism*. Hillsdale, NJ: Lawrence Erlbaum Associates.

Stork, D. G., Walker, S., Burns, M., & Jackson, B. (1990). Preadaptation in neural circuits. *Proceedings of the International Joint Conference on Neural Networks*, Washington, DC (Vol. I, pp. 202-205). Hillsdale, NJ: Lawrence Erlbaum Associates.

Stork, D. G., & Keesing, R. (1990). Interaction of learning and evolution: Principles illustrated by neural networks. Paper presented to workshop on Principles of Organization in Organisms, Santa Fe Institute, June, 1990.

Stork, D. G., & Keesing, R. (1991). Evolution and learning in neural networks: The number and distribution of learning trials affect the rate of evolution. In R. P. Lippmann, J. E. Moody, & D. S. Touretzky (Eds.), *Advances in Neural Information Processing Systems 3 (NIPS-3)* (pp. 804-810). San Mateo, CA: Morgan Kauffman.

Stork, D. G., & Keesing, R. (1992). Preadaptation, learning and evolution. In J. Mittenthal (Ed.), *Principles of Organization in Organisms*. Santa Fe Institute Studies in the Sciences of Complexity, Proc. Vol. XIII. Redwood City, CA: Addison-Wesley, 1992.

Treistman, S. N., & Grant, A. J. (1993). Increase in cell size underlies neuron-specific temperature acclimation in *Aplysia*. *American Journal of Physiology*, **264**, C1061-C1065.

Wilkins, A. S. (1988). *Genetic Analysis of Animal Development*. New York: Wiley.

Wills, C. (1989). *The Wisdom of the Genes: New Pathways in Evolution*. New York: Basic Books.

Wine, J. J. (1971). Escape reflex circuit in crayfish: Interganglionic interneurons activated by the giant command neurons. *Biological Bulletin*, **141**, 408.

5

Optimality and Strategies in Biological and Artificial Neural Networks

Wesley R. Elsberry
Texas A&M University

Wesley Elsberry's chapter, **Optimality and Strategies in Biological and Artificial Networks,** *attempts to relate optimality considerations to classical evolutionary theory. He defines both optimality and satisficing in terms that primarily have to do with function, emphasizing that the same behavioral function can be performed with different neural structures. An example given here is that speed of neuronal transmission is enhanced in mammals by myelination of the axon, but in squid by a large axonal diameter.*

Elsberry also gives an example from courtship behavior of the wrasse fish to show that optimal functioning can sometimes occur in the nervous system. But he emphasizes that Darwinian natural selection only serves to select for traits that enhance **reproductive success of the organism,** *not optimal functioning in any other sense. This is a point also made in other chapters in this section (those of Levine, Werbos, and Leven). And the chapter by Stork points out that in neural functioning as in any other biological functioning, there is a time lag between changes in environmental conditions and adaptations; hence, traits that evolved to perform one function can persist, and be suboptimal, at a later epoch wherein another function is required. In spite of all this, Elsberry argues, a large part of neurobehavioral function in many organisms can both be explained by natural selection and interpreted as optimal. Much of neural function which is not optimal can be understood as satisficing — a concept that originally arose in economics but fits naturally in a framework of evolutionary biology.*

As far as artificial neural networks (ANNs) are concerned, Elsberry says, to the extent that biological functioning is in fact optimal, the goal of biological realism should be striven for in ANNs. The method of genetic algorithms enables some fairly detailed study of the evolutionary process within ANNs themselves. He cites an article by Parisi et al. about ANN whereby connection weights that evolved to perform one function (food gathering) actually enhanced the performance of an unrelated function (learning the logical "exclusive-OR" relationship).

The principal mark of genius is not perfection but originality, the opening of new frontiers.   Arthur Koestler

ABSTRACT

Some general considerations are outlined on how optimality is related to Darwinian natural selection. The position is taken that optimality relates to overall function, rather than structure. Specifically, optimal functioning is defined in terms of what maximizes the possibility of survival to a period of successful reproduction.   Some neural network experiments on evolution are described in which enhanced ability at one skill actually increases the ability to develop a totally unrelated skill.   An evolutionary perspective on satisficing is also outlined.

1. THE BASIC ARGUMENT

In a vastly undersampled multidimensional feature space and essentially unlimited variability, Koestler's dictum just given could have great utility.   However, that's not how biological systems work.   The constraint of history implies that rats will not give rise to furred invertebrate omnivores.   The future directions in which descendant populations will develop depend critically on the set of features of the ancestor.   Variation for novelty's sake thus has a limited payoff, in that the possible variation is constrained by the ancestor's history.

Biological neural networks perform sensory integration and motor control functions in metazoan animals.   The function of biological neural networks, in the sense defined by Millikan (1993), is the emission of appropriate behavior of the animal under normal conditions.   I view these functions as having arisen by means of differential reproduction of animals expressing behavioral traits.   This scenario of natural selection influencing behavior leads to a set of hypotheses regarding the existence and role of optimality in biological neural networks.   First, the appropriate evaluation function for determination of optimality is that of reproductive success.   Representation of alleles in future generations is the only criterion of evaluation in nature, and thus the one that should be used for determining whether an organism has an optimal response to environmental conditions.   Second, a distinction between optimal function and optimal structure needs to be recognized.   Biological neural networks permit the performance of cognitive functions that have been abstracted away from direct sensory encoding.   In these cases, the behavior emitted may not have any direct mapping to a specific neural architecture.   Analysis at the structural level may be misleading, because a variety of structures may support the same or similar behaviors.

The linkage between biological and artificial neural networks is not anywhere near direct, proceeding more by means of analogy than anything else.   Artificial neural network (ANN) systems are even then abstractions away from our current understanding of neuroscience, as evidenced by the paucity of work which takes into account some representation of the actions of even one or two neurotransmitters.   However, the dependence of ANN advances on insights from biology suggests to me that a further strong statement may be made by extension of the already existing analogy.

Determining the optimality of a particular artificial neural network should be an exercise in multivariate analysis. Too often, performance concerning a narrowly defined problem has been accepted as prima facie evidence that some ANN architecture has a specific level of optimality. Taking a cue from the field of genetic algorithms (and the theory of natural selection from which genetic algorithms (GAs) are derived), I offer the observation that optimality is selected in the phenotype; that is, the level of performance of an organism or an ANN is inextricably bound to the system of which it is a part. The context in which the evaluation of optimality is performed will influence the results of that evaluation greatly. While compart-mentalized and specialized tests of ANN performance can offer insights, the construction of effective systems may require additional consideration to be given to the assumptions of such tests. Many benchmarks and other tests assume a static problem set, while many real-world applications offer dynamical problems.

An ANN that performs "optimally" in a test may perform miserably in a putatively similar real-world application. Recognizing the assumptions that underlie evaluations is important for issues of optimal system design; recognizing the need for "optimally suboptimal" response in adaptive systems applied to dynamic problems is critical to proper placement of priority given to optimality of ANNs.

This concept requires some explication of the biological and genetic algorithm background. I next set forth some common concepts, some specific findings, and outline what I see as necessary conclusions that must be drawn from them. Along the way, I treat the terms *optimality, function,* and *satisficing* to a modicum of analysis. The relation to ANNs remains implied by analogy.

The physical structure of biological neural systems is dependent on both historical and contingent factors. The theory of common descent assures us that as far as traits are heritable, descendants derive adaptations from the basis of already existing structures and traits.

Although common descent is accepted in all but a few fringe groups, it is worthwhile to reiterate some of the ways in which we can see the effect that common descent has had upon populations. All organisms in all kingdoms share most of the genetic code. This statement is even stronger than it sounds, because the exceptions are both few in number and incomplete in variance. The mapping of genetic codons to amino acids is identical in all organisms examined save a few protist species, and also certain mitochondrial lines. Even in those groups, the variance only extends to a few different pairings of codon to amino acid.

Respiratory cycles are highly conserved. The Krebs cycle is found widely, and in aerobic organisms, the cytochrome system is a well-studied example of a widespread metabolic pathway.

Patterns of embryological development separate lineages in animals. Protostome development is observed in many invertebrate phyla, whereas deuterostome development occurs in a few invertebrate phyla and chordates.

These various broad attributes link ancestors and descendants in lineages. At points where novel adaptations are incorporated into a lineage, these adaptations arise from or are derived from already existing structures, processes, or traits.

At the level of structural organization, there is not an expectation of optimal design in the sense that a knowledgeable engineer could design an optimal structure from scratch. The optimality that remains to be discovered in biological systems, including neural systems, is the kind of optimality which is conditioned on both current context and past history. The theory of common descent provides the past history, and issues in ecology provide current context.

The important part to remember about optimality in biological neural systems is that it is bound up with the whole organism. We may ignore the phenotype only at the peril of embracing erroneous views of subsystems. The evaluation of components of such a system can yield only limited, and sometimes misleading, insights. Nevertheless, problem decomposition will remain one of our most useful tools in attacking the complexity of biological systems. The resolution of this dilemma is to recognize the limitations of problem decomposition as we employ it, and to concurrently employ synthetic approaches to analysis using knowledge gained from specific studies to confirm validity or determine areas of inconsistency. The inconsistencies will represent regions of nonlinear responses in systems, which will be ripe areas for research and characterization.

By *optimality*, I intend to convey the meaning as having the properties or attributes which contribute to the best possible functioning under a specified set of conditions. Now I should also indicate a meaning for *function*. I here adopt Millikan's (1993) definition of proper function. It is extremely important to recognize that the concept of optimality requires a contextual grounding for any specific instantiation. This is exercised in comparing optimality to the possibly competing concept of satisficing.

Proper function, according to Millikan (1993), is any attribute of utility preserved in an organism and its descendants by the operation of selection. Biological functions do not exist without a history, and the history defines the function. This leads to certain philosophical tag ends, such as the conclusion that the original instantiation of an adaptation cannot thus be considered as having a proper function, but the advantages of consistency that this definition yields outweigh the disadvantages.

The function of biological neural systems, then, can be deciphered by making comparisons between organisms that have neural systems and those that do not, or whose neural systems differ in substantial ways. Biological neural systems provide for rapid transmission of sensory or motor control information through the organism. For some organisms, the mere hardwiring of sensory systems to motor systems is the sum total of neural organization. More interesting cases involve neural systems that perform processing of sensory input before emitting motor control signals. In either case, the primary function of the neural system is to provide the basis for behavior, the sum total of activity of the organism.

The means by which we recognize what an organism is, and what it is not, are by no means settled merely by finding an integument, cell wall, or bilipid layer. As Millikan (1993) pointed out, there is an inadequacy in trying to make an exact and absolute distinction between organism and environment. The organismic system to which Millikan referred includes the organism and that part of the environment which is necessary for the normal function of the organism. In other words, the organism is not a universe to itself, but must be considered in context when issues of behavior are studied. The environment provides the context of resources and other organisms in which and on which the organism operates.

Organisms live, reproduce, and die in a world unconnected to and generally indifferent to the meanings of their existence. This process continues apace, with many individual organisms accessing the same limited pool of resources, often in the same way. This establishes a competition, one without rules, but with numerous constraints. The constraints are imposed both from extrinsic influences and from intrinsic programming, the influence of an inherited program. Viewing these programs as teleonomic processes enables a clearer insight into how organisms approach optimality. Optimality here is seen as having a better approach to the unconnected, inchoate, continuous competition we call life.

Resource utilization has been extensively treated. Many theories in economics explicitly cover this topic. Of recurring interest in the literature is Herbert Simon's concept of *satisficing* (Simon, 1979). As the term finds use elsewhere in this volume, this seems a good opportunity to explore Simon's usage and how the term fits into discussion of optimality and biological neural systems. As Simon used the term in his book, *Models of Thought* (1979), satisficing can be seen to incorporate two distinct concepts. For one, satisficing is the concept of a selection among projected outcomes, and acting on such a selection which is expected to yield an acceptable payoff. On the other hand, satisficing is also used for the concept of an individual terminating an activity when some specific need has been met sufficiently well. I use *projective satisficing* to refer to the first concept, and *activity satisficing* to refer to the second.

Neither projective satisficing nor activity satisficing is strictly applicable as a description of natural selection. Natural selection is an a posteriori process operating strictly on the principle of differential reproductive success, and whose effects are best understood at the population level. Natural selection is not projective in nature, nor does it incorporate intermediary evaluation, such as would be necessary for the application of activity satisficing.

Ernst Mayr referred to natural selection as an optimization process, but a very special one (1989). Mayr gave about ten different reasons why natural selection cannot be expected to give perfect adaptation. For this discussion, his listing of the property of adaptability is a key element. I would prefer to term this consideration *accommodation*, as in the optical literature for the process of adjustment of the eye to new light levels. My preference stems from the fact that adaptation already has the meaning of adjustment to environmental conditions in a population over a period of generations. Mayr noted that the range of accommodation that individuals of a population possess will pose a bar to further selection in refinement toward an optimal solution. However, I wish to explore the notion that this principle of accommodation is, itself, an approach to optimality.

Being able to accommodate, that is, to adapt the individual to the needs of more than one specific environment or situational context, provides the basis for escaping some of the other constraints that hinder the search of natural selection for optimal solutions. In one sense, the accommodation brings the action of natural selection to a halt, yet in another, this accommodation may provide the means by which organisms can produce optimal behavior.

Behavior is, after all, the property of real interest to us in studies of animals, human and nonhuman. In cases of parallel evolution, we can see that a variety of "solutions" exist for common problems in adaptation. Many ecological niches are not defined by the anatomy of organisms, but rather by the behavior of those organisms. Wrasses and gobies can fill the niche of "cleaner fish" in reef systems, and history has settled the issue for the Pacific cleaner wrasse and the Atlantic neon goby. The niche, in each case, has far more to do with the similar behaviors seen in each than with the anatomical details of each species.

That optimal behavior exists is not at issue. Marian Dawkins reported on preliminary findings in bluehead wrasse (*Thalassoma* spp.) that indicate an optimal behavioral response with respect to environmental conditions (1994). In bluehead wrasse, a "supermale" defends his territory from other males and courts females within that territory. Chromatophores in the skin of the supermale enable it to behaviorally adjust its coloration, where the range of change goes from brilliant blue to bright green. The territorial threat display has bright blue with prominent black banding. The courting display is a bright green with dark circles on the pectoral fins. The bright blue coloration reflects light at the optimal wavelength for penetrance in seawater, which results in the territorial display being visible for the maximum possible distance. The courting display coloration is displaced in wavelength from the courting display wavelength in such a manner that it is maximally distant without entering a region of increased falloff in penetrance. A satisficer would have displaced the courtship wavelength the minimal distance necessary to reliably distinguish courtship coloration from territorial coloration. That is not what is seen in the bluehead wrasse.

Even at the level of structural features, adaptation processes seek solutions that provide workarounds to historical constraint. An example is found in comparison of vertebrate and cephalopod neurons. Vertebrate neurons have myelinated axons: Schwann cells envelop axons with insulation, and axonal depolarization jumps from gap (node of Ranvier) to gap in the coverage. The speed at which depolarization progresses down the axon is greatly enhanced relative to axons that are unmyelinated. Invertebrates, including cephalopods, do not have myelinated neurons, but must make do without. Making do, in this case, still involves being able to have fast responses to stimuli, both for predator avoidance and for prey capture. Squid are raptorial animals, often taking fish as prey, and being challenged by larger fish and marine mammals as prey items themselves. The means by which squid get around the limitation on speed of propagation in their axons is to have large-diameter axons. The speed of propagation increases with the diameter of the axon, so squid manage to behave appropriately even though, at a component level, it would seem that a relative disadvantage had been established.

The behavior of organisms is, in many ways, like the result of the interaction between hardware and software in computer systems. That this is the case is not very illuminating or surprising, but it is something that bears repetition in the interest of illustration. Alan Turing developed his ideas about stored program computers as a working out of concepts concerning the instantiation of the human mind. The action of analogy can be seen to result in productive synergy between biological and computational research, starting with Turing and continuing through Hebb, Grossberg, Holland, Hopfield, Alkon, Pribram, and others involved at the interface between biology and computation.

Some of the more interesting recent results in learning have come from studies of evolutionarily stable strategies (ESS) and genetic algorithms. Psychological studies of development provide data that can be explained rather neatly by application of ESS. Naive self-centered strategies may be discarded for more sophisticated modes of stable interaction as individuals learn about relationships with others. Of even greater import are studies of learning in systems with inheritance, as exemplified by genetic algorithms.

A thought-provoking study of learning and genetic algorithms is found in Parisi, Nolfi, and Cecconi (1991). I intend an extended discussion of this paper, and thus begin with quoting the abstract:

We present simulations of evolutionary processes operating on populations of neural networks to show how learning and behavior can influence evolution within a strictly Darwinian framework. Learning can accelerate the evolutionary process both (1) when learning tasks correlate with the fitness criterion, and (2) when random learning tasks are used. Furthermore, an ability to learn a task can emerge and be transmitted evolutionarily for both correlated and uncorrelated tasks. Finally, behavior that allows the individual to self-select the incoming stimuli can influence evolution by becoming one of the factors that determine the observed phenotypic fitness on which selective reproduction is based. For all the effects demonstrated, we advance a consistent explanation in terms of a multidimensional weight space for neural networks, a fitness surface for the evolutionary task, and a performance surface for the learning task.

Parisi et al. presented a very interesting series of simulations, and an even more interesting set of conclusions. Along the way, they noted the existence of what appears to be inheritance of acquired characters. As the authors pointed out, though, a consistent Darwinian explanation is available.

The baseline simulation involves a genetic algorithm operating on a population of simulated organisms. The organisms are controlled by feedforward multilayer neural networks. The neural networks accept as input the angle and distance measures to the nearest bit of simulated food, and also the two outputs from the previous time step. The outputs are interpreted as stay/move forward and turn left/turn right. In the initial generation of organisms, each organism is outfitted with randomized weights in the neural networks. Each organism interacts with a simulated environment alone for 5,000 time steps. Each organism has the possibility of encountering 1,000 pieces of food in this lifetime. The genetic algorithm's evaluation function is simply the number of pieces of food encountered, and thus "eaten," by the organism.

Unsurprisingly, the genetic algorithm evolves populations of organisms that do better at the task of finding food. After 50 generations, the population food acquisition average is around 250, up from about 10 for the initial population. No weight changes occur during the organisms' lives; the only changes occur by mutation in copying the weight values of successful organisms to succeeding generations. (Parisi et al. were careful to note that speaking of assigning fitness values to genotypes is inaccurate, but in the context of the simulations, this simplification will suffice.)

Then Parisi et al. changed a few things. The neural networks were modified to have two more output units, which were trained by back-propagation to predict the angle and distance to the nearest food item at the next time step. Weights within the neural network change over the organisms' lives, but those changes are not used in copying to further generations; the changes due to learning are not made available in reproduction. The populations with the prediction learning task do better over the same number of generations as compared to the baseline population: after 50 generations, the population average food acquisition is about 400. Because prediction of the next sensory input is obviously related to the task of finding food, it is easy to explain the difference in performance of the two sets of simulations. If we imagine two organisms represented as different points on a fitness surface, the effect of the prediction learning task will be to possibly shift the position of each organism on that fitness surface. Even if the two organisms start out with the same relative fitness, learning will accentuate differences in the regions of the fitness surface that each occupies. An organism already at a local maximum will be likely to end up with a lesser fitness than an organism that starts in a region with a higher local maximum. The well-correlated task implies that most changes in weight over the organisms lifetime will tend to explore local areas of higher fitness on the fitness surface.

Using the same network structure, another simulation run was performed. This time, however, the network outputs were trained to pseudo-random numbers. The result: the average food acquisition at 50 generations was about 330. This falls between the baseline simulation, where there was no weight change during life, and the prediction learning simulation, which had correlated weight change. Parisi et al. explain this by noting that any perturbation of weights will tend to differentiate between organisms whose overall fitness starts out as being equal, but which occupy regions of the fitness surface with differing "potential." The effect is not as pronounced as when learning is correlated.

Parisi et al. next explored "indirect inheritance of acquired characters." In this case, they modified the organism's neural network to have an output, this time trained during life to compute the exclusive-or (XOR) of the sensory inputs. The XOR problem is assumed not to be correlated with the food acquisition task. Because the tasks are not correlated, and the fitness evaluation only looks at food acquisition, Parisi et al. expected that no improvement in ability to learn the XOR task would be seen over time. However, the results showed that organisms of later generations did have an improved capacity for learning the XOR task. How might this be? Parisi et al. forwarded the explanation that the performance surface for the XOR and the fitness surface for the food acquisition task will have regions that vary in correlation of slope. Those organisms which have better correlation of the two slopes will be more likely to end with a high fitness value than those organisms with a poor correlation of performance and fitness slopes. This explains the apparent inheritance of an acquired character with a completely Darwinian mechanism.

Again, though, it should be stressed that organisms do not pass on the weights as adjusted by lifetime learning. The performance of new organisms has much the same starting point: descendant organisms do not acquire the ability to perform the learning task, they acquire the ability to learn the learning task more quickly.

Stepping away from simple description of Parisi et al.'s methods and conclusions, a few general comments on the linkage to biology are in order. With the use of one neural network as the basis for learning and motor action for each organism, it is easy to see that changes in weights due to learning at some output units will necessarily change the feedforward response of the other output units, due to change in the node output values for hidden layer nodes. In actual biological organisms, this interaction between systems and consequent perturbation due to learning would be much more difficult to establish in detail. However, the mere fact of some learning task having a perturbing influence on fitness-defining tasks will set the basis for the mode of evolutionary change discussed by Parisi et al., and their mechanism of evolutionary acquisition of learning capability would then be ready to go to work.

As mentioned before, optimality in biological systems is not a matter seen from a global, top-down perspective. As biological neural systems are the basis for behavior, it makes sense to evaluate the behavior of organisms in a theoretical framework that makes specific predictions about that behavior, based on considerations of optimal function. *Optimal foraging theory* (OFT) is one such theoretical framework. OFT asserts that organisms will behave in a predictable manner: they will acquire resources at maximal rates commensurate with circumstances. There exists quite a range of opinion over what is or is not part of OFT, but this catches the major thrust of usage in studies invoking OFT.

OFT has its most notable success as a description of the behavior of organisms where energy constraints are tight; organisms whose energy budget requires high levels of input and whose metabolism, correspondingly, is high, tend to match the predictions made under OFT most closely.

Hummingbirds and bees are favorite subjects in OFT studies, followed closely by various ungulate herbivores and seed and nut gathering birds. In hummingbirds and bees, issues in information theory come to the fore. For hummingbirds, the information of interest involves how hummingbirds might decide to pattern visits to food sources based on knowledge of food source quality. In bees, the issue of information transfer has been the single largest source of controversy within bee foraging studies.

A study of hummingbird foraging had two food sources of differing quality separated by a substantial distance (substantial by hummingbird standards, at any rate). The frequency of visitation to the two feeding stations was characteristic of rate maximizers utilizing the marginal value theorem — except in the case where one feeding station had very poor quality food compared to the other. The prediction, according to marginal value theory, would be that a rate maximizer would never revisit the poor quality feeding station. The hummingbirds, however, continued to visit the feeding station at a very low frequency. The resolution, of course, is to step back and reconsider the system. Hummingbirds feed on nectar from flowers, which may be found on shrubs and trees. The quality of food found in association with a particular shrub or tree will depend on the number and condition of flowers on that shrub or tree. This value will tend to fluctuate, such that a shrub might go from nonproductive status to highly productive status in a short period of time, and this value may fluctuate in either direction over time. Hummingbirds have been shown to be capable of some very interesting cognitive processing, but no one has suggested that precognition is among those capabilities. The hummingbird must engage in scouting behavior in order to learn the current status of food sources, and this includes not totally ignoring potential food sources just because those sources were, at some time past, not of good quality. Information about the environment must be collected by real-world organisms, which is unfortunately not the case for the imaginary theoretical individuals which seem to consistently be the jumping-off point for development of various theories in economics and biology.

Optimal biological neural networks will be identified not by their optimal structure, but rather by their optimal behavior. Differences in structure will not necessarily lead to differences in behavior. Because behavior is a whole-organism ensemble, it is important to make evaluations in light of overall function and fitness, and not merely on specific features in isolation.

Artificial neural networks currently have a simple relationship between structure and function. As ideas derived directly or indirectly from biological study are incorporated, we can be assured that this simplicity will also disappear. The relevant information that will be available for evaluation, and that ultimately matters most, is the behavior of the ANN system. This behavior does not exist in isolation from the problem context, and should not be evaluated without reference to that context.

There is a common perception of a scale of function leading from poor to adequate to better and best. The codification of this scale leads us to consideration of the concept of optimality in general, and for us, the intersection of optimality and function in biological and artificial neural networks. The issue of whether we can expect to find optimality in biological neural networks or to achieve optimality in artificial neural networks should have a serious effect on research carried out in each field. If we are justified in assuming that optimality is a possible attribute of biological neural networks, then we have the basis of an existence proof for extending that expectation to the artificial neural systems whose design and operation are premised on biological neural networks. If, on the other hand, we cannot justify the expectation of optimality in biological neural networks, then our work in striving toward proving or designing optimal artificial neural networks becomes just that much more difficult.

In order to explore this field properly, a good working set of definitions is critical. As the excursion that I plan involves stops at optimality, optimization, optimal foraging theory, satisficing, and other scenic points of interest, I here attempt to give a consistent set of meanings so that the reader may construct a consistent road map. To overextend the metaphor, there are a number of chuckholes and dips posing obstacles, in that the terminology is often given specific usage without the usage meant being unambiguously and forthrightly stated.

1.1. Natural Selection, Optimality, and Satisficing

As the primary proposed mechanism of adaptive evolutionary change, natural selection has undergone rigorous scrutiny and testing since its formal proposal under that name in 1858. In between 1900 and 1930, natural selection was widely regarded as being contradicted by the newly rediscovered principles of Mendelian genetics, so much so that textbooks and articles referred to Darwin as establishing well the fact of evolution, but having forwarded an incorrect theory of its mechanism (natural selection). The Neo-Darwinian theory of natural selection formulated by researchers including Dobzhansky, Fisher, Wright, and Mayr demonstrated that the Mendelist objections to natural selection were misplaced.

The question of the meaning of natural selection for natural adaptation, however, remains open. Certainly, some researchers go too far in asserting an adaptive component for all features of an organism. On the other hand, claiming that natural selection plays no role in the adaptation of features in organisms seems equally naive. Somewhere in between the two extremes lies the middle ground of biological reality. This middle ground denies the naysayers of both camps, who cannot tolerate a pluralistic, complex reality.

The question of whether biological neural networks are or can be optimal probably cannot be answered with either a definitive yes or no at this time. Our understanding of even the normal functioning of biological neurophysiology is woefully inadequate to the task of analysis that the question poses. Even that small amount of data refers to an even comparatively smaller set of populations of research animals, when measured against total biodiversity. Still, we can consider other questions that are inclusive of our question of interest, where a definitive "no" answer would short-circuit our search for an answer. The most inclusive question would be, can optimality arise through evolutionary change? Assuming that the answer is not "no," we may then consider further questions. The answer to the question, can optimality arise through evolutionary change, is most emphatically "yes." Various results demonstrate that various aspects of sensory mechanisms and social behaviors occur with optimal specifications or at optimal rates.

Given that current neurophysiological knowledge does not give us the basis for direct analysis of optimal function, what approach can be taken to give an indication of whether optimality in fact exists in biological neural networks?

First, a recognition of what comprises a proper evaluation function is needed. Evaluation functions which ignore the basic facts of evolutionary biology may be safely ignored. So what, precisely, forms the real, canonical evaluation function for biological organisms? That evaluation function is simply differential reproductive success. Phenotypic characters of an organism which do not either give a reproductive success advantage, or a reproductive success disadvantage, are not preferentially retained or eliminated from the population. This evaluation function can only operate on the composite phenotype of the organism, which is the complex interaction of genotype, environment, and contingent history. Evaluation functions which remove the biological neural network from its context in determining the phenotype of an organism are inherently susceptible to bias, and should either be rejected out of hand, or should have empirical support showing that they are unbiased estimators of the actual biological evaluation function.

REFERENCES

Dawkins, M. (1994). Plenary talk, Annual Conference of the Animal Behavior Society, Seattle, WA.

Mayr, E. (1989). *Towards a New Philosophy of Biology*. Cambridge, MA: Harvard University Press.

Millikan, R. (1993). *White Queen Psychology: Essays on Biology for Alice*. Cambridge, MA: MIT Press.

Parisi, D., Nolfi, S., & Cecconi, F. (1991). *Learning, behavior, and evolution* (Tech. Rep. No. PCIA-91-14). Department of Cognitive Processes and Artificial Intelligence, Institute of Psychology, C.N.R., Rome.

Simon, H. A. (1979). A behavioral model of rational choice (1955). In H. Simon (Ed.), *Models of Thought* (pp. 7-19). New Haven: Yale University Press.

6

Properties of Optimality in Neural Networks

Mark R. DeYong
Intelligent Reasoning Systems, Inc.

Thomas Eskridge
Intelligent Reasoning Systems, Inc. and New Mexico State University

*Mark DeYong and Thomas Eskridge's chapter, **Properties of Optimality in Neural Networks**, is one of two chapters in this book (the other being Elsberry's) that focuses on the variety of ways in which networks could be interpreted as being optimal. DeYong and Eskridge's focus is on the design of artificial neural networks (ANNs), not networks in living animals, although they include a discussion of how faithful one should be to biological detail in ANN design.*

*DeYong and Eskridge make a particularly cogent distinction between three levels of potential optimality in ANNs: the **performance** level, the **attribute** level, and the **design** level. These different levels have some degree of analogy to the biological "triunity" of habits, reason, and affect as discussed in Leven's chapter in this book. Also, they are related to the distinction drawn in both Levine's and Werbos' chapters between optimality of detail versus optimality of overall function. There is also an echo of Werbos' chapter in their statement that optimality at the overall design level need not correspond to a maximum or minimum of a mathematical utility function.*

This discussion is a first step toward a taxonomy of different "optimalities" for ANNs (and perhaps, by analogy, for biological neural networks as well). DeYong and Eskridge go on to deal with the potential significance of other common issues and distinctions, such as linear versus nonlinear; digital versus analog; hardware versus software implementations, and more. They acknowledge that the field needs some general organizing principles about what should be optimized, and how that is to be done, but state we are still in the "fourteenth century" on our quest for such principles. Articles such as this one are valuable for helping us see where the field stands.

ABSTRACT

Ongoing research into the nature of optimality in artificial neural networks (ANNs) is presented. We discuss the relationship of optimality criteria and critical design decisions for the development of ANNs, with the final goal of producing a process model for the optimal design of artificial neurons and ANNs.

1. OPTIMALITY AND OPTIMALITY CRITERIA

Optimality arguments have often been used to describe and predict empirical regularities in many scientific domains (Schoemaker, 1991; Levine, Chapter 1). Optimality arguments describe or explain these regularities as the maximization or minimization of some objective function. The exact form of this optimality function is the subject of much of the discussion in the literature and in this volume. Particularly in the study of artificial neural networks (ANNs), the discussion of optimality is often obscured by the differing ways in which explanations of optimality can be used (e.g., normative or predictive), the different forms of the optimality explanation (e.g., teleological, causal, or process), and by the different criteria on which networks can be optimized.

Optimality arguments are used for both normative and predictive reasons. The assumption of rationality that underlies microeconomic theory is an example of a normative optimality argument that allows economics to be described so that the choices of rational agents can be explained. The rationality assumption basically states that the best option in a given situation should be chosen. However, the assumption is not predictive of the economic behavior of people, because they do not always act rationally. Thus, the predictive capabilities of the normative model are more limited than a process model explanation of observed behavior. A process model explanation is a model that is defined in terms of the observed entities and behaviors. Predictions made from this model can be directly applied to the entities in question, because the information used in developing the prediction is explicitly observed in the entity or its behavior.

The form of optimality arguments is also varied in the study of optimality in ANNs. Teleological explanations of optimality are generally applied when the observed phenomena are so complex that a causal or process model cannot be obtained. Teleological explanations anthropomorphize the entities under study by projecting "wants" and "intentions" on them. Thus, light "wants" to take the shortest amount of time to get from a point A above water to a point B under water. However, teleological explanations fail to provide reliable predictive power in the same way that the (teleological) rationality assumption does. A causal model attempts to rectify this by providing a causal chain of events starting from a set of plausible or possible initial conditions to the observed event. Although this form of optimality argument allows predictions to be made with more confidence than a teleological argument, it is still based on subjective constructions and expectations. A process model takes the causal argument one step further by working from "first principles" so that the subjective matter is confined to (at most) the definitions of terms. This is the most powerful form of optimality argument, because it all but eliminates the biases that adversely affect the other two forms.

Teleological and causal optimality arguments are useful in a number of fields, but in the study of ANNs, more basic process model optimality arguments can be applied. With ANNs, it is possible for the investigator to "open up" the network, and (to some extent) inspect the functionality of each element of the system. Thus, we would like to apply a "design science" approach to the study of optimality in ANNs; that is, we explicitly attempt to design optimal systems (Simon, 1981). By the term *optimal systems* we do not mean to suggest that the design is perfect with respect to some specified goal and no better solution could possibly be found, but that it is preferred over alternatives that attempt to optimize the same goal. This is similar to the use of optimality in biological systems, where inclusive fitness is the optimality criteria that is maximized. Helweg and Roitblat (1991) stated that:

> Natural selection is perforce an optimizing function in that it selects from a set of alternatives those that are more successful at reproducing their genetic copies. Natural selection ranks genotypes in order of their fitness, and selects those that are most fit. An individual that failed to maximize its fitness would quickly (in evolutionary time) be replaced by an individual that was more successful at optimizing its fitness.

This is, of course, a teleological explanation of optimality: natural selection does not actually "rank" and "select." Although this explanation does provide a useful means of reasoning about evolution, it is, like most other teleological explanations, primarily a *postdictive* explanation. This is to say that teleological explanations do not generally have the power to enable predictions to be made, but instead provide a satisficing explanation. What is needed is a process explanation: an explanation embodied by a model that can be used to predict future behavior with respect to optimality criteria.

Optimality arguments in ANN design can take on normative or predictive uses, and fall into any of the three forms mentioned above. With respect to optimizing objective functions, many researchers have focused on optimizing a single criterion. Although we do believe that some criteria are more important than others, we agree with Elsberry (Chapter 5) that optimality in ANNs should be examined with respect to a number of optimization criteria. We have created a hierarchy of criteria divided into levels that illustrates our view of the importance and relationships between optimality criteria for ANNs (Fig. 6.1). This is by no means a complete taxonomy of all criteria, but is meant to serve as a basis for discussion of optimality issues and as a framework for the study and development of a process model of optimal ANN design. We briefly discuss the optimality issues at levels one, two, and four before discussing in more detail the design issues at level three.

1.1. Level One Optimality

The primary issue in determining optimality at the first level is how well the network performs at the task it is designed to do. Performance can be measured by the accuracy in classifying the training set, accuracy in classifying the test set, and accuracy in generalization to new examples.

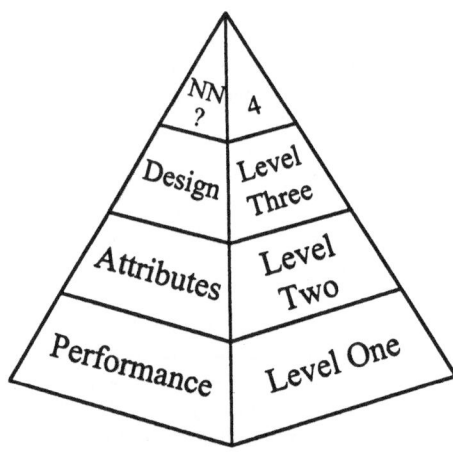

Fig. 6.1. Levels of optimality. The levels of optimality are shown in a pyramid, with the top (fourth) level being the decision to use a neural network, the third level being optimality criteria used in the design of the neuron element and network structure, the second level being optimal attributes of the network, and the bottom (first) level being optimal performance issues.

Optimality criteria such as accuracy on the training set presuppose a particular type of neural network, namely, the common paradigm of supplying a uniform-architecture, feedforward, logistic activation function network with a set of examples on which to train (Rumelhart, McClelland, & the PDP Research Group, 1986). For these networks, training typically consists of supplying a set of input/output examples that, taken together, provide a representative description of the data space. The results of the network on the example are compared with the desired output, and any error is used to modify the weights in the network to reduce the error the next time the training example is presented. Training consists of minimizing the difference between the network output and the desired output. Optimality under the criterion of accuracy on the training set judges competing networks on their performance in producing the desired output for each training instance. The measure of performance can be cumulative error, average error, or the smallest maximum error for any instance in the training set.

Similarly, the criterion of best performance on a test set would judge competing networks on their ability to respond correctly to instances are not in the training set. The training set is typically chosen as a random subset of the all instances available for training, and thus is assumed to representative of the data in the problem space. Because of this, the test set may be significantly different from the data characteristics of the network in actual operation. Thus, the criterion of best generalization performance is an optimality criteria that is sometimes used. For example, a neural network for stock prediction can be developed using training and test sets meant to ensure that the network has adequate ability to respond to the wide variety of market conditions that may arise. But when put into actual practice, the conditions of the market may not exhibit the behaviors on which the network was trained, or may demonstrate several of them at once. In this situation and others like it where the task environment contains some uncertainty, it is important to judge network optimality on its ability to consistently perform well. In cases such as these, the performance metric could be based on either the post-facto decisions humans make on the data presented to the network (i.e., with the same data, a human would have made a similar decision), or on the extent to which the decision made was on the basis of "graceful degradation" (i.e., an extrapolation of the solutions for known training instances.)

1.2. Level Two Optimality

There are a number of properties that are desirable for neural networks once adequate performance (level one optimality) is obtained. In general, the smaller the network is, the more optimal it is. The size of a network can be measured in the number of layers, nodes, or interconnections it has. These criteria for optimality can be significantly biased by the type of neuron element and training procedures used in the development of the network. For example, Cascade Correlation methodology generates networks with a large number of layers, but with few nodes (Fahlman & Lebiere, 1990).

This training of neural networks suggests a number of possible criteria that can be the focus of an optimization process, aimed at reducing the amount of time needed to train the network: the number of times the training set is presented to the network (i.e., the number of epochs), the number of mathematical operations that must be performed, and the number of training instances required. Any principled method of reducing the training time of a network will find wide acceptance within the neural network community for the simple reason that the longer networks take to train, the fewer experiments can be run, and the slower results will be to come. This point is underscored by the current research on catastrophic forgetting, where two or more training instances force a competition for the weights on a number of interconnects, changing the network after one training instance to respond correctly to it, but forcing it to then respond incorrectly to a training instance that was previously responded to correctly (French, 1991; French & Jones, 1991; Hetherington & Seidenberg, 1989). This cycle drastically increases the training time of the network, sometimes making training impossible.

The other optimality criteria mentioned reflect the development of ways to skirt the number of training instances problem. When the investigation is not directly studying the number of training epochs, it is feasible to meet the desired goal of minimizing training time by reducing the number of operations that must be performed, or reducing the number of training instances required to achieve acceptable performance during test and actual use. If the number of mathematical operations performed during the training of the network on one instance can be minimized, then the shortened processing time will allow the training instances to be presented many more times. This can also be achieved by using a simplified neuron model, reducing the number of nodes in the network, or by reducing the number of interconnections between nodes (cf. the randomized interconnection schemes of Minai & Levy, 1993.)

Similarly, if there is a deep understanding of the problem space in which the network will work, training examples can be selected that convey the maximal amount of information.

Obviously, this is a tortuous method of reducing training time, because if the problem space was understood so deeply, a neural network solution would likely not be necessary.

1.3. Level Four Optimality

The criterion at level four basically encompasses the decision to attempt a solution to the problem using neural networks instead of a conventional software, mechanical, or manual approach. Because there have been numerous articles written on the benefits of neural networks over conventional approaches (Eskridge & Barnden, 1992; Rumelhart et al., 1986; Smolensky, 1988), we mention this only for completeness.

2. ISSUES OF OPTIMALITY AT LEVEL THREE

In designing optimal ANNs, we wish to know general principles by which optimal ANNs for other systems can be generated (i.e., a process model of optimal ANN design). Unfortunately, the optimality criteria at levels one and two permit only postdictive explanations of optimality. One ANN is better than another because it has better performance with respect to the level one and two criteria. Although the optimality criteria at levels one and two give a means of comparing two nets and selecting the more optimal, they tell nothing about the process factors that make one net more optimal than the other. It is the issues that are addressed at level three that begin to provide a process explanation for optimal ANN design. This is also the level where optimization for behavioral characteristics that are not precisely defined by the minimization of some mathematical function is performed. In particular, the development of ANNs for the control of the object recognition and reasoning tasks of autonomous systems defies the precise specification of input/output characteristics. The issues investigated at optimality level three are also the types of issues that will allow a process model for optimal behavior in these unstructured task environments (Eskridge & Fields, 1989; Fields, Eskridge, Hartley, Coombs, 1989; Fields & Dietrich, 1987, 1988.)

- o Biological versus nonbiological basis of neuron model
- o Model dynamics
  - Time-dependent versus time-Independent
  - Continuous versus discrete
  - Linear versus nonlinear
- o Model implementation
  - Hardware versus software
  - Analog versus digital
- o Task specific versus generally configurable system architecture

Fig. 6.2. Design tradeoffs basic to optimal neural network design. The competing design decisions form the basis for developing a process model of optimal neural network design.

The optimality issues that we feel are important as a basis for developing a process model for the design of optimal ANNs are shown in Fig. 6.2. These issues are directed principally at the neuron model used by the ANN, and have been motivating factors behind our development of the Hybrid Temporal Processing Element (HTPE, patent pending) (DeYong, 1992; DeYong, Eskridge, & Fields, 1992; DeYong & Fields, 1992; DeYong, Findley, & Fields, 1992). By focusing on the neuron model, the issues of network topological structure can be treated with respect to the problem being solved rather than as a structure into which problems must be forced.

2.1. Biological Versus Nonbiological Basis of Neuron Model

The development of biologically realistic artificial neuron models has been of interest for the last four decades and has diverged in several directions ranging from the purely mathematical parallel conductance model introduced in Hodgkin and Huxley (1952), to the symbolic network simulator GENESIS of Wilson and Bower (1989). These models, however, concentrate solely on biological realism and tend to lose sight of their hardware implementation and application to real-world engineering problems.

Biological realism in neural modeling is motivated both by the goal of understanding the behavior of biological nervous systems and by the realization that biological neurons are complex, versatile signal processing devices that are evidently well suited to a very large variety of computational tasks (Hopfield, 1982; Kohonen, 1988; Lippmann, 1987). Neurons are hybrid analog/digital devices, in which inputs are processed by the time-dependent convolution of relatively slowly varying postsynaptic potentials (PSPs), and outputs are transmitted over long distances by fast, relatively loss-free action potentials (APs). This integration of analog input processing with digital, pulse-encoded communication allows neurons to make use of time and phase differences between signals arriving in real time to represent both temporal and, for example, in the visual system, spatial information (DeYoe & Van Essen, 1988). It also allows neurons to exchange information in times much smaller than their internal processing times, thereby breaking the communications bottleneck that hobbles many massively parallel computers. The combination of high-speed, digitized communication and versatile analog computation makes neurons ideal for many time-dependent signal processing applications.

Although modeling the details of biochemistry down to the ion channels is in all likelihood unnecessary, it is expected that a neuron that does not lump the influence of the axon and collaterals, synapse, and dendrite into a single weight on a single time scale will better reproduce the desirable computational characteristics of biological neurons. This not only aids in the study of biological neural networks, but also provides a means for investigating computational mechanisms clearly not available in the standard neuron model.

Other approaches to the design of neurons trade biological realism for simplicity and traditional computational power. These systems are generally time dependent and digitally based, and therefore negate several beneficial aspects of biological neurons. What these models gain in idealized efficiency (i.e., computed maximum connection updates per second), they lose in actual applicability to real-world science and engineering problems through the inability to distinguish temporal characteristics of the input. The utility of a non-biologically based neuron can be lessened further, depending on the form of the activation function used (e.g., logistic, step function, or linear).

There are a number of problems that can be solved quite well by a network of nonbiological neuron processing elements. However, with respect to the optimality of this solution, in many of these cases, a biologically based implementation can perform better. Because the hybrid analog/digital nature of the biologically based approach can respond to signals encoding information in frequency and phase as well as by accumulating pulses to duplicate the behavior of nonbiological neuron models, it will be applicable in more problem situations. Because biologically based neuron model behaviors subsume the nonbiologically-based behaviors, choosing them for the neuron design can only allow more optimal ANN solutions to be generated.

2.2. Model Dynamics

2.2.1. Time Dependent Versus Time Independent. Many artificial neural networks assume that each input vector is independent of other inputs, and the job of the neural network is to extract patterns *within* the input vector that are sufficient to characterize it. For problems of this type, which amount to spatial pattern recognition, a network that assumes time independence will provide acceptable performance. However, there is a large class of problems where the input vectors cannot be assumed to be independent and the network must process the vector with respect to its temporal characteristics and context. Network architectures that assume time independence are typically unwieldy when applied to a temporal problem, and require additional inputs, neuron states, and/or feedback structures. As mentioned above, networks that assume time dependence have the advantage of being able to handle both time-dependent and time-independent data.

The ability to handle temporal as well as spatial information is particularly important when optimal behavior means survival within a task environment. Miall (1989) stated that "... the current forms of neural networks, while suitable for some computational tasks, have an impoverished temporal repertoire and so are unsuited to many time dependent operations faced by animals." Examples of such time-dependent operations include real-time obstacle avoidance, object tracking, feature and object recognition in the presence of ambiguity and uncertainty, and contextualized problem solving. These operations are critically important to the development of autonomous vehicles for use in the private, public, and defense market sectors.

In the types of networks Miall (1989) referred to, the intrinsic time dependence of the postsynaptic potentials generated by synaptic activity is replaced by the summation of weighted input levels, and the streams of action potentials are replaced by the output of the model neuron. All computation is assumed to occur within a short increment:

$$y(t+\tau) = s\left(\sum \alpha_i x_i(t)\right)$$

where y is the output, x_i are the inputs, α_i are the input weights, and τ is the time increment.

This model has two main shortcomings. First, although the signals are time dependent, the processing is not, limiting its applicability, and second, it provides no insight into computation in the biological network at the scale of the individual cell. This model can process time-varying signals, but it cannot be directly applied to the class of distributed decision making (DDM) problems involving the temporal relationships between processing elements (e.g., temporal winner take all (temporal WTA) problems, as in Barnden, Srinivas, & Dharmavaratha, 1990). It is now reasonably clear that such relationships are routinely used to solve real-time signal processing problems, such as visual object identification and tracking (e.g., DeYoe & Van Essen, 1988).

   2.2.2. Continuous/Asynchronous Versus Discrete/Synchronous. Many problems amenable to solution by ANNs require that the temporal aspects of the problem be taken into account. To accomplish temporal dependency, the ANN must be continuous and asynchronous. Both of these conditions are required because if a system is discrete or synchronous, errors will be introduced into the system by missing signals in the digitization process or by shifting signals in both the digitization and synchronizing processes. Figure 6.3(a) illustrates the information losses caused by digitizing a pulse stream representing a continuous input signal and by enforcing synchronicity on the neuron elements. At the first input spike, a low-amplitude, long-duration PSP begins in the continuous/asynchronous (CA) case, which is only registrable at time T_1 in the discrete/synchronous (DS) case. When the second spike arrives, the amount of current that it adds to the PSP in the CA case pushes it over the threshold V^{th}, causing an axon hillock to fire at time T_2. Voltages V_1 and V_2 show two additional possibilities for the PSPs after the second spike. With voltage V_1, the current level at the time of the next digitization will be lower than the threshold, and therefore the digitizing system will have missed the axon hillock firing altogether. With voltage V_2, the current level remains at a high enough level so that the digitization will recognize the current as being enabled. It could then command the axon hillock to fire at time T_3. However, important temporal properties of the input signals have been changed by delaying the order to fire by T_d. This delay could be as much as the digitizing window T_s.

   The first input AP in Fig. 6.3(b) has a leading edge at time T_1 and a falling edge at time T_2. The rise time of the low-amplitude, long-duration PSP V_1 contained in the time interval TD_1 could be considered a transient, with time TD_2 being an intermediate steady state. When the second input AP arrives it generates a PSP that adds with the previous PSP. If the resulting combined PSP follows the solid line, the period from T_3 until the PSP completely dissipates is considered a transient period with rest as the steady-state value. It is possible to view the entire period from T_1 until the dissipation of the PSP as transient, if TD_2 is viewed as transient. If the PSP follows V_2, the period TD_3 is a transient, with V_2 as the steady state. If the PSP follows V_1 the period TD_4 is a transient, with V_1 as the steady-state value. In any of these cases, the time delay between AP1 and AP2, as well as the transient PSP amplitude, is assumed to contain no information.

   As already mentioned, information can be carried in the transient signals based on the delay between the input pulses. Although these pulses can cause significant changes to the PSP, they can be considered to be in the transient region of a conventional analog system and will be ignored by such a system. In conventional analog systems, where all computation is carried out by level-to-level transformations, no information about the temporal characteristics of the input signals can be used to distinguish the input.

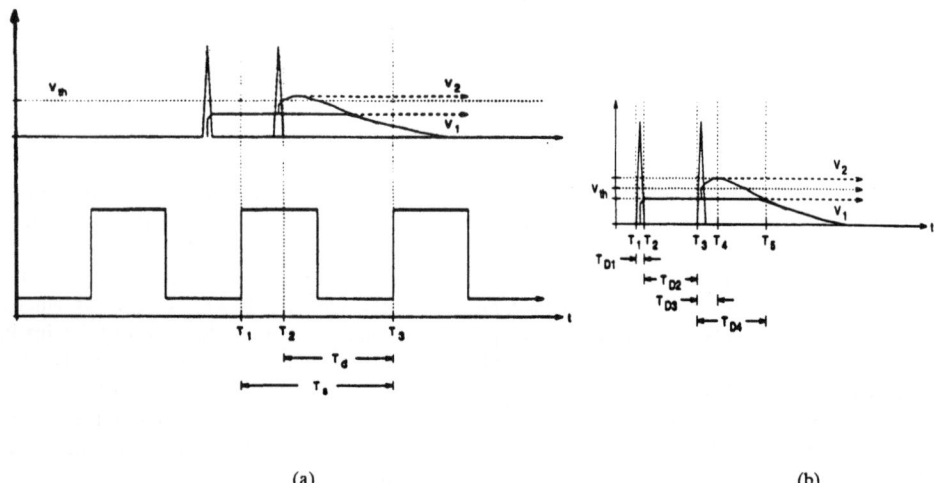

(a)   (b)

Fig. 6.3. Continuous, asynchronous versus discrete, synchronous. (a) Time delay or signal loss that can result from digitizing continuous signals. (b) Related difficulties using a conventional analog approach, which discretizes the signals into transient and steady-state behaviors.

2.2.3. Linear Versus Nonlinear. It has been shown that neuron models with nonlinear input/output transform components are able to solve nonperiodic dynamical systems, whereas those with strictly linear elements are able to solve only periodic dynamical systems (Pineda, 1988; Rumelhart et al., 1986). In conventional, nonbiologically inspired neuron models, the nonlinearity is typically exemplified by the *logistic function* (Rumelhart et al., 1986). This function transforms the real-valued sum of inputs into a real-valued output. In terms of a biologically-based neuron model, this property is realized in a nonlinear transformation from input APs to PSPs. It is important to note that because the nonlinearity applies to the pulse streams as they arrive at a neuron, this nonlinearity transforms the frequency and phase domains of the input, as well as the (pulse stream equivalent of) input strength.

The choice of linear versus nonlinear input/output transformation functions is another case where one option subsumes the other: A network that is capable of implementing nonlinear input/output transformations is also, by definition, able to handle linear transformations. If the optimality task is survival in a task environment, then the extra options afforded by the choice of nonlinear transformation may be critical to the effectiveness of an autonomous system.

2.3.1. Hardware Versus Software. The primary advantage of a hardware solution to a software solution is the speed of solution. Hardware implementations typically provide at least several orders of magnitude improvement in the speed of solution over software implementations. Depending on the design choices made in the consideration of the preceding optimality issues discussed, the implementation of the neuron element in hardware may be the only feasible solution. In particular, hardware support is necessary to process continuous/asynchronous pulse streams. A software simulation of this process will lose the critical temporal information contained in the pulse stream simply because it is being executed on a discrete, synchronous processing unit. As a practical matter, the presence of feedback and multiple inputs can make a software simulation of pulse stream neural elements unusably slow.

2.3.2. Analog Versus Digital. Given that the design choices may force a hardware implementation of the neuron model, the next choice is the implementation technique. This choice will play a large role in determining the efficiency of the resulting system. There are three principal implementation methodologies used in VLSI neural networks: conventional analog, digital, and a hybrid of analog and digital. Analog devices have an advantage over digital devices of generally affording faster processing at a lower hardware overhead. Analog signals also naturally convey temporal information in the input. However, digital devices provide greater noise immunity and a building-block approach to system design. The hybrid approaches generally take a form in which the internal computation of the neuron is implemented in analog, and the extracellular communication is performed digitally. This approach is based explicitly on the electrophysiology of spiking neurons, and gives the best of both worlds: the speed and low hardware overhead of analog, and the noise immunity and building-block nature of digital components.

2.3.3. Conventional Analog. Conventional analog systems are systems in which the range of permissible amplitudes of input and output signals is continuous. Conventional analog systems follow an axiomatic framework that asserts, among other things, that the transient is linear and can be discounted as containing no information. However, transient behavior can be used in some instances to characterize the steady state. For example, in a conventional analog implementation of multiplication, the transient can be used to characterize how long the system will take to settle to steady state, or if it will settle. However, the transient does not contain any information relevant to the solution of the multiplication problem. Transients are used to characterize the physical behavior of the analog system and are not used in the functional behavior of the system. This neglect of the transient can have a detrimental effect on the processing power of the analog system.

Analog circuits are commonly used because of their speed and small implementation size. Analog waveforms also provide a natural means for modeling biological signals such as PSPs. Analog circuits have a fairly constant power dissipation due to the fixed current biases required to keep the devices in the proper operation mode. This is an advantage over digital circuits which have power surges due to circuit switching.

2.3.4. Conventional Digital. There has been interest in developing digital neural network systems and processing elements due to the availability of development tools for digital circuits and the expertise to use them. Digital neural processing elements also have the advantage of conforming to automated testing procedures, which analog systems lack. Because of the difficulty in implementing the inherently analog aspects of biological neurons, many digital neural processing elements implement instead the idealized mathematical expression of the multiplication of synapse weight by the activation value produced in the axon hillock. These circuits do not have particular circuitry for the standard components, but rather implement a simple form of microprocessor, complete with arithmetic logic units (ALUs) and on-board random access memory (RAM). There are several networks that conform to the above description, such as in Melton, Phan, Reeves, and Van den Bout (1992), Hammerstrom, (1990), and reviewed in DeYong (1992). These implementations all have the same problems, namely that the range of behaviors of digital models is limited with respect to biologically inspired neurons, temporal aspects of the input signals could be missed or shifted, and numeric information is stored with limited resolution, which may prevent adequate solutions from being found.

2.3.5. Hybrid Analog/Digital. Hybrid Analog/Digital systems are proposed to circumvent the problems associated with either analog- or digital-only implementations of VLSI neural networks. The idea is to compensate for the weaknesses of one technology with the strengths of the other. For example, communication with analog signals has the disadvantage of being susceptible to noise. This can be overcome by using digital communication techniques.

An example of a system that uses this technique can be found in Waller, Bisset, and Daniell (1991). The synapse weights are stored digitally, then are converted via a digital-to-analog converter to an analog value. This is then passed through a synapse, which converts the analog value into a number of pulses. These pulses are then counted by digital circuitry and produced as output. Although this is clearly a design motivated by engineering considerations, it does illustrate the types of integration that can be made using a hybrid approach.

Other types of hybrid systems generally attempt to perform internal computations in analog and external communication digitally. However, in the examples given above, which are prototypical of the field, this amounts to implementing digital circuits in analog. As such, these methods will have many of the same problems as conventional analog and digital circuits.

In terms of developing a process model of designing optimal ANNs, the design choice of implementation method is not completely clear. The digital implementation strategy cannot be considered as an optimal design choice, because it would preclude the use of other optimality issue options presented earlier. The conventional analog implementation strategy (with the allowance for some unconventional operating procedures such as not always settling into steady state) can be made to encompass both the digital and hybrid implementation techniques. So in a strict sense, the conventional analog implementation method would lead to the greatest amount of design flexibility for the development of new ANNs. However, as a practical matter, the hybrid approach combines the strengths of the analog and digital approaches, using the strengths of one approach to all but eliminate the weaknesses of the other. And because the hybrid approach can opt to use only analog implementation techniques, fabrication under hybrid construction rules can produce any system producible by the analog only method. It is only when digital design elements are introduced that the hybrid method becomes more specific than the analog. But, with careful consideration of how the detailed implementation plan for an ANN will be carried out (specifically, banning the use of digital techniques where it would interfere with other design choices), the hybrid implementation methodology can provide at least as optimal a solution as could be constructed using an analog only approach.

2.4. System Architecture

2.4.1. Task-Specific Versus Generally-Configurable. General-purpose architectures in which identical neurons are fully interconnected make design negligible; the network merely needs to be adapted via weight modifications to a new problem. However, the performance of a network built by this approach can suffer due to inefficiencies caused by unneeded and/or redundant nodes. Similarly, adaptation and convergence may be slow, unpredictable, and never fully complete unless the range of variation of both input and output is well understood. Redundancy in neural networks promotes robust behavior only if the redundant nodes are performing properly. Different learning algorithms produce different results, but most training is considered complete when the output layer is performing within a certain tolerance; the proper operation of all internal nodes is not guaranteed.

Task-specific architectures bypass the dangers of unproductive internal node redundancy, and many of the issues regarding training and convergence. The benefit of a task specific architecture is that the network designer can reduce the space of possible solutions (defined by the network topology and weight values) by building in knowledge of the domain for which the network is to operate.

The issue of optimality in the case of the system architecture is particularly difficult to come to grips with. A general-purpose multilayered, feedforward network has the ability to be trained to work on a wide range of problems. However, it is currently unknown in general whether any given problem of interest will be in the set on which the network will be able to learn. Task-specific architectures provide exceptionally fast response and virtually no training time, because the network is tuned to provide exactly the number of nodes and node connections necessary to perform the desired task.

3. TOWARD A PROCESS MODEL OF OPTIMAL ANN DESIGN

Schoemaker (1991) presented a historical account of the principle of least time. This principle arose in response to the problem of explaining why, when we place a stick in water, it seems that the angle above the water is different from the angle below. The ancient Greeks spent a great deal of time measuring the incident and refracted angles of the stick between air and water. The purpose of this was to gather enough data so that an induction of a relationship between the angles could be generated and then supported. However, it was not until the seventeenth century that Snell algebraically linked the two angles. However, Snell's algebraic law relied on the assumption of a constant that was different for each media (i.e., water or oil) used. Fermat later developed the teleological optimality explanation of the phenomena, stating that the light travels not necessarily the path of least distance, but the path of shortest time. This explanation allowed Fermat to specify a formula for Snell's constant, which was the speed of light in air divided by the speed of light in water. With Fermat's contributions, the explanation of the relationship between the incident and refracted angles was made significantly more useful. Now, a researcher could simply determine the speed of light in the media of interest and, without experimentation, determine the incident angle from the refracted angle and vice versa. The development of the explanation of the principle of least time culminates in the explication of the quantum-mechanical view of Snell's law, which reduces the problem to one of the sum of photon travel times on probabilistic pathways. This causal/process explanation allows the phenomena to be explained at the lowest, most complete level of detail

In comparison, the development of a process model of optimal ANN design is about in the fourteenth century. As a group, we neural network researchers have made an extremely large number of measurements: the development of ANNs for hundreds of different applications, and the development of hundreds of different ANNs to experiment with. Teleological explanations of network structure and behavior are beginning to be presented (see Levine, Chapter 1, and Stork, Jackson, & Walker, Chapter 4, for a review). What has been lacking from this set of data are the "Snells" and "Fermats" of the neural network community. It is hoped that the increased visibility and structures provided in this collection will provide some inspiration to the neural network community to lessen the efforts spent on taking measurements, and increase the effort put into the search for causal and process relationships between network structure and optimal behavior.

REFERENCES

Barnden, J., Srinivas, K., & Dharmavaratha, D. (1990). WTA networks: Time based versus activation based mechanisms for various selection goals. *Proceedings of the IEEE International Symposium on Circuits and Systems*, New Orleans, LA, pp. 215-218.

DeYoe, E. A., & Van Essen, D. C. (1988). Concurrent processing streams in the monkey visual cortex. *Trends in Neuroscience*, 11, 219-226.

DeYong, M. R. (1992). *VLSI hybrid temporal neural networks for signal processing.* Doctoral dissertation, New Mexico State University, Las Cruces.

DeYong, M. R., Eskridge, T. C., & Fields, C. A. (1992). Temporal signal processing with high-speed hybrid analog-digital neural networks. *Journal of Analog Integrated Circuits and Signal Processing*, 2, 367-388.

DeYong, M. R., & Fields, C. A. (1992). Applications of hybrid analog-digital neural networks in signal processing. *Proceedings of the IEEE International Symposium on Circuits and Systems*, San Diego, CA, pp. 2212-2215.

DeYong, M. R., Findley, R., & Fields, C. A. (1992). The design, fabrication, and test of a new hybrid analog-digital neural processing element. *IEEE Transactions on Neural Networks*, 3, 363-374.

Eskridge, T. C., & Barnden, J. A. (1991). *Application of connectionism to analogical reasoning.* Paper presented at 3rd Annual Midwest Artificial Intelligence and Cognitive Science Society Conference, Carbondale, IL.

Eskridge, T. C., & Fields, C. A. (1989). Representing strategic knowledge in continuous, dynamic control functions. In Z. Ras (Ed.), *Proceedings of the International Symposium on Methodologies for Intelligent Systems.*

Fahlman, S. E., & Lebiere, C. (1990). The cascade-correlation architecture. In D. S. Touretzky (Ed.), *Advances in Neural Information Processing Systems 1* (pp. 177-185). San Mateo, CA: Morgan Kaufmann.

Fields, C. A., Eskridge, T. C., Hartley, R. T., & Coombs, M. J. (1989). Experimental analysis of dynamic control strategies for the MGR architecture: Simulation environment and initial results. *Proceedings of the European Conference on Artificial Intelligence*, pp. 165-173.

Fields, C. A., & Dietrich, E. (1987). Multi-domain problem solving: A test case for computational theories of intelligence. *Proceedings of the Second Rocky Mountain Conference on Artificial Intelligence*, University of Colorado, pp. 205-223.

Fields, C. A., & Dietrich, E. (1988). Engineering artificial intelligence applications in unstructured task environments: Some methodological considerations. In D. Partridge (Ed.), *Artificial Intelligence and Software Engineering*. Norwood, NJ: Ablex.

French, R. M. (1991). *Using semi-distributed representations to overcome catastrophic forgetting in connectionist networks* (Tech. Rep. No. 1991-61). CRCC.

French, R. M., & Jones, T. C. (1991). *Differential hardening of link weights: A simple method for decreasing catastrophic forgetting in neural networks* (Tech. Rep. No. 1991-50). CRCC.

Hammerstrom, D. (1990). A VLSI architecture for high-performance, low-cost, on-chip learning. *Proceedings of the International Joint Conference on Neural Networks* (Vol. II, pp. 537-544). Piscataway, NJ: IEEE.

Helweg, D. A., & Roitblat, H. L. (1991). Optimality and constraint. *Behavioral and Brain Sciences*, **14**, pp. 222-223.

Hetherington, P. A., & Seidenberg, M. S. (1989). Is there *catastrophic interference* in connectionist networks? *Proceedings of the 11th Annual Conference of the Cognitive Science Society* (pp. 26-33). Hillsdale, NJ: Lawrence Erlbaum Associates.

Hodgkin, A. L., & Huxley, A. F. (1952). Currents carried by sodium and potassium ions through the membrane of the giant axon of *Loligo*. *Journal of Physiology*, **116**, 449-472.

Hopfield, J. J. (1982). Neural networks and physical systems with emergent collective computational abilities. *Proceedings of the National Academy of Sciences*, **79**, 2554-2558.

Kohonen, T. (1988). An introduction to neural networks. *Neural Networks*, **1**, 3-16.

Lippmann, R. P. (1987). An introduction to computing with neural networks. *IEEE ASSP*, **3**, 4-22.

Melton, M. S., Phan, T., Reeves, D. S., & Van den Bout, D. E. (1992). The TINMANN VLSI chip. *IEEE Transactions on Neural Networks*, **3**, 375-384.

Miall, C. (1989). The diversity of neuronal properties. In R. Durbin, C. Miall, & G. Hutchinson (Eds.), *The Computing Neuron*. Wokingham, UK: Addison-Wesley.

Minai, A. A., & Levy, W. B. (1993). Predicting complex behavior in sparse asymmetric networks. In S. J. Hanson, J. D. Cowan, & C. L. Giles, (Eds.), *Advances in Neural Information Processing Systems 5* (pp. 556-563). San Mateo, CA: Morgan Kaufmann.

Pineda, F. J. (1988). Dynamics and architecture for neural computation. *Journal of Complexity*, **4**, 216-245.

Rumelhart, D. E., McClelland, J. L., & the PDP Research Group (1986). *Parallel Distributed Processing: Explorations in the Microstructure of Cognition Volume 1: Foundations*. Cambridge, MA: MIT Press.

Simon, H. A. (1981). *The Sciences of the Artificial*. Cambridge, MA: MIT Press.

Schoemaker, P. J. H. (1991). The quest for optimality: A positive heuristic of science? *Behavioral and Brain Sciences*, **14**, 205-245.

Smolensky, P. (1988). On the proper treatment of connectionism. *Behavioral and Brain Sciences*, **11**, 1-74.

Waller, W. A. J., Bisset, D. L., & Daniell, P. M. (1991). An analogue neuron suitable for a data frame architecture. In J. G. Delgado-Frias & W. R. Moore (Eds.), *VLSI for Artificial Intelligence and Neural Networks* (pp. 195-204). New York: Plenum Press.

Wilson, M. A., & Bower, J. M. (1989). The simulation of large-scale neural networks. In C. Koch & I. Segev (Eds.), *Methods in Neuronal Modeling* (pp. 291-333). Cambridge, MA: MIT Press.

II

QUANTITATIVE FOUNDATIONS OF NEURAL OPTIMALITY

7

Optimality and Options in the Context of Behavioral Choice

Paul S. Prueitt
Highland Technologies

Paul Prueitt's chapter, **Optimality and Options in the Context of Behavioral Choice**, suggests a general approach for understanding the structure of biological neural networks as collections of subnetworks. The larger networks calculate utility functions based on primary parameters of these subnetworks. All else being equal, these subnetworks optimize the values of their utility functions. Optimization of one subnetwork can compete with optimization of another subnetwork, leading to suboptimal global dynamics, as other authors in this volume also note. However, even local behavior is not always optimal because optimality sometimes competes with another useful organizing principle that Prueitt calls **optionality**. In other words, for dealing effectively with novel or unpredictable situations, the networks have to generate a diversity of responses. This calls for exploratory behavior (see also the chapter by Ogmen and Prakash), in cognitive as well as motor dimensions. Exploratory behavior is seldom optimal because it typically occurs in situations where the optimal response is not yet known.

How might the balance between optimality and optionality be mediated in the brain? Based in part on laboratory experiments of Karl Pribram and in part on his own joint modeling work with Levine, Prueitt suggests that in primates, the "executive" functions of the frontal lobes are important to this mediation process. As suggested also in Levine's chapter, the prefrontal cortex serves as an interface between different subcortical systems involved in processing familiarity and novelty. Roughly, the familiarity system tends toward maximizing utility and the novelty system toward maximizing choices, whereas the frontal lobes are involves in deciding which subsystem to activate.

Prueitt's chapter suggests general mathematical concepts that could provide the bases for these interactions. These include dissipative systems, help-suppression networks, planar rotators, and Gabor functions along with on-center off-surround interactions and opponent processing. He suggests a synthesis between two types of models in the literature. One type of model, *developed in the conference volumes edited by Pribram (see the references in this chapter to Pribram, Kugler, and Hameroff), aims at understanding how cellular and subcellular structures form into typical neural network "nodes" or conceptual units. The other type, developed by Grossberg and his colleagues (see the chapters in this volume by Carpenter, Levine, and Öğmen and Prakash), aims at understanding how typical neural network "nodes" (conceptual units) form into functional networks to perform cognitive tasks.*

ABSTRACT

This chapter speculates about the emergence of processes that give rise to adaptive recognition, expectation and choice by biological systems. This speculation focuses on the creation and annihilation of process compartments whose behavior is localized into subsystems. Each subsystem's behavior may not be optimal for two reasons. First, individual subsystems are in competition for scarce resources and optimize by undercutting the needs of each other. Second, the behavior of subsystems is governed by exploratory activity. This suggests that non-optimizing behavior occurs in conjunction with a structured generation of diversity. The mechanisms that provide this structure are described by an analogy to our common sense notion of self image. This analogy describes the formation of a "system image" arising as a product of system interactions with the environment.

1. SPECULATION

Most connectionist models assume the existence of *in vivo* transformations through which a neural system builds internal representations. For example, artificial neural networks seek a minimization of some chosen measure of the difference between stimulus representation and representation of acquired categories. Structures emerge from the threshold separating chaos and order to produce subsystems that are necessary to these representations. This involves the formation of compartments.

A compartmentalized process, or process compartment, is a process that is localized in space and time. The sum of energy transfer in and out of its boundary remains almost constant, and thus the theory of dissipative systems provides us a partial model. However, compartmentalization has not been fully reduced to the conservation laws of physics. This failure results in a denial of some essential properties of living systems, or the introduction of unexplained vital forces that act counter to the forces of entropy.

For example, simple inhibition and suppression properties widely used in artificial neural network research describe only a few surface characteristics. Neurons work cooperatively within a neuro-ecosystem that has active circuits at quantum, chemical, neural ensemble, anatomical and electromagnetic levels of observation. And, to date, there is no formal theory that accounts for all of these levels and their interaction.

The purpose for most of artificial neural networks research is to extend the artificial intelligence paradigm within a single scientific view that is reducible, in principle, to classical physics. History bears out a resilient effort to explain biology in terms of simple electronic computation. In the formative period of artificial intelligence (1942-1972) the excitatory networks were favored over ones having some inhibition, because inhibition implies something more than mere dissipation of energy. This problem can be solved by subsystems that produce an agent that successfully competes for limited resources and thus decreases the system's production of some other agent. A reduction of the property of inhibition to a slightly more complex interaction does not, however, describe how the selection of specific chains and circuits of reactions is initiated.

We may begin this description by increasing the degree of complexity to the point where compartments emerge from cooperative/competitive interaction. However, systemic problems remain. There is an abundance of explanatory proposals for attacking these problems, but each fails to be consistent with our folk psychology. Thus, they are discouraged.

The practice of science is materialistic and formal, which may or may not be correct, as well as anthropocentric which is a simple, pervasive bias. These properties are well integrated into the theoretical foundation. Clearly, Gödel's and Church's theorems on consistency and completeness imply something about the realism of a unified, formal, and consistent foundation for all scientific investigations. However, a unified foundation is an ideal that is very appealing and there is no way to predict how mid twenty-first century science will address this issue. Meanwhile, a large class of potential explanations are not investigated (Rosen, 1985) and certain other areas are overinvestigated. Science depends on a widely accepted and self-consistent theoretical basis for pursuing investigations (Kuhn, 1970), so we can only be mildly critical of the current scientific paradigm. However, science is also empirical and should account for observations made about real living systems.

2. ECOLOGY OF COMPARTMENTS

This chapter only provides a broad description of process compartments, and not a complete theory. The description addresses the core issues in cognitive and information science, and these issues are far from being resolved. It is important to note, however, that compartments, as they are defined below, are abstractions that generalize the properties of phenomena that occupy multiple levels of organization (see Bradley and Pribram, Chapter 21).

A biological cell is a good example of a process compartment. There is energy and matter flow through its boundary, but this flow is controlled by the properties of the membrane and the spatial/temporal envelope that serves as the cell's boundary in four dimensions. A more complex example is a simple model of the visual process. Photon stimulation of the protein complexes in the retina induce emergent patterns at several junctions between the retina and visual cortex. Repeated stimulation results in pattern clustering and featural recognition at these junctions. Over time, a distributed feature space is constructed through joint expressions of genetic, chemical, and network information at junctions and characteristic optic flow input at the retina. Optic flow patterns then fall under the influence of a set of top-down feature space response templates and visual recognition takes place. Resonance mechanisms, electromagnetic (EM) phase coherence, and phase locking control underlie biochemical circuits

Experimental neuroscience provides firm neurological evidence that retinal images are (nearly) a Fourier transform of temporal invariances available in optical flow (Pribram, 1991, 1993). The evidence suggests that internal representations are encoded in the isophase contours of electromagnetic (EM) activated phase coherence, or resonance, in dendritic ionic fields at the lateral geniculate nucleus. It is thus theorized that EM encoded representations are algebraically isomorphic to structures emergent in the optical flow. The encoding is, however, virtual in the sense that it exists as a set of probability distributions and must be instantiated through subfeatural combinatorics (Prueitt, 1995, 1996a, 1996b). Such virtual encoding becomes an event when conditions elicit local invariance in the form of the compartment. As the influence of stimuli passes through several process compartments, an isomorphism is preserved across a span of time. The invariance achieved by this isomorphism is observable as the event.

For the visual system, encoding is achieved through fine variations of the functional properties of cells and ensembles of cells. The result is a repertoire of subfeatural elements that are evoked in the formation of visual events. Most likely there is a permanent manifestation of the system's repeated reliance on selection from a repertoire. For example, commonality in re-presentations of input patterns of activity has resulted in spatial and temporal selectivity of cells in the posterior visual cortex. This selectivity has been measured as a correlation between interspike intervals and stimulus orientation (spatial stimulus) and frequency of presentation (Berger & Pribram, 1993; Daugman, 1988).

Neural selectivity does not mean that feature recognition is a function only of cell morphology. Information is separated and distributed across multiple regions of the brain and is resident at multiple levels of organization: at the synaptic level, the microtubulin level, the protein complex level, and son on. Whenever information is transformed by a compartment, there is, by necessity, a core component of that transformation that must be measure preserving, or isomorphic, to the component input. Since the composition of several isomorphisms is an isomorphism, representational transformations by multiple compartments preserve relevant information on the category of stimuli and the characteristics of correlated electromagnetic resonance. In this way, stimuli are re-presented while being distributed across and into the brain.

Given the focus of this volume, I next present preliminary work on a model where interaction gives rise to transient systems capable of exercising a higher order control over their subsystems. As stated in the abstract, my intent is to establish a metaphor based on biological process compartments, such as cells in the visual cortex, to organize work on computer based process compartments. The result is a computational model describing the formation of compartments within an ecosystem. The ecosystem is necessary to provide commonality and higher order organization for subsystem compartments.

The study of fluids under conditions far from thermodynamical equilibrium suggests steady-state structures form as internal dynamics evolve toward local optimization. In fact, the availability of complex biochemical machinery is representative of an infrastructure whose computations generate, propagate, and interpret internal representations. In biological systems the same phenomenon creates compartments as a result of symmetries in the distributions of energy and matter. The compartment provides a location and the physical substrate for these algorithms to link cognition, awareness, and intentionality to these chemical computations.

For reasons rooted in an evolutionary selection of simpler, as opposed to more complex, systems, structures are created in nonlinear interactions and produce linear separation and combination of data to and from separate information streams. Linearity allows the most efficient fusion of independent information streams, whereas nonlinear interaction is wonderfully sensitive to perturbation by the environment. Using these two principles, individual compartments produce fused outcomes as a cooperative phenomenon involving multiple sources.

3. STIMULUS REPRESENTATION

Systems designed to optimally match patterns are essential to stimulus-response events in living systems. Pribram and his colleagues have modeled biological signal processing as a measure of spatial-temporal pattern matching, using a convolution in the form of a Gabor elementary function. The set of Gabor elementary functions of time for a given Δt are defined over an infinite index set $\{n, m$ where n and m are integers$\}$ (Gabor, 1948) as:

$$\gamma_{nm}(t) = \exp[-p(t - n\ \Delta t)/2(\Delta t)2]\ \exp(2pimt/\Delta t). \tag{1}$$

MacLennan (1993) summarized several decades of experimental evidence that Gabor functions of space and time are representative of functional primitives in visual cortex. The first factor, $\exp[-p(t - n\ \Delta t)/2(\Delta t)2]$, provides a temporal window defined by the harmonics of the length Δt. For t large, the factor is close to zero. For t finite, the expression has local maxima at $n\ \Delta t$.

The second factor of the Gabor transform is similar to the Fourier elementary function. This factor can be analyzed in terms of principal components and has wide application in military, industrial, and commercial products. The analysis of principal components has led to new signal recognition technologies with direct application to circumventing instrumentation limitations constraining single- and multi-channel microelectrode recording (Bankman, 1993). Such technology moves us closer to understanding biological signal processing.

A generalization is in order that admits the formation of process compartments. This generalization makes a distinction between the spectral composition of stable process compartments and the chaotic transitions from which they arise and dissolve. In later sections of this chapter, this distinction is further formalized as internal and external metaphors with which to view process compartments.

Given a one-dimensional observational trace, $m(t)$, of a stimulus, its within-episode manifestation may be expressed as a finite linear combination of density (probability) measures c_{nm} times elementary functions, $\gamma_{nm}(t)$:

$$m(t) = \Sigma c_{nm}\gamma_{nm}(t). \tag{2}$$

For EM events that correlate with neurocomputation, the computation of coefficients c_{nm} could be achieved via dendritic fast adaptation via gradient descent methods, involving phase alignment during a brief initial period of electromagnetic driven synchronicity (as suggested by Pribram). Observational traces linked to the values for c_{nm} would then exist in the form of Eq. (2) for a finite period of time and would be measurable.

Following a tradition going back at least to Hull, the coefficients c_{nm} are thought of as the stimulus trace (for historical review see Levine, 1991). These traces have not been directly observed in experimental data, but are widely hypothesized as correlating to fine mental events such as mental verbalization and goal formation.

Signal generation expressed between neuron bodies presumably consist of a propagated Δt, a fundamental harmonic, as well as algebraic patterns expressed between coefficients c_{nm}. In the simplest case, the harmonics of Δt are oscillatory signals whose frequencies have integer multiples of Δt (higher harmonics) and whose frequencies divide Δt evenly (lower harmonics). Phase locking and nonlinear transitions near critical points provides synchronization of phenomenon into nearest neighbor harmonics.

It is not required that Δt and c_{nm} be propagated by the same mechanism, nor that mechanisms exist at the same time scale. In fact, the architecture supporting global process is a key determinant in what becomes selected during emergence and what does not. For this to work, as here theorized, the system must be "stratified in time"; that is, it must have multiple levels of organization that are organized by the speed at which things happen. Process compartments are created and annihilated within levels that are observable only at specific time scales. Once created, a compartment separates phenomena that once had a direct interface. Once a compartment has been annihilated, two phenomena that were separated now interact.

The emergence and reinforcement of fundamental harmonics produce coherence and phase synchronization within and between time scales. For example, the competitive enhancement of significant patterns can be experimentally observed in the on-center off-surround cellular dynamics of retinal cells (Ellias & Grossberg, 1975; Grossberg & Levine, 1975; Hubel & Wiesel, 1962) or in machine intelligence by adaptive resonance theory (Carpenter & Grossberg, 1987). These are examples of phase synchronization and coherence. It is important to note, however, that attentional focus and goal formation are subjected to additional architectural principles such as intersystem modulation and opponent processing (Levine, Parks, & Prueitt, 1993).

Both the elemental functions $\gamma_{nm}(t)$ and the coefficients arise, in an as yet unexplained fashion, during the brief formative period early in the life of the compartment. Once these are selected, they act collectively as a process transformation. According to this view, biology creates a medium that produces the Gabor functions, which then transform visual stimulus. Other types of transformations are employed in other brain regions and in other media. A simpler (matrix) linear association may occur in the cerebellum (Houk, 1987), and wavelet transformations are conjectured to form the basis for microtubulin signal propagation (Hameroff, 1987; Hameroff et al., 1993). In any case, the computation of transform coefficients are presumed to be accomplished by fast adaptation of structural constraints, like dendrite spine shape, protein conformation, or neurochemical agents reflecting neurotransmitter concentration. We expect that this is done with the flexibility required to select from multiple (optional) responses in ambiguous situations (see Pribram, 1991, pp. 264-265). This "response degeneracy" (see Edelman, 1987) implies that the process compartment itself, which arises by imposing constraints where none before existed, is nevertheless left underconstrained during its lifetime. This characteristic, if experimentally confirmed, might serve to differentiate a living system from a simpler system arising from fluids forced through phase changes under thermodynamic pressure.

Before moving to the next section, we need to discuss the rationale for regarding response degeneracy as complementary to optimality. The creation of viable options would be complementary to optimality if all but a few selective response potentials were somehow held ineligible for deterministic evolution toward a local basin of attraction. This occurs in two stages. First, there is an enfolding of the implicit properties of the system and its environment during a brief non-algorithmic process (the quantum metaphor: Penrose, 1989; Shaw, Kadar, & Kinsella-Shaw, 1994). Second, a final selection of one from a small number of high potential paths in the probability space created during the first stage (see the notion of multiple drafts versus the Cartesian theater, in Dennett, 1991, pp. 101-138).

The "deselection" of all but a few response potentials is true of frontal/limbic interaction involved in selective attention (Levine et al., 1993) and is likely to be an explanation for other phenomena. The result is the creation of a small number of realizable options, formation of goals, and exercise of free will. Thus a distinction is made between the creation of options by a living system and the deterministic evolution of systems toward local minima. This distinction implies that biological systems, although relying on classical laws, are not reducible to them. The origin of control is placed in a system image as a result of system/not-system interaction.

In the next section, I briefly restate the description of process compartments.

4. THE MODEL IN BRIEF

In biological systems, response categories are acquired through experience and expressed via the mechanisms of immune and neurobiology. Most often, functional expression of these mechanisms is shaped by prior experience into an accommodated internal store capable of producing mental representation. The system image arises when this store is stimulated by external interaction. The stimulation involves recognition or response to novelty. As a result, substructural processes are fused to produce mental representations within context. Functional specificity is supplied by an organized substrate — such as the network of neurons, the conformational space of proteins, or the anatomical organization of the brain.

The model has transient compartments with internal evolution determined by interaction between a system and its environment. Specific instantiation of evolution rules within the compartment is then jointly determined by initial conditions on system variables, by external perturbation, and by sensitivity to symmetry breaking. If the interaction is weak, then behavioral characteristics of external systems are propagated between compartments. Intercomponent frequency entrainment in the electromagnetic (EM) spectrum results in signal generation, propagation, and interpretation between steady state compartments. Intercompartmental interaction allows local/global signaling during restructural transitions and provides periodic interactions with various parts of an complex external world. This interaction extracts information expressed in a common subfeatural set.

In the first stage of a compartment's existence, a brief event, perhaps noncomputational in nature, produces a system image and nominates a finite set of observables. The system image aligns subsystems to serve the needs of higher order phenomena. For example, a protein newly formed from RNA will undergo one or more near chaotic energy/state transitions before settling down into a metastable state. Each transition creates a compartment. RNA transformation to a metastable state can be envisioned as a series of process compartments with a fairly uniform system/not-system interface governed by an invariant system image.

This intuitive notion of system image describes how systems generate anticipatory actions serving intrinsic as well as extrinsic needs. Do proteins express anticipatory actions during phase changes? Any judgment about this is very difficult because of the scale of observation required to be intimate within a protein's world. It is even difficult to formalize the notion of intention when we deal in the world with other humans (Gurwitsch, 1977/1979). An individual acting in the interest of the nation would presumably be influenced by the system image of the nation, as well as by his or her own self image. Note that "need" is better described as "ecological affordance," which is broadly interpreted to be a phenomenon arising from a nonconserved quantity (Gibson, 1979; Shaw et al., 1994). System image can be described by providing examples and by appealing to our common notion of human self-image.

4.1. Representational Computation

There are two major caveats that have been associated with models that rely on stimulus traces. In spite of these two caveats the explanatory power of behavioral theories, based on an assumption of stimulus traces, is considerable.

First, transformation mechanisms may act in a single very fast event or over a longer period of time, marked by sudden transformations of substructure. During transformations, variation of time scale may not be uniform for all observables. Thus representational correlation is more than a static match between values in two vectors. When attempts are made at interpreting real spike train data as simple vector processing, the results are system models with no apparent information. A more reflective model involves the coefficients of a Gabor type transform windowed within episodic events of positive duration (McGuinness & Pribram, 1980) and the selection of the terms of the transform from a set of transformationally similar set of elementary functions, for example, $\gamma_{nm}(t)$ as in Eq. (2).

Second, it is reasonable to assume that the composition of a finite set of elementary functions (principle components) is periodically restructured by locally relative least action principles. During representational binding, utility functions emerge to reflect least action principles subject to ecological affordance at the time scale under consideration. An affordance positively biases the evolution of emergent phenomenon and shapes stimuli into Gaussian envelopes to reflect global properties of an environment. In the same way, ecological niches are formed in large-scale ecosystems. However, in spite of the existence of stimulus clustering into envelopes, environmental affordances are often nonstationary (in time) because the world itself is undergoing change. Thus, principal components selected to fit one set of circumstances may not fit changed circumstances. A locally catastrophic restructuring of substructure is the result.

These two caveats are met with two observations. First, a theory of natural kind is essential to understanding intelligence, and any theory of natural kind must point to the creation of reliable reflexes, and a means for fusing reflexes into coordinated action. Complex responses arise from the selective attention of a system image, while internally processing priorities and expectation and maintaining a hold on the pragmatics of real-time, and often nonstationary, situations. Thus, although the mechanisms that support cognition are subject to adaptation toward optimal processes (measured with respect to several time scales), these processes must also satisfy flexibility in response to a nonstationary world (Prueitt, 1994, 1995).

A second observation involves a common basis for data fusion. Because efficient response to expected occurrences is a function of natural selection, the common availability of a class of natural kind is essential to coexistence. Conjecturally, common substructural constraints provide the basis for expectation, generation of messages, and interpretation of messages. The mechanisms that produce representational transformations act independently as weakly linked systems with behavior shaped by internal affordance and by weak interaction with its environment. To be viable, these closed systems must adapt themselves to primitives that provide commonality and the basis for cooperative processes with other closed systems. These primitives are the external affordance provided by the environment to each of the systems.

In an analog to distributed computer systems, the hardware of the sensory and cognitive mechanisms of the human brain responds to sensory input by making small adjustments in network connectivity, while monitoring the expression of learned (or inherited) constructs, and re-presentation of the trace of the stimulus. When one of a class of internal states is achieved then the system must respond within a limited range of options. However, if the mismatch cannot be compensated by small adjustments, then local restructuring dissolves the process compartment and a new compartment arises with a new set of options.

4.2. Stimulus Processing Within Process Compartments

I suggested above that preprocessing within compartments involves the systemic identification of a finite set $\{\gamma_{nm}\}$ selected from a much larger repertoire of subfeatural elements. The selection may involve activating sequences of events, such as neurotransmitter circuits, that in turn make available the necessary resources for subfeature expression. The subfeatures themselves are emergent phenomenon that form at a faster time scale (see Eq. (5)). These become the ground on which the system depends. Identification is then correlated with the creation of transient compartments that then change state based on the exact condition of the stimulus, acting in a classical if-then (response/stimulus) fashion.

Assuming a set of learned categories having internal representation $d_{nm}\gamma_{nm}(t)$ and incoming composite stimulus having representation $c_{nm}\gamma_{nm}(t)$ then

$$\varepsilon(t) = \Sigma(c_{nm} - d_{nm}) \tag{3}$$

is a measure of stimulus mismatch. This stimulus mismatch can be expected to play a critical role in maintaining, or not, the coherence that supports the process compartment. Behavioral action is represented in a similar language with the notion of a "reflex arc" correlating with the expression of one of the subfeatural elements. Sensory and cognitive representations are shaped by internal constrains designed to optimize an accommodated representation in terms of elementary components, while exchanging information with auxiliary processes. Conjecturally, the fused representations result in indirect accommodation of experience within a memory store and involves both local phenomena (stimulus traces) and a global phenomenon (the memory store).

Accommodation results from the synthesis of stimuli and internal affordance into an singular event that higher order processes can address directly. This must involve an "enfolding of experience" with long-term plans and objectives. Although such a notion does not give a definition of self-image, the phenomenon of self-image and David Bohm's notion of implicate order seem to be related. Both the enfolding of experience and its accommodation depend on measures of coherence and cohesiveness correlative to emergent processes formed from the constraints of an interface between internal and external representations.

Dissipation is the parent of the process compartment. During the first part of a formative phase of the compartment, a choice between viable sets of options is created. During the second part, a specific option, or subset of the potential options, is selected by what appears in the first part as an agent. Its morphology is not dependent on awareness or consciousness but on dynamic interaction between help and suppression processes (Eisenfeld & Prueitt, 1988; Prueitt, 1988; see Figs. 7.1 and 7.2).

System image sits as a higher order mediator of creations, annihilation, and interactions of process compartments. If there is no such mediator, there is no choice and no intentionality. Higher order mediation is found at precisely the point where optimality's complementary principle, optionality, is most clearly identified. When we watch others it can be easy, given the proper technical background, to believe that human behavior is merely experience-determined responses from a highly integrative, genetically determined collection of mechanisms. This is what the reductionists believe. However, when we observe ourselves, we perceive responses to be choices between alternatives. We perceive from our personal experience the periodic generation of a diversity in expectations that depend critically on system image. From this diversity, we choose specific actions as the result of our intention to achieve these expectations.

Our notion of self-image cannot be stated without reference both to internal experience and the world experienced (Bandura, 1978). Through the self image, the individual brings a nondeterministic and nonoptimizing element into a process that is governed by laws that are deterministic. Self-image is generally constant throughout life. It links together experiences into a meaningful whole. However, the description of a system image does not solve some ultimate problem in cognitive psychology. Consciousness is a higher order expression of self-image, but need not be its only manifestation. For example, regulatory processes, occurring over a short period of time, can be shaped by a higher order system image that is constant over a much longer period. The images of subsystems may influence each other as well as higher order (existing over long periods of time) and lower order systems (existing over short periods of time). The relationship is delineated by observational time scale.

Broad anatomic interaction reflects the internal dynamics of the human brain. Transient dissipative structures allow a multiplexing of information, derived from posterior cortex and limbic regions, while perception and action jointly maintain these transient structures through nonconservative induced gradients. As a result of induced changes in anatomic regions, isolated compartments form and maintain a relative stability for finite periods of time. For example, multiple compartments could form in the hippocampus and interact with compartments that arise in parallel in the amygdala. Those compartments not inhibited by frontal lobe mediation would then induce specific plans to be formulated.

This model requires a computational approach. However, the science community does not share a set of conventions under which computational investigation of biological intelligence should be pursued. A deeper explanation cannot be achieved within the current foundation.

Meanwhile, we cannot act as though intentional systems do not introduce teleological elements into a mechanistic world. This simply denies common experience. We should not be surprised if new computational models uncover what are essentially a class of nonalgorithmic events. The mere recognition of such a class could bring behavioral science in line with common sense and reject the folk psychology of twentieth century psychology (Churchland, 1989), which does not address the role of anticipation (Rosen, 1985), or intentionality (Kugler & Turvey, 1987), nor address other essential questions (Werbos, 1993).

System image is instrumental in any full description of the central issues of intelligence and intentionality, and we need a scientific foundation for its discussion. Establishing this foundation may not be far away. As suggested by Eccles (1993), intentionality could be expressed during brief nonlinear restructural transitions of process compartments. Eccles pointed out that the geometrical arrangement of synaptic boutons supports a femtosecond process controlling the release of the transmitter vesicle mediated by Ca^{++} influxes. The control is provided by a process selected through evolution to conserve neurotransmitter. The process has the effect of creating a homogeneous probability distribution measuring the likelihood that any one of six boutons will carry the gradient field interchanges between presynaptic and postsynaptic events. Eccles saw this mechanism as the interface in which the mind couples to the brain by changing the probability and timing of synaptic events. These events are then seen to play a role in the formation of network connections at a high level of organization. At critical locations, even distributions are maintained while the gradient increases to a high level. Symmetry and increased gradients result in a barrier and thus in the formation of a trigger. The trigger is released when self organization at a higher level can be effectively influenced (Eccles, 1993).

Hameroff identified similar symmetry generating mechanisms in the geometric structures of the microtubulin assembly as well as in the temporal dynamics of microtubulin formation. Microtubulins play important roles in cell mitosis and have the potential to control the connectivity of neuronal ensembles through the fine alteration of dendritic arborization, as well as to influence second messenger cascades guiding long-term potentiation (Hameroff, 1987).

4.3. Immune-Neural Interface

The notion of human system image, although itself not well defined, is a metaphor for an higher order "agent" that mediates the formation of compartments and shapes the compartments' evolution. In this model, the transformation of stimulus traces are shaped in this fashion by immunological/neural interactions that are distributed in nature. Of course, this is a huge simplification that is speculative and exploratory. System image itself need not be a nonmaterial mind/body interface as perceived by Eccles (1993), but it could alter probabilities during phase transitions of compartments whose existence is brief when compared to the agent and thus have many of the same properties as envisioned by Eccles. The primary distinction is that system image is a phenomenon that interacts indirectly as intentionality in probability space.

A network model of help-suppression circuits (see Figs. 7.1 and 7.2), motivates a model of self-image formation based on distributions of idiotype classes (Eisenfeld & Prueitt, 1988; Jerne, 1967; Prueitt, 1988; Richter, 1979). Reactions induced by compartments within help-suppression circuits connecting processing levels have nonzero reaction rates and thus produce a driving force for iterative (cross-level) transformation of representational input. Such reentrant processing (Edelman, 1987) is necessary for any adaptive mechanism of invariance recognition (see also Pribram, 1991). In analogy to the immune system, it is also possible for associated features to become functionally paralyzed by help-suppression reaction circuits producing low zone tolerance (LZT) (Eisenfeld & Prueitt, 1988).

Equations that I developed to model help-suppression circuits involve indicator functions, distribution functions, and transition state tables, and their presentation is beyond the scope of this chapter (Prueitt, 1988). Moreover, the model was developed to reflect global characteristics of clonal and natural selection as described in Jerne (1967) and Richter (1979), and may not reflect the internal dynamics of any real immune system as it reacts to any specific antigen. The computational model is therefore only another metaphor. It places one part of a very complex puzzle into a tentative position while other pieces are examined.

The help-suppression circuits of immune networks provide a different view of the ubiquitous phenomenon of opponent processing seen in the dipole field (Grossberg, 1972a, 1972b). The gradient developed by a process compartment drives associated reaction circuits in the ecological substrata, and thus produce self/not-self interactions involving any emergent structures that have become linked with the dipole stimulus. Self/not-self interactions in the immune system establish help-suppression circuits that form control mechanisms influencing production rate variation in neurochemical isomorphs. Similar circuits would be capable of encoding input within the framework of a protocritic process responsible for individual system image. Emergent phenomenon derived from circuits of this type could explain the selective interpretation of experience that characterizes human psychology.

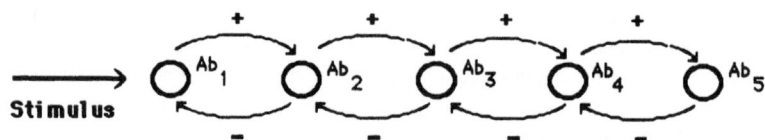

Fig. 7.1. Help-suppression circuits develop a hysteresis that produces an active inability to respond to certain types of immunological challenges.

Now for the details of how emergence leads to self-similar phenomena. Consistent with the ecological approach to learning behavior, the transitions involved in creating a process compartment are open to "information" (meaningful temporal and spatial invariances) from many different sources. The formation of information involves degeneracy and is "nonoptimal" in this sense. Thus, the emergence of a finite set of preestablished reaction circuits from subcellular and cellular environments is predicted to result in one-to-many and many-to-one transformations across levels of organization. Selection of members of the set of possible transitions results from an evening of the probability that each of the possible transformation paths will occur.

Affordance is then reflected in probability distribution as several heightened sets of self-similar potentials are created. Small variations between these are enhanced to become the subfeatures of the compartment.

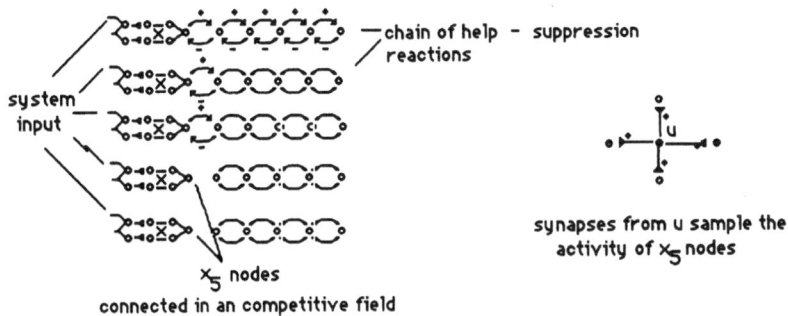

Fig. 7.2. Help-suppression circuits coupled to dipole fields produce a control of subfeature selection that depends on accommodated system image.

4.4. Process Compartments and Prefrontal Involvement in Mediating Choices

The processes particular to prefrontal cortex modulation of sensory input, as well as limbic events, provide keys to understanding the complementarity of optimality and the generation of choice (optionality) by biological systems. First, neural processes satisfy a least action principle in distributed, and often virtual, state spaces. Second, these state spaces exist simultaneously at different time scales (see the next section) and often collapse after a period of transient stability. Third, the periods of stability are correlated with mental phenomena such as are involved in mental search. Mental search, associated with planning, is initiated by neuronal groups in the prefrontal cortex in response to a stimulus, often originating from the posterior cortex. Mental search may also be initiated by limbic events. A small number of dominant foci could be maintained by frontolimbic inhibition of interregionally linked dendritic field coherence through variation of neuromodulators and perhaps though EM signals. The initiated search is ended when there is a resolution of some uncertainty or a specific goal is achieved.

The global force involved in constraining the neural ecosystem is a wave guide that produces discrete localization and a temporal invariance. The discussion of this force has a long history. An exposition of the two competing paradigms of localization and antilocalization can be found in John (1971), Lashley (1929), and Pribram (1973). That memory has localized points of concentration is suggested by a number of experiments. For example, a long line of research supports the notion that feature extraction is the primary functional process supporting recognition systems. Other work points out that the activation or encoding of memory involves cells in many brain regions. Still other research reminds us that activation or encoding of memory is dependent on genetic information contained in cells. In *Brain and Perception*, Pribram's presentation of far frontal cortex function followed an extensive neuropsychological presentation of the full range of sensory and motor control systems, as well as cognitive functions associated with the posterior cerebral convexity, amygdala, hippocampus, and limbic forebrain:

> ... familiarization and innovation occur within a processing space that defines an episode. Episodes provide the context within which perceptions are valued. Context (processing space) and content (stimuli sampled) interact in a reciprocal fashion. (Pribram, 1991, p. 226)

The frontolimbic forebrain, a complex of brain regions including the limbic forebrain, basal ganglia, and frontal cortex, is involved in the creation and maintenance of a processing space capable of placing stimulus into a context from which ambiguity or incompleteness can be addressed:

> Ordinarily, input from sensory or internal receptors preempts allocation ... by creating a "temporary dominant focus" of activation within one or another brain system. ... However when input competition, incompleteness, or ambiguity place extra demands on the routine operations of allocation, envisioning proprieties and priorities, and practical inference become necessary. (Pribram, 1991, p. 239)

The role of process compartments in modeling information fusion is suggested by the above passages. The neurological facts are clear:

1. The anterior (front) portion of the frontal lobe develops a temporary dominant focus that draws energy and information from auxiliary processes.
2. Each local dominant focus enslaves neighboring activity and activity at faster time scales in the production of emergent phenomenon, modeled as a thermodynamically closed (isolated) compartment.
3. The function of envisioning proprieties and priorities (see Pribram, 1991, Chapter 10) serves the same function as does top down expectancy in category classification.
4. The executive control by the prefrontal cortex exposes the operation of intentional inputs into the process compartment.
5. Signal integration depends on measures of coherence and cohesiveness of emergent complex processes from an ecologically sensitive interface between the visual cortex and limbic system. This interface is mediated by the frontal lobes.
6. The propagation of signals from individual neurons to that neuron's area of innervation is accompanied by the modulating influence of electromagnetic events, chemical modulation, and what might reasonably be called second messenger events in the form of protein based dynamics.
7. Although emergent process compartments arising in the posterior cerebral convexity are constrained by the need to know exactly (optimally) what is sensed, the process compartments of the frontolimbic circuit demand flexibility and options.

5. THE CHALLENGE IS CLEAR

Phenomena are localized in space with parallel events occurring during the same period of time but at distinct time scales. This provides science with an enigma, because it is counter to the naive notions of the law of entropy. For a specific time scale and for a specific space localized phenomenon, the path in state space that traces a selection of choices is governed by an implicit order, described as a "system image." Localization is either brought about by this system image or system image emerges as the result of localization. This provides science with an enigma, since it is counter to the naive notions of the law of entropy.

Sir Roger Penrose stated a strong form of the anthropic principle:

> ... preferable, to my way of thinking, would be a rather more scientific version ... namely the anthropic principle, which asserts that the nature of the universe that we find ourselves in is strongly constrained by the requirement that sentient beings like ourselves must actually be present to observe it. (Penrose, 1989, pp. 405-406)

Penrose's previous arguments make a case for admitting both algorithmic as well as nonalgorithmic phenomena as an integral part of human consciousness. The reason why sentient beings are required, it may be argued, is to perform nonalgorithmic tasks in a world where everything else is governed by algorithmic laws (the laws of classical physics). It seems to me that this principle is stated backward, because nonalgorithmic processes also characterize quantum physics. Thus an alternative perspective is that nonalgorithmic events at the quantum level can produce artifacts that serve to enhance the survival of sentient beings. Natural selection would then be predicted to propagate and refine subsystems that make efficient use of quantum phenomena directly or the artifacts that form as longer events.

It is possible that quantum artifacts exist at a nonquantum time scale as (a) the noncomputational phenomena involved in the creation and annihilation of compartments, (b) network properties constraining the interaction between components, (c) the property of an energy manifold expressed within a process compartment, or (d) a message to be generated or interpreted by a mechanism.

It has become increasingly clear to me that a mature computational mathematics of interacting process compartments is necessary to more fully understand the neural ecosystem.

6. THE ROLE OF INTERCOMPONENT LANGUAGE

Isolation of compartments is complemented by the development of intercomponent language (Prueitt, 1994). The conditions for the creation of process compartments arise under constrained dynamics, initially without boundary conditions, initial conditions, or a fundamental set of observables. At the present time, how this occurs is still one of the great mysteries of life. However, there can be no question that intentional systems take willful advantage of episodic emergence during periods in which the system's further evolution has multiple, but finite, possible futures. Such systems are deterministically underconstrained and have nonuniform probability distributions with variance associated to multiple potential evolutionary paths.

Underconstrained systems provide the degeneracy needed for an interactive specification of the proper observables that arise as anticipatory measurements are made (Rosen, 1985). The resulting inferential actions produced by episodic constraints have a direct analog to the motion of a trajectory constrained to exist in a higher dimensional "solution set" as defined by an energy manifold.

The need to cooperate with other compartments and achieve subsystem optimality involves the generation and interpretation of signals. The need for language is an invariant (law) in the behavior of biological systems (Pattee, 1963). In semantic theories of information, symbolic structures are manipulated with mathematical logic through the examination of organizational principles as well as inferential action.

An assumption that internal dynamics becomes isolated is justified because biological membranes, or more generally system boundaries, exchange structural constraints (messages) without measurable associated energy perturbations. This conclusion requires experimental verification (or disverification) of the following generic principles.

1. The isolation of internal dynamics can be accomplished by initial conditions during the formation of the compartment.
2. Phase coherence and synchronization give rise to the process compartment, with energy sinks and sources shaping the potential manifold.
3. Decisive events within the compartment are sensitive to symmetry-forming mechanisms.

Encoding of intercomponent signaling as structural constraints provides an additional challenge for mathematical neurodynamics. Conjecturally the levels of activity in biological systems have a morphology that arises from the stratification of fluids under conditions far from equilibrium. Stratification forces the formation of boundaries (membranes) in which kinematics and kinetic forces are linked together to adaptively encode, via a class of transformations, information about the organism and its environment. Once a stratification has occurred, the origin of episodic stability is jointly derived in conjunction with the environment, from microprocesses (observed as nonholonomic variables), and from macroprocesses (observed with holonomic variables). This view allows a natural statement of entailment (causal description) viewed in an evolutionary perspective consistent with Edelman's neuronal group selection (Edelman, 1987; Edelman & Finkel, 1984) and Changeux and Dehaene's (1989) theory of cognitive function based on multilevel processing and interactions (see also Dehaene & Changeux, 1991).

In Beek, Turvey, and Schmidt (1992), the hypothesis that "the phenomena of movement are understandable as the outcomes of nonlinear dissipative dynamics" (p. 67) is extended to an examination of autonomous and nonautonomous dynamics. Although it is true, perhaps, that self-organizing processes are best expressed in autonomous form (Haken, 1977), the role of nonautonomous variation is the key to capturing the internal dynamics of an open environmentally embedded interface. An autonomous system is one defined by the absence of time as an essential observable and is generally written in the generic form,

$$dx/dt = f(x) \text{ with initial condition } x(t_0) = x_0 \qquad (4)$$

whereas the generic form of a nonautonomous system is

$$dx/dt = f(x,t) \text{ with initial condition } x(t_o) = x_o. \tag{5}$$

In our model, the within episodic dynamics are closely associated with a time-independent model whereas the transitions are modeled with a time dependent model in the form of Eq. (5). A note of caution should be addressed as there may be more severe complications than temporal autonomy to modeling restructural transitions, because the number and nature of observables will undergo fundamental changes.

7. SUMMARY

I have suggested that local coherence in the EM spectrum is produced by systems that are stratified into numerous levels and that produce compartmentalized energy manifolds. These process compartments, and not merely networks of neurons, are prime candidates for the proximal causal mechanisms producing behavior. This view is consistent with the views expressed in Changeux and Dehaene (1989):

> A given function (including a cognitive one) may be assigned to a given level of organization and, in our view, can in no way be considered to be autonomous. Functions obviously obey the laws of the underlying level but also display, importantly, clear-cut dependence on the higher levels. At any level of the nervous system, multiple feedback loops are known to create reentrant mechanisms and to make possible higher-order regulations between the levels. (Changeux & Dehaene, 1989, pp. 71-72)

At all levels, in anatomical regions and across time scales, generic mechanisms appear to operate. More complex models of prefrontal cortex interaction with other cortical systems and with limbic systems, require a formal model of intentional processes (Rosen, 1985; Kugler, Shaw, Vincente, & Kinsella-Shaw, 1990). It is important to introduce the issue of boundary formation in nonautonomous transitions between episodes. Of particular interest is the nonstationary response to symmetry breaking that accompanies these transitions. The model presented in this chapter provides a framework for unification of the neuronal model, process components operating at faster and slower time scales, as well as clarifying the natural role for structural constraints to control signal production and interpretation between compartments (Pattee, 1972). Stratified processing within and between transient compartments can then be seen in ecological terms.

ACKNOWLEDGMENT

This research was supported by National Science Foundation grant IRI-9216401.

REFERENCES

Bandura, A. (1978). The self system in reciprocal determinism. *American Psychologist*, **33**, 344-358.

Bankman, I. (1993). Automated recognition of action potentials on extracellular recording. In K. H. Pribram (Ed.), *Rethinking Neural Networks: Quantum Fields and Biological Data* (pp. 69- 92). Hillsdale, NJ: Lawrence Erlbaum Associates.

Beek, P. J., Turvey, M. T., & Schmidt, R. C. (1992). Autonomous and nonautonomous dynamics of coordinated rhythmic movements. *Ecological Psychology*, **4**, 65-95.

Berger, D., & Pribram, K. H. (1993). From stochastic resonance to Gabor functions: An analysis of the probability density function of interspike intervals recorded from visual cortical neurons. In K. H. Pribram (Ed.), *Rethinking Neural Networks: Quantum Fields and Biological Data* (pp. 47-68). Hillsdale, NJ: Lawrence Erlbaum Associates.

Carpenter, G. A., & Grossberg, S. (1987). A massively parallel architecture for a self-organizing neural pattern recognition machine. *Computer Vision, Graphics, and Image Processing*, 37, 54-115.

Changeux, J.-P., & Dehaene, S. (1989). Neuronal models of cognitive functions. *Cognition*, **3**, 63-109.

Churchland, P. M. (1989). *A Neurocomputational Perspective, The Nature of Mind and the Structure of Science*. Cambridge, MA: MIT Press.

Daugman, J. G. (1988). Complete discrete 2-D Gabor transforms by neural networks for image analysis and compression. *IEEE Transactions on Acoustics, Speech, and Signal Processing*, **36**, 1169-1179.

Dehaene, S., & Changeux, J.-P. (1991). The Wisconsin Card Sorting Test: Theoretical analysis and modeling in a neuronal network. *Cerebral Cortex*, **1**, 62-79.

Dennett, D. C. (1991). *Consciousness Explained*. Boston: Little, Brown.

Eccles, J. C. (1993). Evolution of complexity of the brain with the emergence of consciousness. In K. H. Pribram (Ed.), *Rethinking Neural Networks: Quantum Fields and Biological Data* (pp. 1-28). Hillsdale, NJ: Lawrence Erlbaum Associates

Edelman, G. M. (1987). *Neural Darwinism*. New York: Basic Books.

Edelman, G. M., & Finkel, L. H. (1984). Neuronal group selection in the cerebral cortex. In G. M. Edelman, W. E. Gall, & W. M. Cowan (Eds.), *Dynamic Aspects of Neocortical Function* (pp. 653-695). New York: Wiley.

Ellias, S. A., & Grossberg, S. (1975). Pattern formation, contrast control, and oscillations in the short-term memory of shunting on-center off-surround networks. *Biological Cybernetics*, **20**, 69-98.

Eisenfeld, J., & Prueitt, P. S. (1988). Systematic approach to modeling immune response. In A. Perelson (Ed.), *Proceedings of the Santa Fe Institute on Theoretical Immunology*. Reading, MA: Addison-Wesley.

Gabor, D. (1948). Theory of communication. *Journal of the Institute of Electrical Engineers*, **93**, 429-441.

Gibson, J. J. (1979). *The Ecological Approach to Visual Perception*. Boston: Houghton Mifflin.

Grossberg, S. (1972a). A neural theory of punishment and avoidance. I. Qualitative theory. *Mathematical Biosciences*, **15**, 39-67.

Grossberg, S. (1972b). A neural theory of punishment and avoidance. II. Quantitative theory. *Mathematical Biosciences*, **15**, 253-285.

Grossberg, S., & Levine, D. S. (1975). Some developmental and attentional biases in the contrast enhancement and short-term memory of recurrent neural networks. *Journal of Theoretical Biology*, **53**, 341-380.

Gurwitch, A. (1979). *Human Encounters in the Social World*. Pittsburgh: Duquesne Press. (Original work in German, *Die Mitmenschlichen Begegnungen in der Milieuwelt*, published 1977.)

Haken, H. (1977). *Synergetics: An Introduction*. Heidelberg, Germany: Springer-Verlag.

Hameroff, S. R. (1987). *Ultimate Computing: Biomolecular Consciousness and Nanotechnology*. Amsterdam: Elsevier-North Holland.

Hameroff, S., Dayhoff, J., Lahoz-Beltra, R., Rasmussen, S., Insinna, E., & Koruga, D. (1993). Nanoneurology and the cytoskeleton: Quantum signaling and protein conformational dynamics as cognitive substrate. In K. H. Pribram (Ed.), *Rethinking Neural Networks: Quantum Fields and Biological Data* (pp. 317-376). Hillsdale, NJ: Lawrence Erlbaum Associates.

Houk, J. C. (1987). Model of the cerebellum as an array of adjustable pattern generators. In M. Glickstein, C. Yeo, & J. Stein (Eds.), *Cerebellum and Neuronal Plasticity*. New York: Plenum.

Hubel, D. H., & Wiesel, T. N. (1962). Receptive fields, binocular interaction, and functional architecture in the cat's visual cortex. *Journal of Neurophysiology*, **26**, 106-154.

Jerne, N. K. (1967). Antibodies and learning: Selection versus instruction. In G. E. Stemach & J. Requin (Eds.), *Tutorial in Motor Behavior II*. Amsterdam: Elsevier.

John, E. R. (1971). Brain mechanisms of memory. In J. L. McGaugh (Ed.), *Psychobiology*. New York: Academic Press.

Kugler, P. N., Shaw, R. E., Vincente, K. J., & Kinsella-Shaw, J. (1990). Inquiry into intentional systems I: Issues in ecological physics. *Psychological Research*, **52**, 98-121.

Kugler, P. N., & Turvey, M. T. (1987). *Information, Natural Law, and the Self-assembly of Rhythmic Movements*. Hillsdale, NJ: Lawrence Erlbaum Associates.

Kuhn, T. S. (1970). *Structure of Scientific Revolutions*. Chicago: University of Chicago Press.

Lashley, K. S. (1929). *Brain Mechanisms and Intelligence*. Chicago: University of Chicago Press.

Levine, D. S. (1991). *Introduction to Neural and Cognitive Modeling*. Hillsdale, NJ: Lawrence Erlbaum Associates.

Levine, D. S., Parks, R. W., & Prueitt, P. S. (1993). Methodological and theoretical issues in neural network models of frontal cognitive functions. *International Journal of Neuroscience*, **72**, 209-233.

MacLennan, B. (1993). Information processing in the dendritic net. In K. H. Pribram (Ed.), *Rethinking Neural Networks: Quantum Fields and Biological Data* (pp. 161-197). Hillsdale, NJ: Lawrence Erlbaum Associates.

McGuinness, D., & Pribram, K. H. (1980). The neuropsychology of attention: Emotion and motivational controls. In M. C. Wittrock (Ed.), *The Brain and Psychology* (pp. 95-139). New York: Academic Press.

Pattee, H. (1972). The nature of hierarchical controls in living matter. In R. Rosen (Ed.), *Foundations of Mathematical Biology, Vol. I, Subcellular Systems* (pp. 1-22). New York: Academic.

Penrose, R. (1989). *The Emperor's New Mind.* New York: Penguin.

Pribram, K. H. (1973). *Languages of the Brain: Experimental Paradoxes and Principles in Neuropsychology.* New York: Wadsworth.

Pribram, K. H. (1991). *Brain and Perception: Holonomy and Structure in Figural Processing.* Hillsdale, NJ: Lawrence Erlbaum Associates.

Pribram, K. H. (Ed.) (1993). *Rethinking Neural Networks: Quantum Fields and Biological Data.* Hillsdale, NJ: Lawrence Erlbaum Associates.

Pribram, K. H. (Ed.) (1994). *Origins: Brain and Self Organization.* Hillsdale, NJ: Lawrence Erlbaum Associates.

Prueitt, P. S. (1988). *Mathematical techniques in modeling biological systems exhibiting learning.* Unpublished doctoral dissertation, University of Texas at Arlington.

Prueitt, P. S. (1993). Network models in behavioral and computational neuroscience. In M. Kapis (Ed.), *Non-Animal Models in Biomedical and Psychological Research, Testing and Education.* New Gloucester: PsyETA.

Prueitt, P. S. (1994). An ecological approach to cognition. *World Congress on Neural Networks,* San Diego (Vol. 2, pp. 485-490). Hillsdale, NJ: Lawrence Erlbaum Associates.

Prueitt, P. S. (1995). System needs, chaos, and choice in machine intelligence. In F. D. Abraham & A. R. Gilgen (Eds.), *Chaos Theory in Psychology* (pp. 233-245). Westport, CT: Greenwood.

Prueitt, P. S. (1996a). A theory of process compartments in biological and ecological systems. In J. Albus, A. Meystel, D. Popoelov, & T. Reader (Eds.), *Seminotic Modeling and Situational Analysis in Large Complex Systems.* Bala Cynwyd, PA: Ad Rem Press.

Prueitt, P. S. (1996b). An implementing methodology for computational intelligence. In J. Albus, A. Meystel, D. Popoelov, & T. Reader (Eds.), *Seminotic Modeling and Situational Analysis in Large Complex Systems.* Bala Cynwyd, PA: Ad Rem Press.

Richter, P. H. (1979). Pattern formation in the immune system. *Lectures on Mathematics in the Life Sciences,* 11, 89-107.

Rosen, R. (1985). *Anticipatory Systems, Philosophical, Mathematical and Methodological Foundations.* Elmsford, NY: Pergamon Press.

Shaw, R., Kadar, E., & Kinsella-Shaw, J. (1994). Modelling systems with Intentional dynamics: A lesson from quantum mechanics. In K. H. Pribram (Ed.), *Origins: Brain and Self Organization.* Hillsdale, NJ: Lawrence Erlbaum Associates.

Werbos, P. J. (1993). Chaotic solitons, computation, and quantum field theory. In K. H. Pribram (Ed.), *Rethinking Neural Networks: Quantum Fields and Biological Data* (pp. 299-314). Hillsdale, NJ: Lawrence Erlbaum Associates.

8

Knowledge, Understanding, and Computational Complexity

Ian Parberry
University of North Texas

*Ian Parberry's chapter, **Knowledge, Understanding, and Computational Complexity**, deals with an issue that is not discussed elsewhere in this book, but is vital for optimality considerations both in artificial and biological neural networks. This is the issue of how extensive computational resources are needed for particular cognitive tasks. Parberry argues that many of the discussions about the ability of computers to emulate human or animal cognitive function have focused solely on whether computers can **in principle** replicate a particular class of functions or behaviours. In addition, he adds, they need to consider whether replications that are realizable in theory would in fact consume such a large amount of computational power as to be unrealistic in practice. He argues that computers could in principle be designed to replicate anything the brain does, including intentionality, but that it may be hideously expensive to do so.*

There are implications of computational complexity theory for optimality that go beyond what is discussed in this chapter. Computational complexity is analogous in some sense to the cognitive effort expended by biological organisms when confronted with unexpected or novel events. Brain loci for effort have been extensively studied by another of this book's authors, Karl Pribram (see, e.g., Pribram & McGuinness, 1992, and Section 3.2 of Bradley and Pribram, Chapter 21 in this volume). The arguments by several authors in this book (notably Stork, Werbos, Leven, and Levine) that much of neural and cognitive functioning is not optimal do not consider the expenditure of effort in the process of making a decision. The possibility still exists that satisficing decisions (Simon, 1979) can be interpreted as optimal with respect to some complex utility function that increases with satisfaction and decreases with effort. Hence, Parberry's work may point the way to a broader basis for interpreting network behaviour, both biological and artificial, in an optimality framework.

ABSTRACT

Searle's arguments that intelligence cannot arise from formal programs are refuted by arguing that his analogies and thought-experiments are fundamentally flawed: he imagines a world in which computation is free. It is argued instead that although cognition may in principle be realized by symbol processing machines, such a computation is likely to have resource requirements that would prevent a symbol processing program for cognition from being designed, implemented, or executed. In the course of the argument the following observations are made: (1) A system can have knowledge, but no understanding. (2) Understanding is a method by which cognitive computations are carried out with limited resources. (3) Introspection is inadequate for analyzing the mind. (4) Simulation of the brain by a computer is unlikely not because of the massive computational power of the brain, but because of the overhead required when one model of computation is simulated by another. (5) Intentionality is a property that arises from systems of sufficient computational power that have the appropriate design. (6) Models of cognition can be developed in direct analogy with technical results from the field of computational complexity theory.

1. ARGUMENT

Penrose (1989) stated, "I am inclined to think (though, no doubt, on quite inadequate grounds) that unlike the basic question of computability itself, the issues of complexity theory are not quite the central ones in relation to mental phenomena." On the contrary, I intend to demonstrate that the principles of computational complexity theory can give insights into cognition.

Searle (1980) published a critique of artificial intelligence (AI) that almost immediately caused a flurry of debate and commentary in academic circles. The paper distinguished between *weak* AI, which uses the computer as a tool to understand cognition, and *strong* AI, which has as its main goal the recreation of cognition in a computer by means of a formal symbol-processing program. Searle professed to prove by thought-experiment, analogy, and introspection that no formal program can think, and thus deduced that strong AI is misguided.

Despite the flood of criticism and countercriticism that has been published, Searle seemed to have changed his opinions little over the next decade (Searle, 1984, 1990). As a theoretical computer scientist I do not find his arguments convincing. I propose here to expose some fundamental misunderstandings in his arguments. I do not directly refute his claim that strong AI is misguided, but I propose to show that his demonstration of this proposition is deeply flawed. I believe that strong AI cannot be dismissed on purely philosophical grounds. However, in the course of my argument I raise some of my own doubts about strong AI.

The three main weapons that Searle uses against strong AI are introspection, reasoning by analogy, and *gedankenexperiment*. Introspection can be highly unstable pedagogical ground, because in using the mind to observe and reason about itself, one risks running afoul of the Heisenberg Uncertainty Principle: the process of self-analysis may change the mind to the extent that any conclusions are cast into serious doubt. Nonetheless, I am prepared to allow introspection within certain bounds: I will allow Searle to look within himself and state that he understands English and does not understand Chinese.

I am suspicious of reasoning by analogy primarily because one needs little experience to realize that an analogy can be inappropriate if not properly subjected to the scrutiny of logic. Similarly, the *gedankenexperiment*, despite its illustrious history, can be seriously misguided. Because a *gedankenexperiment* is carried out purely in the mind, the conductor of the experiment is free to construct a fictional world in which reality does not apply, and hence runs the risk of coming to conclusions that have no basis in the real world. This is the fundamental flaw in Searle's reasoning: he carries out his *gedankenexperiment* in an imaginary world where computation costs nothing.

Many academics from outside the field of computer science who like to publish papers in the field appear to suffer from the misguided belief that computer science is a shallow discipline (if nothing else, because it has the word "science" in its name). Searle, like many critics of computer science, does not appear to be aware of current tends in research. Searle's arguments are limited to the theoretical computer science before the 1970s, which is based on the concept of *computability*, and the Church-Turing thesis that all models of symbolic computation are essentially the same.

Such a computational model assumes that computation is free. Unfortunately, just because a function is computable in the Church-Turing sense does not automatically mean that it is computable in the real world. Computation consumes resources, including time, memory, hardware, and power. A theory of computation, called *computational complexity theory*,[1] has grown from this simple observation, starting with the seminal paper of Hartmanis and Stearns (1965). The prime tenet of this technical field is that some computational problems intrinsically require more resources than others. The resource usage of a computation is measured as a function of the size of the problem being solved, with the assumption that we can solve small problems with the computers available to us now, and we will wish to scale up to larger problems as larger and faster computers become available.

The crux of Searle's argument is the following: just because a computer can compute something does not imply that it understands it. This is a reasonable hypothesis in the light of 1950s computer science: a function being computable is not sufficient reason to believe that something that computes it truly understands it. According to Searle, proponents of strong AI, in contrast, believe the opposite. The Turing[2] test (Turing, 1950) pits a human being against a computer. If an independent observer cannot tell in conversation with the two via some anonymous medium such as a teletype which is the computer and which is the human being, then the computer is said by proponents of strong AI to be "intelligent."

---

[1] Computational complexity theory should not be the confused with the more recent science of complexity popularized by physicists.

[2] Alan Turing made fundamental contributions to both theoretical computer science and AI, which is not surprising because the two fields were at the time inexplicably intertwined by the fact that the only computational device upon which to model a computer was an intelligent one: the brain.

Searle's *gedankenexperiment* consists of the following. Program a computer to converse in a natural language by providing it with a table of all possible inputs and their corresponding outputs. When given an input, the computer looks up the correct response in the table, and outputs that response. Searle reasoned that this computer passes the Turing test, but cannot be said to really *understand* what it is doing. He justified the latter observation with an analogy. A human being can be given such a lookup table for a language that he or she does not understand, for example, Chinese. This person can pass the Turing test in Chinese, *despite the fact that he or she does not understand Chinese.* Unlike many of Searle's critics, I am quite comfortable with this line of argument, and quite willing to concede that a computer programmed in this manner does not understand what it is doing in any reasonable sense of the word. However, Searle missed an important point early in his argument. He assumed that such a computer program is *possible.* I believe that such a program is *not* possible, for the simple reason that it requires too much in the way of resources.

Because the number of legal utterances in a natural language is uncountable (Langendoen & Postal, 1984), it is impossible to compile a complete look-up table of a language such as English or Chinese. However, this is not a serious barrier to the experiment. It would be sufficient for the purposes of passing the Turing test to compile a table of commonly used statements and legitimate responses. Whilst the number of commonly used questions and statements is a matter of some debate, a conservative lower bound is easy to obtain by considering questions of a particular form.

Consider queries of the form

Which is the largest, a $<noun>_1$, a $<noun>_2$, a $<noun>_3$, a $<noun>_4$, a $<noun>_5$, a $<noun>_6$, or a $<noun>_7$?

where $<noun>$ denotes any commonly used noun. Seven nouns were chosen rather than any other number because that appears to be the number of concepts that a typical human being can grasp simultaneously (Miller, 1956). How many queries are there of this form? There is little difficulty in constructing a list of 100 commonly known animals (see, e.g., Fig. 8.1). Therefore there are 100 choices for the first noun, 99 for the second, etc., giving a total of $100 \times 99 \times 98 \times 97 \times 96 \times 95 \times 94 = 8 \times 10^{13}$ queries based on Fig. 8.1 alone.

This is a very large number that requires grounding in everyday experience. The *Science Citation Index*[3] is a publication that approaches the human limit for usable information crammed into the smallest amount of space. Each page contains approximately 275 lines of 215 characters each, and each inch thickness of paper contains 1000 pages (over 5.9×10^7 characters). Assuming we could fit two queries of the above form and their responses on each line, each inch of paper would contain 5.5×10^5 queries. Therefore, if a look-up table for queries of the above form were constructed, and all the pages were stacked up, they would be 1.45×10^8 inches, that is, 2300 miles high. This would require a volume of paper almost 200 feet long, 200 feet wide,

---

[3] Published by the Institute for Scientific Information.

and 200 feet high. In contrast, the Great Pyramid of Cheops was (at the time of construction) over approximately 760 feet square and 480 feet high (see Fig. 8.2).

aardvark	crocodile	guinea pig	orangutan	shark
ant	deer	hamster	ostrich	sheep
antelope	dog	horse	otter	shrimp
bear	dolphin	hummingbird	owl	skunk
beaver	donkey	hyena	panda	slug ·
bee	duck	jaguar	panther	snail
beetle	eagle	jellyfish	penguin	snake
buffalo	eel	kangaroo	pig	spider
butterfly	ferret	koala	possum	squirrel
cat	finch	lion	puma	starfish
caterpillar	fly	lizard	rabbit	swan
centipede	fox	llama	raccoon	tiger
chicken	frog	lobster	rat	toad
chimpanzee	gerbil	marmoset	rhinoceros	tortoise
chipmunk	gibbon	monkey	salamander	turtle
cicada	giraffe	mosquito	sardine	wasp
cockroach	gnat	moth	scorpion	weasel
cow	goat	mouse	sea lion	whale
coyote	goose	newt	seahorse	wolf
cricket	gorilla	octopus	seal	zebra

Fig. 8.1. One hundred animals.

A reasonable defense against this objection is that computers can store data more efficiently than the printed word. It is possible in principle to construct a hard-disk array capable of storing our example lookup table. If we extrapolate slightly from the current state of the art, a disk capable of storing 2.5×10^9 characters takes on the order of 100 cubic inches of volume and costs on the order of \$1,000. Therefore, 8×10^{13} queries at 100 characters per query requires 3.2 million disks, which would take up a volume of 1.85×10^5 cubic feet (or a cube 57 feet on a side), and cost \$3.2 billion.

It is clear that our toy example only scratches the surface of the true size of a lookup table for a natural language. It is not too difficult to compile a list of 1400 fairly common concrete nouns (see the Appendix). It is not unreasonable to expect computers to be able to match the highest human ability, which would be nine nouns per query (Miller, 1956). The total amount of storage required for $1400^9 = 2 \times 10^{28}$ queries, with 100 characters per query, 5 bits per character, is 10^{31} bits.

If we were to store this on paper, it would require a stack almost 10^{10} light years high. In contrast, the nearest spiral galaxy (Andromeda) is 2.1×10^6 light years away, and the distance to the furthest known galaxy in 1988 was 1.5×10^{10} light years (Emiliani, 1988). If we were to use hard disks, it would take 5×10^{20} drives, which would occupy a cube 580 miles on a side. Even if we were to extrapolate wildly beyond the limits of foreseeable technology and conjecture that each bit could be stored on a single hydrogen atom, it would require almost 17 tons of hydrogen. Yet our query set is still relatively small compared to the true number of reasonable natural language queries. It is not too difficult to compile 30 different adjectives or adjectival clauses to replace "largest"), which multiplies the resource requirement by 30. Increasing the list of nouns to 2,000 increases it by a further factor of 20. Increasing the list of nouns to 10,000 increases it by a further factor of almost 2 million.

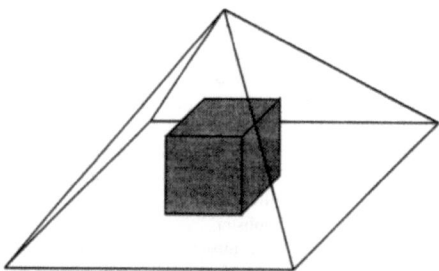

Fig. 8.2. The Great Pyramid of Cheops and the lookup table.

Therefore, it is fairly safe to conclude that it is not possible to pass the Turing test by simply using a lookup table. Where does this leave Searle's Chinese Room *gedankenexperiment*? A lookup table certainly contains *knowledge*, but no *understanding*. Searle's *gedankenexperiment* illustrates that understanding enables us to perform computations with a reasonable amount of resource usage; certainly less memory than is required to store a lookup table, and less time than is required to access one. This is a purely operational definition of understanding, and thus may not be satisfactory to a philosopher such as Searle who is more interested in a denotational definition, but I believe that any theory of cognition that does not take this into account rests on unstable foundations.

Naturally, understanding is not a Boolean trait; one can have a little understanding, rather than being limited to *no understanding* or *complete understanding*. With a little understanding of the concept of size, one can reduce the lookup table for the example queries simply by sorting the list of objects in increasing order of size. We appear to understand such things not by memorizing lists of facts, but by grounding the abstract concepts of the objects involved in everyday experience, from which information we compute facts such as their relative size. I believe that understanding evolved as the most efficient way of storing, cross-referencing, and reasoning about large quantities of environmental data (that is, the most efficient way that can be realized within the design parameters of evolution).

One point on which Searle and I agree is that a digital computer can, in principle, simulate a human brain. The electrical behaviour of a single neuron is far from being well understood, but I would be surprised if it could only be described using continuous mathematics. My first objection is on general principle: most phenomena in the Universe appear to be discrete, although in many cases the quanta are so small that continuous mathematics is a good approximation to reality. My second objection comes from the experimental observation that the brain often continues to function when large numbers of neurons are damaged, and under conditions in which a large number of them misfire. I find it difficult to believe that this robustness would be possible if it were *essential* that every neuron compute a real value to infinite precision. Fixed precision is almost certainly enough, and probably not too large a precision. Any fixed precision computation can be realized by a discrete computation.

Searle felt uncomfortable with the consequences of the Church-Turing thesis. Computers can be realized with any medium that can represent Boolean values and compute binary conjunction and complement, including water pipes. In principle, a plumber could devise a sewer system that can simulate a human brain. Searle found this absurd, but not for the same reasons that I do. There is far too much computational power in the brain to implement it as a sewer system.

Can we make a rough estimate as to how much computational power is contained in the human brain? Barrow and Tipler (1986) gave a range of 10^{10} to 10^{12} floating point operations per second, but they assumed that the computation is taking place purely within the soma of the neuron. Conventional wisdom currently conjectures that a significant amount of the computation actually takes place within the synapses. Turing (1950) made an estimate in the 1950s that with the benefit of modern knowledge seems optimistically low.

It is difficult to obtain reliable estimates of the number of neurons in the human brain. Shepherd (1988) estimated that the human cortex has a surface area of about 2,400 square centimeters, and Rockell, Hiorns, and Powell (1980) reported a uniform density of about 8×10^4 neurons per square millimeter, from which we can conclude that the number of neurons in the cortex alone is of the order of 10^{10}. I assume that the bulk of the information passed from one neuron to another passes through the synapses; the number of such connections per neuron varies with the type of neuron in question, and is somewhat difficult to estimate, but a figure of 10^3 connections per neuron is probably conservative. It is probably optimistic to assume that a pair of inputs to a neuron can be combined using a single floating-point operation; even so, this implies that each neuron computes the equivalent of 10^3 floating-point operations to combine the information input to it across its synapses. Combining these naive estimates with a firing time of 10^2 seconds per neuron, we see that the brain appears to have a processing power equivalent to at least 10^{15} floating-point operations per second.

Searle's water-pipe brain simulator is clearly something that can be imagined, but not constructed. Even under high pressure, water would flow so slowly in the pipes that in order to achieve 10^{15} floating-point operations per second it would require on the order of 10^{15} floating point operations to be computed simultaneously at different parts of the sewer. Even if these results could be combined in a meaningful way in a fast enough manner, the sheer size of the system makes it so unreliable that it would stand little hope of passing the Turing test. For that matter, could a computer do a better job? Current supercomputers can execute 10^{10} floating-point operations per second, and it is estimated that we might reach 10^{12} by 1994 (Bell, 1989). The brain appears to have more available computational power than a thousand of these hypothetical supercomputers.

This type of argument rests on shaky pedagogical ground because it is impossible to make an accurate assessment of the brain's computational power given our current almost complete lack of understanding of the principles of brain style computation. Our estimate may well be too high or too low by several factors of ten. A second weakness is that technology is advancing so rapidly that, if Bell is correct and 10^{12} floating-point operations per second are achievable by 1994, and advances in technology double computing speed annually, then computers may reach the 10^{15} floating-point operations per second needed to rival the brain by as early as 2004.

One thing that we can be fairly certain of, however, is that the brain's architecture is in a sense optimized for the type of computation that it is to perform. I say "in a sense" because there is little reason to believe that it is the absolutely optimum architecture (for evidence that biological computing systems are suboptimal, see, e.g., Dumont & Robertson, 1986; Stork, 1992; Stork, Jackson, & Walker, 1991). Rather, it is reasonable to believe that evolution has led to a locally optimal solution to a complicated optimization problem whose constraints include such factors as computational efficiency, heat loss, weight, volume, and nutritional requirements. Current computers, on the other hand, have architectures that are optimized within the constraints of current technology for the types of symbol processing problems for which they are used. It is hardly surprising that the architectures of the brain and the computer are radically different.

The simulation of the brain on a computer, then, is the task of simulating one model of computation on a second, architecturally different, model. The concept of one computer model simulating another is a key one in the theory of computational complexity. The Church-Turing thesis states that any reasonable model of computation can simulate any other one. Computational complexity theory has similar theses that state that these simulations can be carried out with a fairly small overhead in resource use; there is the *sequential computation thesis* (Goldschlager & Lister, 1983), the *parallel computation thesis* (Goldschlager, 1977, 1982), the *extended parallel computation thesis* (Dymond, 1980; Dymond & Cook, 1980), and the *generalized parallel computation thesis* (Parberry & Schnitger, 1988).

Nonetheless, each simulation requires an overhead in either hardware or time, often by as much as a quadratic in amount of that resource used by the machine being simulated. Therefore, any computer doing a neuron-by-neuron simulation of the brain need not only be as computationally powerful as the brain, but dramatically more so. For example, contrast our figures on raw computing power above with experimental figures in the DARPA study (1988) on simulating synaptic weight updates in current neuron models (summarized in Table 8.1). The reason why the computer figures are so poor (capable of simulating neural capacity somewhere between a worm and a fly, see Table 8.2) is that the raw computing power figures that we gave earlier completely ignored the extra overhead involved in the simulation. This is the real reason that we should abandon any hope of simulating cognition at a neuron-by-neuron level, rather than any philosophical or pedagogical objection.

Searle's reasoning by analogy that there is little reason to believe that a simulation of cognition is not the same as cognition is unconvincing. Certainly a simulation of a fire is not a fire, but for some purposes it does just as well. Pragmatically, if a simulation of cognition is not possible by reason of the fact that such a simulation carries too much overhead, then it is merely a matter of definition whether one calls it true cognition. Nonetheless, Searle has raised an important point that has deep ramifications. Strong AI proceeds by construction of a model of how the mind performs a task (such as understanding short stories, Schank & Abelson, 1977), and then implementing (Searle would say "simulating") that model on a computer. But what is introspection, if it is not *simulating cognition*? When one introspects, one constructs a conscious model of mind process, in essence a *simulation* of the mind. What right have we to believe that the products of introspection, which is no more than the construction of an internal simulation of mind, bear any real resemblance to the mind?

Computer	Synapses	Updates
PC/AT	1.0×10^5	2.5×10^4
Symbolics	1.0×10^7	3.5×10^4
VAX	3.2×10^7	1.0×10^5
SUN3	2.5×10^5	2.5×10^5
MARK III, V	1.0×10^6	5.0×10^5
CM-2 (64K)	6.4×10^7	1.3×10^6
Butterfly (64)	6.0×10^7	8.0×10^6
WARP (10)	3.2×10^5	1.0×10^7
Odyssey	2.6×10^5	1.0×10^7
CRAY XMP 1-2	2.0×10^6	5.0×10^7
MX-1/16	5.0×10^7	1.3×10^8

Table 8.1. Number of Synapses, and Synaptic Weight Updates per Second for Some Common Computers to Simulate a Neural Network (from DARPA study, 1988). The measurements for the MX—1/16 are projected performance only.

Creature	Synapses	Updates
Leech	7×10^2	2×10^4
Worm	5×10^1	2×10^5
Fly	8×10^7	1×10^9
Aplysia	2×10^8	2×10^{10}
Cockroach	9×10^8	3×10^{10}
Bee	3×10^9	5×10^{11}
Man	1×10^{14}	1×10^{16}

Table 8.2. Number of Synapses, and Synaptic Weight Updates per Second for Some Common Creatures (from DARPA study, 1988).

A crucial part of Searle's argument is the concept of *intentionality*, which describes directed mental states such as beliefs, desires, wishes, fears, and intentions. Intentional states are related to the real world, but are not in one-to-one correspondence with it. One can have beliefs that are false, desires that are unfulfillable, wishes that are impossible, fears that are groundless, and intentions that cannot be realized (Searle, 1979a, 1979b). There are conscious intentional states, and unconscious intentional states, yet Searle devised a logic of intentional states that is mirrored in linguistics (Searle, 1979b). Searle came to this conclusion from introspection, that is, by constructing a conscious and therefore *by its very nature* symbolic simulation of intentionality. If a simulation of intentionality is fundamentally different from intentionality itself, and if a conscious model of intentionality is merely a simulation of intentionality (rather than the real thing), then we are drawn to the inevitable conclusion that Searle's logic of intentionality tells us little about intentionality itself. The inadequacy then is not in strong AI, which can take any consciously generated symbol-based model of cognition and turn it into a computer program, but rather with the analytical tools of cognitive psychology.

Searle argued that a formal program cannot have intentionality, and that intentionality is a crucial part of cognition. I am in agreement with the latter hypothesis, but in the former hypothesis Searle exhibited a strong anti-machine bias that he did not defend to my satisfaction. He was willing to accept that an animal has intentionality because it is the simplest explanation of its behaviour, but only because it is made of the same "stuff" as we are; apparent intentional behaviour from a robot was insufficient for him because (Searle, 1980, p. 421) "as soon as we knew that the behavior was the result of a formal program, and that the actual causal properties of the physical substance were irrelevant we would abandon the assumption of intentionality."

Searle made the assumption that intentionality is a property of the "stuff" of biological organisms, and cannot arise from the "stuff" of computers by execution of a formal program. We do not know enough of how intentional states are realized in human beings (Searle, 1979b, considered the question briefly and dismissed it as irrelevant) to be able to say with confidence that formal programs can never exhibit them. It is reasonable to hypothesize that intentional states can arise in computational systems that are both sufficiently powerful and properly organized. There is no reason to believe that intentional states arise in simple machines such as thermostats, and it is a reasonable hypothesis that they do occur in higher primates and human beings. A reasonable hypothesis is that simple machines lack the computational power to have intentional states. That proper organization is necessary for intentional states is a properly conservative view; it is too optimistic to believe that they occur naturally in any computational system that is powerful enough to exhibit them. The real reason why computers do not have intentional states is that they are too simple to have them.

Searle was offended by the thought that a *mere* computer program could have intentional states, or think. But there is nothing "mere" about a computer program. The task of producing correct, fast, and robust software for even simple tasks (such as an airline reservation system) is incredibly difficult, as anyone who has attempted to write more than "toy" programs will agree. I have no philosophical problem with the hypothesis that there exists a program that when executed gives rise to cognition. However, there is a great chasm between that belief and strong AI. It may be that the program is far too complicated for us to understand. It may be that, when written as a formal program, it has resource usage that is beyond the power of the human race to provide. Simply because cognition can be realized in the brain (in some fashion that we do not yet fully understand) with a reasonable resource requirement is no reason to believe that its resource requirements as a formal program will be reasonable too. We have already seen that there is a large overhead in simulating one machine by another; it is often the case in computational complexity that a program for machine A requires large overhead to implement on machine B, regardless of whether machine B simulates the program for A directly, or executes a completely unrelated program that produces the same results. The overhead of achieving cognition on a computer may be so large as to render the task impossible.

For example, it is clear that one cannot simulate intentionality by a Chinese Room algorithm, because such a lookup table must have entries for questions of the form

Would you believe that a <noun>$_1$ could be larger than a <noun>$_2$, a <noun>$_3$, a <noun>$_4$, a <noun>$_5$, a <noun>$_6$, or a <noun>$_7$?

or of the form

Which would you like to see most of all, a <noun>$_1$, a <noun>$_2$, a <noun>$_3$, a <noun>$_4$, a <noun>$_5$, a <noun>$_6$, or a <noun>$_7$?

The previous arguments about the size of the lookup table apply equally well here.

In summary, I believe that intentionality and cognition can *in principle* be obtained by executing the appropriate formal symbol manipulation program, but that there are other barriers that prevent intentionality and cognition from being realized that way in practice. To draw an analogy, the *Principia Mathematica* (Whitehead & Russell, 1910-1913) reduces mathematics to symbol manipulation, yet this is not how mathematicians do mathematics. Whilst they freely acknowledge that it is a necessary condition for any "proof" to be in principle expressible in formal logic, it is not necessary that it be so expressed. Mathematicians reason informally principally for the purposes of communication: a human being simply cannot understand a proof of any great depth and difficulty if it is expressed in symbolic logic. I believe that in the same sense, the mind can in principle be reduced to a symbol manipulation program, but the program would be far too long and complicated for human beings to understand (see also Campbell, 1989, p. 109), and that the reason why we don't see thinking beings that are "mere symbol processors" is that the mind reduced to a symbol processing program may be too greedy of resources to be realized in the physical world.

We must also face the fact that it may not be possible to build a computer that matches the brain in speed, size, reliability, portability, power consumption, and ease of fabrication. It may be, as some biologists believe, that biology is the only way to achieve these goals simultaneously. But perhaps not. Perhaps the brain is the only way that such a computational device could evolve. It is an open question whether we can devise one ourselves, independent of the constraints of evolution. It is still an open question as to whether we could make such a device sentient. I believe that it is not possible given our current state of technology and current state of knowledge about cognition, but Searle's arguments have failed to convince me that such a thing is in principle impossible.

Many believe that neural networks adequately refute Searle's Chinese Room *gedankenexperiment* (see, e.g., Churchland & Churchland, 1990). Searle dismissed neural networks and parallel computation as not bringing anything new to the concept of computation as it applies to cognition. In a sense he was right; they bring nothing new to the 1950s style of computability theory that he used to bolster his arguments. However, parallel computers are more efficient at solving some problems than sequential computers are (see Parberry, 1987), and the same can be said of neural networks (see Parberry, 1990, 1992).

The prime contribution of neural networks is not their mode of computation. The fact that they use a computational paradigm that differs from the traditional Church-Turing one is self-evident in some cases, but this is not the death knell for computer science as many of the proponents of neural networks would have us believe. Theoretical computer science has dealt with unconventional modes of computation for decades, as we examine later in this chapter.

The prime contribution of neural networks is the capacity for efficient computation of certain problems. The first computers were created in rough analogy with the brain, or more correctly, in rough analogy with a carefully selected subset of what was known about the brain at the time. Although technology has advanced greatly in recent decades, modern computers are little different from their older counterparts. It is felt by some scientists that in order to produce better computers we must return to the brain for further inspiration.

I believe that it is important to determine which features of the brain are crucial to efficient computation, and which features are by-products or side effects of these (see Parberry, 1990, 1994). I do not believe that a computer that is comparable in computing power to the brain can be obtained by merely simulating its observed behaviour, simply because the overhead is too great. The general principles of brain computation must be understood before we try to implement an artificial system that exhibits them.

Computational complexity theory is a powerful technique that can be used to divine some of the general principles behind brain computation. However, the theory is in its infancy. Surprisingly, many apparently simple questions about efficient computation turn out to be difficult and deep. Whilst computational complexity theorists equate exponential resource usage with intractability and polynomial resource usage with tractability, in real life any resource usage that grows more quickly than log-linearly in problem size is probably too large to be of any real use. It remains to develop the tools that can make that fine-grained a distinction in resource requirements; for example, we cannot distinguish between problems with time requirements that intuitively grow exponentially with problem size from those that do not (see, e.g., Garey & Johnson, 1979).

Nonetheless, computational complexity theory often gives insights that may have profound philosophical ramifications. For example, many neural network researchers use a continuous model (i.e., one in which the neurons compute a continuous value). It can be shown in certain technical senses that if one assumes that neuron outputs are robust to small errors in precision, then their model is essentially the same as a discrete one within a "reasonable" overhead in resources (Obradovic & Parberry, 1990a, 1990b). More importantly, the same is true *even without the assumption of robustness* (Maass, Schnitger, & Sontag, 1991).

The general framework used by neural network researchers is a finite network of simple computational devices wired together similarly to the network shown in Fig. 8.3 so that they interact and cooperate to perform a computation (see, e.g., Rumelhart, Hinton, & McClelland, 1986). Yet there is little attention paid to how these finite networks scale to larger problems. When one designs a circuit to solve a given task, such as performing pattern recognition on an array of pixels, one typically starts with a small number of inputs, and eventually hopes to scale up the solution to real life situations. How the resources of the circuit scale as the number of inputs increases is of prime importance. A good abstraction of this process is to imagine a potentially infinite series of circuits, one for each possible input size, and to measure the increase in resources from one circuit in the series to the next (see Fig. 8.4).

Inputs from sensors

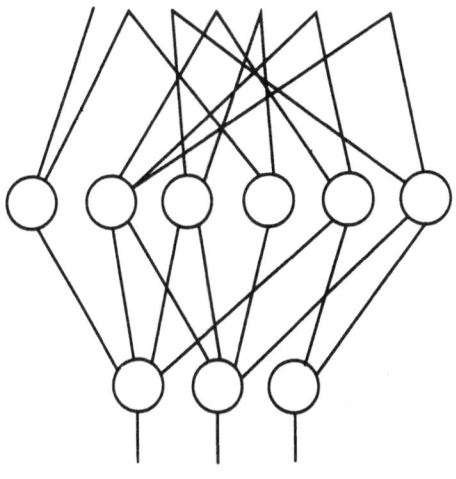

Outputs to effectors

Fig. 8.3. A finite neural network with nine nodes and two layers.

There is an apparent flaw in this abstraction, however. Because for every natural number n, every Boolean function with n inputs can be computed by a finite Boolean circuit (essentially by using a lookup table, a formalization of Searle's Chinese Room), our infinite-family-of-finite-circuits model can compute any Boolean function, and so violates the Church-Turing thesis. This can be remedied in one of three ways (among others). First, we could insist that each finite circuit in the infinite series be similar to its predecessor in the series in the sense that a Church-Turing computer can compute the differences between the two. This type of circuit is called a *uniform* circuit (whereas the former is called a nonuniform circuit). Second, we could insist that the number of processing units in the finite circuits grows only polynomially with input size. This would avoid the embarrassment of being able to compute every Boolean function, because it is easy to show by a counting argument that there are Boolean functions that require exponential size. Third, we could insist that the structure of each circuit be different from its predecessor, but that there exists a computable set of circuits in which substantially more than half of the alternatives compute the correct function. The first solution satisfies the Church-Turing thesis, and the other two do not. This need not necessarily be a problem: the Church-Turing thesis is a *model* of reality, and is not inviolate. The second solution is not particularly desirable, since it can make the design of the circuits difficult in practice. The third solution is more appealing because although we may not be able to compute the layout of each circuit, a subset of circuits chosen randomly from the computable set of alternatives has high probability of turning out a circuit that works.

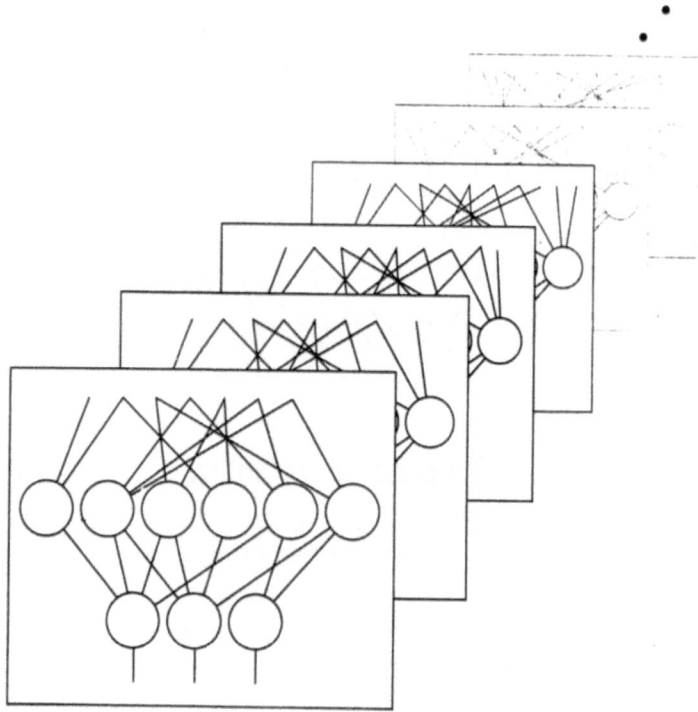

Fig. 8.4. A neural network family.

Allowing computers access to a random source appears to make them more efficient than a plain deterministic computer in some circumstances (see, for example, Cormen, Leiserson, & Rivest, 1990, Section 33.8). (Note that this is different from randomly choosing a deterministic algorithm.) In this case, it is sufficient for the algorithm to compute the correct result with high probability, say 0.999. Surprisingly, such a randomized algorithm can be replaced with a nonuniform one with only a small increase in resources (Adleman, 1978). This principle can even be applied to probabilistic neural networks such as Boltzmann machines (Parberry & Schnitger, 1989).

Randomness and nonuniformity are two methods for reducing the resource requirements of algorithms, both of which reach outside the confines of the Church-Turing thesis. The use of randomness occurred to Turing (1950), as did the possibility that the program of the mind (if it exists) may be far too complicated for us to analyze. Turing also raised the possibility that a computer could learn, as does a child. Computational complexity theory has started to ask questions in this domain with the recent development of *computational learning theory* (see, e.g., Natarajan, 1991). Of prime importance is the *probably-approximately-correct*, or PAC model of learning, in which it is sufficient for the system to learn a response that is with high probability close to the correct answer. Valiant (1984) proposed the original distribution-free PAC learning, and more recent versions include the Universal distribution (Li & Vitanyi, 1989).

The theoretical results described above demonstrate that randomness and continuous computation do not offer a large increase in efficiency because they can be simulated with a small increase in resources by discrete, deterministic computation. This does not, of course, mean that we should use discrete computation to realize neural networks. As discussed above, what is small overhead for a theoretician often becomes overwhelming in practice. Nonetheless, the theoretical results indicate that there is nothing new and alien in probabilistic and continuous computation, and that any philosophy based in them does not necessarily differ radically from a philosophy based on discrete, deterministic computation, contrary to all appearances.

ACKNOWLEDGMENTS

I would like to thank George Mobus, Tim Motler, Kathleen Swigger, and John Young for discussions on the subject of this paper and comments on a draft of the manuscript.

REFERENCES

Adleman, L. (1978). Two theorems on random polynomial time. In *19th Annual Symposium on Foundations of Computer Science* (pp. 75-83). IEEE Computer Society Press.

Barrow, J. D., & Tipler, F. J. (1986). *The Anthropic Cosmological Principle.* Oxford, UK: Clarendon Press.

Bell, G. (1989). The future of high performance computers in science and engineering. *Communications of the ACM, 32,* 1091-1101.

Campbell, J. (1989). *The Improbable Machine.* New York: Simon and Schuster.

Churchland, P. M., & Churchland, P. S. (1990). Could a machine think? *Scientific American, 262,* 32-37.

Cormen, T. H., Leiserson, C. E., & Rivest, R. L. (1990). *Introduction to Algorithms.* Cambridge, MA: MIT Press.

DARPA Neural Network Study (1988). Alexandria, VA: AFCEA International Press, 1988.

Dumont, J. P. C., & Robertson, R. M. (1986). Neuronal circuits: An evolutionary perspective. *Science, 233*, 849-853.

Dymond, P. W. (1980). *Simultaneous resource bounds and parallel computations.* Doctoral dissertation (Tech. Rep. #TR145). Department of Computer Science, University of Toronto.

Dymond, P. W., & Cook, S. A. (1980). Hardware complexity and parallel computation. In *21st Annual Symposium on Foundations of Computer Science* (pp. 360-372). IEEE Computer Society Press, 1980.

Emiliani, C. (1988). *The Scientific Companion.* New York: John Wiley and Sons.

Garey, M. R., & Johnson, D. S. (1979). *Computers and Intractability: A Guide to the Theory of NP-Completeness.* San Francisco: W. H. Freeman.

Goldschlager, L. M. (1977). *Synchronous parallel computation.* Doctoral dissertation (Tech. Rep. # TR114), Department of Computer Science, University of Toronto.

Goldschlager, L. M. (1982). A universal interconnection pattern for parallel computers. *Journal of the ACM, 29*, 1073-1086.

Goldschlager, L. M., & Lister, A. M. (1983). *Computer Science: A Modern Introduction.* Englewood Cliffs, NJ: Prentice-Hall.

Hartmanis, J., & Stearns, R. E. (1965). On the computational complexity of algorithms. *Transactions of the American Mathematical Society, 117*, 285-306.

Langendoen, D. T., & Postal, P. M. (1984). *The Vastness of Natural Languages.* Basil Blackwell.

Li, M., & Vitanyi, P. M. B. (1989). A theory of learning simple concepts under simple distributions and average case complexity for the universal distribution. In *30th Annual Symposium on Foundations of Computer Science* (pp. 34-39). IEEE Computer Society Press.

Maass, W., Schnitger, G., & Sontag, E. D. (1991). On the computational power of sigmoid versus Boolean threshold circuits. In *32nd Annual Symposium on Foundations of Computer Science.* IEEE Computer Society Press.

Miller, G. A. (1956). The magic number seven, plus or minus two: Some limits on our capacity for processing information. *Psychological Review, 63*.

Natarajan, B. K. (1991). *Machine Learning: A Theoretical Approach.* San Mateo, CA: Morgan Kaufmann.

Obradovic, Z., & Parberry, I. (1990a). Analog neural networks of limited precision I : Computing with multilinear threshold functions. In *Advances in Neural Information Processing Systems 2* (pp. 702-709). San Mateo, CA: Morgan Kaufmann.

Obradovic, Z., & Parberry, I. (1990b). Learning with discrete multi-valued neurons. *Proceedings of the Seventh Annual Machine Learning Conference* (pp. 392-399).

Parberry, I. (1987). *Parallel Complexity Theory. Research Notes in Theoretical Computer Science.* London: Pitman.

Parberry, I. (1990). A primer on the complexity theory of neural networks. In R. Banerji (Ed.), *Formal Techniques in Artificial Intelligence: A Sourcebook,* Volume 6 of Studies in Computer Science and Artificial Intelligence, pp. 217-268. Amsterdam: North-Holland.

Parberry, I. (1992). Circuit complexity and neural networks. In P. Smolensky, M. Mozer, & D. Rumelhart (Eds.), *Mathematical Perspectives on Neural Networks, Developments in Connectionist Theory*. Hillsdale, NJ: Lawrence Erlbaum Associates.

Parberry, I. (1994). *Circuit Complexity in Neural Networks*. Cambridge, MA: MIT Press.

Parberry, I., & Schnitger, G. (1988). Parallel computation with threshold functions. *Journal of Computer and System Sciences*, **36**, 278-302.

Parberry, I., & Schnitger, G. (1989). Relating Boltzmann machines to conventional models of computation. *Neural Networks*, **2**, 59-67.

Penrose, R. (1989). *The Emperor's New Mind*. Oxford: Oxford University Press.

Pribram, K. H., & McGuinness, D. (1992). Attentional and para-attentional processing: Event-related brain potentials as tests of a model. In D. Friedman & G. Bruder (Eds.), *Annals of the New York Academy of Sciences*, **658**, 65-92.

Rockell, A. J., Hiorns, R. W., & Powell, T. P. S. (1980). The basic uniformity in structure of the neocortex. *Brain*, **103**, 221-244.

Rumelhart, D. E., Hinton, G. E., & McClelland, J. L. (1986). A general framework for parallel distributed processing. In D. E. Rumelhart & J. L. McClelland (Eds.), *Parallel Distributed Processing: Explorations in the Microstructure of Cognition* (Vol. 1, pp. 282-317). Cambridge, MA: MIT Press, 1986.

Schank, R., & Abelson, R. (1977). *Scripts, Plans, Goals and Understanding*. Hillsdale, NJ: Lawrence Erlbaum Associates.

Searle, J. R. (1979a). The intentionality of intention and action. *Inquiry*, **22**, 253-280.

Searle, J. R. (1979b). What is an intentional state? *Mind*, **88**, 74-92.

Searle, J. R. (1980). Minds, brains and programs. *The Behavioral and Brain Sciences*, **3**, 417-457.

Searle, J. R. (1984). *Minds, Brains and Science*. Cambridge, MA: Harvard University Press.

Searle, J. R. (1990). Is the brain's mind a computer program? *Scientific American*, **262**, 26-31, 1990.

Shepherd, G. M. (1988). *Neurobiology*. New York: Oxford University Press.

Simon, H. (1979). A behavioral model of rational choice (1955). In H. Simon (Ed.), *Models of Thought* (pp. 7-19). New Haven: Yale University Press.

Stork, D. G. (1992). Preadaptation and principles of organization in organisms. In A. Baskin & J. Mittenthal (Eds.), *Principles of Organization in Organisms* (pp. 205-224). Addison-Wesley, Santa Fe Institute.

Stork, D. G., Jackson, B., & Walker, S. (1991). "Non-optimality" via preadaptation in simple neural systems. In C. G. Langton, C. Taylor, J. D. Farmer, and S. Rasmussen (Eds.), *Artificial Life II* (pp. 409-429). Addison-Wesley, Santa Fe Institute.

Turing, A. M. (1950). Computing machinery and intelligence. *Mind*, **59**, 433-460.

Valiant, L. G. (1984). A theory of the learnable. *Communications of the ACM*, **27**, 1134-1142.

Whitehead, A. N., & Russell, B. A. W. (1910-1913). *Principia Mathematica*. Cambridge, UK: Cambridge University Press.

APPENDIX: 1400 CONCRETE NOUNS

aardvark	antelope	axe	bathtub	blackberry	branch	burette	cap	chalkboard	closet
abacus	antenna	axle	baton	blackbird	bratwurst	bus	cape	chameleon	cloud
abalone	anteroom	axolotl	battalion	bladderwort	brazier	bush	capillary	chamois	coach
abbey	anther	axon	battery	blade	breadboard	bustard	capstan	chandelier	coat
abdomen	anvil	azalea	battlefield	blanket	breadfruit	buteo	capstone	chapel	cobweb
abscess	aorta	baboon	bayonet	blastula	breakwater	buttercup	capsule	chariot	cochlea
accipiter	ape	baby	bayou	blister	bream	butterfly	car	chateau	cockatoo
acropolis	apostrophe	bacillus	beach	blossom	breast	buttock	caravan	check	cockle
aileron	apple	backpack	beacon	blowfish	breastplate	button	carbine	checkbook	cocklebur
aircraft	appliance	backyard	bead	blueberry	breastwork	buttonhole	carbuncle	checkerboard	cockleshell
airedale	apron	bacterium	beak	bluebird	brewery	buzzard	carburetor	cheek	cockpit
airfield	apse	badge	beam	bluebonnet	brick	buzzsaw	carnation	cheekbone	cockroach
airfoil	aquarium	bagel	bean	bluefish	bridge	cabbage	carnival	cheesecake	cocktail
airplane	aqueduct	bagpipe	bear	blueprint	bridle	cabin	carp	cheesecloth	coconut
airport	arachnid	balcony	beaver	boa	brief	cabinet	carriage	cheetah	cod
airstrip	arena	ball	bed	boar	briefcase	cable	carrot	cherry	codpiece
aisle	ark	ballfield	bedbug	board	brigade	cactus	cart	cherub	coffeecup
alarm	arm	balloon	bedpost	boardinghouse	broadloom	cafe	cartwheel	chest	coffeepot
albatross	armada	ballroom	bedroom	boat	bronchus	cake	cashew	chestnut	coin
album	armadillo	balustrade	bedspread	boathouse	brontosaurus	calf	casino	chickadee	colander
alcove	armature	banana	bedspring	boatyard	brook	calliope	cask	chicken	collarbone
alder	armchair	bandage	bee	bobcat	broom	callus	casket	chigger	collard
allele	armhole	bandstand	beech	bobolink	brush	camel	casserole	child	college
alley	armoire	bangle	beefsteak	boil	brushfire	camellia	cassette	chimney	collie
alligator	armpit	banjo	beehive	bolster	bubble	camera	cassock	chimpanzee	colon
almanac	army	bank	beet	bomb	buckboard	campfire	castle	chin	colt
almond	arrow	bannister	beetle	bongo	bucket	campground	castor	chipmunk	column
aloe	arrowhead	bar	bell	bonito	buckle	campus	cat	chloroplast	comb
alpenstock	artery	barbecue	belt	bonnet	buckthorn	can	catalpa	church	comet
alsatian	artichoke	barbell	bench	bookcase	bud	canal	catapult	churn	comma
altar	ashtray	barge	berry	bookend	buffalo	canary	cataract	cicada	computer
altimeter	ass	barn	bezel	bookshelf	buffet	candelabra	catbird	cigar	concertina
alveolus	asteroid	barnacle	bib	bookstore	bug	candle	caterpillar	cigarette	conch
amaranth	atlas	barnyard	bible	boomerang	bugle	candlestick	catfish	cilia	condominium
amethyst	atom	barometer	biceps	boson	bulb	candlewick	catheter	cinema	coneflower
amoeba	atrium	barracuda	bicycle	botfly	bulkhead	cane	catkin	cinquefoil	coney
amphibian	attic	barrel	bikini	bottle	bull	canine	cauldron	circus	conifer
amplifier	auditorium	barrow	bilge	boulder	bulldog	canister	cauliflower	citadel	continent
anaconda	auger	baseball	billboard	boulevard	bulldozer	canker	civet	cookie
anchor	auk	basin	billfold	bouquet	bullfinch	cankerworm	cave	clam	cork
anchovy	aurochs	basket	bin	bowl	bullfrog	cannery	cell	clarinet	corkscrew
andiron	aurora	basketball	biplane	bowstring	bullock	cannister	cellar	classroom	cormorant
anemone	autoclave	bass	birch	box	bumblebee	cannon	cemetery	claw	cornfield
angel	automobile	bassinet	bird	boxcar	bun	cannonball	centaur	cleat	cornflower
angelfish	aviary	bat	birdbath	bracelet	bungalow	canoe	centimeter	cliff	corridor
angstrom	avocado	bath	biscuit	brad	bunk	cantaloupe	centipede	clipboard	corset
ant	avocet	bathrobe	bison	braid	bunny	canteen	cerebellum	clock	cortex
anteater	awl	bathroom	bittern	brain	buoy	canyon	chaise	cloister	cotoneaster

cottage
cottonseed
cottonwood
country
courtyard
couscous
cow
cowbell
cowbird
cowslip
coyote
crab
crabapple
cradle
crane
cranium
crankcase
crankshaft
crater
cravat
crayfish
crayon
credenza
creek
crewel
cricket
crocodile
crocus
crossbow
crow
crown
crucible
crucifix
crumb
crypt
cuckoo
cucumber
cudgel
cufflink
cup
cupboard
cushion
cutlass
cuttlebone
cuttlefish
cutworm
dachshund
daffodil
dagger
dahlia
daisy
dam
dandelion
dashboard
deer
deerstalker
desk
dewdrop
diaper

diary
dictionary
dingo
dinosaur
discus
dish
dishwasher
distillery
doberman
dockyard
dog
dogfish
doghouse
doll
dolphin
donkey
doorbell
doorknob
doorstep
dormitory
doughnut
dove
dragon
dragonfly
drake
dromedary
drosophila
drum
duck
duckling
dusthin
eagle
ear
eardrum
earphone
earring
earthworm
earwig
easel
echidna
eel
egg
eggplant
egret
electron
elephant
elk
elm
eucalyptus
ewe
eye
eyeball
eyebrow
eyeglass
eyelash
eyelid
falcon
farm
farmhouse

faucet
fawn
feather
featherbed
fern
ferret
fiddle
fiddlestick
finch
finger
fingernail
fingerprint
fingertip
fir
firecracker
firefly
firehouse
fireplace
fish
fist
flashlight
flatiron
flea
fledgling
flounder
flowchart
flowerpot
flute
fly
flycatcher
foal
foot
footpath
footprint
footstool
forest
fork
forklift
fort
fountain
fowl
fox
foxglove
foxhole
foxhound
foxtail
freeway
frigate
frog
fruit
fungus
furlong
gadfly
galaxy
gallberry
gallstone
galvanometer
gander
gannet

gannet
garage
garden
gardenia
garter
gasket
gate
gauntlet
gavel
gazelle
gear
gecko
gene
geranium
gerbil
germ
geyser
gibbet
gibbon
giraffe
girdle
glacier
gnat
gnome
gnu
goat
goldenrod
goldfinch
goldfish
goose
gooseberry
gorilla
gourd
grackle
grape
grapefruit
grapevine
gravestone
graveyard
greatcoat
greenhouse
greyhound
grosbeak
guillotine
guineapig
guitar
gull
gun
gyrfalcon
gyroscope
hackberry
hacksaw
hailstone
hairpin
halibut
hammer
hammock
hamster
hand

handbag
handgun
handkerchief
hangar
hare
harp
harpoon
harpsichord
hat
hatchet
hawk
haystack
hazelnut
headboard
headlight
headstone
hedgehog
hen
heron
hippopotamus
hollyhock
honeybee
honeycomb
honeydew
hoof
hornet
horse
horsefly
horsehair
horseshoe
hourglass
house
houseboat
housefly
huckleberry
human
hummingbird
hurricane
hyacinth
hydra
hydrangea
hydrant
hyena
ibex
iceberg
icebox
icicle
inch
infant
inn
insect
intestine
iris
jackass
jackboot
jackdaw
jacket
jackknife
jaguar

javelin
jawbone
jellyfish
jockstrap
jug
kaleidoscope
kangaroo
keeshond
kerchief
kestrel
ketch
kettle
key
keyboard
keyhole
keystone
kid
killdeer
kingfisher
kite
kitten
kiwi
knee
kneecap
knife
knot
koala
labrador
lacewing
ladle
lagoon
lake
lamb
lamprey
landfill
lapel
larkspur
larva
larynx
lasso
lathe
laundry
lavatory
leaf
leaflet
leash
lectern
leghorn
legume
lemming
lemon
library
lifeboat
ligament
ligature
limousine
limpet
lion
lip

lizard
llama
lobster
lock
locknut
locomotive
locust
lodge
log
loincloth
lollipop
longhorn
loudspeaker
louse
lovebird
lowboy
lute
mace
magazine
magnolia
magpie
mailbox
mallet
mandrake
mandrill
manometer
manor
mantelpiece
maple
marble
mare
marionette
marketplace
marmoset
marmot
marquee
marrowbone
marten
martini
mastiff
mastodon
mattock
mattress
mausoleum
meadow
medal
menu
metronome
metropolis
midge
millipede
minefield
minesweeper
minibike
minicomputer
minnow
mitten
moat
mobcap

moccasin
mockingbird
mole
mollusk
mongoose
monkey
moon
moor
moose
mop
mosquito
moth
mothball
mother
motor
motorcycle
mound
mount
mountain
mouse
moustache
mouth
mouthpiece
mudguard
muffin
mulberry
mule
mushroom
musket
muskmelon
muskox
muskrat
mussel
mustang
muzzle
myrtle
nanometer
napkin
nautilus
navel
nebula
neck
necklace
necktie
nectarine
needle
nest
net
nettle
neuron
neutrino
neutron
newt
nickel
nightcap
nightclub
nightdress
nightgown
nightingale
nightshirt

nit
noose
nose
nosebag
notebook
notocord
nuthatch
oak
oar
oasis
ocarina
ocean
ocelot
octopus
odometer
omelet
onion
orange
orangutan
orchard
orchestra
orchid
organ
oriole
oscilloscope
osprey
ostrich
otter
owl
ox
oxcart
oyster
pacemaker
paddle
paddock
padlock
pail
paintbrush
palace
palette
palfrey
palm
pamphlet
pan
panda
pansy
panther
paperback
paperweight
papoose
parachute
parakeet
parrot
parsnip
patio
pawn
pea
peach
peacock

peanut	postmark	salmon	snowflake	tarantula	truck	wolf
pear	pot	samovar	snowmobile	tarpon	trunk	wombat
pearl	potato	sandbag	snowshoe	taxi	truss	woodpecker
pebble	pothole	sandpiper	sock	teacart	tub	worm
pecan	propeller	sandwich	sofa	teacup	tuba	wrist
peccary	proton	sardine	soybean	teahouse	tulip	wristband
pelican	pterodactyl	sausage	spade	teakettle	tuna	wristwatch
pen	puffball	sawfly	sparrow	teal	tunic	xylophone
penguin	puffin	sawmill	speedboat	teapot	turban	yacht
penny	pug	saxophone	spider	teardrop	turnip	yak
periscope	puma	scabbard	spiderwort	teaspoon	turnpike	yam
periwinkle	pumpkin	scallop	spinnaker	tee	turtle	yardstick
petal	pumpkinseed	scarecrow	spinneret	telephone	turtleneck	yarmulke
petticoat	pupil	scarf	spleen	telescope	twig	zebra
pheasant	puppet	scorpion	spleenwort	television	typewriter	zucchini
photon	puppy	screw	sponge	tepee	tyrannosaurus	
piano	purse	sea	spoon	termite	umbrella	
pickaxe	pushpin	seagull	spore	tern	unicorn	
pie	pussycat	seahorse	squash	terrapin	vacuole	
pier	python	seal	squirrel	textbook	valley	
pig	quail	sealion	stamen	thermostat	vertebrate	
pigeon	quark	shallot	stamp	thesaurus	viaduct	
pigeonhole	quill	shark	starfish	thigh	videotape	
pigpen	quilt	sheep	steamboat	thimble	village	
pigtail	quince	sheepskin	stegosaurus	thrip	vine	
pike	rabbit	shinbone	sternum	throat	viola	
pill	raccoon	shoe	stethoscope	throne	violet	
pimple	racetrack	shoehorn	stickpin	thrush	violin	
pin	radish	shoelace	stool	thunderstorm	virus	
pinafore	rainbow	shotgun	stopwatch	tick	vise	
pinball	raindrop	shovel	stove	ticket	vixen	
pincushion	raisin	shrew	strawberry	tiger	volcano	
pine	rake	shrimp	streetcar	titmouse	volleyball	
pineapple	rapier	shrub	streptococcus	toad	vulture	
pinhead	raspberry	shuttlecock	sunfish	toe	wagon	
pinhole	rat	silkworm	sunflower	toenail	waistcoat	
pinion	rattlesnake	skate	sunspot	toilet	wallaby	
pinpoint	razor	sketchbook	swallow	tomato	wardrobe	
pinto	reindeer	ski	swallowtail	tomb	washbasin	
pinwheel	retina	skiff	swan	tombstone	washboard	
pion	rhinoceros	skirt	sweatband	tongue	washbowl	
pistol	ribbon	skittle	sweater	tonsil	wasp	
piston	rickshaw	skullcap	sweatshirt	tooth	wastebasket	
pitchfork	rifle	skunk	swimsuit	toothbrush	watch	
pizza	ring	skyjack	switchblade	toothpick	watchband	
plane	river	skylark	switchboard	tornado	watchdog	
planet	robe	skylight	sword	torpedo	waterfall	
platelet	robin	skyscraper	swordfish	torso	watermelon	
plowshare	rock	sled	swordtail	tortoise	wattle	
plum	rook	sledgehammer	sycamore	tortoiseshell	weasel	
polecat	rosary	sleeve	synapse	town	whale	
poncho	rose	sleigh	syringe	trachea	wharf	
pond	rosebud	slingshot	tablecloth	tractor	wheelchair	
pony	rosebush	sloop	tablespoon	train	whip	
poodle	roundworm	sloth	tadpole	tram	wick	
poppy	rowboat	slug	tamarind	tray	widgeon	
porch	rudder	snail	tambourine	treadmill	wiener	
porcupine	rutabaga	snake	tampon	tree	wigwam	
porpoise	sailboat	snapdragon	tanager	triceratops	wildcat	
possum	sailfish	snorkel	tapeworm	trilobite	windmill	
postcard	salamander	snowball	tapir	trombone	window	

9

Optimal Statistical Goals for Neural Networks Are Necessary, Important, and Practical

Richard M. Golden
University of Texas at Dallas

Richard Golden's chapter, **Optimal Statistical Goals for Neural Networks Are Necessary, Important, and Practical,** *provides a systematic mathematical theory, based on probability theory, for optimization in a wide variety of neural networks for pattern classification. The networks that Golden fits into his framework are the Hopfield, brain-state-in-a-box (see Chapter 16 by Abdi, Valentin, and O'Toole), harmony theory, and Boltzmann machine networks. (Editor's note: the Cohen-Grossberg network, which is a generalization of the Hopfield and has a global Lyapunov function, fits into this framework as well.)*

The framework Golden uses includes an activation update rule, a learning rule, and an objective function. It also includes a probability density function specifying the likelihood of a particular weight vector, and a probability density function specifying the likelihood of a particular activation vector given the weight vector. The network's probabilistic assumptions are interpreted in terms of **logically consistent** *(idealized) belief structure and therefore obey standard laws such as transitivity. The computational goal of the learning rule is to find a weight vector which is most probable given a sequence of training patterns presented to the network. One result of this theory is to determine the class of statistical environments that a given network can represent.*

In the discussion, Golden deals with several questions that are key to the overall ideas of this book. These include "Why assume models of human behavior should be rational?," "Does an energy function always exist?," and "What if the learning dynamics of a model are not optimal?" Golden recognizes deviations of human inference and decision making from strict rationality but seeks to explain them under Herbert Simon's rubric of "satisficing," which he interprets as follows: "humans try to achieve their computational goals but are limited by the algorithms that they are forced to use." This statement is a good description of a locally optimizing neural network whose actual behavior may not always be optimal for many reasons. Nonoptimal network behavior could result from interactions with other subnetworks that are also trying to optimize different objective functions, with neural networks in other humans that also have conflicting computational goals, and with a complex, nonstationary environmental context. The distinction between local and global optimization is a theme of many of this book's other chapters (e.g., DeYong, Leven, Levine, Öğmen/Prakash, and Werbos).

ABSTRACT

The description of a neural network's behavior in terms of its attempts to achieve specific computational goals is necessary to obtain a complete understanding of the network's behavior. A practical procedure for constructing the computational goal for a given neural network architecture is described. To illustrate the approach, specific computational goals for a variety of neural networks are then formulated as the solutions to a set of nonlinear statistical optimization problems. Some reasons for using a statistical formulation are that (a) rational decision makers should use statistical inference, and (b) a statistical formulation permits an objective evaluation of a neural network's computational goals using goodness-of-fit tests. Finally, some commonly asked questions about this approach are noted and answered.

1. INTRODUCTION

Marr (1982) proposed that at least three levels of description are necessary to understand the behavior of a complex information processing system. These three levels of description are the implementational theory level, the algorithmic theory level, and the computational theory level. The computational level of description is designed to specify the goal of the information processing task. Moreover, the *relevance* of the goal for the information processing task must also be specified at this level and its uniqueness must be justified. The algorithmic level of description is designed to specify the algorithms required to carry out the computations necessary to achieve the computational goal with sufficient detail so that the algorithm can be implemented as a computer program. Finally, the implementational level of description is designed to specify how the procedures at the algorithmic level are actually implemented.

An *artificial neural network* (Levine, 1991; McClelland, Rumelhart, & the PDP Research Group, 1986) is a collection of simple neuron-like computing *units*. Each unit has a real-valued state that is referred to as the unit's *activity level*. The list of the activity levels of all units in the model can be represented as a vector $\mathbf{X}$ which is referred to as the *activation pattern*. The *activation updating rule* or *classification dynamics* of the neural network indicates how the activation pattern over the units will change as a function of the current activation pattern and the current pattern of connections among the neuron-like computing units. The free parameters of the model which explicitly specify the interaction coefficients of the neural network model are usually referred to as the *connection weights* and can be represented as a weight vector $\mathbf{W}$. The *learning rule* specifies how the elements of the weight vector $\mathbf{W}$ are updated while the neural network model is learning.

Because artificial neural networks are complex information processing systems, Marr's (1982) theoretical framework is relevant for evaluating our current understanding of the behavior of these systems. The purpose of this chapter is to propose a particular formal theoretical framework for precisely and unambiguously expressing the computational goal of a neural network. This theoretical framework is expressed in terms of the language of statistics following proposals by a variety of researchers (Ackley, Hinton, & Sejnowski, 1985; Golden, 1988a, 1988b, 1988c; Marroquin, 1985; Smolensky, 1986; Tishby, Levin, & Solla, 1989; White, 1989).

First, the proposed probabilistic framework is defined and justified. Next, a useful method for the construction of a probabilistic computational level of description for neural networks is presented in conjunction with some sample applications of the theory. Finally, additional commonly asked questions about the approach are noted and answered.

2. A STATISTICAL COMPUTATIONAL FRAMEWORK FOR NEURAL NETWORKS

2.1. Formal Presentation of the Framework

A theoretical framework for the description of neural network models is now provided following the development of Golden (1988a, 1988b, 1988c, in press).

2.1.1. Implementational level of Description. The implementational level of description is expressed by explicitly writing down the dynamical systems associated with the activation updating and learning rules of the neural network model.

Activation Updating Rule. Let $\mathbf{W}$ be a constant vector indicating the connection weights and other free parameters of the neural network model. A neural network model is assumed to be defined by an *activation updating rule* which is explicitly specified by either a discrete time-invariant deterministic (or stochastic) dynamical system of the form:

$$\mathbf{X}(t + \delta t) = \mathbf{F}(\mathbf{X}(t), \mathbf{W}) + \mathbf{U}_t \qquad (1)$$

where $U_1, ..., U_t, ...$ is a zero-mean stochastic process indexed by t, or a continuous time-invariant deterministic dynamical system of the form:

$$d\mathbf{X}/dt = \mathbf{F}(\mathbf{X}(t), \mathbf{W}). \qquad (2)$$

Note that Eq. (1) may be updated only once, as in the activation updating rule of a multilayer feedforward backpropagation network (Rumelhart, Hinton, & Williams, 1986). Also note that the above definition of the activation updating rule is not applicable to neural networks where the learning takes place during the activation updating process. It may be possible to extend the proposed theoretical framework to handle this latter situation in specific cases, but the above theoretical framework is sufficient for handling enough interesting cases to illustrate the general characteristics of the theory.

Learning Rule. The learning rule describes how the weight vector $\mathbf{W}$ is adjusted based upon the network's experiences with its environment. Let the set $\tau_N = \{\mathbf{X}_1, \mathbf{X}_2, ..., \mathbf{X}_N\}$ be defined as the *training sequence*. To keep things simple, assume that the elements of τ_N are independently and identically distributed according to an unobservable *environmental probability distribution* p_e.

The general class of neural network learning rules that are considered are assumed to have one of the following forms. Neural network *batch* learning rules are specified by either discrete-time dynamical systems of the form

$$\mathbf{W}(t + \Delta t) = \mathbf{H}(\mathbf{W}(t), \tau_N) \tag{3}$$

or continuous-time dynamical systems of the form:

$$d\mathbf{W}/d\mathbf{t} = \mathbf{H}(\mathbf{W}, \tau_N). \tag{4}$$

Note that if the neural network is required to demonstrate adaptive learning, Eqs. (3) and (4) are not appropriate because they require that each weight update is functionally dependent on the neural network's past, present, and future experiences. For this reason, a third type of dynamical system for the weight update dynamics is now considered. An *online* neural network learning rule is specified by a discrete-time dynamical system of the following form:

$$\mathbf{W}(t + \Delta t) = \mathbf{H}(\mathbf{W}(t), \mathbf{X}_t) \tag{5}$$

where the stochastic process, $\mathbf{X}_1$, $\mathbf{X}_2$, ... is a sequence of independent and identically distributed random vectors whose distribution is given by the environmental probability distribution p_e. Thus, $\mathbf{X}_t$ is the tth activation pattern observed by the neural network learning rule during the training process. An online neural network learning rule can presumably "track" slowly changing (i.e., nonstationary) statistical environments, but the current version of the theoretical framework presented here does not directly deal with the nonstationary learning problem.

2.1.2. Algorithmic and Computational Levels of Description. Although the dynamical system representation of a neural network model is sufficient to specify the network's dynamics, the dynamical system representation is not an adequate description of the information-processing dynamics of the network model. The computational level of description specifies the information processing goals of the model, whereas the algorithmic level provides the necessary conceptual linkage between the implementational level and the computational level of description. That is, the algorithmic level of description provides a useful intermediate level of representation that explains how the implementation of the neural network is related to the achievement of the neural network's computational goals. Although all three levels of description (implementational, algorithmic, and computational) are equally important, this chapter focuses upon the computational level which is concerned with the optimal statistical goals of the neural network.

Computational Goal of the Activation Updating Rule. Let the activation pattern $\mathbf{X}$ be partitioned into a *response* subvector $\mathbf{R}$ and *stimulus* subvector $\mathbf{S}$. It is assumed that the elements of $\mathbf{S}$ are known, and that the elements of $\mathbf{R}$ are unknown and must be estimated. The computational goal of the activation updating rules in Eqs. (1) or (2) is to find a response pattern $\mathbf{R}^*$ which is most probable given stimulus $\mathbf{S}$, the current knowledge state of the neural network $\mathbf{W}$, and the network's belief structure (i.e., probabilistic assumptions). That is, the computational goal of Eq. (1) or (2) is to generate a sequence of response vectors $\mathbf{R}_1$, $\mathbf{R}_2$, ... that converges in some sense to a global maximum of a probability mass (or density) function p. The probability distribution or *belief structure* p indicates the neural network's belief that $\mathbf{R}$ will occur with probability (belief) $p(\mathbf{R}|\mathbf{S},\mathbf{W})$ given that the neural network has observed stimulus $\mathbf{S}$ in knowledge state $\mathbf{W}$.

Four important comments must be made at this point. First, the activation updating rule of the neural network is essentially optimal *by definition*, because a belief structure p is constructed so that the activation updating rule is seeking a global maximum of p. Second, the precise definition of p for a given neural network model is equivalent to a strong claim about the class of statistical environments that the neural network model is incapable of completely learning. Third, it is not obvious that a belief structure p can be found such that the activation updating dynamics for a given neural network architecture is seeking a global maximum of p. On the other hand, some guidelines for the construction of such belief structures are discussed later in this chapter. And fourth, the parametric form of the belief structure, p, *implicitly* incorporates knowledge of the network's biases, architectural assumptions, and activation updating rule. The sample space of p specifies the set of permissible activation patterns that are relevant to the neural network's *information processing dynamics*, and p also implicitly provides a "similarity metric" for judging the similarities and differences between activation patterns.

Computational goal of the learning rule. The computational goal of the classes of learning rules such as Eq. (3), (4), or (5) is to find a weight vector $\mathbf{W}^*$ that is most probable given a training sequence, τ_N, of N training patterns with respect to the probability distribution $p(\mathbf{X}|\mathbf{W})$. Or in other words, find a $\mathbf{W}$ that maximizes $p(\mathbf{W}|\tau_N)$. This goal can be achieved if it can be proved that Eq. (3) and (4) is an optimization algorithm which is seeking a $\mathbf{W}^*$ that is a global minimum of the *cross-entropy* or Kullback-Leibler information criterion (Kullback & Leibler, 1951; White, 1989), $E_N(\mathbf{W})$.

A heuristic derivation of the error function $E_N(\mathbf{W})$ is now provided in order to construct an explicit relationship between $E_N(\mathbf{W})$ and $p(\mathbf{X}|\mathbf{W})$. Let $p(\mathbf{W})$ be a probability density function that specifies the likelihood of a particular weight vector before the training sequence, τ_N, has been observed. Let $p(\tau_N)$ be the likelihood of a particular training sequence consisting of the N training stimuli observed by the network. These N training stimuli are denoted by X_1, X_2, ..., X_N. Note that $p(\tau_N)$ is not functionally dependent on $\mathbf{W}$. Then the likelihood of a given weight vector $\mathbf{W}$ given τ_N is expressed by

$$p(\mathbf{W}|\tau_N) = p(\tau_N|\mathbf{W})\, p(\mathbf{W}) \,/\, p(\tau_N) \qquad (6)$$

where

$$p(\tau_N|\mathbf{W}) = p(\mathbf{X}_1|\mathbf{W})\, p(\mathbf{X}_2|\mathbf{W})\, p(\mathbf{X}_3|\mathbf{W}) \cdots p(\mathbf{X}_N|\mathbf{W})$$

because $\mathbf{X}_1$, $\mathbf{X}_2$, ..., $\mathbf{X}_N$ are independently and identically distributed.

Unfortunately, $p(\mathbf{W}|\tau_N)$ does not uniformly converge to a fixed function of $\mathbf{W}$ as the number of elements, N, in the training sequence, τ_N, is increased. To address this problem, define

$$E_N(\mathbf{W}) = -(1/N) \log [p(\mathbf{W}|\tau_N)] + K \qquad (7)$$

where K is a constant that is not functionally dependent on $\mathbf{W}$ so that $E_N(\mathbf{W})$ is a monotonically decreasing function of $\mathbf{W}$. The minima of $E_N(\mathbf{W})$ are thus equivalent to the maxima of $p(\mathbf{W}|\tau_N)$. Now, substituting Eq. (6) into Eq. (7), the following expression is obtained:

$$E_N(\mathbf{W}) = -(1/N) \log [p(\mathbf{W})] - (1/N) \Sigma_i \log[p(\mathbf{X}_i|\mathbf{W})] + K_1 \qquad (8)$$

where K_1 is a constant that is not functionally dependent on $\mathbf{W}$ and the summation ranges from $i=1$ to $i=N$. If $p(\mathbf{W})$ has a strict lower bound when $\mathbf{W}$ is sufficiently close to the set of global minima of $-\log [p(\mathbf{W}|\tau_N)]$ (this assumption can be considerably relaxed), then as N becomes large the $-(1/N) \log [p(\mathbf{W})]$ term in (8) will approach zero. Thus, for sufficiently large sets of training patterns the error function is given by the log-likelihood function, $E_N(\mathbf{W})$, which converges to the cross-entropy or Kullback-Leibler Information Criterion (KLIC) (Kullback & Leibler, 1951; White, 1982, 1989) as the length, N, of the training sequence increases. That is,

$$E_N(\mathbf{W}) = -(1/N) \Sigma_i \log[p(\mathbf{X}_i|\mathbf{W})] \qquad (9)$$

where the summation ranges from $i=1$ to $i=N$.

3. JUSTIFICATION FOR A PROBABILISTIC REPRESENTATION

3.1. Neural Networks Are Inductive and Not Deductive Machines

Most researchers in the field of neural networks would not consider a network of logic gates in the CPU of their computer to be a neural network. But of course, such a network is formally equivalent to a network of McCulloch-Pitts formal neurons (McCulloch & Pitts, 1943). One reason that neural network researchers are reluctant to refer to logic gate networks as neural networks is that logic gates are solving deductive logic inference problems using the Boolean algebra instead of inductive logic problems. Most researchers would agree that a distinguishing characteristic of neural networks is their ability to generalize from experience and extract the "regularities" from their environment. Or, in other words, the computational goal of neural networks is to solve the inductive logic problem where the network is forced to make appropriate generalizations from its experiences in the world.

It should be emphasized at this point that because inductive logic necessarily entails "going beyond" the data, the criteria for what constitutes a "good" inductive inference must be carefully considered. For example, Cox (1946) presented a very interesting argument concerned with the justification of probabilistic inference. Cox (1946) proved an assertion of the following type. If (a) the belief in a response $\mathbf{R}$ given a stimulus $\mathbf{S}$ can be represented as a real-valued function of $\mathbf{S}$ and (b) the computation of new beliefs from old beliefs is assumed to be consistent with the deductive logic (i.e., the Boolean algebra), then without any loss in generality the belief of $\mathbf{R}$ given $\mathbf{S}$ can be represented as the conditional probability of $\mathbf{R}$ given $\mathbf{S}$. Similar arguments for using probability theory as an inductive logic that specifies the computational goals of a rational decision maker have been made by other researchers, including von Neumann and Morgenstern (1944), Ramsey (1988), and Savage (1954).

Note that the "representation" of the belief of **R** given **S** as a conditional probability means that the following axioms of probability theory must be satisfied. Let Ω be a set referred to as the *sample space* of all possible elementary events (i.e., responses) for a given *S*. To keep the discussion simple, assume that Ω contains a finite number of elements. Now, let events *R* and **Q** be events in Ω. If:

1. $0 \leq p(\mathbf{R}|\mathbf{S}) \quad \forall \mathbf{R} \in \Omega$,
2. $p(\Omega|\mathbf{S}) = 1$,
3. $p(\mathbf{R} \cup \mathbf{Q}|\mathbf{S}) = p(\mathbf{R}|\mathbf{S}) + p(\mathbf{Q}|\mathbf{S})$ if $\mathbf{R} \cap \mathbf{Q} = \varnothing$,

then $p(\mathbf{R}|\mathbf{S})$ is the *conditional probability* of **R** given **S**.

Any system of inductive logic that calculates and represents belief according to the above three axioms is equivalent to the probability theory or is logically inconsistent with respect to the above three axioms. To provide some additional insights into the plausibility of the above three axioms, these axioms of probability theory are now presented as axioms of rational decision making. To simplify the presentation, slightly "stronger" axioms of probability theory will be presented than is required. Golden (in press) provides a useful introduction to these axioms from the perspective of fuzzy measure theory on crisp sets.

1. Axiom 1: $0 \leq p(\mathbf{R}|\mathbf{S}) \leq 1 \quad \forall \mathbf{R} \in \Omega$. This axiom states that the decision maker's "belief" that a particular event will occur can be represented as a real number between 0 (event will not occur) and 1 (event will occur). Intermediate belief values represent intermediate degrees of belief. Thus, a belief of 0.5 for a particular event (i.e., $p(\mathbf{R}|\mathbf{S}) = 0.5$) indicates the decision maker is completely uncertain about the occurrence of the event. This assumption implies that (a) the decision maker assigns a belief to every event which he, she, or it believes could occur in the environment, (b) the beliefs of the decision maker can always be rank-ordered, and (c) belief values have maximum and minimum values.

2. Axiom 2: $p(\Omega|\mathbf{S}) = 1$. The decision maker assumes that at least one event in the sample space will occur in its environment.

3. Axiom 3: $p(\mathbf{R} \cup \mathbf{Q}|\mathbf{S}) = p(\mathbf{R}|\mathbf{S}) + p(\mathbf{Q}|\mathbf{S})$ if $\mathbf{R} \cap \mathbf{Q} = \varnothing$. This assumption states that the beliefs associated with mutually exclusive events combine in an additive manner. Although the additivity assumption is arbitrary, the additive method of computing beliefs has the following desirable properties. Suppose events **R** and **Q** are mutually exclusive (i.e., either **R** or **Q** can occur but events **R** and **Q** can not simultaneously occur). First, since the belief measure is nonnegative, the belief that either event **R** or event **Q** will occur is always greater than or equal to the belief that event **R** will occur (monotonicity). Second, the order of combination of beliefs is irrelevant (commutativity and associativity).

3.2. Objective Statistical Tests of Generalization Properties

Consider a rational decision maker who has a belief $p(\mathbf{X}|\mathbf{W})$ that one of a finite number of outcomes, $\mathbf{X}$, will occur in the environment. This belief is functionally dependent on the decision maker's knowledge state $\mathbf{W}$. Thus, different values of $\mathbf{W}$ will cause the decision maker to change his, her, or its beliefs about the likelihood of occurrence of the outcome $\mathbf{X}$. Let $p_e(\mathbf{X})$ indicate the relative frequency that event $\mathbf{X}$ will occur in the decision maker's environment. Then, a necessary but not sufficient condition for correct generalization is that a $\mathbf{W}$ exists such that:

$$p(\mathbf{X}|\mathbf{W}) = p_e(\mathbf{X}) \quad \forall \mathbf{X} \in \Omega \tag{10}$$

where Ω is the sample space. If this condition is not satisfied, then statisticians say that the parameterized probability model (belief structure) $p(\cdot|\mathbf{W})$ is *misspecified* with respect to the environmental distribution p_e. A similar definition can be developed for probability density functions.

This definition of generalization is straightforward, insightful, and useful. Given that the belief structure, p, of a particular neural network model is known, the class of statistical environments for which that neural network can make correct generalizations and inferences is explicitly specified. Moreover, goodness-of-fit tests such as White's Information Matrix Test (White, 1982) or the chi-squared goodness-of-fit test (Manoukian, 1986, pp. 86-88) may be used to test for the presence of model misspecification. Such statistical tests are based on comparing the model's predicted probability of event $\mathbf{X}$ with respect to the estimated relative frequency of event $\mathbf{X}$ in the model's statistical environment. Objective statistical tests for deciding which of several alternative neural network (i.e., statistical) models "best fit" a given statistical environment can also be developed (Vuong, 1989; see Golden, 1995, in press). Thus, the construction of p provides a compact summary of the class of statistical environments that the neural network model will never completely learn!

4. CONSTRUCTION OF A PROBABILISTIC COMPUTATIONAL DESCRIPTION

In this section, a particular method of providing a given neural network model architecture with a probabilistic computational level of description is suggested following Golden (1988a, 1988b, 1988c, in press). The reason for proposing this method is not to argue that this is the only correct method for constructing such probabilistic computational level descriptions. Rather, the reason for providing this procedure is to illustrate a useful procedure for constructing probabilistic computational goals for a large class of artificial neural network architectures.

The method consists of two steps and is based on the development of Golden (1988a, 1988b; see Golden, in press, for a review). The end result of the method is an explicit parametric probability mass (or density) function that the neural network model uses to find a "most probable" activation pattern as a consequence of the activation updating rule as in Eq. (1) or (2). Once this probability distribution is constructed, the computational goal of the learning process as specified by Eq. (3), (4), or (5) is determined. In other words, the neural network model's probabilistic representation is intentionally constructed so that the neural network model's activation updating rule is trying to make "optimal" inferences. Then, the computational adequacy of the learning process is evaluated using the probabilistic representation that was assumed for the classification (i.e., testing as opposed to training) process.

4.1. Construction of a Computational Goal for Classification

The first step consists of constructing an energy function, $V(X;W)$, that has the property that the activation updating rule in Eq. (1) or (2) can be viewed as a heuristic optimization algorithm that is seeking a global minimum of $V(X;W)$ for some fixed constant weight vector W. For example, a multilayer feedforward back-propagation neural network (Rumelhart et al., 1986) produces an activation pattern $R = \phi(S;W)$ for a given input activation pattern S and a weight vector W. Thus, a possible energy function for this neural network would be

$$V(R,S;W) = (1/2)|R - \phi(S;W)|^2. \tag{11}$$

As a second example, the activation updating rules of the Brain-State-in-a-Box neural network model (Anderson, Silverstein, Ritz, & Jones, 1977; Golden, 1986, 1993), the Hopfield (1982) model, the Boltzmann machine, and Harmony theory may be viewed as heuristic algorithms for seeking a global minimum of $V(X;W)$ with

$$V(X;W) = -(1/2) \, X^T W X \tag{12}$$

where W is a symmetric matrix whose ij^{th} element indicates the "connection strength value" between units i and j of the network.

The second step consists of constructing a probability mass (or density) function, p, that is a monotonically decreasing function of the energy function V. That is, find a function g whose domain and range are the real numbers such that $p = g(V)$ where: (a) $g(x+k) < g(x)$ for all positive k, and (b) p is a valid probability mass function with respect to some sample space Ω.

The reason why p must be a monotonically decreasing function of V is that the activation updating rule Eq. (1) or (2) will then be searching for an activation pattern X^* that is most probable with respect to p. This assumption is referred to as the *optimal classification dynamics* assumption (Golden, in press). Because the range of p is the closed interval $[0,1]$ and the sum (or integral) of p over all disjoint subsets of the sample space Ω must be equal to one, it is usually convenient to construct p using the following approach. For the case where p assigns a probability mass, $p(X|W)$, to each X in sample space Ω, Golden (1988a, 1988b, 1988c, in press) suggested:

$$p(X|W) = (1/Z) \exp[-V(X;W)] \quad \text{where} \quad Z = \Sigma_Y \exp[-V(Y;W)]. \tag{13}$$

For the case where p is a probability density function, the normalization constant Z is computed by integrating as opposed to summing over the sample space Ω as follows:

$$p(\mathbf{X}|\mathbf{W}) = (1/Z) \exp[-V(\mathbf{X};\mathbf{W})] \quad \text{where} \quad Z = \int_{\mathbf{Y}}\exp[-V(\mathbf{Y};\mathbf{W})] \, d\mathbf{Y}. \tag{14}$$

Golden (1988a, 1988b, 1988c), following Smolensky (1986), suggested a general argument regarding the uniqueness of the above mapping given certain restrictions on the parametric form of $V(\mathbf{X};\mathbf{W})$. The probability distribution constructions in (13) and (14) can also be motivated using a Markov random field framework (Besag, 1974; Marroquin, 1985; also see Tishby et al., 1989, for a statistical mechanics perspective).

To illustrate the procedure describing how p can be constructed from V, substitution of Eq. (11) into Eq. (14) with $\mathbf{X} = (\mathbf{R},\mathbf{S})$ and the assumption that the sample space is a real vector space with dimensionality equal to the dimension of $\mathbf{R}$ yields the following belief function for a multilayer neural network:

$$p(\mathbf{R}|\mathbf{S};\mathbf{W}) = (1/Z) \exp\left(-(1/2)|\mathbf{R} - \phi(\mathbf{S};\mathbf{W})|^2\right) \tag{15}$$

which is a multivariate Gaussian density with mean vector $\phi(\mathbf{S};\mathbf{W})$ and covariance matrix equal to the identity matrix. Note that if the sample space of Eq. (15) was not a real vector space but was limited to vectors in a binary vector space, then $E_N(\mathbf{W})$ would be mathematically intractable because Z would be functionally dependent on $\mathbf{W}$. On the other hand, if the sample space of Eq. (15) is a binary vector space, then a computationally tractable belief function is given by using a different energy function to construct $p(\mathbf{X}^i|\mathbf{W})$ which when substituted into Eq. (13) yields the cross-entropy error learning function (Golden, 1988c). Thus, the construction of an appropriate probabilistic computational goal for classification is based on (a) the appropriateness of the energy function $V(\mathbf{X};\mathbf{W})$ as a distance metric for classification (see Golden, 1988c, for additional discussion), (b) the sample space of the data generating process which the neural network is attempting to represent, and (c) the computational tractability of the belief function.

4.2. Construction of a Computational Goal for Learning

Within the framework proposed by Golden (1988a, 1988b, 1988c), the probabilistic computational goal for learning is specified once the probabilistic goal for the classification dynamics of the neural network is specified. As discussed in the previous section of this chapter, the goal of learning is to choose the *most probable* value of $\mathbf{W}$ with respect to the network's belief structure p. The error function, $E_N(\mathbf{W})$, in Eq. (9) obtains its global minima at precisely those values of $\mathbf{W}$ which are most probable with respect to p for N sufficiently large. For example, substituting the belief function p in Eq. (15), for a multi-layer feedforward network with continuous-valued targets, into Eq. (9) yields:

$$E_N(\mathbf{W}) = C + (1/N) (1/2) \Sigma_i | \mathbf{R}^i - \phi(\mathbf{S}^i;\mathbf{W})|^2 \tag{16}$$

where C is a constant that is not functionally dependent on $\mathbf{W}$ and the summation index i ranges from 1 to N. The error function in (16) is typically minimized by the back-propagation learning algorithm (see Rumelhart et al., 1986, for additional details) during the learning process. Thus, the classification dynamics of the back-propagation network are optimal in the sense that the network is seeking the most probable response **R** given **S** and **W** with respect to (15). In addition, the learning dynamics of the back-propagation network are optimal in the sense that the network is seeking the most probable set of weights, **W**, with respect to (15). Finally, the parametric form of (15) explicitly indicates which class of statistical environments can never be learned by the back-propagation learning algorithm.

As a second example, suppose the energy function has the form of Eq. (12) as in the Hopfield (1982), Brain-State-in-a-Box (Anderson et al., 1977; also see Golden, 1986, 1993, for an analysis of the Brain-State-in-a-Box model), Harmony theory (Geman & Geman, 1984; Smolensky, 1986), Boltzmann machine (Geman & Geman, 1984; Ackley et al., 1985) neural networks. Substitution of (12) into (13), and then (13) into (9) leads to the error learning cost function:

$$E_N(\mathbf{W}) = -(1/2)(1/N) \, \Sigma_i \, [\mathbf{X}_i]^T \mathbf{W} \mathbf{X}_i + \log[Z(\mathbf{W})] \tag{17}$$

where Z is a function of **W**. Connectionist gradient descent algorithms for minimizing this error learning cost function were described by Smolensky (1986) and Ackley et al. (1985).

Thus, despite the seemingly great differences in the nature of the Hopfield (1982), Harmony theory, Brain-State-in-a-Box, and Boltzmann machine neural network models, all of these network models have the same computational goals for classification and learning. True, it can be shown that the Harmony theory model and Boltzmann machine model under certain conditions can find a global maximum of (13) using energy function (12) more effectively than the deterministic neural models, but all of these models will be unable to internally represent the same statistical environments.

5. COMMONLY ASKED QUESTIONS ABOUT THE APPROACH

5.1. Why Assume Models of Human Behavior Should Be Rational?

With respect to the proposed theoretical framework, neural network models are viewed as "heuristic" algorithms that attempt to achieve their computational goals. Simon's (1979) view of satisficing algorithms is similar to the viewpoint proposed here. The pattern of errors associated with the network model can then be interpreted as a set of "clues" regarding the specific heuristics used by the model to achieve its goals. If one did not assume the computational level of description was consistent with the goals of a rational decision maker, then a theory of why the model's behavior yields "reasonably correct" answers would not be available. For example, it is well known that in certain situations people will violate the "transitivity" axiom. It is easy to imagine an individual who states that they prefer: (i) peach ice cream to strawberry ice cream, (ii) strawberry ice cream to vanilla ice cream, and (iii) vanilla ice cream to peach ice cream! The transitivity axiom is a fundamental assumption of probability theory and directly follows from the assumption that beliefs can be represented by real numbers and thus "ordered" along a single dimension. Although such violations of the transitivity axiom seem to clearly indicate that a probabilistic specification of computational goals is not appropriate for people or mathematical models of human behavior, it is worthwhile to carefully consider and assess the situation before one "throws the baby out with the bath water."

True, people *do* make inferences that violate the transitivity axiom. On the other hand, these inferences are made relatively rarely and only in certain contrived situations. Consider the problem of walking through a door. This is a difficult motor control problem that is still not well understood. An intelligent system has decided that (a) it is more likely that the door is directly ahead than that the door is directly behind the robot, and (b) it is more likely that the door is directly behind the robot than above the robot. It seems that a good "engineering design" would be to build in a transitivity axiom so that it could make the inference that (c) it is more likely that the door is directly ahead than above the robot.

It is very difficult to imagine that animal perceptual and cognitive processing systems that are capable of making inferences far beyond the capabilities of modern computer systems could work so well on a fundamentally flawed design. Instead, it seems more reasonable to view human behavior as "satisficing" following Simon (1979). That is, humans try to achieve their computational goals but are limited by the algorithms that they are forced to use. From this perspective, it is quite appropriate to assume the existence of rational computational goals. And finally, if we assume humans are using heuristic algorithms to achieve their idealized computational goals, then models of human behavior should use heuristic algorithms designed to achieve those same idealistic goals.

5.2. Does an Energy Function Always Exist?

The existence of an energy function that is minimized by the activation updating rule is not guaranteed. On the other hand, the claim is made that the construction of such an energy function that is minimized by the activation updating rule in some sense is usually possible. In fact, appropriate energy functions for almost every neural network model in the literature can be constructed (see Golden, 1988a, 1988b, 1988c, in press, for a review).

5.3. Why Use a Probabilistic Interpretation if Learning is Ignored?

It is sometimes argued that if the activation updating rule of a particular neural network such as the Brain-State-in-a-Box (BSB) model is known to be minimizing an energy function and the weights of the model are known, a probabilistic computational level of description is not required. One could simply state that the computational goal of the BSB model is to converge to an attractor that represents a particular category or that the computational goal of the BSB model is to attempt to find a global minimum of its energy function.

Assuming that one accepts the assumption that the computational goal of a neural network is inductive in nature, then the above computational goals are inadequate because their relationships to the problem of inductive inference have not been explicitly identified. Moreover, if a probability distribution is not constructed for the neural network model, then a precise statement regarding the suitability of a particular neural network model architecture for a particular statistical environment cannot be made.

5.4. Is the Probabilistic Interpretation Unique?

Let $p = G(v)$ where v is an energy function and p is the model's belief function. Then, within the framework of the theory proposed here, any function G such that the minima of v correspond to the maxima of p can be used provided that p is a valid probability mass (or density) function. Moreover, some neural network models may possess multiple energy functions! Thus, from this perspective, the probabilistic interpretation is not unique.

On the other hand, suppose that one considers the following alternative definition of a neural network model. Let a neural network model be a triplet: (L,A,p) where L is the learning rule, A is the activation updating rule, and p is the model's belief function, which is maximized in some sense by the algorithm A. From this perspective, the probabilistic interpretation is unique because the probabilistic interpretation is a fundamental component of the model! The theoretical framework proposed here assumes that the specification of a neural network model is not complete until the neural network model engineer explicitly identifies the class of statistical environments that are consistent with the neural network's classification and learning dynamics.

5.5. Why Restrict the Above Framework to Map Estimation?

MAP (maximum a posteriori) estimation of a response vector $\mathbf{R}$ given a stimulus vector means choosing that response vector $\mathbf{R}$ which is a global maximum of some conditional probability distribution $p(\cdot|\mathbf{S})$. The theoretical framework presented here was developed within a MAP estimation framework for expository reasons. A more general theoretical development involving expected utility functions could be developed in a straightforward manner. That is, it is assumed that in addition to having a representation of uncertainty given by a probability distribution $p(\cdot|\mathbf{S})$, the decision maker also possesses a representation of utility (i.e., subjective cost of making a particular response $\mathbf{R}$). Let the decision maker's cost of choosing $\mathbf{R}_l$ when the "correct" response was $\mathbf{R}_k$ be given by $U(\mathbf{R}_l, \mathbf{R}_k)$. Then, the decision maker seeks a response $\mathbf{R}_l$ such that the function:

$$V(\mathbf{R}_l) = \Sigma_k \, U(\mathbf{R}_l, \mathbf{R}_k) \, p(\mathbf{R}_k|\mathbf{S})$$

is minimized. Note that if $U(\mathbf{R}_l, \mathbf{R}_k) = 1$ for $\mathbf{R}_l \neq \mathbf{R}_k$ and $U(\mathbf{R}_l, \mathbf{R}_k) = 0$ for $\mathbf{R}_l = \mathbf{R}_k$, then selecting the $\mathbf{R}_l$ so that $V(\mathbf{R}_l)$ is minimized is formally equivalent to MAP estimation. Von Neumann and Morgenstern (1944) and Savage (1954) have argued that minimizing an expected utility function is the optimal strategy for a certain class of rational decision makers who are embedded in an environment characterized by uncertainty.

5.6. What if the Learning Dynamics of a Model Are Not Optimal?

The proposed theoretical framework prescribes the construction of a belief structure p such that the classification dynamics of the neural network are optimal. That is, so that the neural network is seeking a global maximum of p. If one decides the goal of the learning process is to find the most probable set of weights $\mathbf{W}$, then the optimal cost function for learning is determined by p. It is also quite likely that the learning rule will not be minimizing the optimal cost function for learning. For example, it is not obvious that the Hebbian learning rule suggested by Anderson et al. (1977) and Hopfield (1982) is a good learning rule in general for searching for the global minima of Eq. (14).

On the other hand, this type of learning rule might be viewed as optimal or suboptimal for certain restricted classes of statistical environments. The proposed framework *does not* recommend "throwing out" learning rules such as the Hebbian learning rule simply because they are nonoptimal. Instead, the proposed framework simply asks that researchers recognize the suboptimality of such learning rules with respect to a given belief structure p.

5.7. What About Fuzzy Measures?

If one accepts the axioms of fuzzy measure theory as constraints on the behavior of a rational decision maker, then fuzzy measures may be used as a language to express the computational goal of a neural network. Probability measures may be viewed as special cases of "fuzzy" measures (for a review see Golden, in press, and Klir and Folger, 1988, Chapter 4). Although such generality may seem initially desirable, the generality of fuzzy measures may not provide a sufficient number of constraints on the inductive decision making process.

For example, if probability theory is used as a theory of belief, then $p(A) = 1-p(\neg A)$ where $\neg A$ denotes the complement of the set A. That is, as one's belief that A will not occur decreases, then one's belief that A will occur must increase. On the other hand, fuzzy measure theory only requires that: $b(A) \leq 1-b(\neg A)$ where b designates a *fuzzy measure* as opposed to a probability measure p. Thus, it is possible to change one's belief that A will not occur without affecting one's belief that A will occur!

6. SUMMARY AND CONCLUSIONS

This chapter began by arguing that neural network models are complex information processing systems. Because neural network models are complex information processing systems, Marr's (1982) computational level of description for understanding such systems is relevant to the analysis of neural networks. Accordingly, the description of a neural network's behavior in terms of its attempts to achieve specific computational goals is necessary for understanding the behavior of neural networks according to Marr (1982). Specific computational goals for neural networks were then formulated as the solutions to specific nonlinear statistical optimization problems. Some reasons for using a statistical formulation were that (a) rational decision makers should use statistical inference, and (b) a statistical formulation permits an objective evaluation of a neural network's computational goals using goodness-of-fit tests. Finally, some commonly asked questions about this approach were noted and answered.

REFERENCES

Ackley, D. H., Hinton, G. E., & Sejnowski, T. J. (1985). A learning algorithm for Boltzmann machines. *Cognitive Science*, **9**, 147-169.

Anderson, J. A., Silverstein, J. W., Ritz, S. A., & Jones, R. S. (1977). Distinctive features, categorical perception, and probability learning: Some applications of a neural model. *Psychological Review*, **84**, 413-451.

Besag, J. (1974). Spatial interaction and the statistical analysis of lattice systems. *Journal of the Royal Statistical Society, Series B*, *36*, 192-236.

Cox, R. T. (1946). Probability, frequency, and reasonable expectation. *American Journal of Physics*, **14**, 1-13.

Geman, S., & Geman, D. (1984). Stochastic relaxation, Gibbs distributions, and the Bayesian restoration of images. *IEEE Transactions, PAMI*, **6**, 721-741.

Golden, R. M. (1986). The brain-state-in-a-box neural model is a gradient descent algorithm. *Journal of Mathematical Psychology*, **30**, 73-80.

Golden, R. M. (1988a). Probabilistic characterization of neural model computations. In D. Z. Anderson (Ed.), *Neural Networks and Information Processing* (pp. 310-316). New York: American Institute of Physics.

Golden, R. M. (1988b). Relating neural networks to traditional engineering approaches. *Proceedings of the Artificial Intelligence and Advanced Computer Technology Conference*, 255-260. Glen Ellyn, IL: Tower Conference Management Company.

Golden, R. M. (1988c). A unified framework for connectionist systems. *Biological Cybernetics*, **59**, 109-120.

Golden, R. M. (1993). Stability and optimization analyses of the generalized brain-state-in-a-box neural network model. *Journal of Mathematical Psychology*, **37**, 282-298.

Golden, R. M. (1995). Making correct statistical inferences using a wrong probability model. *Journal of Mathematical Psychology*, **38**, 3-20.

Golden, R. M. (in press). *Mathematical Methods for Neural Network Analysis and Design*. Cambridge, MA: MIT Press.

Hopfield, J. J. (1982). Neural networks and physical systems with emergent collective computational abilities. *Proceedings of the National Academy of Sciences, USA*, **79**, 2554-2558.

Klir, G. J., & Folger, T. A. (1988). *Fuzzy Sets, Uncertainty, and Information*. Englewood Cliffs, NJ: Prentice Hall.

Kullback, S., & Leibler, R. A. (1951). On information and sufficiency. *Annals of Mathematical Statistics*, **22**, 79-86.

Levine, D. S. (1991). *Introduction to Neural and Cognitive Modeling*. Hillsdale, NJ: Lawrence Erlbaum Associates.

Manoukian, E. B. (1986). *Modern Concepts and Theorems of Mathematical Statistics*. New York: Springer-Verlag.

Marr, D. (1982). *Vision*. San Francisco: W. H. Freeman.

Marroquin, J. L. (1985). *Probabilistic solution of inverse problems* (A.I. Memo No. 860). Cambridge, MA: MIT Press.

McClelland, J. L., Rumelhart, D. E., & the PDP Research Group (1986). *Parallel Distributed Processing. Volume 2: Psychological and Biological Models*. Cambridge, MA: MIT Press.

McCulloch, W. S., & Pitts, W. (1943). A logical calculus of the ideas immanent in nervous activity. *Bulletin of Mathematical Biophysics*, **5**, 115-133.

Ramsey, F. P. (1988). Truth and probability. In P. Gardenfors & N. Sahlin (Eds.), *Decision, Probability, and Utility: Selected Readings* (pp. 19-47). Cambridge, UK: Cambridge University Press.

Rumelhart, D. E., Hinton, G. E., & Williams, R. J. (1986). Learning internal representations by error propagation. In D. E. Rumelhart & J. L. McClelland (Eds.), *Parallel Distributed Processing. Volume 1: Foundations* (pp. 318-362). Cambridge, MA: MIT Press.

Savage, L. J. (1954). *The Foundations of Statistics*. New York: Dover Publications.

Simon, H. (1979). A behavioral model of rational choice (1955). In H. Simon (Ed.), *Models of Thought* (pp. 7-19). New Haven, CT: Yale University Press.

Smolensky, P. (1986). Information processing in dynamical systems: Foundations of harmony theory. In D. E. Rumelhart & J. L. McClelland (Eds.), *Parallel Distributed Processing. Volume 1: Foundations* (pp. 194-281). Cambridge, MA: MIT Press.

Tishby, N., Levin, E., & Solla, S. (1989). Consistent inference of probabilities in layered networks: predictions and generalization. *Proceedings of the International Joint Conference on Neural Networks* (Vol. 2, pp. 403-409). Piscataway, NJ: IEEE.

von Neumann, J., & Morgenstern, O. (1944). *Theory of Games and Economic Behavior*. Princeton, NJ: Princeton University Press.

Vuong, Q. H. (1989). Likelihood ratio tests for model selection and non-nested hypotheses. *Econometrica*, **57**, 307-333.

White, H. (1982). Maximum likelihood estimation of misspecified models. *Econometrica*, **50**, 1-25.

White, H. (1989). Learning in artificial neural networks: A statistical perspective. *Neural Computation*, **1**, 425-464.

10

Rule Induction and Mapping Completion in Neural Networks

Graham D. Tattersall
University of East Anglia

*Graham Tattersall's chapter, **Rule Induction and Mapping Completion in Neural Networks**, deals with the problem of optimal generalization. The framework in which he studies this problem is feedforward networks such as the multilayer perceptron or radial basis function net. He develops a mathematical method for computing the most statistically likely generalization, and shows that this can be achieved using sinusoidal nonlinear functions.*

The main operation of this type of neural network can be interpreted as learning a mathematical function based on a few training examples. It is desired to have a function of a larger class of data points that not only extends the original function but has the same properties as the original function. This can be done for both continuous and discrete valued functions, but the theory is most fully developed in this chapter for discrete (i.e., Boolean) functions. The statistical properties are, broadly speaking, understood in terms of using a discrete version of the Fourier transform to break up a function into spectral components and thereby extend it to a function that adds as few other spectral components as possible.

*The Fourier type networks proposed here use activation functions that oscillate and hence do not increase monotonically with input strength; for that reason they are not likely to be realistic models of **biological** neural networks. They are likely to have many applications, however, in **artificial** neural networks for both industrial and financial applications, such as the determination of credit worthiness that Tattersall mentions. The concern for finding optimal statistical properties of an artificial network is shared by Golden's chapter in this book.*

ABSTRACT

The problem of generalisation of logical functions is addressed. It is argued that instead of arbitrarily choosing a particular neural net to perform generalisation, the statistically most likely generalisation should be computed and it appears that this computation can be performed by a class of feedforward ANNs using sinusoidal non-linearities. The ANNs are called the *self-organising perceptron* (SOP) and *Fourier multilayer perceptron* (FMLP), and are able to correctly generalise with data on which conventional ANNs fail. The justification for using the SOP and FMLP is based on an analysis of the amount of structure in a function and a reformulation of the Shannon-Hartley Law to show that they generalise in the statistically most likely way.

1. INTRODUCTION

One of the most important functions of neural networks is to generalise from a sparse set of input-output examples of some process and produce a useful output value estimate when given a previously unseen input. The process of generalisation consists of completing the input-output mapping or, in other words, inducing a rule which is consistent with the given training examples and which can then be applied to previously unseen input arguments.

The process of generalisation is central to pattern recognition and rule induction in artificial intelligence. In both application areas, generalisation makes it possible to work without having a complete "lookup table" of all input pattern — output decision pairs. More formally the process of generalisation can be viewed as completely defining a vector transformation or mapping rule denoted by f such that $[y] = f[x]$, where $[x]$ is an input pattern or set of logical attributes and $[y]$ is a pattern classification or decision code. Generalisation involves finding $f[x]$ from a set of examples of $[y]$ and $[x]$. This type of vector mapping is exactly the function provided by artificial neural networks (ANNs) in which an input pattern $[x]$ is mapped to an output vector $[y]$. In ANNs the vectors $[x]$ and $[y]$ are intrinsically continuously valued but there is no reason why they should not represent logical attributes and in this case the network potentially behaves as a rule system.

Many attempts have been made using ANNs to learn the rules underlying sets of data in the hope that the net would then be able to generalise correctly on previously unseen inputs. Most commonly, the ANN is a sigmoidal multilayer perceptron (MLP) or radial basis function (RBF) as described by Lowe (1989), and to a limited extent these experiments have been successful as reported, for example, by Scalia et al. (1989).

Experiments by Tattersall and Foster (1989) suggest that the MLP and RBF are interpolative systems which can only successfully generalise on mappings whose output values change smoothly as the input argument is changed. This property is often well matched to mappings of data derived from the physical world. Such functions often exhibit smoothness, because the mechanisms underlying data production are continuous and have properties such as mass, thermal capacity, and electrical capacitance which lead to inertia and hence functions that change relatively slowly. These kinds of function are often said to contain "first-order structure" because of the high correlation between function values at adjacent points in their domains.

The successful generalisation of physically derived data with first order structure using MLPs and RBFs has led researchers to use them to try to obtain generalisation with symbolic or logical data. Unfortunately, the functions underlying this type of data are not governed by physical inertia, so there is no reason to expect them to be smooth. Consequently, nets which generalise by interpolation will often incorrectly complete these functions because they lack first-order structure. However, the functions may contain "higher order" or "hidden" structure. This means that the values of the function at nonadjacent points in its domain are correlated, and in principle, generalisation should be possible by appropriately exploiting the correlation.

A simple example of incorrect generalisation by an interpolating net such as the MLP is given by the parity function, which contains hidden structure but no first-order structure. This function is perfectly regular, and missing values can be predicted by a human quite easily after examining a few examples of the function. However, the function value does not change smoothly as the input argument is changed, so an MLP is incapable of correctly generalising to unseen arguments of the function, although it can easily learn given examples.

Neural nets such as the MLP and RBF cannot generalise even on functions with very simple hidden structure, and therefore have limited application as replacements for conventional rule induction systems. This chapter describes modifications which provide both supervised and unsupervised forms of neural network which can correctly generalise when presented with examples of logical functions containing hidden structure. The networks use sine functions as the nonlinearities in their hidden units and are called the self organising perceptron (SOP) and Fourier multilayer perceptron (FMLP). They are able to generalise on high-order structure in the data as well as having the interpolation properties of conventional ANNs such as the MLP.

A reformulation of the Shannon-Hartley Law is used to show that the FMLP and SOP generalise in the statistically most likely way and that if the number of hidden units is limited, they develop a mapping rule that gives the minimum possible error rate on both seen and unseen logical data. The FMLP and SOP both operate by finding a function or rule $F(X)$ that is consistent with the training examples and also has the minimum possible frequency bandwidth. In the case of logical functions, which are cyclic, the minimum bandwidth function is the function whose frequency transformation contains the minimum possible number of spectral lines. It is suggested that the minimum bandwidth function which is consistent with the given set of training examples corresponds to a minimum entropy function and hence most probable function completion. The argument is supported by examining the nature of generalisation and its relationship to different types of data structure. Quantitative measures of the amount of structure in logical functions are developed and are used to explain the operation of the SOP and FMLP.

2. EXAMPLES OF THE GENERALISATION PROBLEM

The problem of generalisation of symbolic and continuously valued data can be illustrated by a pair of examples. In the first example we consider generalisation of symbolic (logical) data pertaining to a hypothetical case of credit worthiness. The use of neural networks to deal with this type of problem is currently attracting much attention from large companies which bill individuals after provision of goods or services.

The credit worthiness assessment of an individual is based on a number of attributes of that individual, such as whether he or she is an owner occupier, marital status, post code, sex, and so on. The problem of generalisation is to take a fairly small set of individuals with known credit worthiness and examine their personal attributes. These examples must then be used to infer the credit worthiness of other individuals not in the example set (see Fig. 10.1).

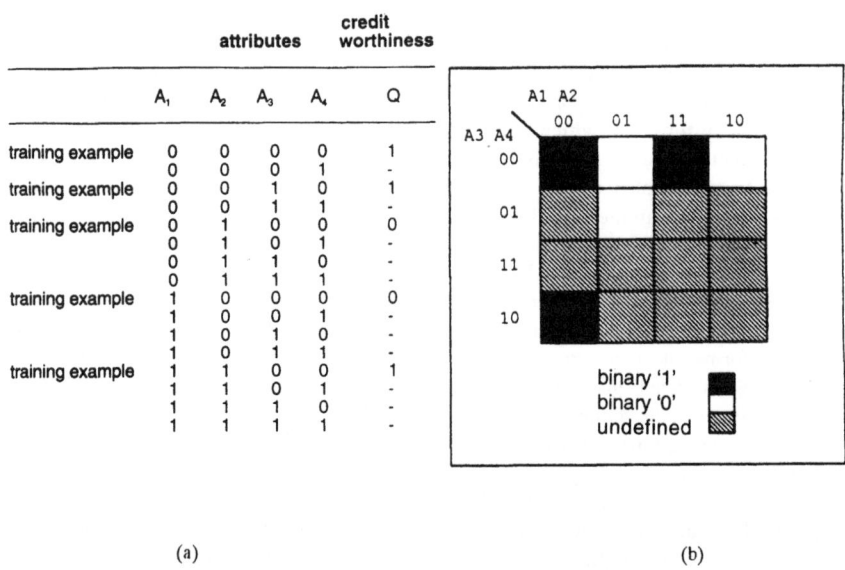

	attributes				credit worthiness
	A_1	A_2	A_3	A_4	Q
training example	0	0	0	0	1
	0	0	0	1	-
training example	0	0	1	0	1
	0	0	1	1	-
training example	0	1	0	0	0
	0	1	0	1	-
	0	1	1	0	-
	0	1	1	1	-
training example	1	0	0	0	0
	1	0	0	1	-
	1	0	1	0	-
	1	0	1	1	-
training example	1	1	0	0	1
	1	1	0	1	-
	1	1	1	0	-
	1	1	1	1	-

(a)   (b)

Fig. 10.1. (a) Truth table representation of hypothetical credit worthiness function. (b) Karnaugh map representation of hypothetical credit worthiness function.

Usually, the symbolic attributes are coded as binary variables such as: A_1 = "owner occupier," A_2 = "marital status," A_3 = "post code in Central London," A_4 = "male." Similarly, the condition of credit worthiness can be denoted by a binary variable, Q, and having assigned this coding to the examples, it is evident that the example set constitute entries in a truth table or Karnaugh map (Karnaugh, 1953) for a Boolean function as shown in Fig. 10.1. In terms of the truth table, generalisation is the process of filling in the blank rows of the truth table or elements of the Karnaugh map with the "best" set of values which are consistent with the example set.

The second example is of generalisation applied to continuously valued data. A common task of this nature is speech recognition, in which a continuously valued spectral description of spoken words is used to classify a sound as a particular word. Typically, the spectral description is a multidimensional continuously valued vector which is input to the classifier, which then produces an output vector that encodes a particular classification. Thus, the classifier is required to act as a vector to vector mapper which implements a desired mapping function. The vector mapper should generalise, so that given a small number of examples of the spectral vector representation of known words, it produces the correct output vector value (classification) when presented with previously unseen spectral vectors as its input.

A simplified description of this process is shown in Fig. 10.2, in which the input and output of the mapper are scalar variables. The function relating input and output is continuous, and this figure shows that the example set of required input-output pairs consists of actual samples of the function. The purpose of generalisation is therefore to produce the continuous function from its samples. This view of generalisation suggests that the body of knowledge on waveform reconstruction and Nyquist sampling is relevant to this problem and, as discussed in the following section and by Tattersall and Foster (1989), RBFs and MLPs have a similar function to the interpolation filters used to recover continuous waveforms from their sampled form.

3. MLPS AND RBFS AS GENERALISING MACHINES

Figure 10.2 shows that if the input-output examples in a training set are samples of a smoothly changing underlying function, then the complete continuous function should be recoverable by passing them through a suitable low-pass interpolation filter. It turns out that this is one interpretation of the action of the radial basis function and multilayer perceptron.

3.1. Radial Basis Functions as Low-Pass Interpolators

The operation of the radial basis function network is illustrated in Fig. 10.3, in which a set of basis functions are added together so that their sum closely fits discrete training examples of a continuous function which is to be learnt by the system. Typically the basis functions are multivariate Gaussians whose amplitude, mean value, and variance can be scaled to make the network output fit the given examples of the function.

The low-pass filter action of the network is easily understood by initially assuming that the function's training samples are regularly spaced and that one basis function is positioned over every sample in the function domain. In this situation the network's output is the convolution of the training samples with the radial basis function. spectrally this is low-pass filtering because the Fourier transform of the multivariate Gaussian is a low pass frequency response.

In practice the samples are not positioned regularly and the bandwidth and amplitude of each of the basis functions must be individually adjusted to match the sample rate in its locality. Moreover, it is usually impractical to place a basis function over every single training sample because of computational load, and in this case, the function is subsampled by using a relatively small number of basis functions for its synthesis. The positions of the basis functions in the pattern space are adapted iteratively to optimise the accuracy of the synthesised function, and the bandwidth of each radial basis function is reduced to reflect the lower effective sample rate of the function. The MLP can similarly function as a low-pass interpolator; details are not given here.

3.2. Deficiencies of the MLP and RBF as Generalising Machines

The values of functions which describe logical processes often change abruptly as a single variable in the argument changes value. Such functions may be governed by very simple rules, and yet the sudden changes in their value means that generalisation by low-pass interpolation will be completely incorrect. Predictably, it will usually be found that radial basis function systems or sigmoidal MLPs fail to generalise correctly when only trained on a subset of possible input-output examples of these kinds of function, as is demonstrated by the practical results presented in Section 9 of this chapter.

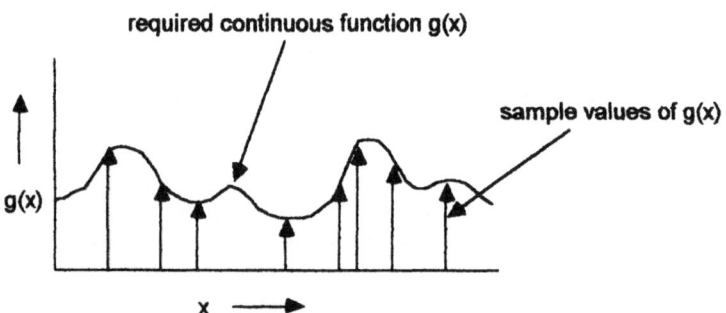

Fig. 10.2. Generalisation by finding a continuous function from its samples.

4. REQUIREMENTS OF AN IDEAL GENERALISING MACHINE

Supervised learning machines such as the MLP are required for learning a mapping function $Y = f(X)$ without exposure to all possible input/output pairs of Y and X values. This is only possible if the system can *generalise* from the sample values of the function which are given during training so that the statistically most likely value of output y is produced when a previously unseen input vector x is input to the machine. This means that the most likely complete function which is consistent with the training samples must be found.

The definition of the *most likely* function is debatable, but in this chapter we use the idea that the completed function must not only be consistent with the given values of the function, but should also have the same statistical properties as the seen parts of the function. This means that the types of feature which are observed with a certain frequency in the given parts of the function should also be present with the same frequency in the completed function. Conversely, features which do not occur in the seen parts of the function should not be introduced into the generalised parts of the complete function. That is, the simplest or least complex function consistent with the given data examples must be found; the evaluation of function complexity is discussed later in the chapter.

A powerful justification for using this approach to generalisation is that it is similar to the principle employed in a Bayesian classifier. In the classifier, an estimate is made of the class conditional probability distribution of a set of data from a set of training examples. It is assumed that *these statistics extend to unseen regions* of the data domain and can be used to classify patterns lying in these regions. In the case of the mapping completion problem, it is assumed that the statistics of the function in seen regions of the input domain also extend to unseen regions and therefore that the completed function must retain the same statistical characteristics.

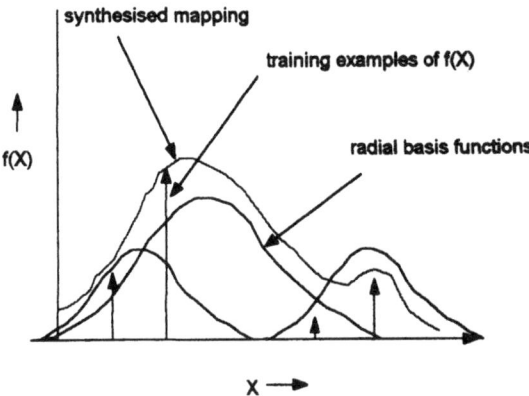

Fig. 10.3. Synthesis of function using radial basis functions.

Generalisation is only possible if the function to be generalised is redundant in some way. For example, if a function has the statistics of white noise, it will be impossible to predict the value of the function at an unseen point in the domain, even if the values of the function at all other points are known. Conversely, a function which has constant value over the entire input domain is predictable from a single given value and is highly redundant. In general, redundancy is present if the values at different points in the domain are correlated and the correlation is manifest as *structure* in the input-output mapping.

5. MEASURING THE STRUCTURE IN LOGICAL FUNCTIONS

The foregoing arguments are applicable both to functions of continuous and discrete valued variables. However, the primary object of this chapter is to show how ANNs can operate with discrete valued logical data, and from this point on, the discussion is related specifically to the problem of using neural nets to find the most likely Boolean function (rule) which is consistent with a sparse set of examples of a logical process.

It is desirable to quantify the amount of structure in a function so that the complexity of different functions can be compared precisely. It is shown in later sections that such a precise comparison is necessary to determine how an incomplete function should best be estimated by a neural network. A natural measure of the amount of structure, or complexity, of a function is the length of the shortest data description required to completely define the function. This approach has been taken by Risanen (1978, 1986), who rigorously applied information theory to determine a lower bound for the necessary description length of a series of data.

In this chapter we use a more intuitive engineering approach based upon the information theory proposed by Shannon (1948). The technique is analogous to the way in which the number of bits required to send a message over a communication channel is computed. In the latter case, the number of required bits is equal to the logarithm of the inverse probability of the message. The idea can be applied to the measurement of function structure by conducting a thought experiment in which two people are connected by a communication channel through which an attempt is made to transmit complete information about the Boolean function whose structure is to be measured. The value of the function at each possible point in its domain constitutes a message whose probability is determined by the topological statistics of the function.

In the case of Boolean functions, only the two messages "1" or "0" are possible. If the function were unstructured such that each value had a probability of .5, then one bit of information would have to be transmitted to define each value of the function. If the function has N input variables then its domain contains 2^N points; hence, 2^N messages (bits) must be sent to completely define the function. This corresponds to transmitting the entire truth table for the function. Conversely, a highly structured function whose statistics make the values of the function at different points in the domain predictable requires much less than one bit of information per function value. This is because the probability of each value (message) being "1" or "0" is greater than .5, and so much less than 2^N bits need to be transmitted to define the function.

A suitable measure of function structure is therefore the number of bits which would be needed to send the entire function truth table, minus the information which actually needs to be transmitted if the statistics of the function are taken into account. This measure has a value of zero bits if the function is unstructured such that the probability of each function value is .5 and a value of $2^N - 1$ bits if the function is completely structured such that the probability of each value of the function being "1" or "0" is 100%.

In practice it is very difficult to evaluate the statistics of anything but very simple functions. However, an alternative technique exists which exploits the relationship between bandwidth and information as formalised by Shannon, and this approach is described in the next subsection.

5.1. Bandwidth and Function Structure

The Shannon-Hartley Law (Shannon, 1948) relates the information-carrying capacity, C, of a communication channel to its bandwidth, B, and the signal-to-noise ratio, S/N, and is given by $C = b \log_2 (1 + S/N)$. This law can be reformulated to define the number of spectral lines required for the frequency-domain encoding of a function which requires a certain number of bits of information for its complete definition. Thus amount of structure can be measured by counting the function's spectral lines rather than by trying to evaluate its topological statistics.

The reformulation of the Shannon-Hartley Law commences by calculating how many different messages may be encoded by a single spectral line (carrier frequency) of variable amplitude in the presence of noise. A plausible coding scheme would be to represent each message by a different amplitude of the spectral line. To prevent the additive noise causing the different levels from being confused during decoding, their separation would have to be greater than the expected value of the noise. Assuming that the RMS (root mean square) value of the spectral line is s_s and that the RMS value of the noise is s_n, then the number of different spectral line amplitudes which could be decoded without confusion would be M:

$$M = (s_s^2 + s_n^2) / s_n^2 \tag{1}$$

Eq. (1) allows the number of messages which can be encoded by a single spectral line to be expressed in terms of the signal-to-noise ratio S/N_q, as $M = 1 + S/N_q$. Assuming each of the messages is equiprobable, their probability is just $1/M$ and so the number of bits of information which can be encoded by a single spectral line in the presence of additive noise is $h = \log_2(1 + S/N_q)$. Extending this to the case in which Q spectral lines are available for encoding information, it is seen that H bits can be encoded by Q spectral lines where:

$$H = Q \times \log_2(1 + S/N_q) \tag{2}$$

This derivation is not mathematically rigorous but gives some insight into the process. Eq. (2) can be used to relate function structure to the number of spectral lines needed to encode the function in the frequency domain as shown in Fig. 10.4. It is assumed that quantisation noise is added in this process by the use of a threshold which sets the value of decoded function to either "1" or "0." The value of quantisation noise can be found by assuming that quantisation error has a uniform probability distribution and is therefore given by $N_q = \Delta^2/12$, where Δ is the function value quantisation interval.

For convenience, the two levels from the quantiser are assumed to be +1 and −1 and so $\Delta = 2$ and the signal power, S, is 1. Thus the signal to noise ratio is $S/N_q = 3$, and the number of bits of information which can be encoded by Q spectral lines is $H = 2Q$. Thus, if it is found that a function requires at least Q spectral lines, in its frequency domain description so that it can be inverse transformed without any errors, it can be deduced that the number of bits of information required to describe the function is $H(F) = 2Q$.

An argument which is often set against the validity of this relationship is that the number of lines depends on the type of basis vectors chosen for the frequency domain representation. For example if the first chosen basis vector (carrier frequency) is actually the same as the function to be encoded, then only one spectral line will ever be needed and the apparent amount of structure in the function will therefore be high even if the function is actually noiselike.

The counterargument is that the chosen basis vectors should not favour any one particular type of function. For this reason it is proposed that the discrete cosine transform should always be used to encode the functions in the frequency domain because the basis vectors of this transform "sample" the N-space of the function at uniform solid angle intervals. In fact, the Fourier transform has similar properties to the discrete cosine transform, and no significant difference in structure value has been observed when using both of these transforms over a diverse range of function types.

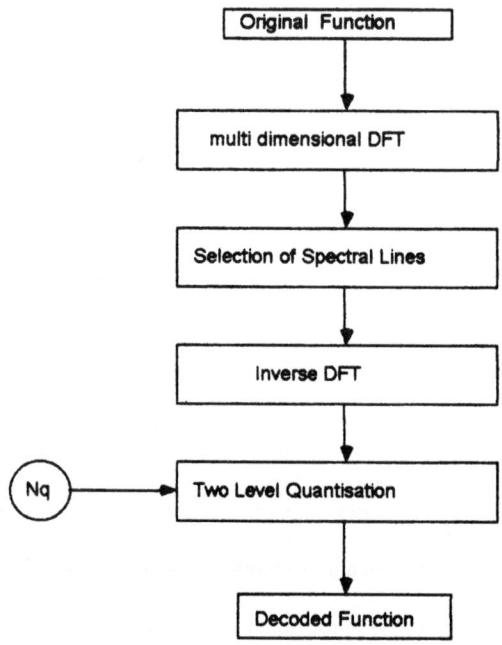

Fig. 10.4. Coding of a function in the frequency domain.

5.2. Calculating Function Structure Using the Bandwidth Measure

The previous section demonstrated a very simple relationship between the amount of structure in a function and the number of spectral lines which its frequency-domain encoding occupies. A simple way to calculate the structure of a function is therefore to evaluate its multidimensional discrete Fourier transform and then find the minimum number of spectral lines which yield the original function when inverse transformed and thresholded. The selection of the spectral lines is done in order of their energy, with the largest being taken first and the smallest last.

Some examples of this evaluation are shown in Fig. 10.5, which uses Karnaugh map notation to show the original function, its multidimensional DFT, and the smallest set of spectral lines which enables the function to be recovered by inverse transforming. For ease of representation, the 16 possible four dimensional DFT coefficients are also laid out in the same format as a Karnaugh map. This means for example that the coefficient listed under coordinates 0110 is the amplitude a of four dimensional frequency given by $\cos (0 \times \pi \times x_1 + 1 \times \pi \times x_2 + 1 \times \pi \times x_3 + 0 \times \pi \times x_4)$, where x_i, $i = 1, ..., 4$, are the function arguments. Fig. 10.5 shows that highly structured functions such as four-bit parity only require one spectral line whereas more random functions require up to eight spectral lines for their representation.

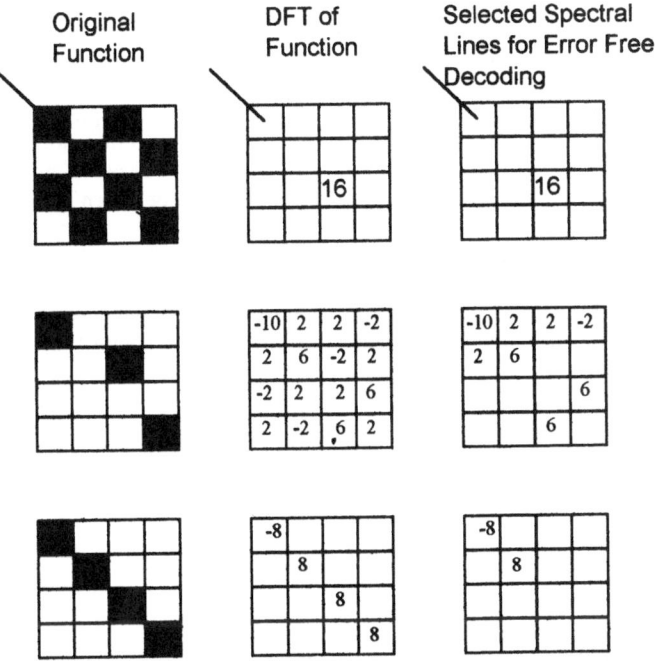

Fig. 10.5. Examples of functions of various entropies and their DFTs.

The size of the subset of highest energy DFT coefficients which enables the original function to be recovered without error is always exactly half of the total number of nonzero spectral lines in the DFT, rounded up to the nearest integer number of lines. Omitting half of the lines in the transform of the function introduces noise into its reconstruction, but the noise is effectively removed by the use of the threshold after the inverse transform which sets the noisy reconstructed value to "1" or "0." The amount of structure in a function can therefore be found directly from its DFT by counting the number of nonzero coefficients, N_c, as $H(F) = N_c$.

It must be emphasised that the Karnaugh map is only used as an aid to visualisation, and that the spectra shown are not a transform of the pattern in the two-dimensional Karnaugh map.

6. THE BEST ESTIMATION OF AN INCOMPLETE LOGICAL FUNCTION

We now return to this chapter's fundamental question: how to find the simplest explanation of an incomplete set of data. In the context of logical functions this means that the system is presented with an incomplete truth table generated by an unknown function and must insert the most probable values in the blank entries of the table. This process is analogous to the process used to classify patterns with a Bayesian classifier: the statistics of the data are estimated from a training set and it is then assumed that these statistics will hold for previously unseen data in the test set. Hence, the basis for choosing the most likely function completion is to ensure that the statistics of the completed function are the same as the statistics of the given parts of the function. Choosing a completion function whose statistics do not match the statistics of the training examples will always reduce the amount of structure in the function because the predictability of each function value will be reduced. Thus the criterion for choosing a particular completion is to find a function which fits all the seen parts of the incomplete function and has the highest possible structure value. Because function structure is related to the number of spectral lines in the DFT or DCT of the function, the criterion for completion can be stated as follows:

The most probable completion of an incomplete function is the function having fewest spectral lines in its DFT which is consistent with the seen parts of the incomplete function.

Examples of this approach are shown in Fig. 10.6, which shows partially defined logical functions along with their most likely and second most likely completions based on the numbers of spectral lines in the DFTs of the set of possible completed functions.

6.1. Overgeneralisation by Bandwidth Constriction

A useful corollary to the bandwidth criterion is that selection of a highest energy subset of spectral lines which is insufficiently large to properly encode the original function, will yield a new function on inverse transformation, which progressively overgeneralises as the number of spectral lines selected is reduced. An example is shown in Fig. 10.7, which shows the Karnaugh maps generated by the inverse transformation of the DFT of functions from which different numbers of spectral lines have been selected. It can be seen that the errors in the reconstituted functions increase monotonically as the number of selected spectral lines diminishes. This property is important in the context of the *self-organising perceptron* (SOP) and *Fourier multilayer perceptron* (FMLP) neural networks to be defined later in this chapter. These networks are used for rule induction because the hidden units of the network synthesise spectral lines, and choosing a certain number of hidden units in the network fixes the complexity of the rule which will be implemented and the generalising ability of the network.

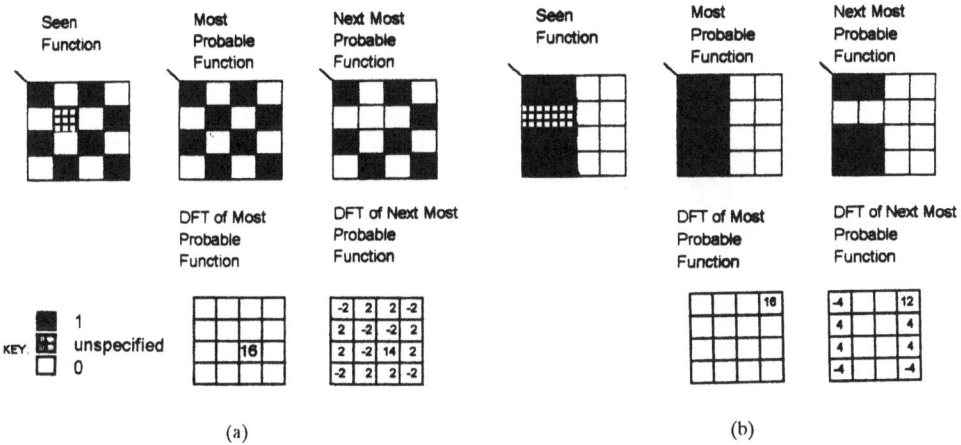

Fig. 10.6. (a) Minimum bandwidth completion of four-bit parity function with one unspecified value. (b) Minimum bandwidth completion of $F(x_1, x_2, x_3, x_4) = x_1$.

6.2. Computational Implications of Using the Bandwidth Criterion

An approach to inducing an optimal rule or function to fit a set of examples is to methodically search through the set of all possible complete functions which fit the given examples, evaluate their multidimensional DFTs, and select the function with the lowest number of spectral lines. However, with rules of practical complexity, this may prove computationally overintensive and other approaches to using the bandwidth criterion may have to be employed. As an example of the computational load incurred by the direct search technique, assume that a rule or function has n binary input attributes and that p examples of the mapping generated by the function are given. An optimal rule must be induced from these p examples. The number of possible function arguments is 2^n of which p are given. The number of function arguments for which the function value is not known is therefore $2^n - p$, and so the number of possible completions of the function is $2^{(2^n - p)}$.

The DFT of each possible completion must be evaluated and each DFT requires 2^n multiplications and additions. The total number of multiply-accumulate operations needed to find the optimal function is therefore $2^n . 2^{(2^n - p)}$. Inserting some test values into this expression gives enormous numbers of multiply-accumulate operations for all but trivial problems and so alternative ways of applying the bandwidth criterion are essential.

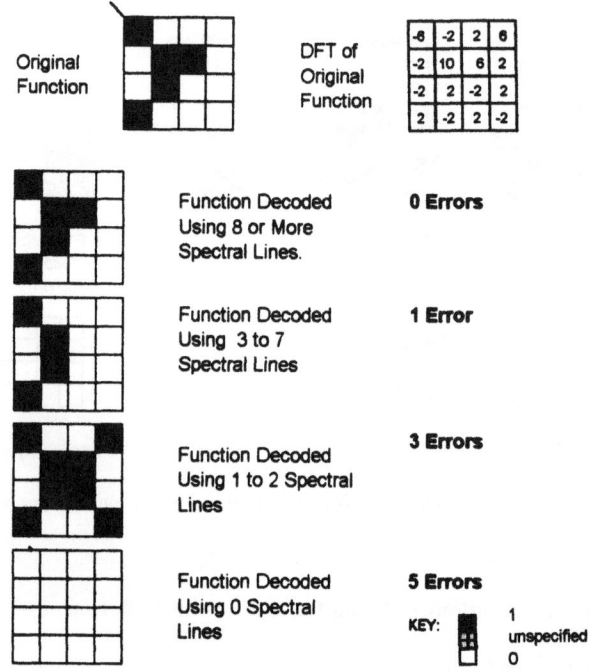

Fig. 10.7. Example showing monotonic increase in function errors as number of spectral lines selected is reduced.

7. RULE INDUCTION BY SPECTRAL PEAK PICKING — THE SELF-ORGANISING FOURIER PERCEPTRON

The previous sections have shown that structure or redundancy in a function is manifest by the existence of spectral energy at only some points in the domain of its multidimensional frequency transform. A highly structured function has few peaks, and an unstructured function has many peaks. We have shown that the highest amplitude half of this set of peaks is required for perfect reconstruction of the function, and the positions and amplitudes of the peaks in this set can be viewed as an encoding of the rule which generates the function.

In practical situations, only a subset of all possible input-output values of the function is defined and the missing values are set to a default value, usually the average of the defined function. The result of using default values for unspecified parts of the function is to cause spectral energy to appear at frequencies which would actually have zero energy if the complete function had been defined.

This is illustrated by the parity function example in Fig. 10.8, which shows the DFT for the completely defined function and for the same function in which five points in the domain have been left unspecified. These have been assigned a default value equal to the average of the defined parts of the function. It can be seen that the DFT of the completely defined function only has energy at one frequency. The incompletely defined function has small amounts of energy at other frequencies, as well as a large amount of energy at the frequency point defining the complete parity function.

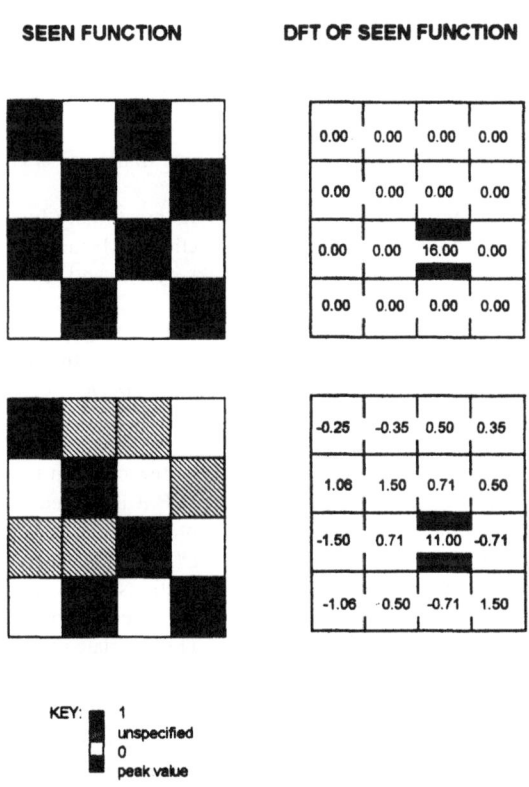

SEEN FUNCTION   **DFT OF SEEN FUNCTION**

KEY: 1
      unspecified
      0
      peak value

Fig. 10.8. Spectral peak of parity function surrounded by spectral noise floor due to incomplete definition of function.

This example illustrates the general effect of not defining some parts of a function: a "spectral noise floor" is created by the missing values, and its level may rise to submerge significant peaks if insufficient points in the function are defined. however, if the spectral peaks associated with the underlying function are unambiguously visible above the noise floor, the complete function can be estimated from the partially defined function by picking these peaks and finding their inverse transform, assuming all other frequency coefficients are set to zero.

It has already been shown in Section 6.1 that the error rate in the reconstruction of a function increases monotonically as the number of spectral terms used in the inverse transformation is reduced, so a method of estimating a complete function from a partially defined function is as follows:

1. Evaluate the DFT of the partially defined function, setting undefined function values to the average of the defined part of the function.
2. Pick the largest spectral peak in the transform and inverse transform, setting the energy at all other frequency values to zero.
3. Test to see if the function generated by the inverse transform is consistent with the defined parts of the original function.
4. If the outcome of the test in (3) is positive then the most likely function completion has been found in accordance with the criterion argued in Section 6. If the test outcome is negative, then go to (2) and increase the number of selected peaks by one and repeat the process.

This algorithm can be cast in a neural network framework, which in principle can be trained continuously, and can update its estimate of the complete function underlying the training data as it is trained. The network is called the *self-organising Fourier perceptron* (SOP) and is shown schematically in Fig. 10.9.

The SOP learns incrementally as follows: a process governed by some unknown rule generates a set of training examples defining parts of the function to be learnt by the SOP. As the nth training example, $F^n(X)$, is received by the SOP it is passed through a multidimensional discrete Fourier transformer. Each training example only represents a single impulse in the domain of the function and therefore its transform, $D(F^n(X))$, is easy to compute. The transform is loaded into a leaky accumulator which contains fractions of all past transform vectors, and as more training examples are received, the accumulator begins to build up an estimate of the spectrum of the rule/function underlying the data. The estimated spectrum is $D'(w_1, w_2, ..., w_N)$.

A function which has structure will start to exhibit some spectral peaks in the accumulated transform. Selecting the highest energy subset of these peaks defines the DFT, $\hat{D}(w_1, w_2, ..., w_N)$, of a rule which fits the seen data with minimum error given the number of peaks in the subset. The peak picking is done by "self-organisation" such that connections are made to those outputs of the leaky accumulator which exhibit large signal amplitudes. these connections can be thought of as the hidden units of the system and can have values of one or zero.

The system is used in "recognition mode" by applying the estimated rule to the current input argument, X^m. This is simply done by evaluating the inverse transform (IDFT) at the position of the required function argument, X^m, to produce an estimated output, $\hat{F}(X^m) = IDFT\{\hat{D}(w_1, w_2, ..., w_n, X^m)\}$. If the error between this function and the training set is too high, then the number of hidden units, and hence selected spectral peaks, is increased. Conversely, if the error is zero, the number of hidden units is reduced to ensure that the simplest possible function consistent with the training set is generated.

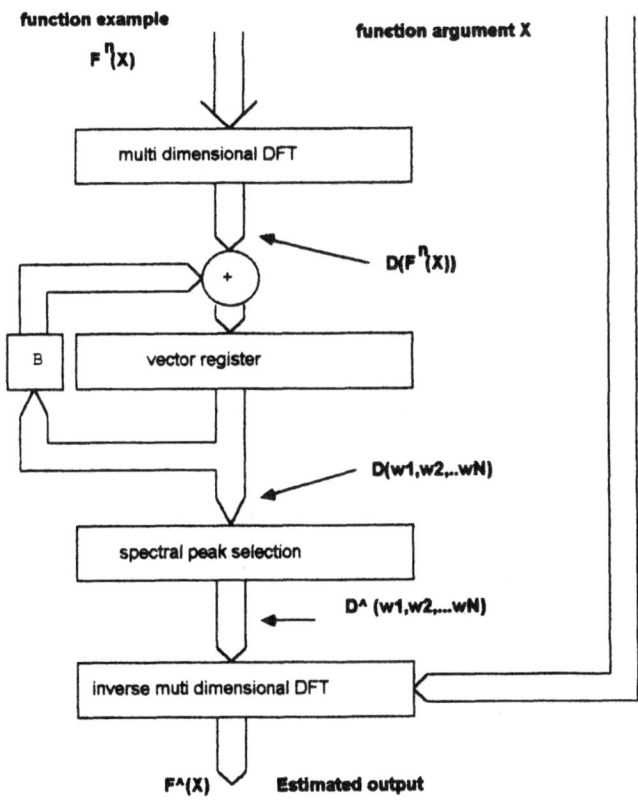

Fig. 10.9. Structure of self-organising Fourier perceptron.

This system has been tested by training on many four-variable functions in which some values are not defined. The estimated functions generated by the SOP using different numbers of spectral lines or hidden units are presented in Fig. 10.10. As expected, an SOP with just one hidden unit produces a very simple function estimate which does not always fit the seen data. This is because the simplest function which fits the data has an entropy which cannot be carried by a single spectral line. Increasing the number of hidden units allows functions to be generated which are consistent with the seen data. Eventually, further increase in the number of hidden units causes the SOP to generate functions which are not the simplest consistent with the seen data because the system has too much information capacity and stops generalising.

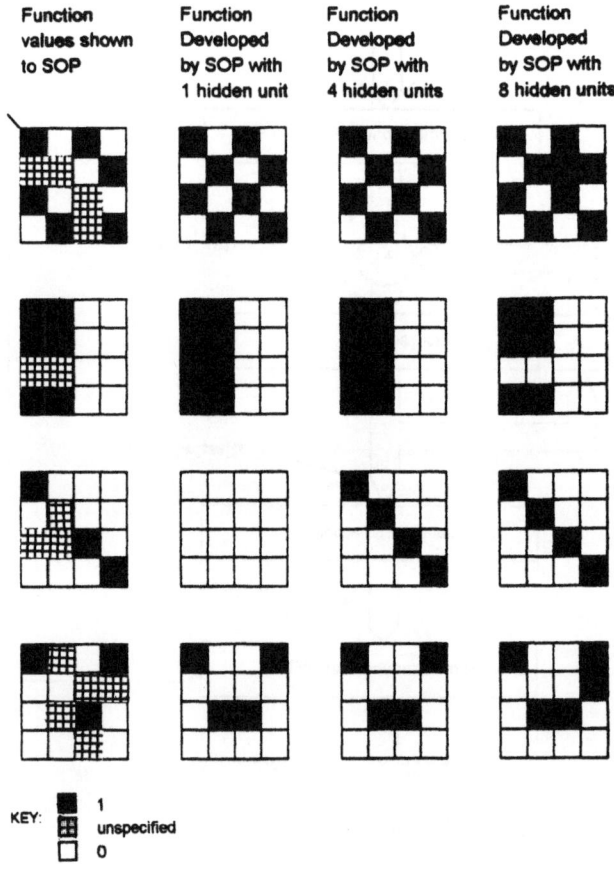

Fig. 10.10. Functions developed by self-organising perceptron.

In view of the necessity of not overproviding the SOP with hidden units, an appropriate strategy is to start the system with just one unit and add more if it is found that the estimated function values are frequently incorrect. This approach will ensure that the most general rule possible will be initially developed and that it will only become more complex if the error rate is unacceptably high.

8. RULE INDUCTION USING THE FOURIER PERCEPTRON (FMLP)

The Fourier multilayer perceptron (FMLP) has been used by several neural net researchers as an alternative to the multilayer perceptron (see, e.g., Lapedes & Farber, 1987). The architecture of the FMLP is identical to the MLP except that the sigmoidal nonlinearities associated with each neural element are replaced by sinusoids as shown in Fig. 10.11. In an FMLP with one hidden layer, the weights from the input units into each hidden unit control the complex frequency generated by the unit, whilst the weights summing into the output units add together the complex frequency terms from the hidden units to form the required output value. The FMLP is therefore structured to generate an input-output map by multidimensional Fourier synthesis.

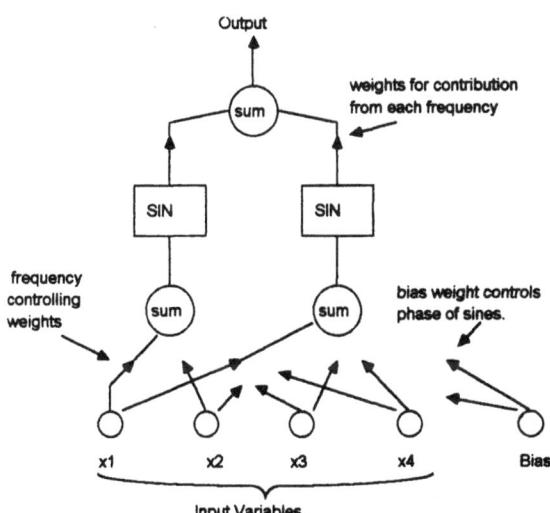

Fig. 10.11. The Fourier perceptron (FMLP).

Learning in the FMLP is usually done using error gradient descent with the back propagation algorithm described by Rumelhart, Hinton, and Williams (1986). Ideally the FMLP minimises its mean square output error by adjusting the complex frequencies synthesised by the available hidden units to match the highest energy multidimensional frequencies in the function which it is being taught to synthesise.

At first sight this action appears ideal for inducing the optimal rule or function from a subset of training examples. Each hidden unit in the network synthesises one spectral line in the frequency-domain description of the function, so a network with $H(F)/2$ hidden units should be capable of perfectly learning a function whose information content is $H(F)$ bits. An FMLP with a restricted number of hidden units will therefore synthesize a complete function whose bandwidth is the minimum possible while still fitting the training examples. As shown earlier, minimum bandwidth implies minimum entropy, and hence the most probable complete function. Another important property of the ideal FMLP is that if the number of hidden units is less than $H(F)/2$ then the network will be forced to overgeneralise. However, the learning algorithm minimises the mean square error, and consequently the available hidden units will be used to model the highest energy spectral lines in the function and thereby minimise the number of errors in the resultant function as shown by the examples in the previous section on over-generalisation.

The practical use of the FMLP is more difficult. First, the system is prone to becoming trapped in local energy minima like the MLP. A second more important problem is that the frequencies generated by the hidden units are not necessarily orthogonal over the domain of the function, and this may cause several hidden units of different frequencies to jointly model variations in the function which are really due to a single complex frequency.

The latter problem can be reduced by encouraging the hidden units to take on frequency values which are harmonics of the reciprocal of the width of the domain of the function. This can be done by artificially extending the domain of the training examples by repeating the given function periodically in intervals beyond the actual domain of the function. For example, a one-dimensional logical function whose values are specified at $x = 0$ and $x = 1$ can be extended to generate new training examples such that $F(x) = F(x + 2n)$ where n is an integer.

In spite of this kind of trick, optimal learning in the FMLP is very unreliable on all but simple logical functions, and an improved learning algorithm is required. However, to demonstrate the fundamental suitability of the FMLP architecture, a number of experiments have been performed to confirm the claim of optimal rule induction. In these experiments an FMLP has been presented with a subset of the truth table of a number of functions and the complete function generated by the FMLP compared against the optimal completion as defined by the DFT evaluation of entropy. It is found that the completions are optimal as long as the FMLP does not become trapped in local minima; a sample of these results is shown in Fig. 10.12.

9. COMPARISON OF THE FMLP, SOP, SIGMOIDAL MLP, AND RBF

In Section 3 it was argued that the radial basis function net and multilayer perceptron both generalise by low-pass interpolation. If this is true, these nets are unsuited to data which exhibits structure not manifest as function smoothness. In particular, these nets cannot deal with data representing symbolic variables whose functional relationship may be very simple, yet are subject to sudden changes in value as the argument value is changed.

To illustrate these problems, and attempt to show that the SOP and FMLP behave more appropriately than the MLP and RBF when faced with symbolic data, the ability of each type of network to generalise on a variety of functions has been tested. It is impossible to exhaustively test the performance of each network, so various function types with different properties have been chosen. The experiments consist of defining a binary function and finding the minimum complexity of each neural network which can learn the complete function without error. A network of the same complexity is then trained using a partially defined version of the function and its ability to complete the function is examined. The chosen functions and the function completions provided by each type of network are shown in Fig. 10.13 in the form of Karnaugh maps. The left-hand column shows the function on which the network has been trained. Subsequent columns show the functions actually learnt by an MLP, RBF, SOP, and FMLP respectively. Each network has one layer of hidden units, and its complexity is defined by the number of hidden units recorded under the Karnaugh mp for each network.

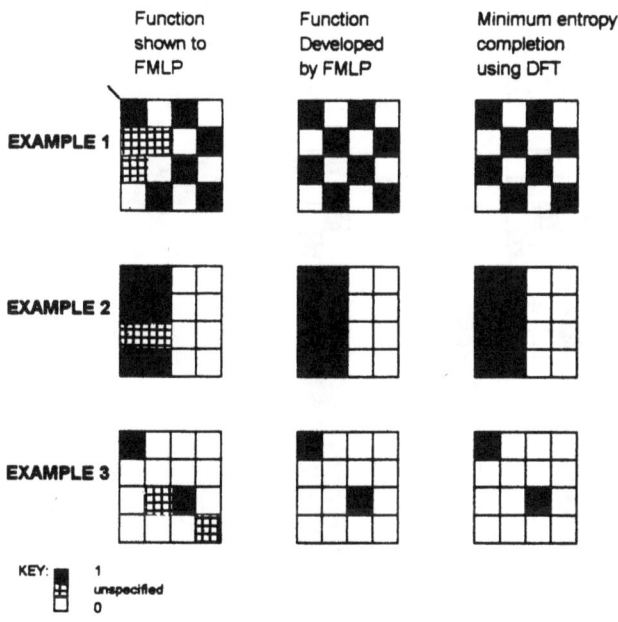

Fig. 10.12. Examples of function completion by the Fourier perceptron.

· *The parity function.* The first function is four-bit parity, and it can be seen that each type of network is able to learn this function perfectly if trained on all 16 possible points in the domain. It is worth noting that the radial basis function net has great difficulty in dealing with the function and requires 16 hidden units for perfect learning. When each network is trained by the same function, with the value at just one point undefined, both the MLP and RBF fail to generalise in a way which maintains the regularity of the function. The FMLP and SOP generalise to the most regular (simplest) function consistent with the training set.
· *The "F(A,B,C,D) = A" function.* This function is shown in the third and fourth rows of Karnaugh maps in Fig. 10.13. Experiment showed that the numbers of hidden units required for the MLP, RBF, SOP, and FMLP to perfectly learn this function when trained on all 16 possible values are 1, 5, 1, and 1, respectively. The function contains first-order structure. That is, the

value of the function is generally the same at adjacent points in the domain. It is thus to be expected that all of the network types should correctly generalise on this function. As predicted, when trained on the function, excluding two of its values, each network correctly generalises.

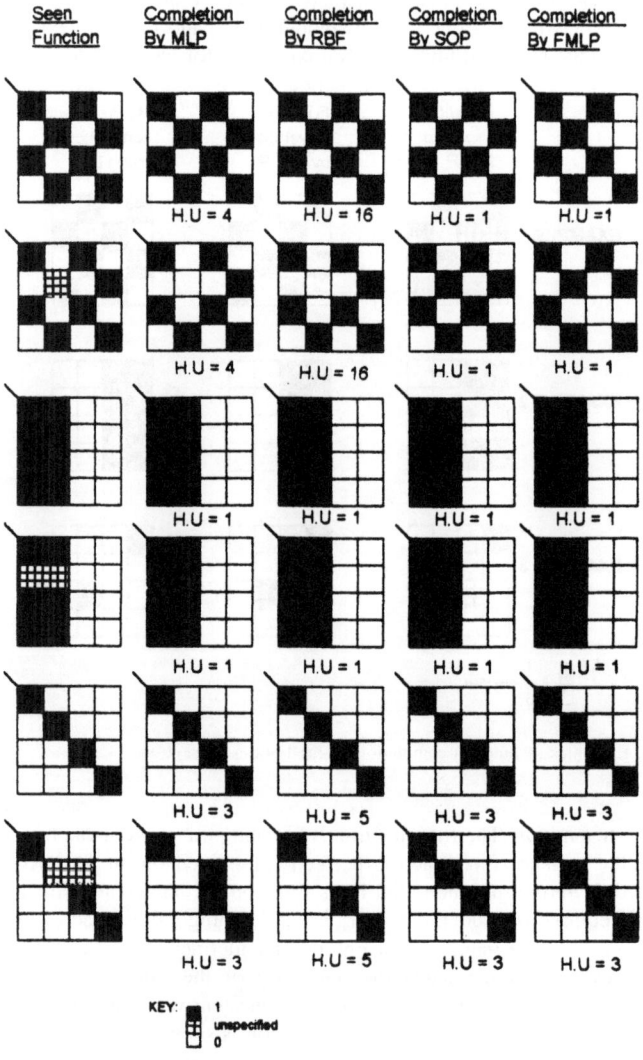

Fig. 10.13. Comparison of the function completions produced by various neural networks.

· *The "diagonal" function.* Results pertaining to this function are shown by the Karnaugh maps in the fifth and sixth rows of Fig. 10.13. Again, all networks were found capable of learning the function, given enough hidden units, and when trained on all of its values. However, both the MLP and RBF fail to find the simplest underlying function when trained on a subset of the values. This example illustrates how the SOP and FMLP can complete a function without introducing any new features, whereas the MLP and RBF introduce arbitrary new features.

10. CONCLUSIONS

This chapter has attempted to demonstrate that a rule system can be viewed as a mapping between a set of input attributes and a set of output actions or decisions. Such mappings are characterised by a function, $F(X)$, where X is the set of inputs and $F(X)$ is the output decision. Functions of this type frequently contain structure which means that the values of the function for one set of inputs can be estimated from the function values at other sets of inputs. It has been shown that this structure is manifest as the repeated occurrence of certain topological features over the domain of the function and that the statistical probability of these features can be used to derive an overall measure of structure in the function in terms of information entropy.

This idea has been extended by a reformulation of the Shannon-Hartley Law to enable the entropy and hence structure of a function to be evaluated from the number of spectral lines occupied by the Fourier transform of the function. It has also been shown that the best estimates of unseen parts of a function can be obtained by searching for a function consistent with the seen data and with the minimum possible number of spectral lines in its transform.

A variant of the multilayer perceptron which uses sinusoidal instead of sigmoidal nonlinearities has been shown capable of optimal function completion or, in other words, rule induction. Unfortunately, this type of MLP, known as the Fourier perceptron (FMLP), is hard to train reliably since it often gets trapped in local minima, and so a new perceptron structure has been developed. The new perceptron is known as a self-organising Fourier perceptron (SOP) and does not suffer from any learning difficulties. The SOP can produce very general rules at the start of training and can progressively make the rules more complex if its current rule makes too many errors. Moreover, if the SOP is restricted in size it will provide the optimal rule to fit a set of incomplete function values and can therefore be said to perform optimal rule induction.

REFERENCES

Karnaugh, M. (1953). The map method for the synthesis of combinational logic circuits. *Transactions of the American Institute of Electrical Engineers (Communications on Electronics)*, **72**, 593-599.

Lapedes, A., & Farber, R. (1987) *Nonlinear signal processing using neural networks: Prediction and system modelling.* Los Alamos National Laboratory (Rep. No. LA-UR-87-2662).

Lowe, D. (1989). Adaptive radial basis function non-linearities, and the problem of generalisation. *Proceedings of the First International Conference on Artificial Neural Networks* (pp. 171-175), London.

Risanen, J. (1978). Modelling by shortest data description. *Automatica*, **14**, 465-471.

Risanen, J. (1986). Stochastic complexity and modelling. *Annals of Statistics*, **14**, 1080-1100.

Rumelhart, D. E., Hinton, G. E., & Williams, R. L. (1986). Learning internal representations by error propagation. In D. E. Rumelhart, J. L. McClelland, and the PDP Research Project, *Parallel Distributed Processing* (Vol. 1, pp. 318-362). Cambridge, MA: MIT Press.

Scalia, F., Marconi, L., et al. (1989). An example of back propagation: diagnosis of dyspepsia. *Proceedings of the First International Conference on Artificial Neural Networks* (pp. 332-335). London: IEEE.

Shannon, E. C. (1948). Mathematical theory of communication. *Bell System Technical Journal*, **27**, July and October.

Tattersall, G. D., & Foster, S. (1989). Single layer look up perceptrons for speech recognition. *Proceedings of the First International Conference on Artificial Neural Networks* (pp. 148-152). London: IEEE.

11

Evolution of Dynamic Reconfigurable Neural Networks: Energy Surface Optimality Using Genetic Algorithms

Robert E. Dorsey and John D. Johnson
University of Mississippi

Robert Dorsey and John Johnson's chapter, **Evolution of Dynamic Reconfigurable Neural Networks: Energy Surface Optimality Using Genetic Algorithms,** *poses the problem of optimal connection structures between neurons for performance of specific tasks. Networks with dynamically reconfigurable connection structures are becoming increasingly popular, as the problems are becoming sufficiently complex that function does not immediately suggest form.*

Dorsey and Johnson, building on some previous work of Harold Szu (a speaker at the Optimality conference but not a contributor to this book), develop a novel methodology that uses genetic algorithms to select structures that are optimal in the sense of minimizing an energy function. The examples that they use involve feedforward networks with small numbers of nodes, which Dorsey and Johnson set to solve classic problems such as the exclusive-OR problem. The weights within a network can either be zero or positive, and the genetic algorithm allows these weights to evolve. The dependence of energy on the weights is complex enough that hill-climbing algorithms on the energy surfaces have apparently proved ineffective.

Dorsey and Johnson speculate that their methodology might prove helpful in determining neuronal structures for encoding specific concepts or performing specific behaviors. This could provide a synthesis between the approaches of Hebb that focus on specific neural connections, and those of Lashley that focus on collective properties. The chapter in this book by Bengio et al. also uses genetic algorithms to optimize synaptic weights (in conditioning models). If this methodology proves to be useful in neuroscience, it might also be usable in the study of optimizing social structures, as discussed in Bradley and Pribram's chapter.

ABSTRACT

This chapter investigates the use of the genetic algorithm for training the Dynamic Reconfigurable Neural Network (DRNN). The assumption of energy efficiency provides the biological motivation for the reconfiguration of synaptic strengths and activity of connections. The DRNN, trained with the genetic algorithm, offers researchers a new tool to model neural activity. In addition, it requires modelers to reexamine what is meant by optimal and also whether or not the collective network performance rather than the individual neuron action should ultimately be the focus of neurophysiological research.

1. INTRODUCTION

Recent advances in neurosciences have identified active hairlike neurofilaments that can expand and contract forming synapses among neurons. As Freeman (1991) pointed out, "This ability of a neuron to reach out and form new synaptic links earns it the name 'hairy neuron.'" These structures have caused a resurgence of interest among some researchers (see Szu, 1985, 1991) in developing neural network models that are able to dynamically reconfigure themselves as needed. Although important in theory, the ability of the hairy neuron to dynamically restructure itself while solving the current problem has posed considerable difficulty for traditional training methods. The inability of conventional training strategies to yield acceptable results on this "dynamic reconfigurable neural network" (DRNN) led to Harold Szu's challenge to one of the authors at the M.I.N.D. (Metroplex Institute for Neural Dynamics) conference to identify an algorithm that would be able to train and restructure the DRNN. Previous to this challenge Szu (1991) wrote:

> Thus, writing the neural system operation specifications for a general-purpose neural network is a challenging task that requires a constraint-relaxing optimiza- tion in the admixture of algorithm and architecture — namely, a dynamic optimization that is still to be developed.

This chapter represents a response to Szu's training challenge, as well as an exploration of alternative architectures that can be used to solve the well-known XOR operation.

Because the growth action of the neurofilaments of the hairy neuron appears to be independent of long-term synaptic potentiation, both the neuronal core and the surrounding active neurofilaments may serve as independent computing elements for the selection of pathways and strengths of synaptic connections. This discovery points to a twofold mode of modification for connections in an artificial neural model: first, a conventional capability to change synaptic weights, and second, an unconventional potential to dynamically create and annihilate active connections. Models of these fully interconnected dynamic networks have the potential to meet the goal discussed by Pribram (1991) to "emphasize a minimum of constraints in the processing wetware or hardware," and thus move beyond the currently highly constrained neuroscience modeling framework.

In the current chapter, the assumption of energy efficiency provides the biological motivation for the reconfiguration of synaptic strengths and activity of connections. Thus, the Hebbian tuning and adaptive topology become cooperative, as well as competitive, in order to minimize overall energy. An algorithm based loosely on genetic evolution provides the means to a parsimonious selection of pathways and strengths of synaptic connections. In contrast to backpropagation, or other gradient methods, the genetic algorithm does not rely on a slope-directed search from the current point but rather searches *globally* for a solution by iterating from one population of points to another. The probability of convergence to false optima is thereby reduced. Additionally, because the genetic algorithm is a direct search method, the network objective function is not required to be differentiable. This allows for flexibility in the selection of alternative objective functions.

Dorsey and Mayer (1994) provided a detailed description of the properties and dynamics of this algorithm. They showed that for a wide variety of complex optimization problems, the genetic algorithm is able to consistently achieve the global solution. The genetic algorithm is ideal for the DRNN because it provides a global search and allows a complex and nondifferentiable objective function.

The application of genetic algorithms to neural networks has followed two separate but related paths. First, genetic algorithms have been used to find the optimal network architectures for specific tasks. Todd (1988) and Miller, Todd, and Hegde (1989) represented various network architectures as connection constraint matrices that were mapped directly into a bit-string genotype. Modified standard genetic operators were then used to act on populations of these genotypes to produce successively higher fitness levels. This methodology was applied to three problems: the XOR, the Four Quadrant Problem (two-dimensional XOR), and the Pattern Copying Problem (a direct mapping between a binary input vector and itself). Guha, Harp, and Samad's (1989) network architectures were represented using groups of units with probabilistic projections between them. Although interesting, this work seems to sidestep the issue of a universal functional mapping because it leaves unresolved the question of whether the model's architecture performs poorly on a given task due to the appropriateness of the given architecture, or, rather to the inability of the backpropagation learning rule to achieve a global solution on the said architecture.

Alternatively, in a recent paper Dorsey, Johnson, and Mayer (1994) demonstrated the superior performance of the genetic algorithm across a large class of problems with generic feedforward network architectures. In fact, the problems typically used one hidden layer and six hidden layer neurons. The genetic algorithm appears to overcome the shortcomings of backpropagation. There have also been previous attempts by Whitley and Starkweather (1990), Whitley, Starkweather, and Bogart (1990), Miller (1988), Montana and Davis (1989), and Whitley and Hanson (1989) to use genetic search instead of gradient-descent learning to establish the appropriate weight matrices in fixed architectures.

This chapter, on the other hand, takes a general, fully interconnected, and dynamic network and uses a modified genetic procedure to evolve not only the connection weights but also the underlying connection structure.

2. GENETIC ALGORITHMS

In learning the appropriate weights and connection structure, the genetic algorithm uses a process that in ways parallels the Darwinian process of natural selection to iterate toward a solution (see Goldberg, 1989, and Holland, 1975). Given a specific dynamic neural network energy function to be optimized, the genetic algorithm starts with an initial population of candidate solutions (the first generation) and then selects a subset of the population to act as progenitors to contribute offspring to the next generation of candidate solutions. As in natural systems, the new offspring inherit a combination of the traits from their parents whose traits were either the result of a crossover between vectors in a prior generation or bestowed as an initial condition set by the organizer of the initial population or mutation. The key to this process is selectivity. Not all population members from the previous generation are given an equal chance of surviving and thus contributing to the population of potential solutions. Thus, it is likely that only a select few will actually contribute. In particular, the population members with the highest probability of surviving are those with traits favorable to solving for the optimum of the specific energy function. In contrast, members of the current population least likely to contribute candidate solutions to the next generation are those possessing unfavorable traits. In this way, a new population of candidate solutions (the second generation) is built from the most desirable traits of the initial population. As iteration continues from one generation to the next, traits most favorable in finding an optimal solution for the objective function thrive and grow, whereas those least favorable die out. Mutation may also occur at any stage of the progression from one generation to the next. By randomly introducing new traits into the natural selection process, mutation tests the robustness of the population of current potential solutions. As with traits bestowed on members of the initial population, if these newly introduced traits add favorably to the ability of their recipients to optimize the specific objective function, then the new trait will thrive and grow. Otherwise, the effect of the mutation will die out. Eventually, the initial population evolves to one that contains an optimal solution and the evolutionary process terminates. As Szu (1991) wrote:

> ... bioevolution may begin by chance, but it proceeds by dynamics. Initial chance and dynamics could together provide the statistical and the correlational information, the preferential and the synergistic boundary conditions, and the desirable and the realizable performance measures, which are all expressible in terms of a constrained free-energy principle in a top-down design of the minimum-maximum pattern-classification principle in this article. This energy principle can describe a comprehensive learning principle, including the topological dynamics of the cellular morphology.

3. METHODOLOGY

In order to optimize the DRNN shown in Figs. 11.1 and 11.2, we adopted certain protocols. These are clearly not unique, and alternative procedures could be followed and still be consistent with Szu (1991). Nonetheless, we felt these procedures to be the most logical.

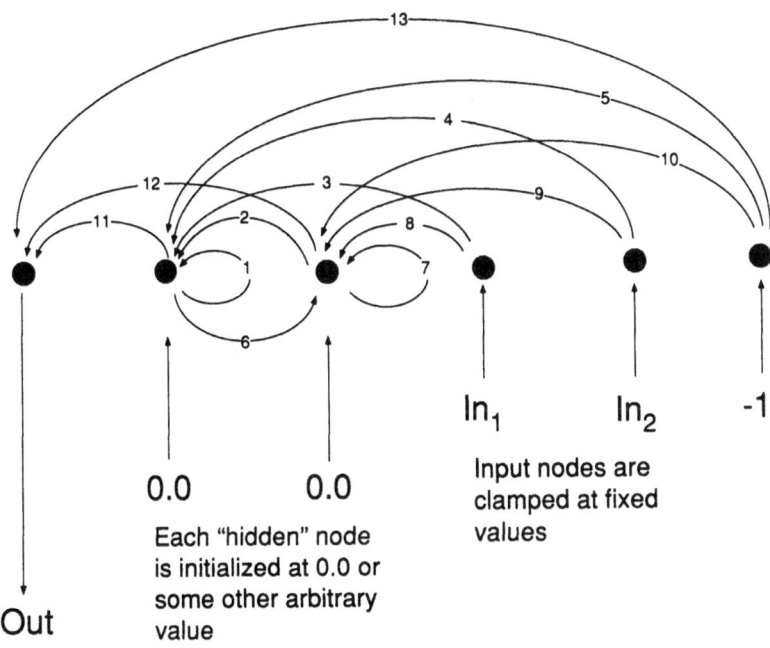

Fig. 11.1. The dynamic reconfigurable neural network with two recursive nodes.

The DRNN is operationalized in the following manner:

· The DRNN consists of k input or data nodes x_i ($k = 2$ for the XOR problem), m recursive nodes (2 and 4 were used) and one bias term node. The data nodes are fixed at the data values for each observation. The bias term node is fixed at -1.0.
· The k input nodes are each connected to the m recursive nodes (km connections). Each recursive node is connected to all recursive nodes including itself (mm connections). The bias term node is connected to each recursive node (m connections). The output node is connected to each recursive node and to the bias term node ($m + 1$ connections). The total number of connections is therefore $m(k + m + 2) + 1$.
· The output of each of the recursive nodes is the sigmoid function

$$o_{jt} = \frac{1}{1 + \exp\left(-\sum_{i=1}^{m} \omega_i o_{i,t-1} - \sum_{i=m+1}^{m+k} \omega_i x_{i-m} + \omega_{m+k+1}\right)}$$

where w_i are the connection weights.

· The initial input to the recursive nodes is initialized to an arbitrary value in the range [0,1]. We use 0 for the initial weights reported in this chapter.

· The output of the network for the tth iteration of the jth observation is the value

$$\hat{Y}_{jt} = \frac{1}{1 + \exp\left(-\sum_{i=1}^{m} \omega_i o_{i,t-1} + \omega_{m+k+1}\right)}$$

and the converged output for the jth observation is denoted Y_{jt}.

· The DRNN is said to converge if the output from the output node varies by less than 10^{-9} for five consecutive iterations[1].

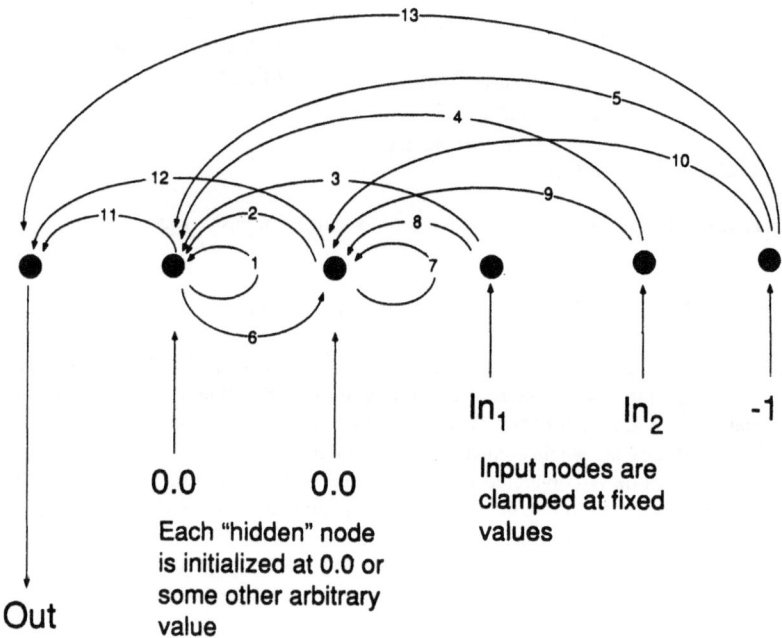

Fig. 11.2. The dynamic reconfigurable neural network with four recursive nodes.

---

[1]Two or even three consecutive iterations are not sufficient. Numerous solutions were found where this criterion was met for three consecutive iterations only to be following by a output values significantly different than previous outputs.

To optimize the DRNN we use the genetic algorithm with weights selected randomly from a uniform distribution over the range [-100, 100]. Two objective functions were used. For the first series of optimization runs the objective function

$$Min \sum_{j=1}^{N} (\hat{Y}_{jT} - Y_j)^2$$

is used where Y_{jT} is the converged forecast of the jth observation and Y_j is the jth observation. For the other series of runs, the objective function

$$Min \left(\gamma \sum_{j=1}^{N} (\hat{Y}_{jT} - Y_j)^2 + \sum_{i=1}^{m(k+m+2)+1} IND(\omega_i \neq 0) \right)$$

is used where $\gamma > m(k + m + 2) + 1$, $IND(\cdot) = 1$ when the argument is true and equals 0 otherwise and, as in Eq. (2), Y_{jT} implies the converged forecast of the jth observation. This objective function is referred to as the energy function.

The first objective function simply selects a set of weights that will minimize the sum of squared errors, whereas the second objective function requires this to be achieved with the minimum number of total connections.

Both objective functions are used with the XOR problem. For the XOR problem, the DRNN consists of two input nodes, either two or four recursive nodes, the bias term node and the output node. This results in a total of 13 (2 recursive nodes) or 33 (4 recursive nodes) connections. There are four observations for the XOR problem.

4. RESULTS

The genetic algorithm quickly identifies a set of weights to solve the first objective function, and because no priority is given to solutions with fewer connections, hard zeros almost never occur. Several typical solutions are provided in Table 11.1. When the objective function is modified to incorporate a penalty for the existence of superfluous connections, the genetic algorithm readily achieves the solution of eight connections discussed in Szu (1991) and in Rumelhart, Hinton, and Williams (1986). Their solution is shown at the top of Fig. 11.3. In fact, for this orientation of nodes, both connections to the bias term are not necessary. Only seven connections are needed for this architecture. This compact solution is shown in Fig. 11.4. This solution has both recursive nodes connected to the two input nodes and one is connected to the bias term. In addition, the recursive node not connected to the bias term serves as input to the other recursive node.

Some of the other solutions found by the genetic algorithm incorporating seven connections are listed in Table 11.2 and shown in Fig. 11.4. As can be seen, with two recursive nodes, there are two structures that can solve the problem and contain seven connections. To see whether or not the genetic algorithm would be able to solve the XOR problem with a more complex DRNN, the four recursive nodes were used. The genetic algorithm found a solution with the same fundamental structure as one of the seven connection solutions. This solution from the four recursive node DRNN is provided in Table 11.3.

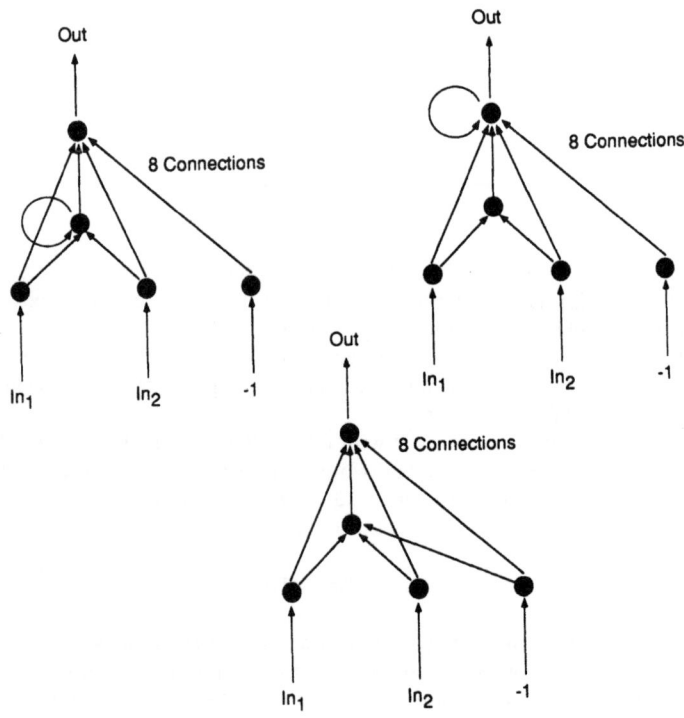

Fig. 11.3. Alternative eight-connection structures for the XOR problem.

This seven-connection structure is not unique. The other solution shown in Fig. 11.4 is also able to solve the XOR problem with seven weights. In this case the reduced structure has one recursive node connected to the two data nodes and to the bias node. This recursive node also has input from the second recursive node. The second recursive node only receives input from the first recursive node and from itself.

Weights	Solution 1	Solution 2	Solution 3
1	54.097	5.1580	19.1757
2	66.357	−80.1753	1.2389
3	97.767	89.7624	5.0123
4	−63.581	85.5918	30.0780
5	54.870	97.2459	40.5659
6	−52.713	−22.4606	−86.2190
7	−51.417	−21.2104	6.6190
8	−86.065	62.0763	32.4577
9	74.879	61.9885	40.8300
10	19.988	14.5472	33.6546
11	1.000	−1.0612	−47.2223
12	1.015	1.0000	1.0046
13	−78.455	45.0968	−21.4239

Table 11.1. Solutions Based on Minimizing Sum of Squared Errors.

Weights	Solution 1	Solution 2	Solution 3
1	0.0000	0.0000	0.0000
2	0.0000	61.2176	58.3973
3	9.5374	28.7965	52.2191
4	−84.9750	28.7965	−6.1705
5	0.0000	85.6920	48.3934
6	67.8922	−75.2506	0.0000
7	0.0000	58.3850	0.0000
8	−20.5624	0.0000	−66.1518
9	47.3249	0.0000	10.8252
10	42.3825	0.0000	0.0000
11	0.0000	1.0000	1.0000
12	1.0000	0.0000	0.0000
13	0.0000	0.0000	0.0000

Table 11.2. Seven-Connection Solutions Based on Minimizing the Energy Function.

Weights	Solutions	All other weights are zero
18	−99.9171	
19	74.0024	
20	−23.5243	
21	−38.3391	
26	57.7830	
27	−23.6824	
31	1.0000	

Table 11.3. Sample Solution for the Four Recursive Node DRNN.

This use of the one recursive node simply to modify the signal of the other node is not uncommon among the solutions found. As the DRNN is optimized with the genetic algorithm, it will often reach a plateau with a larger number of connections prior to discovering how to eliminate the next connection. A typical genetic algorithm optimization run is shown in Fig. 11.5. This figure shows the value of the error function for the four-hidden-node DRNN as the genetic algorithm searches and ultimately converges to the seven-node structure shown in Table 11.3. The scale used is log-log because the genetic algorithm very quickly achieves a close solution but takes much longer to achieve the final solution. It is also apparent from the figure that intermediate solutions are achieved during the search process that solve the XOR problem, but with more than the minimum connections.   Figure 11.3 shows, for example, several of the structures formed to solve the XOR problem where eight connections were used. As can be seen, there are several configurations that use one of the recursive nodes for signal modification.

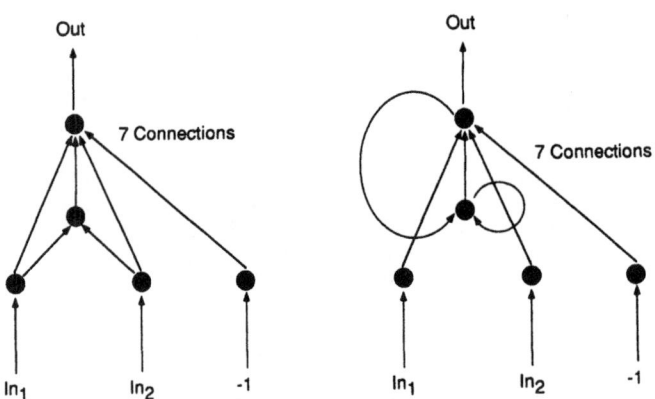

Fig. 11.4. Alternative seven-connection structures.

As mentioned earlier, the structure discussed by Szu (1991) and by Rumelhart, Hinton, and Williams (1986) is not the minimum energy function found by the genetic algorithm. The energy of their solution can be reduced by eliminating one of the connections to the bias node. In addition, an architecture with even fewer connections was found with the genetic algorithm. Figure 11.6 shows the lowest energy structure found that will solve the XOR problem. This six connection structure has one recursive node receiving input from the two data nodes as well as from the other recursive node. The second recursive node receives input from the bias node and from the other recursive node. This is again not unique because either recursive node can take on either function.

The values for the weights are shown in Table 11.4. This solution was not easily found by the genetic algorithm and required extensive search. To help understand the difficulty in finding this solution, Figs. 11.7 through 11.10 show the energy surface as the weights are varied. In each of these figures the weights were varied in .02 increments over the range ±5.0 around the optimal value of the weight. In some cases the optimal value was zero corresponding to the connection optimally not existing. All other weights were held at their optimal values and the value of the energy function was plotted as the two weights were allowed to vary. For each figure the neural network was evaluated at 10,000 points. Figure 11.7 shows the energy function as weights 1 and 2 are varied, Fig. 11.8 has weights 3 and 4, Fig. 11.9 shows 5 and 6, and Fig. 11.10 allows weights 7 and 8 to vary. As can be seen, the energy surface is quite complex. In addition, for those weights that are optimally zero, weights 1, 3, 4, and 7 in the figures, discontinuities exist as the energy function crosses the value zero. At that value the energy function drops by an increment of 1 corresponding to the elimination of that connection. In Fig. 11.7 both weights are optimally zero so there is a point value at (0,0) that represents the optimal value of the energy surface.

5. CONCLUSION

This study has shown that a global search algorithm, such as the genetic algorithm, has the ability to identify a set of weights that will solve the training problem of the dynamic reconfigurable neural network and thus meet Harold Szu's challenge. Furthermore, if the objective function is structured in a manner to generate penalties for extraneous nodes and connections, the algorithm has the capacity to identify a parsimonious structure to solve the candidate problem.

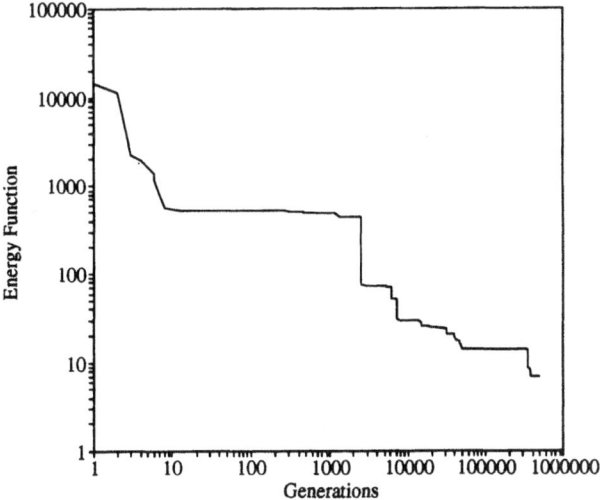

Fig. 11.5. Changes in the value of the energy function during search.

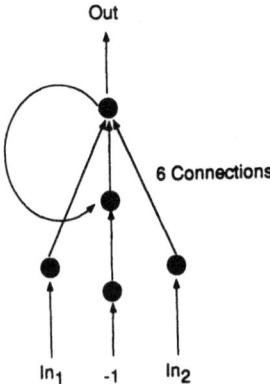

Out

6 Connections

In_1 -1 In_2

Fig. 11.6. A six connection structure for the XOR problem.

Weights	Solution
1	0.0000
2	100.2900
3	0.0000
4	0.0000
5	0.7142
6	−97.0725
7	0.0000
8	45.7195
9	48.7554
10	0.0000
11	0.0000
12	1.0000
13	0.0000

Table 11.4. Six-Connection DRNN Weights.

The reason that the DRNN is difficult to train can be seen from the complex nature of the energy surface. The existence of an extremely convoluted energy surface in conjunction with the discontinuities associated with the removal of nodes and connections makes the DRNN a virtually intractable problem for hill-climbing-based algorithms.

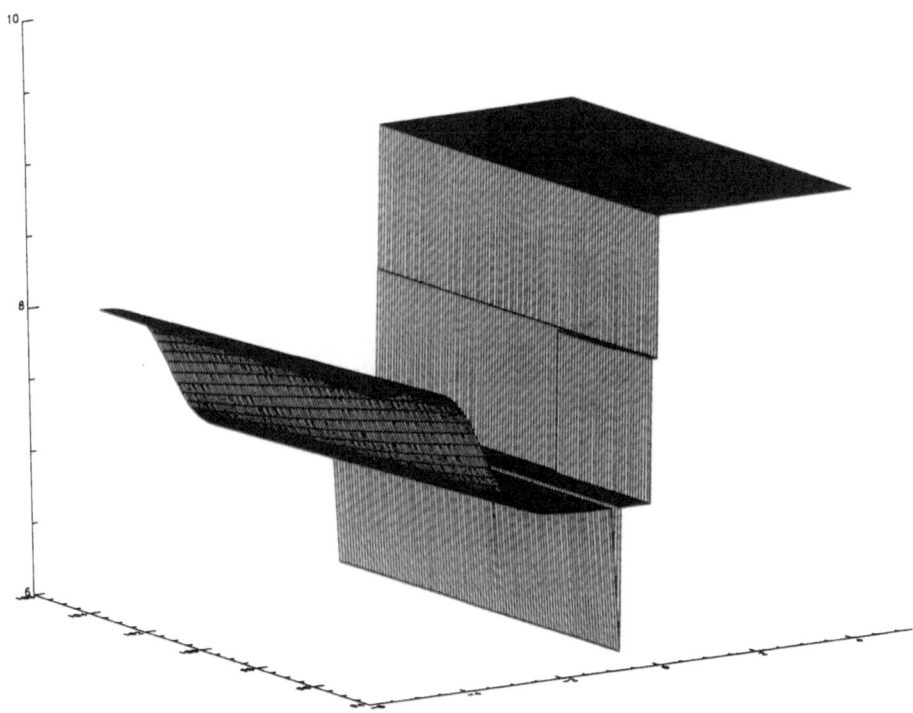

Fig. 11.7. Energy function as weights 1 and 2 are varied.

A further result of this study is the identification of a wide variety of architectures for each energy level that is able to solve the problem. Although each structure can solve the problem, the individual characteristics of convergence for each solution have yet to be explored. In addition, each of the structures mentioned above can be formed in a multitude of ways depending on the starting point (or initial condition) of the genetic evolution in an almost random way.

Because there is clearly a large number of ways to solve any particular problem, the issue of what is optimal becomes more interesting. The idea of a multitude of structures for solving the same problem resulting from different initial conditions is possibly what Karl Pribram (1991) had in mind when he wrote:

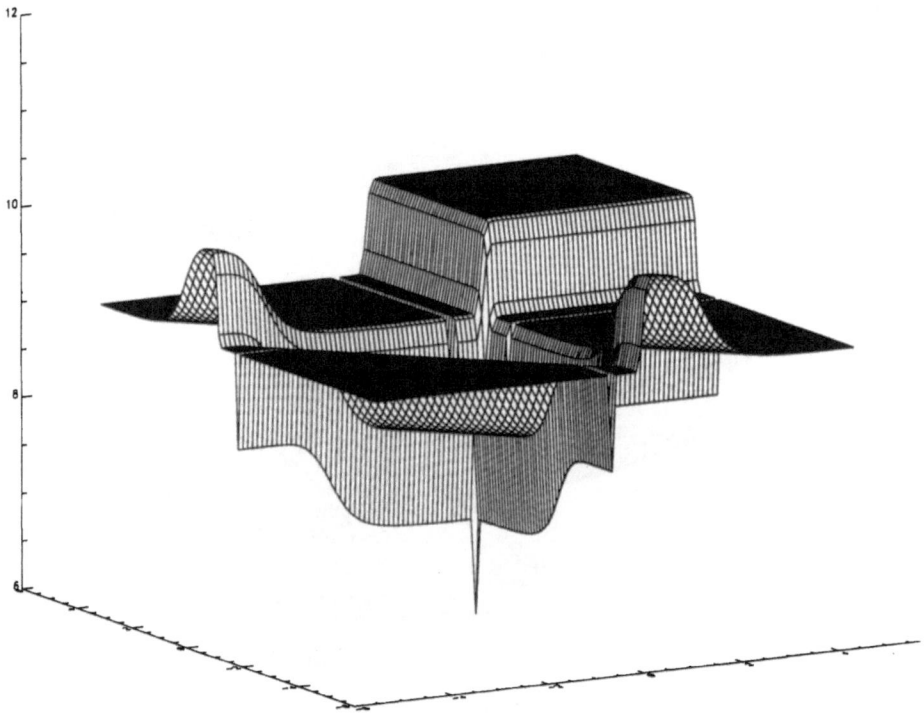

Fig. 11.8. Energy function as weights 3 and 4 are varied.

Einstein was wrong in expression if not in intent when he stated his view that God does not play dice with the universe. Indeed he does, and has six-sided cubes (numbered at that), or perhaps 10-dimensional superstrings with which to play. Playing marbles would only get him Hamiltonians: The marbles would accumulate in equilibrium structures composed of sinks of least energy. In my evening's search for relevant information, in Einstein's search for determinant structure, the books and dice are constraining initial conditions as it is of the process of shuffling the books or throwing the dice.

The genetic algorithm is inherently a random process and thus identifies different solutions depending on the particular string of random numbers that are called. In this light it is important to return to questions that have concerned researchers in the past but that have been of less concern in recent studies.

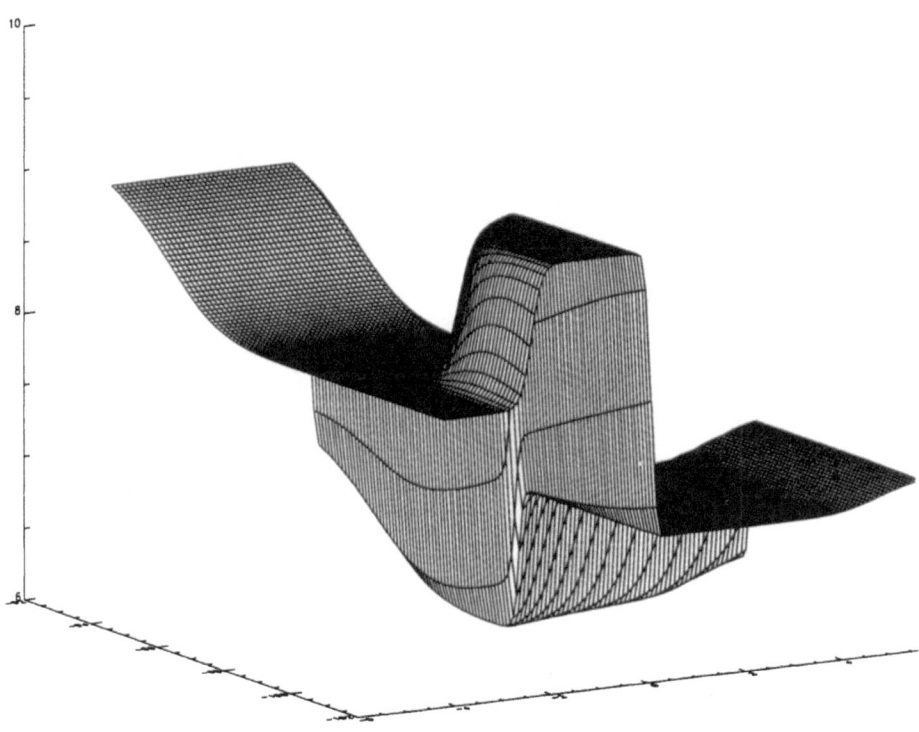

Fig. 11.9. Energy function as weights 5 and 6 are varied.

What is perceived as disorder with respect to some particular activity ordinarily results, however, from the shuffling and throwing process. On closer scrutiny, randomness could be seen to reflect the structure of the initial conditions as they become processed in shuffling, throwing, or selecting (Pribram, 1972, 1986).

If the starting point of the conditioning process matters and if a multitude of equally valid structures will yield acceptable (if not always optimal) solutions, then do these DRNN structures offer us insight in to Donald Hebb's (1949) dilemma of whether perception is to depend on the excitation of specific cells, or on a pattern of excitation whose locus is unimportant? Hebb chose the former alternative: "A particular perception depends on the *excitation of particular cells* at some point in the central nervous system."

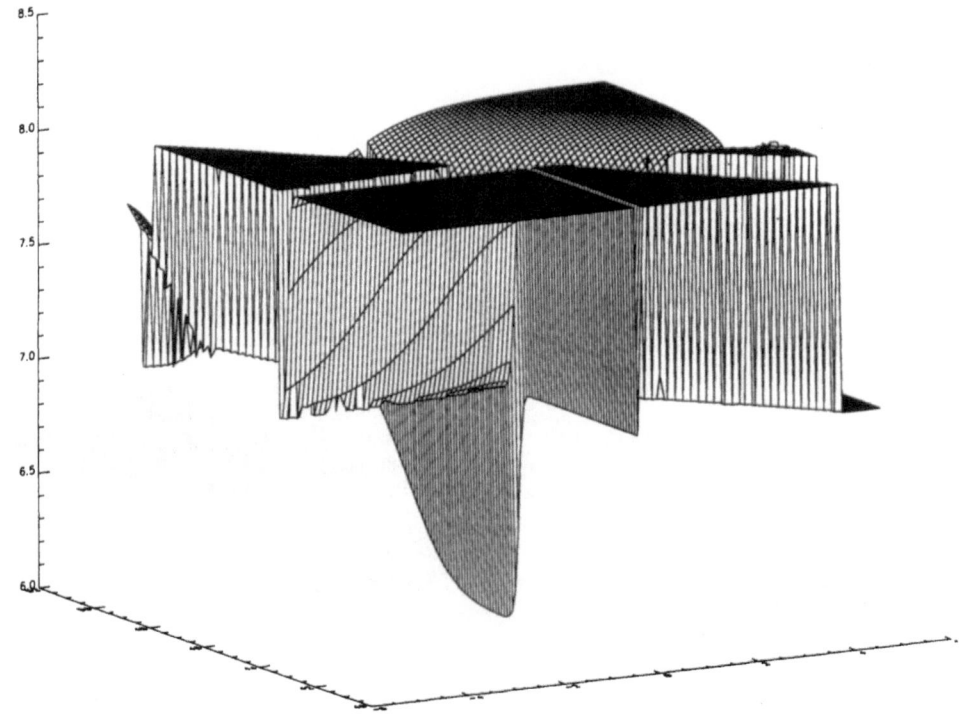

Fig. 11.10. Energy function as weights 7 and 8 are varied.

As neurophysiological evidence accumulated ... this choice, for a time, appeared vindicated: Microelectrode studies identified neural units responsive to one or another feature of a stimulating event such as directionality of movement, tilt of line, and so forth. Today textbooks in psychology, in neurophysiology, and even perception, reflect this view that one percept corresponds to the excitation of one particular group of cells at some point in the nervous system. (Pribram, 1991)

Does concentration on *specific cells* cause us to dismiss other equally plausible DRNN structures? Or should we be troubled even as in Lashley (1942):

Here is the dilemma. Nerve impulses are transmitted over definite, restricted paths in the sensory and motor nerves and in the central nervous system from cell to cell through definite inter-cellular connections. Yet all behavior seems to be determined by masses of excitation, by the form or relations or proportions of excitation within general fields of activity, without regard to particular nerve cells. It is the pattern and not the element that counts.

Because multiple DRNN structures solve the XOR problem equally well should this lead us to concentrate more on the *pattern of excitation* rather than on the excitation of specific cells? However, because there is an ultimate structure, the locus of processing is not necessarily unimportant. Is this what Pribram had in mind?

According to the views presented here and in keeping with Lashley's intuitions, this computational power is not a function of the "particular cells" and the conducting aspects of the nervous system (the axonal nerve impulses), nor is it necessarily carried out within the province of single neurons. At the same time, the theory based on these views does not support the notion that the locus of processing is indeterminate. Rather the locus of processing is firmly rooted within regions of dendritic networks at the junction between neurons.

The DRNN, trained with the genetic algorithm, offers researchers a new tool to model neural activity. It also requires modelers to reexamine what is meant by optimal and also whether or not the collective network performance rather than the individual neuron action is what is ultimately the focus.

ACKNOWLEDGMENTS

Drs. Dorsey and Johnson were supported in part by the U.S. Department of Commerce NOAA and the Research Foundation of the Institute of Chartered Financial Analysts. We would like to thank, without implicating, Harold Szu, Daniel Levine, and various participants of the MIND seminar held at the University of Texas at Dallas for helpful insights that led to the development of this chapter.

REFERENCES

Dorsey, R. E., Johnson, J. D., & Mayer, W. J. (1994). A genetic algorithm for the training of feedforward neural networks. In A. Whinston & J. D. Johnson (Eds.), *Advances in Artificial Intelligence in Economics, Finance and Management* (Vol. 1). Greenwich, CT: JAI.

Dorsey, R. E., & Mayer, W. J. (1994). Optimization using genetic algorithms. In A. Whinston & J. D. Johnson (Eds.), *Advances in Artificial Intelligence in Economics, Finance and Management* (Vol. 1). Greenwich, CT: JAI.

Freeman, W. J. (1991). On "A dynamic reconfigurable neural network." *Journal of Neural Network Computing,* **XX**, 21-23.

Goldberg, D. E. (1989). *Genetic Algorithms: in Search, Optimization and Machine Learning.* Reading, MA: Addison-Wesley.

Guha, Harp, & Samad (1989). Genetic synthesis of neural networks. *Proceedings of the 3rd International Conference on Genetic Algorithms.* San Mateo, CA: Morgan Kaufmann.

Hebb, D. O. (1949). *The Organization of Behavior.* New York: Wiley.

Holland, J. (1975). *Adaption in Natural and Artificial Systems.* Ann Arbor: University of Michigan Press.

Lashley, K. S. (1942). The problem of cerebral organization in vision. In *Biological Symposia, Vol. VII, Visual Mechanisms* (p. 306). Lancaster: Jacques Cattell Press.

Miller, G. F. (1988). *Evolution and learning in adaptive networks.* Unpublished manuscript, Psychology Department, Stanford University.

Miller, G. F., Todd, P. M., & Hegde, S. U. (1989). Designing neural networks using genetic algorithms. *Proceedings of the 3rd International Conference on Genetic Algorithms.* San Mateo, CA: Morgan Kaufmann.

Montana, D. J., & Davis, L. (1989). *Training feedforward neural networks using genetic algorithms* (Tech. Rep.). Cambridge, MA: BBN Systems and Technologies Technical Report.

Pribram, K. H. (1972). Book review of J. Monod, *Chance and Necessity. Perspectives in Biology and Medicine.*

Pribram, K. H. (1986). The cognitive revolution and mind/brain issues. *American Psychologist,* **41,** 507-520.

Pribram, K. H. (1991). *Brain and Perception: Holonomy and Structure in Figural Processing.* Hillsdale, NJ: Lawrence Erlbaum Associates.

Rumelhart, D. E., Hinton, G. E., & Williams, R. J. (1986). Learning internal representation by error propagation. In D. E. Rumelhart & J. L. McClelland (Eds.), *Parallel Distributed Processing: Exploration in the Microstructures of Cognition* (Vol. I, pp. 318-362). Cambridge, MA: MIT Press.

Szu, H. (1985). Matched filter spectrum shaping for light efficiency. *Applied Optics,* **24,** 1426-1431.

Szu, H. (1991). A dynamic reconfigurable neural network: Learning principles, paradigms, rules, and morphology. *Journal of Neural Network Computing,* **XX,** 3-19.

Todd, P. M. (1988). *Evolutionary methods for connectionist architectures.* Unpublished manuscript, Psychology Department, Stanford University.

Whitley, D., & Starkweather, T. (1990). GENITOR II: A distributed genetic algorithm. *Journal of Experiment and Theory in Artificial Intelligence,* **2,** 189-214.

Whitley, D., Starkweather, T., & Bogart, C. (1990). Genetic algorithms and neural networks: Optimizing connections and connectivity. *Parallel Computing,* **14,** 347-361.

Whitley, D., & T. Hansen, T. (1989). *The GENITOR algorithm: Using genetic recombination to optimize neural networks.* Working Paper, Computer Science Department, Colorado State University.

12

Optimization by a Hopfield-Style Network

Arun Jagota
University of North Texas

Arun Jagota's chapter, **Optimization by a Hopfield-Style Network,** *is one of two chapters in this book that deal with what are usually considered classical optimization problems — the other being Jayadeva and Bhaumik's chapter which draws analogies between some of those problems and problems in visual perception. In this case, Jagota applies a neural network based on Hopfield's to solve several problems in graph theory: graph coloring, minimum vertex and set cover, constraint satisfaction problems, and Boolean satisfiability.*

Jagota points out that his technique does not always produce **global** *best solutions, though for one of the problem types (vertex cover) the solutions are within a certain factor of optimal. (Neither, of course, does the Hopfield-Tank or any of the other known neural net algorithms for the Traveling Salesman Problem.) Rather, his contribution is in showing that a single simple network can provide solutions for several problems, all of which are sufficiently close to optimal for real-world applications. These could be considered "satisficing" solutions, perhaps, but Jagota's approach is more grounded in formal mathematics than the usual satisficing approaches which settle on the first available satisfactory solutions. He shows that a variety of different graph theoretic problems can all be reduced to the same binary weights network in a manner that is* **goodness-preserving,** *that is, does not change relative orderings of solution qualities in the energy functions. This can be interpreted as not changing preferences based on utility functions.*

ABSTRACT

We map several hard optimization problems, by reduction to MAX-CLIQUE, on to Hopfield network special case called HcN and then approximately solve them via its dynamical algorithms. These problems include graph coloring, minimum vertex and set cover, constraint satisfaction problems (*N*-queens), and Boolean Satisfiability. Approximation performance is experimentally determined on random instances. Our optimizing dynamics are discrete and provably efficient. Our broad contribution is in optimizing several problems in a *single* binary weights network, which, for *all* problems in this chapter, admits *no* invalid solutions. All reductions, except one, are goodness-preserving in a formal sense. We contrast this with the variety of handcrafted energy functions for the same individual problems in the literature, several of which admit invalid solutions also; several employ higher precision weights; and the goodness correspondence is not always clear, in a formal sense.

1. CLIQUES AND MAXIMUM CLIQUE

In a graph with undirected edges, a *clique* is a set of vertices such that every pair is connected by an edge. A clique is *maximal* if no strict superset of it is also a clique. A k-clique is a clique of size k. A clique is *maximum* if it is a largest clique. In Fig. 12.1, the vertex sets $\{a, c, d, e\}$, $\{c, d\}$, $\{a, b\}$, and $\{c, d, e\}$ are, respectively, nonclique, nonmaximal 2-clique, maximal-but-not-maximum clique, and maximum clique (size 3).

MAX-CLIQUE is the optimization problem of finding a largest clique in a given graph and is NP-hard (Karp, 1972) even to approximate (Arora et al., 1992; also see Crescenzi, Fiorini, & Silvestri, 1991; Freige, Goldwasser, Lovasz, Safra, & Szegedy, 1991; Garey & Johnson, 1979). That is, unless $P = NP$, which is highly unlikely, there is no algorithm that, for *every* given graph, can find a maximum clique — or even a clique within, for example, a constant factor of it — tractably (i.e., in time polynomial in the size of the graph).

2. OPTIMIZATION BY REDUCTION

In this section, we show how to map combinatorial optimization problems to MAX-CLIQUE in such a way that maximal cliques correspond to feasible solutions to the original problem and the size of the clique corresponds to the "goodness" of the solution. All the reductions described here have the following property: every maximal clique in a reduced instance corresponds to some feasible solution of the original problem. Our reason for exhibiting these reductions is as follows. In Section 4, we describe a special case of the Hopfield model that we call the Hopfield-clique network (HcN), which encodes the MAX-CLIQUE problem. In particular, given any graph G, we can construct an HcN instance N such that the stable states of N are exactly the maximal cliques of G. We will use this fact to solve other problems in HcN by first reducing them to MAX-CLIQUE and then reducing the resulting MAX-CLIQUE instance to HcN. Because HcN admits only maximal cliques as stable states, every HcN stable state thus represents a valid solution of the original problem.

To describe an optimization problem A we specify the *instances* I, the sets F_1 of *feasible solutions* to I, and a numerical measure $\text{goodness}_A(s:I)$ defined for $s \in F_1$. In the MAX-CLIQUE problem, the instances are undirected graphs G, F_G is the set of subgraphs of G which are cliques, and $\text{goodness}(s:G)$ is the number of vertices in s. A *reduction* from A to another optimization problem B formally consists of a pair of functions (g, h) such that for each instance I of A, $J := g(I)$ is an instance of B, and h is a function from F_J to F_1. The reduction is *goodness preserving* if for all instances I of A and feasible solutions b_1, b_2 to the corresponding instance J of B,

$$goodness_B(b_2:J) > goodness_B(b_1:J) \leftrightarrow goodness_A(h(b_2):I) > goodness_A(h(b_1):I)$$

This captures the idea that finding better solutions to B helps one find better solutions to A. The reduction of MAX-CLIQUE to HcN is goodness preserving (larger maximal cliques $\leftrightarrow$ deeper local minima; see later). All reductions to MAX-CLIQUE in this chapter, except one, are goodness preserving. Consequently, all these reductions are goodness-preserving in HcN: deeper minima formally correspond to better solutions; global minima are exactly the optimum solutions. Many of these reductions are possible only because of the NP-hardness of MAX-CLIQUE. If MAX-CLIQUE were in P, then no reduction from an NP-complete problem to MAX-CLIQUE would be possible unless $P = NP$. A neural network for graph matching such as the one described in Hertz, Krogh, and Palmer (1991) has less "computational power" in this sense (because maximum matching is in P; consequently, no NP-hard problem is reducible to maximum matching unless $P=NP$).

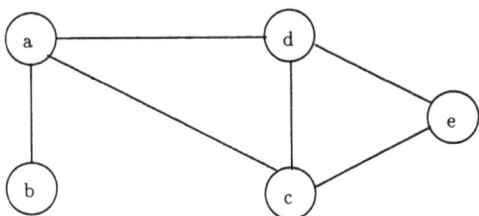

Fig. 12.1. Graph with maximal cliques $\{a, b\}$, $\{a, c, d\}$, and $\{c, d, e\}$.

3. REDUCTIONS TO MAX-CLIQUE

In this section, we present the reductions. Subsequent sections describe HcN, the optimizing dynamics, the experimental results and some theoretical issues, and comparisons with earlier work.

3.1. Set and Vertex Cover

The *minimum vertex cover* (MVC) is an NP-hard optimization problem on graphs (see Garey & Johnson, 1979); the *minimum set cover* (MSC) is its generalization to hypergraphs and thus NP-hard. MSC is also known as the *Hitting Set Problem* (Garey & Johnson, 1979). We describe MSC and then specialize its description to MVC. This terminology is from Motwani (1992). Given a vertex-set $V = \{v_1, \cdots, v_n\}$ and an edge-set $E = \{e_1, \cdots, e_m\}$ such that $e_i \subseteq V$, a subset $C \subseteq V$ such that for all $e_i \in E$: $e_i \cap C \neq \varnothing$ is called a *set cover*. That is, C is a set of vertices that cover all the edges. MSC is the problem of finding the smallest set cover in a given hypergraph instance. When, for all i, $|e_i|$ equals 2, the hypergraph is a graph and the set cover is a vertex cover. MSC then becomes MVC. MSC and MVC model many resource-selection problems. Let V represent people and edge e_i represent the subset of the people who have skill i. Then a set cover C represents a subset of the people such that for every skill, there is at least one person in C who has that skill. MVC also models computer networking (e.g., monitoring a network with the minimum number of processors).

C is a vertex cover in a graph G iff $V \setminus C$ is a clique in G_c, the complement graph of G (edges in G_c are nonedges in G and vice versa). This is one reduction of MVC to MAX-CLIQUE that we shall use on our Hopfield net special case, mainly for empirical comparisons. Our main reduction is different, and is motivated by a well-known property of vertex covers attributed to F. Gavril (see Garey & Johnson, 1979, p. 134). This property also trivially generalizes to set covers. Therefore we describe it in the set cover setting and present the MSC-to-MAX-CLIQUE reduction that exploits it. We then specialize this reduction to the MVC case.

A *matching* M in a hypergraph is a set of edges such that no two share a vertex (i.e., $\forall$ $e_i, e_j \in M$: $e_i \sqcap e_j = \varnothing$). Let $M_v = \bigcup_{e_i \in M} e_i$ denote the vertices in any *maximal* matching M. M_v is a set cover because for every edge e_i in E, $e_i \sqcap M_v \neq \varnothing$. Let $f(n) = \max_{e_i \in E} |e_i|$. A minimum set cover S^* must have at least one vertex from every edge in M. Hence $|M_v| \leq f(n)|S^*|$. That is, the vertices in *any* maximal matching are a set cover of size at most $f(n)$ times that of the minimum set cover. Clearly, the approximation may be good only when $f(n) << n$. For a graph, $f(n) = 2$. Hence the vertices in any maximal matching in a graph are a vertex cover of size at most twice that of the minimum vertex cover. This is the property that F. Gavril observed.

Our MSC-to-MAX-CLIQUE reduction exploits this property by representing maximal matchings of a hypergraph H as cliques of a new graph. The *line graph* G_l of a hypergraph H has a vertex v_i for every edge e_i of H and two vertices v_i, v_j of G_l are adjacent iff the intersection of the corresponding edges e_i, e_j of H is nonempty. Given H, our reduction constructs the line graph of H and then complements it. The k-cliques in the resulting graph G_{l_c} are k-matchings in the hypergraph H. Thus every maximal clique of G_{l_c} represents a distinct set cover of H within factor $f(n)$ of optimum. Clearly, the size of G_{l_c} (number of vertices) may be up to $\Theta(n^{f(n)})$, which can be astronomical unless $f(n) << n$. When H is a graph G and we are interested in the minimum vertex cover, G_{l_c} is the complement of the line graph of G and its size can be up to $\Theta(n^2)$ vertices.

Figure 12.2(b) shows the line graph of the graph of Fig. 12.2(a). Figure 12.2(c) shows its complement. The 3-clique $\{e_4, e_5, e_6\}$ in Fig. 12.2(c) is a 3-matching in Fig. 12.2(a).

3.2. Constraint Satisfaction Problems

The *constraint satisfaction problem* is a key problem in many artificial intelligence applications, especially in computer vision and in natural language processing. For a good overview see Mackworth (1992).

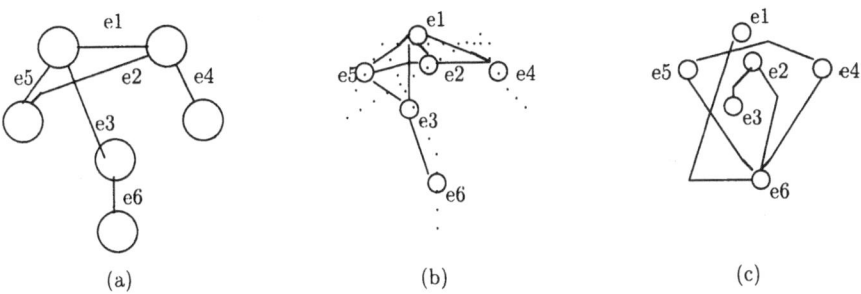

Fig. 12.2. (a) A graph, (b) its line graph, and (c) the complement of its line graph. k-cliques in (c) are k-matchings in (a).

Definition 1. A *binary constraint satisfaction problem (B-CSP)* is a set of N variables $X_1, ..., X_N$ that take values from the sets $D_1, ... , D_N$ respectively, under given binary compatibility constraints. For all variable pairs $X_i, X_j, C(X_i, X_j) \subset D_i \times D_j$ is a binary constraint set and specifies the pairs (d_i, d_j) such that $X_i = d_i$ is *compatible* with $X_j = d_j$. A *solution* of a B-CSP is an N-tuple of values to the N variables so that all pairs of value assignments are compatible.

Because graph k-coloring is a B-CSP, solving a B-CSP problem is in general NP-hard.

N-Queens Example. The N-queens problem is one of placing N queens on an $N \times N$ chessboard so that no queen threatens another. The *standard* formulation of the N-queens problem as a B-CSP uses N variables: $X_i = j$ represents a queen on column i and row j. The value set D_i of each variable is $\{1, ..., n\}$, denoting the n possible row values. The choice of variables enforces the *column constraints* of no more than one queen per column; the similar row and diagonal constraints are represented explicitly as $C(X_i, X_j)$.

Standard Representation of B-CSPs. A B-CSP is commonly represented as a *constraint graph* $G=(V, E)$. There is a vertex v_i for every variable X_i. An edge (v_i, v_j) represents the binary constraint between X_i and X_j, and is labeled by the set $C(X_i, X_j)$ of compatible assignments (see Mackworth, 1992). If we consider only B-CSPs that have solutions, all constraint sets $C(X_i, X_j)$ must be nonempty. We can then choose a convention in which two vertices v_i, v_j are nonadjacent if and only if their constraint set is *complete*. That is, $C(X_i, X_j) = D_i \times D_j$. This convention is space efficient.

B-CSPs as MAX-CLIQUE. Our way of representing a B-CSP is different — we transform it to a graph so that all its maximal feasible solutions are explicitly represented as maximal k-cliques of that graph (solutions as N-cliques). We call this representation a *B-CSP clique-graph.* The reduction is as follows. For an N-variable B-CSP, define an N-partite graph G, partition i representing variable X_i. Partition i contains $|D_i|$ vertices, one for each possible value of X_i and named by the value. We make $C(X_i, X_j)$ the set of edges between vertices in partition i and partition j. Every compatibility constraint of the B-CSP is represented as an edge of the graph.

The fact that the vertex set of a partition is an independent set (no edges) enforces the constraint that a variable may be assigned at most one value.

It is clear that our reduction, in one sense, involves essentially only a language change and that there is a one-to-one correspondence between maximal sets of mutually compatible assignments to k variables of a B-CSP and k-cliques of its clique-graph. Hence the set of N-cliques is exactly the set of the B-CSP's solutions.

Figure 12.3 shows a clique graph representation of the 3-queens problem, according to its standard B-CSP formulation (see the N-queens example).

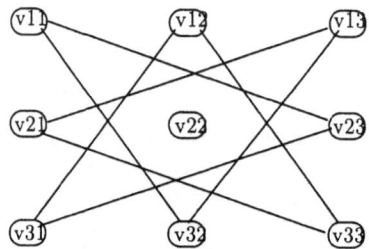

Fig. 12.3. 3-queens B-CSP clique-graph. Edges represent compatible assignments.

Space Requirements. The N-queens constraints are analytical, and need not be stored explicitly in conventional methods. The constraints are $X_i \neq X_j$, $X_i - X_j \neq j - i$, and $X_i - X_j \neq i - j$, for each pair of variables (Nadel, 1989). In our method, however, the constraints must be stored explicitly. HcN is fully-connected; it requires $\Theta(N^4)$ connections (3,123,750 for N=50). HcN has nonzero-valued weights only for the nonedges. We have calculated that the number of non-edges (forbidden pairs) in the N-queens clique-graph is $\Theta(N^3)$ ($\approx$125,000 for $N = 50$). To represent arbitrary B-CSP instances, $\Theta(\Sigma|D_i|)^2$ connections are required.

One attractive feature of our representation is that all maximally compatible partial solutions (includes full solutions) are represented explicitly; further, its incorporation into HcN's parallel architecture potentially provides a constant-time (i.e. independent of N) test for an N-tuple to be a solution. Second, new parallel algorithms for solving B-CSPs emerge from its graph and neural network connection (see our results section).

3.3. Boolean Formulas

Satisfiability of Boolean formulas is in general NP-complete (Garey & Johnson, 1979). MAX-SAT, the problem of finding the maximum number of clauses satisfiable by a value assignment, is NP-hard. MAX-SAT is the optimization problem that is experimentally approximated in this chapter. We employ a recent reduction of conjunctive form (CF) Boolean formulas to cliques in graphs (Crescenzi, Fiorini, & Silvestri, 1991). We view this reduction differently than in Crescenzi et al. (1991) — as a representation of any CF Boolean formula as a B-CSP, followed by the representation of the B-CSP as a B-CSP clique-graph.

As an example of this reduction, consider the CF formula $(v_1 + v_2)(\bar{v}_1 + v_2)(v_1 + \bar{v}_2)$ consisting of a conjunction of three clauses c_1, c_2, c_3. We represent it as a B-CSP with variable X_i for clause c_i. The value set of X_i is the set of literals in clause c_i. In this case, $D_1 = \{v_1, v_2\}$, $D_2 = \{\bar{v}_1, v_2\}$, and $D_3 = \{v_1, \bar{v}_2\}$. The constraint sets are: $C(X_i, X_j) = (D_i \times D_j)\setminus\{(u_i, u_j)\}$ where u_i and u_j represent different values to the same Boolean variable (v_1 or v_2 in this case). That is, the only incompatible assignments are if two clauses assign different values to the same Boolean variable. Figure 12.4 is the B-CSP clique graph for this example. The B-CSP clique graph of a formula of m clauses is m-partite. The k-cliques represent assignments that satisfy k clauses, so m-cliques represent assignments that satisfy the formula. In this example, the 3-clique $\{c_1v_1, c_2v_2, c_3v_1\}$ says that $v_1 = v_2 = T$ satisfies all three clauses in the formula.

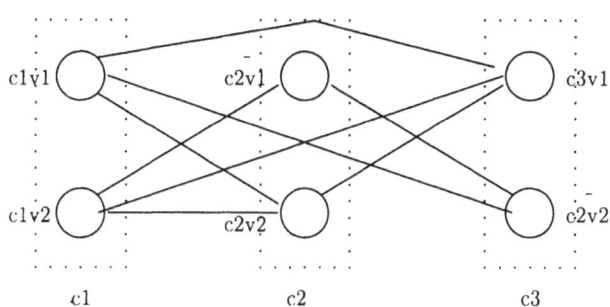

Fig. 12.4. B-CSP clique-graph for the formula $(v_1 + v_2)(\bar{v}_1 + v_2)(v_1 + \bar{v}_2)$.

3.4. Graph K-Coloring

The graph k-coloring optimization problem is to label as many of the vertices of a graph with one of k colors as possible, such that no pair of adjacent labeled vertices has the same color. The problem is NP-hard for $k \geq 3$ (see Garey & Johnson, 1979). The problem has applications to resource partitioning. It is well known that graph k-coloring is a B-CSP. We use this connection to reduce graph k-coloring to MAX-CLIQUE but skip the intermediate B-CSP step in the description. We assume graph-theoretic terminology.

Given an n-vertex graph G and the number of colors k, a new n-partite graph G' with k vertices in each partition is created. Then $V_i = \{v_{i_1}, \cdots, v_{i_k}\}$ is the vertex-set of partition i. v_{i_j} denotes vertex v_i of G having color j. The set of edges between vertices in partitions i and j is $V_i \times V_j$ if (v_i, v_j) is not an edge in G; it is $(V_i \times V_j) \setminus \{(v_{i_l}, v_{j_l})\}$ for $l=1, \cdots, k$ if (v_i, v_j) is an edge in G. There is an exact correspondence between m-cliques, $1 \leq m \leq n$, in G' and sets of m

vertices of *G* properly colored. Figure 12.5 shows a graph and its 2-coloring MAX-CLIQUE reduction.

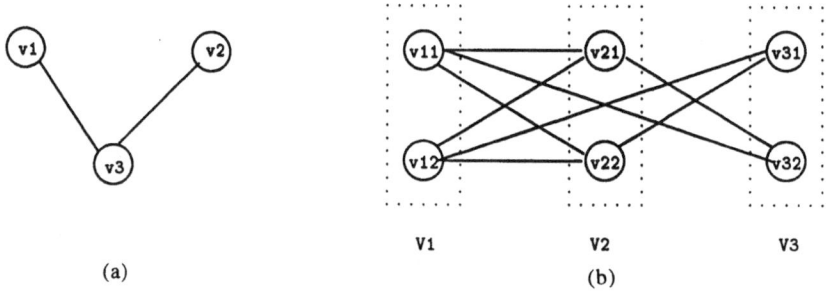

Fig. 12.5. (a) A graph and (b) its two-coloring reduction to MAX-CLIQUE.

3.5. Common Vertex-Induced Subgraphs

Consider a pair of graphs *G* and *H*. A *common vertex-induced subgraph (CVS)* is a subgraph, induced by a set *V'* of vertices, that is common to both *G* and *H* (up to isomorphism). Figure 12.6 shows two graphs and one of their common vertex-induced subgraphs.

Applications of CVS. The problem of graph isomorphism (are *G* and *H* isomorphic?) is a special-case: Find the largest common vertex-induced subgraph. If it is *G* (or *H*) the answer is yes, else no. Common subgraphs are also useful for pattern matching: geometric or otherwise. One example is to find the largest common pattern between an image and a model, with the essential features of both modeled as graphs.

Representing CVS as Cliques. We use the reduction of Barrow and Burstall (1976) to represent common vertex-induced subgraphs of a given pair of graphs *G* and *H* as cliques of a new graph *G H*. We view this reduction as one to a B-CSP. The set of variables is $X = V(G)$. The domain of every variable is $D=V(H)$. The value $x_i = j$ is compatible with $x_k = l$ if and only if

$$(i,k) \in E(G) \leftrightarrow (j,l) \in E(H)$$

The graph *G H* is just the clique-graph of the above B-CSP.

k-vertex common vertex-induced subgraph of *G* and *H*} $\leftrightarrow$ *k*-clique of *G H*.

Fig. 12.7 shows the graph *G H* for the pair of graphs *G* and *H* of Fig. 12.6.

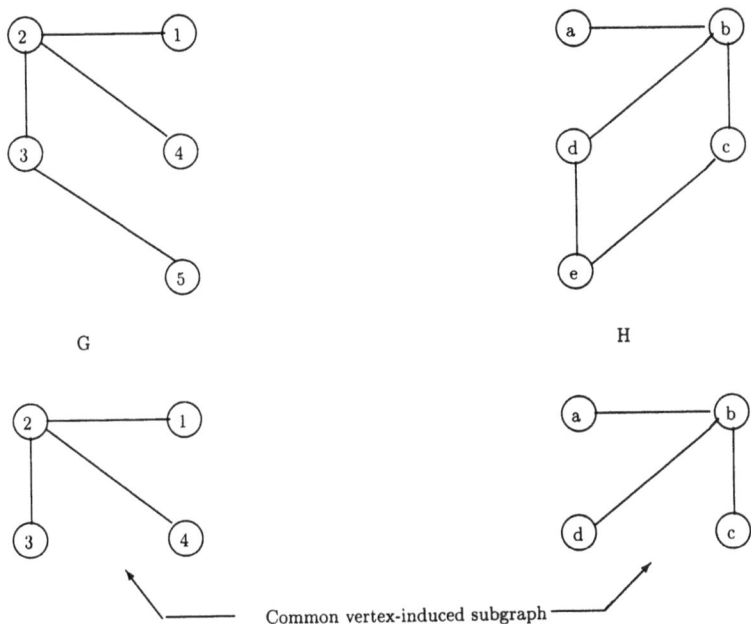

Fig. 12.6. Graphs G and H and their common subgraphs induced by the vertex sets $\{1,2,3,4\}$ and $\{a,b,c,d\}$ respectively.

4. HOPFIELD-CLIQUE NETWORK

The Hopfield network (Hopfield, 1982, 1984) is a recurrent N-unit network closely related to Ising spin models. $\mathbf{W}=(w_{ij})$ is a real, symmetric with zero diagonal, matrix of weights, describing the interconnection weights of all unit pairs i, j. In the discrete version (Hopfield, 1982), unit states are: $\mathbf{S}_i \in \{0,1\}$. For $x \neq 0$, let $\theta(x) = 1$ if $x < 0$ and 0 otherwise. The network state $\mathbf{S}=(S_i)$ is updated serially as follows: at time t exactly one unit i is picked and updated according to the rule $S_i(t) := \theta((\mathbf{WS}(t - 1))_i)$. This prescribes a family of serial-update rules, updates according to any of which monotonically decrease the energy function $E = -(1/2)\mathbf{S}^T\mathbf{WS} - \mathbf{I}^T\mathbf{S}$ (Hopfield, 1982), thus guaranteeing eventual convergence to a discrete local minimum of it. Here $\mathbf{I} = (I_i)$ is the vector of external biases to units.

Hopfield-Clique Network. The Hopfield-clique network (HcN) (Jagota, 1990b, 1992a) is a binary weights special case of the Hopfield network. For all $i \neq j$, $w_{ij} \in \{\rho,1\}$, where $\rho < -n$. For all i, $w_{ii} = 0$. $\mathbf{I} = b^n$ is the vector of external unit biases; $0 < b \leq 1$. $G_N=(V,E)$ is the graph underlying it whose vertices are the units and there is an (undirected) edge between vertices i and j if and only if $w_{ij} = 1$. V is the set of vertices. The discrete HcN has $\mathbf{S}_i \in \{0,1\}$. A set of vertices $V' \subset V$ is an alternative description of the network state $\mathbf{S} = (S_i) \in \{0,1\}^N$ — the elements of V' are the units that are ON. We use these notations interchangeably. **S** is a stable state if and only if $\theta(\mathbf{WS} + \mathbf{I}) = \mathbf{S}$, that is, a local minimum state of the discrete energy function.

    Lemma 1 (Grossman & Jagota, 1992; Jagota, 1990b). The HcN stable states are exactly the maximal cliques of its underlying graph G_N.

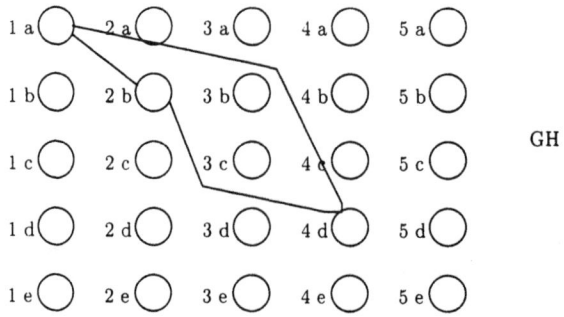

Fig. 12.7. Graph $G\,H$ whose k-cliques represent common vertex-induced subgraphs of G and H of order k. Not all edges are shown. A clique $\{1a,2b,4d\}$ is shown, which represents a common vertex-induced subgraph of order 3 in G and H.

Lemma 1 is straightforward, and closely related versions may also be found in Godbeer, Lipscomb, and Lubey (1988) and Shrivastava, Dasgupta, and Reddy (1990).

    Lemma 2. The energy of a clique of size C is $-\binom{C}{2}$.

    Lemma 2 may also be trivially verified by calculation of energy of a clique. From Lemma 2, the encoding of MAX-CLIQUE in HcN is goodness preserving; it follows as a corollary that maximum cliques are exactly the global minima.

    The associative memory storage rule of HcN may be used to store graphs (of reduced problems). Instances of all problems in this chapter are easier stored in HcN in terms of reductions to maximum independent set, MIS (rather than MAX-CLIQUE). For this reason, we describe a version of the storage rule that complements the graph (to convert an MIS reduction to a MAX-CLIQUE one) before storage. The initial weight state is: For all $i \neq j$, $w_{ij}(0) = 1$. A set S of vertices is stored as follows. For all $i \neq j$, $w_{ij}(t + 1) := \rho$ if units i and j are in the set S; $w_{ij}(t + 1) := w_{ij}(t)$ otherwise. A graph G may be stored by its edges. Because G is complemented, after storage, the stable states are maximal cliques of G_c and hence maximal independent sets of G. G may also be stored by larger clique covers, which is more efficient (in practice we exploit this because some of our reductions conveniently allow us to).

Recall that in serial updates, exactly one unit i is picked and updated according to the rule $S_i(t) := \theta((W S(t - 1))_i)$. Any serial-update (energy-descent) dynamics on discrete HcN converges in $\leq 2N$ unit-switches (Grossman & Jagota, 1992; Jagota, 1994). An equivalent result is in Shrivastava et al. (1990)

Steepest Descent. Steepest descent (SD) is an instance of serial-update dynamics in which the unit i whose switch reduces the energy maximally is picked to switch. Specifically, i satisfies $\Delta E_i \leq \min_j \Delta E_j < 0$ for some choice of $S_i(t)$ and all choices of j and $S_j(t)$. Here, $\Delta E_k = [S_k(t) - S_k(t - 1)] (WS(t - 1))_k$ is the energy change caused by the particular switch of k. Let $SD(V_0)$ denote, starting with $V_0 \subseteq V$ as initial state, updating units via steepest descent until HcN converges to a stable state. $SD(V_0)$ is characterized as a graph algorithm (Jagota, 1990a; also see Grossman & Jagota, 1992; Jagota, 1992a). Efficient convergence in $\leq 2N$ unit switches follows from this characterization (although it follows directly from the more general serial-updates result). $SD(V)$ and $SD(\varnothing)$ emulate two greedy large-clique-finding algorithms (Jagota, 1990a, 1992a). that are well known in the graph algorithms literature (see Griggs, 1983; Karp, 1976).

Stochastic Steep Descent. Stochastic steep descent (SSD) is a stochastic variant of SD. In SSD, the deterministic moves of SD are replaced by *gradient descent* moves that are stochastic but favor the steepest direction. The motivation is, with initial state V, to improve on the already good optimization performance of $SD(V)$ (see Jagota, 1992a) without a significant risk of worsening it. The unit to be updated is selected via a probability distribution P that has zero probability of "uphill" moves and favors large (steepest-descent like) decreases in energy. Specifically, let $C(t) \equiv \{i \mid \Delta E_i(t) < 0\}$. Then $P[S_i$ is switched at time $t] = 0$ if $\Delta E_i(t) \geq 0$; $\delta[\Delta E_i(t)]$ otherwise. $\Sigma_{i \in C(t)} \delta[\Delta E_i(t)] = 1$. P ensures that exactly one unit is switched. Our choice of δ (hence P) that approximates SD is the linear distribution, represented by $\delta[\Delta E_i(t)]$

$$= \frac{\Delta E_i}{\sum_{j \in C(t)} \Delta E_j}.$$ The probability of switching a unit is proportional to the amount of energy the

switch decreases. SSD performs only gradient descent moves for two reasons: (a) to approximate SD and (b) because any serial update gradient descent update scheme on HcN converges in $\leq 2N$ unit switches. Nongradient-descent heuristics like simulated annealing are much slower. The idea behind SSD is that multiple runs on the same input will produce different solutions, and because it approximates SD, it is expected to produce at least one solution that is better. Let $SSD_{max}(V_0, i)$ denote i runs of SSD on the same graph, with V_0 as input for each run. The best clique found is chosen. Much better empirical MAX-CLIQUE approximations than from one run validate this approach (Jagota & Regan, 1992).

5. CONNECTIONS WITH EARLIER NEURAL NETWORK APPROACHES

A word on well-known notation: minimal and maximal are analogous to local optima; minimum and maximum are analogous to global optima.

5.1. Vertex Cover

An algorithm for vertex cover has a *ratio bound f(n)* if on any *n*-vertex graph, $\frac{C}{C^*} \leq f(n)$, where C is the size of the vertex cover obtained by the algorithm and C^* is the size of its minimum vertex cover.

The earliest Hopfield net encoding of the vertex cover problem that we have found is described in Godbeer et al. (1988, p. 19) and attributed to M. Luby (1986) as personal communication. Given a graph $G=(V,E)$, a Hopfield net is constructed with weights $w_{ij} = -1$ for $(i,j) \in E$ and $w_{ij} = 0$ otherwise. Furthermore $w_i = \deg(i)-(1/2)$ where w_i is a bias to each unit and $\deg(i)$ is the degree of vertex i in G. Minimal vertex covers are stable states (Godbeer et al., 1988). It is easy to check that infeasible solutions are not admitted as stable states, for if there is an uncovered edge e, that is, one whose endpoints u and v are both OFF, then both u and v receive positive input and one of them will switch ON. In Godbeer et al. (1988) there were no experimental results with this encoding, and we are not aware of any in the literature. However, we can show that the ratio bound given by this encoding is poorer than the factor of 2 guaranteed both by the matching observation and by its realization in HcN. We apply the following example from Motwani (1992, p. 48) to the encoding of Godbeer et al. (1988). Figure 12.8 shows the situation.

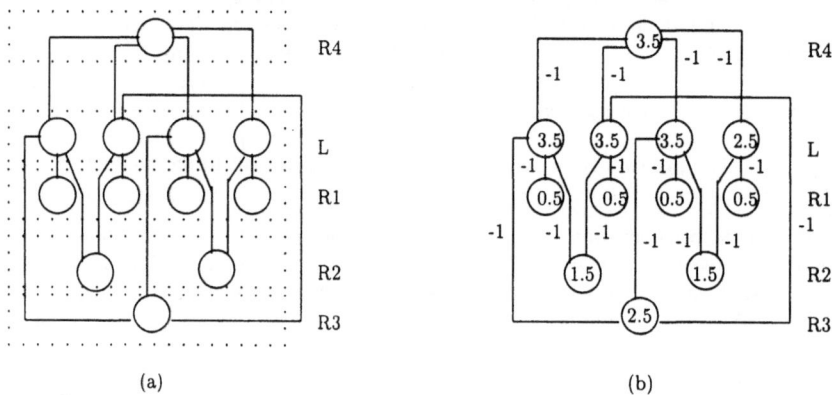

(a)   (b)

Fig. 12.8. (a) Graph and (b) its vertex cover encoding in a Hopfield network. The encoding has a large stable state R, though the minimum vertex cover L is small.

L is a minimum vertex cover, but it is easy to check that R is also a network stable state, which may thus be retrieved (even with serial steepest descent, in which a unit that maximally reduces the energy is switched, depending on how ties break). By generalizing this example to $|L| = r$ and r groups $R_1, \cdots, R_r$ of vertices such that every vertex in R_i is connected to i different vertices in L, it is easy to check that the ratio bound is $\Omega\left(\dfrac{|R|}{|L|}\right) = \Omega(\log r)$, where $R = \sqcup_i R_i$.

A second Hopfield net encoding of the vertex cover problem (Ramanujam & Sadayappan, 1988) is as follows. It employs the energy function

$$E = A \sum_{i,j} V_i V_j + \frac{B}{2} \sum_{i,j} [(1 - V_i)(1 - V_j)a_{ij}]$$

The first term minimizes the number of ON vertices; the second term minimizes the number of edges uncovered by the ON vertices. A and B are nonnegative constants that control the relative importance of these terms. Given a graph $G = (V, E)$, a Hopfield net is constructed with weights $w_{ij} = -2A - Ba_{ij}$ and unit biases $w_i = B \sum_j a_{ij} = B \deg(i)$ where $\deg(i)$ is the degree of vertex i in G. $\mathbf{a} = (a_{ij})$ is the adjacency matrix of G.

It is worth noting that this encoding admits stable states that correspond to infeasible solutions to MVC if the constants A and B are chosen injudiciously. There is no discussion on issues of choosing A and B in Ramanujam and Sadayappan (1988). Consider the following example: a graph consisting of two vertices v_1 and v_2 that are adjacent. It is easy to see that when $A > B/2$, $E(\varnothing) < E(\{v_1\}) = E(\{v_2\})$ and hence $\varnothing$, an infeasible solution, is a stable state. ($\varnothing$ is an infeasible solution because it does not cover any edge, let alone all edges, of the graph.) By similar reasoning, for an arbitrary graph G, if $A > (B/2)\Delta$, where Δ is the maximum degree in G, then $\varnothing$ is a stable state. This is because for any vertex v, $E(\{v\}) - E(\varnothing) \geq A - (B/2) \Delta$ as at most Δ edges get covered by switching v ON.

If $A < B/2$, on the other hand, then this encoding does not ever admit infeasible solutions. This is easy to see. Consider an infeasible solution. There is at least one uncovered edge e, that is, one whose endpoints u and v are OFF. Switching u or v ON increases the first term in the energy E by the amount A and decreases the second term by at least the amount $B/2$, because at least the edge e that was not covered earlier becomes covered. The cumulative energy change is thus negative and the infeasible solution cannot be a stable state.

It is also useful to note that setting A to 0 is problematic. Because the second term is zero-valued for every vertex cover, this setting (a) makes every vertex cover a stable state, (b) with the same energy value. No experiments are reported with this encoding in Ramanujam and Sadayappan (1988), and we are not aware of any in the literature.

The second issue of this encoding, compared to ours, is that the weights and biases are real-valued. In our encoding, the weights are binary-valued: $\{0, -1\}$. Our encoding is thus easier to implement in hardware. A third issue is that the ratio bound of this encoding is not clear.

A third approach (Shrivastava et al., 1990) encodes maximal independent sets (MIS) (the authors use their own term, irredundant isolated graphs) as stable states in the Hopfield model. It is based on the well-known fact that if I is an independent set then W, that is, the set of OFF units in a stable state, is a vertex cover. Their encoding was earlier presented in Godbeer et al.

(1988, p. 17). However, in Shrivastava et al. (1990), a very interesting and initially surprising (to this author) result was proven: using synchronous updates, the MIS-encoded network converges in zero or one steps to a fixed point and two or three steps to a two-cycle. No experiments were reported, however.

We employ two encodings of the vertex cover problem in HcN. Our first encoding (encoding 1) is based on the reduction of minimum vertex cover to MAX-CLIQUE based on the idea of maximal matchings. Encoding 1 has ratio bound of 2 (all stable states represent vertex covers of at most twice the optimum size). This holds regardless of the dynamical rule.

Our second encoding (encoding 2) of vertex cover is based on the following more common reduction of minimum vertex cover to MAX-CLIQUE. A graph G is converted to its complement graph G_c on which MAX-CLIQUE is approximated using HcN. The obtained clique in G_c is an independent set in G; hence the set of OFF units is a vertex cover of G. Encoding 2 is essentially the same as the encoding in Godbeer et al. (1988) and Shrivastava et al. (1990).

It is clear that neither encoding 1 nor encoding 2 admits infeasible solutions. The set of stable states of encoding 2 represents exactly the set of minimal vertex covers. The set of stable states of encoding 1, on the other hand, represents only those vertex covers that correspond to maximal matchings. In fact, the minimum vertex cover may not be in this latter set.

Although encoding 2 (and the one in Shrivastava et al., 1990) works well on uniformly-at-random instances (see Section 6), the same example of Fig. 12.8(a) illustrates that its ratio bound is poor [again, $\Omega(\log r)$]. This time, L, which is a maximal independent set, may be retrieved as the stable state from which the vertex cover $R = V \backslash L$ is obtained. But L is the minimum vertex cover and $\frac{|R|}{|L|} = \Omega(\log r)$. Again, even steepest descent, with initial state V, may retrieve L, depending on how the ties break.

5.2. Constraint Satisfaction Problems (CSPS)

Two early neural net approaches to (a) graph coloring — a constraint satisfaction problem (CSP) and (b) general CSPs are given by Ballard, Gardner, and Srinivas (1987) and Tagliarini and Page (1987), respectively. The eight-queens problem was solved in Akiyama, Yamashita, Kajiura, and Aiso (1989), and a solution of 1000 queens was alluded to in Kajiura, Akiyama, and Anzai (1989) but no details provided.

An efficient discrete near-Hopfield network for CSPs was proposed in Adorf and Johnston (1990). Some parts of the network use asymmetric connections; thus convergence is not guaranteed. However, in practice (Adorf & Johnston, 1990) it worked very well on N-queens [finding exact solution in empirically observed $O(N)$ unit-switches]. The 1024-queens problem was solved in minutes. Our reduction of CSPs to MAX-CLIQUE in this chapter was first presented in a poster (Jagota, 1991). A neural net approach to CSPs using hidden units was proposed in Baram and Dechter (1991). The number of hidden units required scales poorly with the CSP size; hence its use is limited to small CSPs. One of the apparent motivations was to establish a one-to-one correspondence between the CSP solutions and global minima. Our approach provably exhibits the same correspondence. A continuous Hopfield net formulation of N-queens was presented in Takefuji (1992a), but no empirical results were discussed. A more recent paper on neural approach to CSPs is Bourret and Gaspin (1992). The encoding is unclear and no empirical results are given. The authors are apparently unaware of earlier neural approaches to CSPs.

It is useful to examine the energy function in Akiyama et al. (1989). Let v_{rc} denote the vertex at row r and column c in an N-queens encoding. In Akiyama et al. (1989), the units are real-valued in [0, 1]. The energy function in Akiyama et al. (1989), exactly as reported there, is

$$E = A/2 \sum_{r=1}^{n} (\sum_{c=1}^{n} v_{rc} - 1)^2 + B/2 \sum_{c=1}^{n} (\sum_{r=1}^{n} v_{rc} - 1)^2 +$$

$$d/2 \sum_{d=1}^{2} n \sum_{r+c=d} \sum_{R \neq r, R-C=d} v_{rc} v_{RC} + (A+B)/2 \sum_{r,c} v_{rc}(1 - v_{rc})$$

We now show that if the energy function is restricted to the corners of the $[0, 1]^n$-cube, that is, if the states are binary, then it may admit infeasible solutions if the constants are not chosen judiciously. On the $\{0, 1\}^n$ space, the last term may be discarded as it is always zero-valued. Consider the two-queens instance. It is easy to check that $E(\{v_{11}\}) = E(\{v_{22}\}) = (A/2 + B/2)$ and $E(\{v_{11}, v_{22}\}) = D$. Furthermore, for any strict superset S of $\{v_{11}, v_{22}\}$, $E(S) > E(\{v_{11}, v_{22}\})$. Thus with $(A+B)/2 > D$, $\{v_{11}, v_{22}\}$, which is an infeasible solution, is a stable state. The settings suggested in Akiyama et al. (1989) are $A = B = C = D = 1$ without explanation.

Our approach in the current chapter admits only feasible solutions, that is, maximally compatible sets of queens, as stable states. The N-queens formulations in Adorf and Johnston (1990), Akiyama et al. (1989), Jagota (1991), and Takefuji (1992) all use N^2 neurons; the one in Adorf and Johnston (1990) uses N additional "guard" neurons.

5.3. Boolean Satisfiability

Two previous papers that have encoded Boolean formulas in Hopfield networks are Chen and Hsieh (1989) and Pinkas (1991). The former appears only as a short abstract, whereas Pinkas (1991) does not discuss its usage for solving Boolean satisfiability (SAT) per se. The encoding of SAT in HcN used in this chapter differs from this author's own earlier work (Jagota, 1990a). The latter uses the reduction of k-SAT to MAX-CLIQUE indirectly described in Garey and Johnson (1979). This reduction does not lend itself to approximate solution of MAX-SAT. The former is based on a somewhat simpler reduction (Crescenzi et al., 1991), which is also well-suited to approximate solution of MAX-SAT. Both our encodings use binary weights. The weights in Pinkas (1991), although of low precision, are integer-valued.

5.4. Graph K-Coloring

Because graph k-coloring is a CSP, there is some overlap between the early work we list here on graph k-coloring and what we listed earlier on CSPs. One of the earliest neural net approaches to graph coloring appears to be Dahl (1987) and is based on the Hopfield model.

Another approach at about the same time is given in Ballard et al. (1987). They map graph coloring to Hopfield networks by reduction to maximum weighted independent set. This approach is similar — in strategy and in detail — to our mapping in this chapter. One main difference is that their reduction to weighted independent set requires an extra set of control vertices; ours doesn't. The second difference is that their reduction is based on $\Delta + 1$ colors, where Δ is the maximum degree of the original graph G; our reduction inputs the number of colors k as a parameter. What they gain by these differences is the mapping of the graph minimum coloring problem; our mapping is only for the graph k-coloring problem. They prove that global optima in their mapping correspond to minimum colorings. The optimization algorithms in Ballard et al. (1987) are, however, quite different than the dynamics we employ in the current chapter. In Ballard et al. (1987), no graph coloring experiments are reported.

The Boltzmann machine is used for graph coloring in Aarts and Korst (1989, p. 159) (also see Korst & Aarts, 1989). The mapping of graph coloring to the Boltzmann Machine (Aarts & Korst, 1989) is structurally similar to the one in Ballard et al. (1987); Aarts and Korst (1989) were apparently unaware of this somewhat earlier work. Both reductions may be viewed as to maximum weighted independent set and both employ $\Delta + 1$ colors. The approach of Aarts and Korst (1989) differs from the one in Ballard et al. (1987) by not using the extra control vertices; rather, it uses varying thresholds for the other vertices to reward colorings that use smaller number of colors. The mapping in Aarts and Korst (1989) is also similar to our mapping of graph k-coloring to HcN in the current chapter.

The description in Aarts and Korst (1989) of their mapping is somewhat more complicated. First the mapping is described in combinatorial terms, then the weights and biases are derived, and then it is proven (Theorem 9.3), in somewhat lengthy fashion, that their mapping is feasible and order preserving. By contrast, our mapping methodology is different. We had first shown, only once, that maximum clique maps to HcN in feasible and order-preserving fashion. We then map other problems to HcN by reducing them to maximum clique. In doing so, we need only be concerned with properties of their reduction to maximum clique. This simplifies at least the exposition considerably. On the other hand, the mapping of Aarts and Korst (1989) is, like the one of Ballard et al. (1987), for the graph minimum coloring problem. They formally proved that global minima correspond to minimum coloring solutions. Our mapping is for the graph k-coloring problem (k-coloring solutions are the global minima). Consequently their mapping is more significant; in HcN we can only approximately solve the graph coloring problem indirectly by constructing different HcN instances for different k. The graph coloring experimental results reported in Aarts and Korst (1989, p. 173, Table 9.3) are not comparable with our experimental results (Table 12.4 in the following section). First, their test set comprises random graphs with expected degree a constant (independent of number of vertices N); ours comprises random graphs with expected degree $p(N - 1)$ where $p \in [0, 1]$ is a constant. Second, in their exposition, their emphasis is on running time comparisons, not solution quality ones. Indeed, their Table 9.3 results only show the local energy minimum values in their solutions, not the number of vertices properly colored. Finally, their objective is to solve the graph coloring problem, and ours is to approximately solve the graph k-coloring problem (in particular to find the maximum number of vertices colorable with k colors).

A third approach is in Adorf and Johnston (1990), in which experiments on graph 3-coloring are reported. Their test graphs are essentially 3-colorable random N-vertex graphs as in Caspi (1992) (the generation mechanism is similar but differs in detail slightly). Their network does not have guaranteed convergence. It converges in N transitions on one set of graphs, but the empirically observed probability of convergence within $9N$ transitions decreases exponentially with increasing N.

A recent approach is in Caspi (1992) where the Rochester Connectionist Simulator is used to solve graph coloring. Caspi experimentally showed that his neural network can find a full solution to the k-coloring problem (i.e., color all the vertices with k colors) for most k-colorable random graphs. Almost all k-colorable random graphs are uniquely k-colorable and hence k-chromatic (Turner, 1988). Thus Caspi's full solutions to the graph k-coloring problem are almost always optimal solutions to the graph minimum coloring problem. However, these experiments also turn up some pathological cases of k-colorable random graphs that are hard to fully solve by their approach. Their empirical results are not comparable with the results of our chapter because our test set is composed of random graphs, not random k-colorable graphs. Finding optimal colorings in random graphs is probably hard. In any case, our dynamics are all efficiently convergent in time independent of the problem instance (depending only on n) and applied mainly for obtaining quick approximation, whereas Caspi's approach can occasionally be much slower (and does not have a proof of convergence).

6. EXPERIMENTS AND RESULTS

6.1. Vertex Cover

A *p-random* n-vertex graph is one in which each of the $\binom{n}{2}$ pairs of vertices is connected by an edge with probability p. Table 12.1 reports vertex cover experiments using HcN on p-random 50- and 100-vertex graphs. The column labeled "ISVC" is based on loading the complement G_c of the given graph G in HcN. A stable state (maximal independent set in G) is then found using SD(V) on HcN; the vertex cover (of G) is read-out as the set of OFF units. The next three columns are dynamics applied to G_{l_c}, the complement of the line graph of G, stored in HcN. Each dynamics retrieves a stable state (maximal independent set in G_l; maximal matching in G). The vertices in G associated with the vertices (edges of G) in the stable state are a vertex cover. Smaller maximal cliques in G_{l_c} provide smaller vertex covers. Hence our optimization goal here is exactly the opposite — to find the smallest, as opposed to the largest, maximal clique. Somewhat surprisingly, the problem of finding the smallest maximal clique is, in general, also NP-hard (Jagota, 1992b). The dynamics that works best is $SSD_{min}(\varnothing, n)$ — n runs of SSD from initial state $\varnothing$; the smallest size output (worst in the traditional optimization sense) is picked.

The last two columns are expected size of minimum vertex cover ($E[C^*]$) and expected size of maximum minimal vertex cover ($E[|\{MaMiC\}|]$), respectively. These e estimates are very sharp and are obtained indirectly from the theory of random graphs applied to independent sets.

It is well known that by Matula's theorem, the maximum independent set size in a p-random graph can be pinpointed almost exactly (see Palmer, 1985). However, we noted earlier that this estimate is inaccurate for small n and p away from 0.5 (Jagota, 1992a; Jagota & Regan, 1992). We obtain our estimates directly from the formula for the expected number of *maximal* k-independent sets in p-random n-vertex graphs (Bollobás & Erdös, 1976):

$$E_{n,p}(k) = \binom{n}{k}(1-p)^{\binom{k}{2}}[1-(1-p)^k]^{n-k}$$

Test Parameters		VC Via IS	VC Via Linegraph Reduction				Expected Values	
p	n	ISVC	$E[\text{\# units}]$	$SD(V)$	$SD(\varnothing)$	$SSD_{min}(\varnothing,n)$	$E[C^*]$	$E[|MaMiC|]$
0.5	50	42.86	612.5	48.0	48.6	46.2	42	47
0.1	100	72.3	495	95.2	93.46	86.0	66	87

Table 12.1. Size of Vertex Cover Found by Various Algorithms, Averaged Over 15 p-Random n-Vertex Graphs in Each Row.

First, the expected sizes k_1 and k_0 of maximum independent set and minimum maximal independent set, respectively, are obtained by tabulating the distribution $E_{n,p}(k)$ (which is binomial-like) and noting the k for which the values transition from ≈ 1 to $<<1$ and from $<<1$ to ≈ 1, respectively. Both transitions are known to be sharp in theory and are also so in practice. The expected sizes of minimum vertex cover and maximum minimal vertex cover are then $n - k_1$ and $n - k_0$, respectively. This procedure is much more accurate.

From Table 12.1, the independent-set-based approach (ISVC) performs the best; all the line-graph-based approaches are significantly poorer. ISVC finds near-optimal solutions (compare it with $E[C^*]$). ISVC, however, has a poor (worst-case) ratio bound. whereas the other three dynamics have ratio bound of 2. The variances in the results were small, justifying the small sample size.

6.2. N-Queens

In the N-queens encoding in HcN, $|V| = N^2$ and the size of every stable state is $\leq N$. Hence, $SD(V)$ makes $\Theta(N^2)$ unit switches. $SD(\varnothing)$ makes N unit switches. The operation of $SD(V)$ on an N-queens instance may be viewed as follows (cf. Table 12.2). The initial state represents the board configuration in which a queen is placed on every square. From a given configuration, the next state is generated as follows. If no queens on the board are threatened, then a new queen is placed on the first lexicographically empty square on which a queen may be placed without threatening existing queens. If no such square exists, then we are at a stable state. If, on the other hand, at least one queen is threatened, then the queen that is threatened by the largest number of other queens on the board is removed. Ties are broken lexicographically. $SD(\varnothing)$ may be viewed in similar fashion but from the initial state: the board is empty.

6.3. Boolean Satisfiability

An n-variable 3-SAT Boolean CNF formula has $8\binom{n}{3}$ possible clauses. A p-random formula of the above kind independently selects each of these clauses for inclusion with probability p, generating an expected number $8p\binom{n}{3}$ of clauses. Our network uses three units per clause.

Over all the runs, a satisfying assignment was found just once: by $SSD_{max}(V,N)$ with $n=50$; $p=.001$. Here N was both the number of iterations of SSD and the number of units in the network (cf. Table 12.3).

6.4. Graph K-Coloring

Recall that in graph k-coloring, given a graph G and an integer k, we are interested in coloring the maximum number of vertices colorable with k colors. In this chapter we solve this problem approximately via efficient neural network heuristics. We report the number of vertices colorable, by various neural network heuristics, as a function of graph properties and k.

Random Graphs. Our first test set comprises p-random graphs (recall the definition, given earlier). Table 12.4 reports graph k-coloring experiments using HcN on p-random 50-vertex graphs.

N	4	6	8	10	12	16	30	40	50
SD(V)	4	5	7	9	10	14	26	37	45
SD($\emptyset$)	3	5	5	7	9	13	23	31	38

Table 12.2. Number of Mutually Nonattacking Queens Placed Efficiently via SD(V) and SD($\emptyset$). The optimal number in each case is N. SD(V) outperforms SD($\emptyset$) in each instance.

Test Formulae Parameters		Avg # Clauses	Dynamics		
p	n		SD(V)	SD($\emptyset$)	SSDmax($\emptyset,N$)
0.5	5	39.2	35.5	35.6	37.06
0.02	20	178.73	166.33	162.13	168.93
0.001	50	167.73	161.4	160.4	162.06

Table 12.3. Number of Simultaneously Satisfiable Clauses Found by Various Algorithms, Averaged Over Fifteen p-Random n-Variable 3-SAT Boolean CNF Formulas in Each Row.

The chromatic number $\chi(G)$ of a graph G is the minimum k such that G is k-colorable. The last two columns of Table 12.4 numerically tabulate lower and upper bounds on $\chi(G_{n,p})$ that hold for almost every $G_{n,p}$. These estimates are obtained from the theory of random graphs as follows (this parallels our earlier vertex cover estimates). From Spencer (1987, p. 53), $n/\alpha(G_{n,p})$ $\leq \chi(G_{n,p}) \leq n/(\alpha(G_{n,p})/2)$. $\alpha(G)$ is the size of the largest independent set in G. Recall that for sufficiently large n, $\alpha(G_{n,p}) \approx 2 \log_{1/(1-p)} n$. (Also see Spencer, 1987, p. 53.) Also recall that $\alpha-(G_{n,p})/2$ is the expected size of the minimum maximal independent set for sufficiently large n (see Bollobás & Erdös, 1976).

Test Parameters				Algorithms		Bounds on $X(G_{n,p})$	
p	k	SD(∅)	SD(V)	SSDmax(∅,n)	SSDmax(V,n)	Lower	Upper
0.1	1	18.33	20.73	20.66	21.33	2.0	4.0
	2	32.93	34.53	35.13	35.26		
	3	42.53	43.6	44.4	44.2		
	4	48.33	48.13	49.06	48.73		
0.3	1	8.2	10.06	10.3	11.06	4.16	8.32
	2	16.3	19.06	19.6	20.0		
	3	23.6	26.93	27.33	27.66		
	4	30.2	32.93	33.66	34.53		
	5	35.93	39.13	39.2	39.53		
	6	41.06	43.26	43.73	43.53		
	7	45.06	46.46	47.06	46.8		
	8	48.0	48.33	49.13	49.46		

Table 12.4. Number of Vertices Colored by Various Algorithms, Averaged Over 15 p-Random 50-Vertex Graphs in Each Row. For all rows of fixed p, the 15 graphs were identical.

As noted earlier, the above formula for $\alpha(G_{n,p})$ is inaccurate for n in the range of our experiments. We estimate $\alpha(G_{n,p})$ as we did before for obtaining the vertex cover estimates. From Bollobás and Erdös (1976), the expected number of maximal independent sets of size l in $G_{n,p}$ is $E_{n,p}(l) = \binom{n}{l}(1-p)^{\binom{l}{2}}[1-(1-p)^l]^{n-l}$. First, the expected sizes l_1 and l_0 of maximum independent set and minimum maximal independent set respectively are obtained by tabulating the distribution $E_{n,p}(l)$ (which is binomial-like) and noting the l for which the values transition from ≈ 1 to $\ll 1$ and from $\ll 1$ to ≈ 1 respectively. The lower bound and upper bound for $\chi(G_{n,p})$ are then n/l_1 and n/l_0 respectively. Note that we do not use $l_{1/2}$ for the upper bound because our numerical investigations have also shown that for the range of n employed in our experiments the expected size of minimum maximal independent set is better estimated directly by l_0 than by $l_{1/2}$. The lower and upper bounds on $\chi(G_{n,p})$ reported in Table 12.1 are also used to guide the selection of k for the experiments.

From Table 12.4, some judgements of the performance of the dynamics (columns 3-6) and the intrinsic character of the k-coloring random graph instances are possible. For the same number of colors k, more vertices are colored for the $p=0.1$ graphs than for the $p=0.3$ graphs. Equivalently, to color the same number of vertices, larger number of colors are needed for $p=0.3$ graphs as compared with for $p=0.1$ graphs. This is entirely predictable, as higher density graphs, in general, require more colors (also see columns 7 and 8). We may also note that for fixed p, the addition of each new color initially has larger benefits, in terms of additional vertices colorable, but the benefits tail off as the number of colors grows large. Finally, SD(∅) has the worst performing dynamics but the other three exhibit very similar performance, with SD(V) performing just slightly worse than $SSD_{max}(\varnothing,N)$ and $SSD_{max}(V,N)$. $SD(\varnothing)$ catches up with the others, however, as the number of colors allowed is increased.

"*l Random Cliques Graphs*". Our second test set comprises "*l* random cliques" *n*-vertex graphs. These graphs are generated by generating *l* cliques at random and taking their union. First, the size *s* of each clique is computed independently and randomly from 1 to $n/2$. Then *s* vertices are picked at random from the *n* vertices to put in the clique. This process is repeated *l* times to generate the *l* cliques. Then the labeled union is taken of these *l* cliques to form the resulting graph. This graph is then subjected to a *k*-coloring reduction to MIS.

It is worth noting that our associative memory storage rule of this chapter is a convenient way to take the union of cliques and to realize the *k*-coloring reduction to MIS of "*l* random cliques" graphs. (Recall that MIS is the complement problem of MAX-CLIQUE and some problems are more convenient to reduce to MIS than to MAX-CLIQUE.) As an example, let $n=4$, $l=2$, and $k=2$. Let the two random cliques be $\{v_1,v_2,v_3\}$, $\{v_2,v_3,v_4\}$. Then the training set is $\{c_1v_1,c_2v_1\}$, $\{c_1v_2,c_2v_2\}$, $\{c_1v_3,c_2v_3\}$, $\{c_1v_4,c_2v_4\}$, $\{c_1v_1,c_1v_2,c_1v_3\}$, $\{c_1v_2,c_1v_3,c_1v_4\}$, $\{c_2v_1,c_2v_2,c_2v_3\}$, $\{c_2v_2,c_2v_3,c_2v_4\}$. That is, there are cliques for the partitions and *k* copies, one for each color, of each of the *l* random cliques.

"*l* random cliques" graphs are harder for certain MAX-CLIQUE heuristics than *p*-random graphs (Jagota, 1992a). Table 12.5 provides the same conclusion for graph *k*-coloring. In particular, the performance of $SD(\varnothing)$ relative to the other dynamics is poorer on *l* random cliques graphs than on *p*-random graphs. Furthermore, in contrast with Table 12.5, the $SD(\varnothing)$ performance relative to the other dynamics does not appear to improve significantly as the number of colors is increased. Significantly, the performance of the other three heuristics, relative to each other, remains relatively unchanged, with the qualification that $SSD_{max}(V,N)$ is perhaps a more consistent, although still narrow, winner over $SSD_{max}(\varnothing,N)$. From Table 12.5, we may note that the percentage of vertices colored by the dynamics on "20 random cliques" 100-vertex graphs is much less than the same for "10 random cliques" 50-vertex graphs (compare the rows with the same *k*). That is, the former appear to be harder to color.

Test Parameters			Algorithms			
n	*l*	*k*	$SD(\varnothing)$	$SD(V)$	$SSD(\varnothing,n)$	$SSD(V,n)$
50	10	2	8.8	12.2	11.6	12.13
		3	11.2	14.93	15.06	15.53
		4	13.73	17.8	17.86	18.53
		6	18.73	22.6	23.2	23.6
		8	23.4	26.93	27.73	28.33
100	20	2	9.53	11.93	11.8	12.2
		4	14.06	17.13	17.53	17.93

Table 12.5. Number of Vertices Colored by Various Algorithms, Averaged Over 15 "*l* Random Cliques Graphs" in Each Row. For all rows of fixed *n*, the 15 graphs were identical. SSD denotes SSD_{max}.

7. DISCUSSION

These results indicate that our approaches so far are not the best ones for individual problems. Neural network methods designed for N-queens, for even large N, quickly find exact solutions (e.g., Adorf & Johnston, 1990). Conventional methods for N-queens do even better. Recently, one of the simplest local search algorithms has been proven to almost always find a satisfying assignment in 0.5-random 3-SAT formulas (Koutsoupias & Papadimitriou, 1992). Our dynamics found a satisfying assignment only once.

The main contribution of our work is a single simple binary-weights network for approximately solving several optimization problems which does not admit infeasible solutions and in one case (vertex cover) provides an asymptotically optimal ratio bound. All our mappings, except encoding 1 of vertex cover, are goodness preserving. The goal of this chapter has not been to present a superior optimization technique but to exploit the simplicity and massive parallelism of neural networks for sufficiently good cost-effective optimization for real-world purposes.

REFERENCES

Aarts, E., & Korst, J. (1989). *Simulated Annealing and Boltzmann Machines: A Stochastic Approach to Combinatorial Optimization and Neural Computing.* New York: Wiley.

Adorf, H., & Johnston, M. D. (1990, June). A discrete stochastic neural network for constraint satisfaction problems. *International Joint Conference on Neural Networks* (Vol. 3, pp. 917-924). San Diego: IEEE.

Akiyama, Y., Yamashita, A., Kajiura, M., & Aiso, H. (1989). Combinatorial optimization with gaussian machines. In *International Joint Conference on Neural Networks* (Vol. 1, pp. 533-540). Piscataway, NJ: IEEE.

Arora, S., Lund, C., Motwani, R., Sudan, M., & Szegedy, M. (1992). Proof verification and hardness of approximation problems. *Proceedings of the 33rd Annual IEEE Symposium on Foundations of Computer Science* (pp. 14-23). IEEE Computer Science Society Press.

Ballard, D. H., Gardner, P. C., & Srinivas, M. A. (1987). *Graph problems and connectionist architectures* (Tech. Rep. No. 167) Department of Computer Science, University of Rochester.

Baram, Y., & Dechter, R. (1991). *Processing constraints by neural networks.* Department of Computer and Information Science, University of California, Irvine, June, unpublished.

Barrow, H., & Burstall, R. (1976). Subgraph isomorphism, matching relational structures and maximal cliques. *Information Processing Letters,* 4, 83-84.

Bollobás, B., & Erdös, P. (1976). Cliques in random graphs. *Proceedings of the Cambridge Philosophical Society,* 80, 419-427.

Bourret, P., & Gaspin, C. (1992). A neural based approach of constraints satisfaction problem. In *International Joint Conference on Neural Networks, Baltimore, June, 1992* (Vol. 4, pp. 588-593). Piscataway, NJ: IEEE.

Caspi, Y. (1992). The performance of an artificial neural network for graph coloring. In C. H. Dagli, L. I. Burke, & Y. C. Shin (Eds.), *Intelligent Engineering Systems Through Artificial Neural Networks* (Vol. 2, pp. 319-324). New York: ASME Press.

Chen, W., & Hsieh, K. (1989). A neural network for 3-satisfiability problems. In *International Joint Conference on Neural Networks, Washington, DC, June, 1989* (Vol. 2, p. 587). Piscataway, NJ: IEEE. Abstract in Proceedings.

Crescenzi, P., Fiorini, C., & Silvestri, R. (1991). A note on the approximation of the maxclique problem. *Information Processing Letters*, **40**, 1-5.

Dahl, E. D. (1987). Neural networks algorithms for an np-complete problem: map and graph coloring. In *IEEE First International Conference on Neural Networks* (Vol. 3, pp. 113-120). San Diego: IEEE/ICNN.

Freige, U., Goldwasser, S., Lovasz, L., Safra, S., & Szegedy, M. (1991). Approximating clique is almost NP-complete. In *Symposium on Foundations of Computer Science* (pp. 2-12). IEEE Computer Science Society Press.

Garey, M. R., & Johnson, D. S. (1979). *Computers and Intractability: A Guide to the Theory of NP-Completeness*. New York: Freeman.

Godbeer, G. H., Lipscomb, J., & Luby, M. (1988). *On the computational complexity of finding stable state vectors in connectionist models (Hopfield nets)* (Tech. Rep. No. 208/88). Department of Computer Science, University of Toronto.

Griggs, J. R. (1983). Lower bounds on the independence number in terms of the degrees. *Journal of Combinatorial Theory, Series B*, **34**, 22-39.

Grossman, T., & Jagota, A. (1992). On the equivalence of two {Hopfield}-type networks. In *IEEE International Conference on Neural Networks* (pp. 1063-1068). Piscataway, NJ: IEEE.

Hertz, J., Krogh, A., & Palmer, R. G. (1991). *Introduction to the Theory of Neural Computation*. Reading, MA: Addison-Wesley.

Hopfield, J. J. (1982). Neural networks and physical systems with emergent collective computational abilities. *Proceedings of the National Academy of Sciences, USA*, **79**, 2554-2558.

Hopfield, J. J. (1984). Neurons with graded responses have collective computational properties like those of two-state neurons. *Proceedings of the National Academy of Sciences, USA*, **81**, 3088-3092.

Jagota, A. (1990a). *A Hopfield-style network as a maximal clique graph machine* (Tech. Rep. No. 90-25). Department of Computer Science, SUNY at Buffalo.

Jagota, A. (1990b). *A new Hopfield-style network for content-addressable memories* (Tech. Rep. No. 90-02). Department of Computer Science, SUNY at Buffalo.

Jagota, A. (1991). Backtracking dynamics in a Hopfield-style network. In *International Joint Conference on Neural Networks, Seattle, July, 1991*. Piscataway, NJ: IEEE. Abstract in Proceedings.

Jagota, A. (1992a). Efficiently approximating Max-Clique in a Hopfield-style network. In *International Joint Conference on Neural Networks* (Vol. 2, pp. 248-253). Piscataway, NJ: IEEE.

Jagota, A. (1992b). *On the computational complexity of analyzing the Hopfield-clique network* Unpublished manuscript, Department of Computer Science, SUNY at Buffalo.

Jagota, A. (1994). A Hopfield-style network with a graph-theoretic characterization. *Journal of Artificial Neural Networks*, 1, 145-166.

Jagota, A., & Regan, K. W. (1992). *Performance of max-clique heuristics under description-length weighted distributions* (Tech. Rep. No. 92-24). Department of Computer Science, SUNY at Buffalo.

Kajiura, M., Akiyama, Y., & Anzai, Y. (1989). Neural networks vs. tree search in puzzle solving. In *International Joint Conference on Neural Networks, Washington, DC, June, 1989* (Vol. 2, p. 588). Piscataway, NJ: IEEE. Abstract in Proceedings.

Karp, R. M. (1972). Reducibility among combinatorial problems. In R. E. Miller & J. W. Thatcher (Eds.), *Complexity of Computer Computations* (pp. 85-103). New York: Plenum Press.

Karp, R. M. (1976). The probabilistic analysis of some combinatorial search algorithms. In J. F. Traub (Ed.), *Algorithms and Complexity: New Directions and Recent Results* (pp. 1-19). New York: Academic Press.

Korst, J. H. M., & Aarts, E. H. L. (1989). Combinatorial optimization on a Boltzmann machine. *Journal of Parallel and Distributed Computing, 6,* 331-357.

Koutsoupias, E., & Papadimitriou, C. H. (1992). On the greedy algorithm for satisfiability. *Information Processing Letters, 43,* 53-55.

Mackworth, A. K. (1992). Constraint satisfaction. In S. C. Shapiro (Ed.), *Encyclopedia of Artificial Intelligence* (2nd ed., pp. 285-290). New York: Wiley.

Motwani, R. (1992). *Lecture notes on approximation algorithms* (Tech. Rep.) Department of Computer Science, Stanford University.

Nadel, B. A. (1989). Representation selection for constraint satisfaction: A case study using n-queens. *IEEE Expert, 2,* 16-23.

Palmer, E. M. (1985). *Graphical Evolution.* New York: Wiley. (Matula's theorem is on page 76).

Pinkas, G. (1991). Symmetric neural nets and propositional logic satisfiability. *Neural Computation, 3,* 282-291.

Ramanujam, J., & Sadayappan, P. (1988). Optimization by neural networks. In *IEEE International Conference on Neural Networks, San Diego, June, 1988* (Vol. 2, pp. 325-332). Piscataway, NJ: IEEE.

Shrivastava, Y., Dasgupta, S., & Reddy, S. M. (1990). Neural network solutions to a graph theoretic problem. In *Proceedings of IEEE International Symposium on Circuits and Systems* (pp. 2528-2531). New York: IEEE.

Spencer, J. (1987). Random graphs ii. In J. Spencer (Ed.), *Ten Lectures on the Probabilistic Method* (pp. 51-56). Philadelphia: Society for Industrial and Applied Mathematics.

Tagliarini, G. A., & Page, E. (1987). Solving constraint satisfaction problems with neural networks. In *IEEE First International Conference on Neural Networks* (Vol. 3, pp. 741-747). San Diego: IEEE/ICNN.

Takefuji, Y. (1992). Neural network models and N-queen problems. In Y. Takefuji (Ed.), *Neural Network Parallel Computing* (pp. 1-26). Boston: Kluwer.

Turner, J. S. (1988). Almost all k-colorable graphs are easy to color. *Journal of Algorithms, 9,* 63-82.

III

OPTIMALITY IN LEARNING, COGNITION, AND PERCEPTION

13

An Examination of Mathematical Models of Learning in a Single Neuron

David C. Chance
University of Central Oklahoma

John Y. Cheung
University of Oklahoma

Sue Lykins
Oklahoma State University

Asa W. Lawton
University of Oklahoma

The chapter by David Chance, John Cheung, Sue Lykins, and Asa Lawton, **An Examination of Mathematical Models of Learning in a Single Neuron,** *deals with different models in the literature of classical (Pavlovian) conditioning. The authors relate the conditioning problem to optimality on two levels. First, there is optimality at the level of the organism or network that learns. The conditioning process itself is often considered as part of the organism's adaptive responses to the environment — a main theme of all the chapters in this book's first section. Many neural network theorists (among them Barto, Sutton, Werbos, Grossberg, and Levine) have studied conditioning as part of the paradigm of reinforcement learning: how an organism (or artificial network) learns to approach those objects in the environment that are positively reinforcing and avoid those objects that are negatively reinforcing. This is also considered, in a control framework, in this book's chapter by Öğmen and Prakash. However, Chance et al. trace the history of psychological theory: the notion that reinforcement learning is done optimally was a mainstream notion in psychology (with Guthrie the sole important dissenter) in the 1930s, 1940s, and 1950s but is less widely believed now.*

Second, there is optimality at the level of the single network connection, which Chance et al. interpret as a synapse between two neurons. There are trade-offs between the advantages of different learning rules: Hebbian, error-correcting, and the modifications due to various recent researchers or groups such as Sutton-Barto, Klopf, Gelperin-Hopfield-Tank, and Tesauro.

*Although the authors simulated all of these models, the authors draw the clearest comparisons between the Hebbian and Klopf (also known as drive-reinforcement) models. The Hebb rule depends on paired pre- and postsynaptic activities, whereas the Klopf rule depends on paired **changes** in pre- and postsynaptic activities. The Hebb rule does better at reproducing effects of stimulus duration, interstimulus interval, and acquisition curves, whereas the Klopf rule does better at reproducing complex contextual effects such as blocking and conditioned inhibition. The authors speculate that a comprehensive conditioning model from a single learning rule for single neurons may not exist: the whole nervous system is greater than the sum of its parts, and even actual neurons are complex enough to be modeled as networks.*

ABSTRACT

Classical conditioning is used as a means to investigate unsupervised learning in a single neuron. The double advantage of being extensively studied and rich in terms of responses to multiple temporally related stimuli makes for a good benchmark against which learning laws can be tested. This chapter relates the mathematical developments in the area of real time single-neuron learning by looking at the Hebbian, Rescorla-Wagner, Sutton-Barto, Tesauro, Gelperin-Hopfield-Tank, Klopf (drive-reinforcement) and Widrow-Hoff rules for synaptic adaptations. The learning rules are compared based on four classical conditioning simulations: CS duration, blocking, reacquisition, and second-order conditioning. The results of the examination have shown both the strengths and weaknesses of the individual models. A comparison between the Hebbian and drive-reinforcement models brings into question the use of a single learning law as the basis for the optimal development of a multiple neuron system.

1. INTRODUCTION

The research discussed in this chapter was designed to gain a better understanding of single-neuron models that incorporated time as an important variable in the learning process. One of the most sophisticated learning models proposed to date was Klopf's (1988) real-time drive-reinforcement theory. His model appeared to accurately depict a large number of Pavlovian classical conditioning procedures. In addition, he included comparisons with both Hebb (1949) and Sutton and Barto (1981). In our work we have also looked at the mathematical model of Rescorla and Wagner (1972), which assumed time but did not incorporate time explicitly in its learning rule (Levine, 1991). We also looked at Sutton and Barto (1981) who did include time in their learning and activation rules, and followed up on others in the 1980s (Gelperin, Hopfield, & Tank, 1985; Tesauro, 1986) who primarily extended the ideas of Sutton and Barto. We also included Klopf's formulation of the Hebb rule and the Widrow-Hoff model (1960).

A computer simulator was written based on the various real-time models of a single neuron. Various mathematical models were programmed into the simulator and tested using the classical conditioning procedures, many of them presented by Klopf. Our goals were (a) to understand mathematically the activation and learning rules of several different models of single neuron function, (b) to explore the use of mathematical tools to aid in the modeling of weight changes in a single neuron, (c) to examine the role of time as a factor in learning, (d) to compare results with psychological studies in classical conditioning, and (e) to provide the computer simulation as a teaching and research tool for psychology and engineering students.

Our exploration of the models involves several of psychology's theoretical issues. As psychology grew into a science, it brought with it an adaptionist evolutionary concept that emphasized the power of natural selection as an optimizing agent (Gould & Lewontin, 1979; see also Elsberry, Chapter 5). The goal was to better understand human adaptation to his environment (e.g., James, Dewey, Angell Carr, Thorndike, Watson), in other words, human ability to learn in a changing environment. The common assumption was that learning was movement toward an optimal state. Both Thorndike and Skinner also stressed the consequences of action. This led to the equating of reinforcement, reward, and consequences of behavior with a hedonistic explanation. For example, Olds and Milner (1954), Delgado (1955), and others sought to explain hedonism through pleasure and punishment centers in the brain. Klopf (1972, 1979, 1982) talked about the hedonistic versus the heterostatic neuron. Although he dropped this terminology in his 1988 paper, the drive-reinforcement model of that paper is still based on neuronal reinforcement. However, many psychologists have moved away from such theories. As Malone (1990, p. 313) stated, "it is long past time to bury hedonism once and for all."

Of the early learning theorists, Edwin Guthrie was one of the few who disagreed that learning always moved toward an optimal state (Malone, 1990). He discussed meaningless acts, awkwardness, and maladaptive and foolish behavior, and thought it unwise to treat them merely as failures (nonoptimal) in the course of otherwise reasonable goal-directed activity. (For further treatment of nonoptimality examples, see Levine, Chapter 1). Guthrie (1952) also warned against placing too much stress on general characteristics as opposed to attempting to explain the behavior of individuals in particular.

During this century, behavioral research in classical conditioning has been viewed from two different perspectives: behavior of the single individual versus group behavior. The comparisons of models presented in this chapter were based on group behavior.

All the neuronal models we analyzed had at least two things in common. First, they assumed that "essentially *all* behavior can be explained by the optimization of a single variable" (Levine, 1991, pp. 284 and 285). Second, these modelers assume stimulus substitution relative to the conditioned response (CR) and the unconditioned response (UR), which were viewed as similar. However, as early as 1932 Warner said "whatever response is grafted onto the conditioned stimulus (CS), it is not snipped from the unconditioned stimulus (US)" (quoted in Wagner & Brandon, 1989, p. 149). Wagner and Brandon further comment that "in many, if not most, circumstances of Pavlovian conditioning, the CS does not appear to act like a substitute for the US and/or evoke a CR that mimics the practiced UR" (p. 149). The models presented in this chapter combine the CR and UR as a single output, however, making no distinction between the two responses.

The work of Hull, especially in the 1950s and 1960s, had a major impact on mathematical modeling of psychological data. Hull (1947) approached learning in terms of testable postulates that could eventually lead to general truths. This was a part of the shift to viewing psychological phenomena from a physical science perspective. Hull's postulates led to the 1972 model of Rescorla and Wagner, and became the basis for Klopf's drive reduction/induction learning model in the 1980s.

2. NEURONAL LEARNING RULES

In particular, based on the experimental results of classical conditioning, many researchers have postulated rules to demonstrate the learning behavior of a single artificial neuron. In all the single-neuron models examined here, the activation function was generally accepted to be the weighted sum of the inputs, where the weights signify the corresponding synaptic strength, and the output is a nonlinear function of the weighted sum, thus representing the repetition frequency or the firing frequency of a real neuron. However, there is no general consensus on how a single neuron would modify its own synaptic weights in response to external conditions, although some recent research promised insights (Claiborne, Zador, Mainen, & Brown, 1991).

Among the various synaptic weight adaptation (learning) laws that have been postulated, the major differences stem from the perspectives taken by their designers. Some, such as Hebb (1949) and Rescorla (1973), designed the learning rules from a psychological perspective based on association of stimuli. Recent changes of learning rules include refinements such as the Sutton-Barto (1981) and the drive-reinforcement (Klopf, 1988) rules, which come from the perspective of adaptive control theory. Others fashioned their adaptation laws from less biologically plausible origins, using the difference between the neuronal output and the given external reinforcement. (Reinforcement is used here in the most general sense, as any input signal that determines the changes in connection weights.) These are often called *error-based* adaptation laws, and include the back propagation rule (Le Cun, 1985; Parker, 1982, 1985; Rumelhart, Hinton, & Williams, 1986; Werbos, 1974) and the Widrow-Hoff (1960) adaptation rule. Some recent research suggests that back propagation of error may be an actually occurring phenomenon in cases where nitric oxide acts as a retrograde messenger (Kandel & Hawkins, 1993). Still other adaptation laws are based on the input patterns of a cooperative network of neurons, thus achieving pattern-based matching operations. A good example is the adaptive resonance theory (ART) class of neuronal networks. Some adaptations are probability based where the neuronal output is analyzed in the probabilistic sense and the synaptic weights are modified in such a way that the next neuronal output would respond closely to expectation. An example of this category is the Rescorla-Wagner approach.

The most general learning law in the literature, perhaps, is Amari's (1977):

$$w_i(t+1) = w_i(t) + c * r_i(t),$$

where t represents the time step; $w_i(t)$ and $w_i(t+1)$ the ith connection weight at the present time step t and the next time step $t+1$; c the step size; and $r_i(t)$ the external (extracellular) or internal (intracellular) reinforcement for that trial or event. The form of this law indicates that learning is incremental and cumulative. When a trial comes, learning occurs in the presence of direct reinforcement. If there is no reinforcement, there is no learning. Although Amari's adaptation law does not point to the association between input and output stimuli, the formulation nevertheless suggests the notion of neuronal plasticity being incremental and a function of time.

A common early assumption about the underlying mechanisms of learning was that associative changes are properties of complex neural circuits (Cohen, 1985, cited in Abrams & Kandel, 1988; Lashley, 1929). One of the first to challenge this assumption was the Canadian psychologist D. O. Hebb (1949) who suggested a cellular mechanism for classical conditioning. Hebb proposed that learning involved changes in the efficacy of plastic synapses at the neuronal level, and further, that these changes occurred through correlations between approximately simultaneous pre- and postsynaptic levels of neuronal activity. In other words, presynaptic activity, followed directly by postsynaptic activity, was hypothesized to result in a change in the efficacy of the associated synapse. Prior to Rescorla (1966), the most important principles of classical conditioning based on Pavlov's (1928) reflexive model of associative learning were those of contiguity (the bond that occurs between two events which occur closely together in space and time), frequency (how often contiguous events occur together), and intensity (the strength of the contiguous association). Three models — those of Hebb, Widrow-Hoff, and Rescorla-Wagner — were proposed to address the contiguity, frequency, and intensity of the associative bond.

Although Hebb gave no explicit mathematical formulation to his rule, it was commonly accepted (Sutton & Barto, 1981) as $w_i(t+1) = w_i(t) + c * x_i(t) * y(t)$, where $x_i(t)$ is the input signal strength (or presynaptic activity) and $y(t)$ the output of the neuron. The Hebbian model was thus one of the earliest attempts to explain the role of real-time learning mechanisms in terms of the temporal association of signals, albeit a simultaneous one. The reader is cautioned here that the notion of simultaneous time is only one way to indicate the presence of more than one activity "at the same time" on a very gross time scale, perhaps on the time scale of the entire epoch, and hence should not be construed as precise timing order within the same epoch. As Carew, Hawkins, Abrams, and Kandel (1984) noted, Hebb's postulate accounted for temporal specificity, the hallmark of associative learning, using plausible physiological mechanisms, and for stimulus and response specificity, which are common features of classical conditioning, without the need of complex neural circuitry. The pairing of input signals with neuronal output together as reinforcement suggests associative learning as a result of contiguity. The summation of reinforcement accounts for the concept of frequency. The strength of reinforcement as a function of the input and output signal strength reflects the notion of intensity.

Hebb's rule however, has several drawbacks. First, learning is only based on the activity present in the current epoch. There are no associations with historical occurrences. Second, Hebb's rule provides no means of *reducing* synaptic efficacy. Thus, weights could potentially go to infinity. Third, there is no means of dealing with inhibitory connections between neurons (Caudill, 1989). Classical conditioning is too complex to incorporate expectations and stimulus patterns into a simple correlation rule like Hebb's (Sutton & Barto, 1981). Last, the necessity for Hebb's postulate to produce action potentials in a postsynaptic neuron is neither "necessary nor sufficient" to produce the temporally specific change in synaptic efficacy that underlies, for example, the classically conditioned gill withdrawal reflex in *Aplysia* (Carew et al., 1984).

The *Widrow-Hoff* adaptation law (1960) is formulated based on the least mean squares (LMS) principle rather than known learning criteria. The LMS principle seeks to minimize the error, which is taken to be the difference between the target neuronal output response value and the present value. The Widrow-Hoff law can be written as $w_i(t+1) = w_i(t) + c * x_i(t) * [z - y(t)]$ where z is the (constant) target neuronal output response value. Hence the source of reinforcement is external, and learning occurs only when there is an explicit difference between the target and output response.

The goal of this adaptation law is to change synaptic weights in such a way that the output response value will follow the given target value z. In other words, the neuron learns to reproduce the target value based on a linear combination of present active inputs. The form of the Widrow-Hoff rule again suggests learning through contiguity (pairing of $x_i(t)$ and a function of $y(t)$. Similarly to Hebb's rule, learning here is only associated with the activity present in the current epoch. Hence, learning only occurs with overlapping input stimuli; that is, it only occurs when there is an unconditioned response (UR) paired with any active stimuli.

The *Rescorla-Wagner* cognitive theory postulated that the animal learns expectations about events following presentation of a stimulus complex. These expectations are equivalent to associative strengths at any given time and can be changed when events such as unconditioned stimuli differ from the composite expectation, which is the summation of associative strengths of present stimuli. The Rescorla-Wagner learning law is a direct formulation of Hull's (1947) approach based on probability of the animal's response to its input stimuli. In the Rescorla-Wagner formulation there is a synaptic strength associated with each input stimulus. When more than one input stimulus is provided, there is a total synaptic strength analogous to neuronal output. Using the present neural network formulation, if the active input stimuli are a fixed value, then synaptic strength may be taken to represent synaptic weights, and total synaptic strength as neuronal output. With minor modifications one can fit the Rescorla-Wagner learning law into a neuronal formulation, resulting in the learning law $w_i(t+1) = w_i(t) + c_k * x_i(t) * [z_k - y(t)]$ where c_k is a constant that varies according to the different stages of the conditioning paradigm; and z_k the maximum synaptic strength at a particular stage.

The Rescorla-Wagner and Widrow-Hoff laws appear similar although they are derived from totally different paradigms. Although the Widrow-Hoff model is based on minimizing the difference between target values and the neuronal output values, the Rescorla-Wagner model is based on the probability of synaptic strengths in response to input stimuli toward a maximum synaptic weight or strength value. There are two major differences, however, between these models. First, in the Widrow-Hoff model, the learning constant c remains the same, whereas in the Rescorla-Wagner model the learning constant c_k varies throughout the same conditioning period. Second, the Widrow-Hoff model uses a target value that may change from epoch to epoch, whereas the Rescorla-Wagner model uses a constant z_k that represents a maximum synaptic strength for the conditioning interval.

It is historically important to note that in the 1960s, the principles of contiguity, frequency, and intensity were seriously questioned. Rescorla (1966) suggested that contiguity between two events was insufficient for conditioning, because a CS (conditioning stimulus) must not only be contiguous with a US but must also be an accurate predictor of US occurrence. Rescorla called this the *contingency* of events, which he defined as a statistic derived from the probability that US will occur in the presence of a CS, and the probability that it will occur in the absence of the CS. Combining these two probabilities allows for calculation of a contingency coefficient that measures the degree to which CS and US will occur together. Thus positive contingency between a CS and a US produces excitatory conditioning, and negative contingency produces inhibitory conditioning. But both contingency and contiguity are necessary (Lieberman, 1990).

Association by contiguity alone was further questioned by Garcia and Koelling (1966) who challenged the idea that it did not matter what stimulus was chosen as a CS. They found that in studies of taste-aversion learning, for example, nausea could not be conditioned to a noise, nor fear to a taste, although each of these CSs is easily associated with the other US. Thus, some CS-US combinations associate more readily than others. This finding has been variously labeled *preparedness*, *relevance*, *selective association*, *associative bias*, and *belongingness* (Lieberman, 1990). Also, Garcia and Koelling used a delay of at least 20 min between presentation of the taste and onset of illness. Thus conditioning is not simply due to the linking of contiguous events; some additional processes must be involved.

Yet another challenge to the contiguity-alone principle came from Kamin's (1968) experiment on *blocking*, in which prior conditioning of one element of a compound stimulus prevents conditioning to the other element. Blocking shows that conditioning will not take place if another stimulus that already predicts the US is present. These three findings taken together demonstrate that unconditioned stimuli are not associated with every stimulus that precedes them. Instead, conditioning seems to depend on stimuli that are good predictors of a US, that is, CSs that inform the organism that an important event is about to occur (Lieberman, 1990).

Rescorla and Wagner's model demonstrated paradigms such as conditioning, extinction, blocking, overexpectation, overshadowing, conditioned inhibition, superconditioning, discrimination, and pseudodiscrimination. They could not, however, replicate configural learning, latent inhibition, or extinction of conditioned inhibitors. Their model was also unable to make predictions about effects on the conditioning of intratrial temporal relationships between stimuli, and yielded a negatively accelerated acquisition curve, unlike those found in animal research.

In the 1970s and early 1980s, some researchers of animal behavior (e.g., Klopf, 1972; Sutton & Barto, 1981) began to consider the role of time as a fundamental dimension for understanding natural intelligence. (Grossberg, 1967, was already considering time with his outstar model, but his assumptions are very different from those reported in this chapter.) As Klopf said later (1988), "Real-time learning mechanisms emphasize the *temporal* association of signals: each critical event in the sequence leading to learning has a time of occurrence associated with it, and this time plays a fundamental role in the computations that yield changes in the efficacy of synapses" (p. 90).

The first major step in creating a real time neural network model of classical conditioning to go beyond that of Hebb (1949) came from Sutton and Barto (1981); Gelperin et al. (1985), and Tesauro (1986), among others. They agreed that neurons may be reinforcement learning devices different from those previously proposed in neural theories (Klopf, 1972, 1979). In the *Gelperin, Hopfield, and Tank (GHT)* model, the present input stimuli are scaled by the change in output values in the adaptation law. The GHT adaptation law can be described as $w_i(t+1) = w_i(t) + c * x_i(t) * [(y(t) - y(t-1)]$. The major thrust of the GHT adaptation law is that it is not the output values that provide the reinforcement but rather the *change* in output values from the previous time step. Hence, learning occurs only when there is a substantial change in the neuronal output. Using only the input stimulus, the GHT model also behaves in similar ways to the Widrow-Hoff in that the neuron only learns when there are strong input stimuli that cause an unexpected output. Because the output change may be positive or negative, the weight may also be increasing or decreasing.

Tesauro (1986) postulated that learning not only depends on the output change, but also on the input change. Tesauro's earlier learning rule is as follows:

$$w_i(t+1) = w_i(t) + c * x_i(t-1) * (y(t) - f[y(t-1)])$$
$$f[y] = \min(1, y + \eta)$$

where η is a small constant and f is the expectation function that anticipates the present output from the previous output. The output change is not just a change from the previous value, but rather a change from the expected present value. The expected output is the previous output plus some additional amount η, bounded by its maximum value. When there is a substantial difference between present output and expected present output, learning occurs.

A later refinement of Tesauro's rule incorporates the change in the past stimulus, rather than the input stimulus strength:

$$w_i(t+1) = w_i(t) + c * \sigma[x_i(t-1) - x_i(t-2)] * (y(t) - f[y(t-1)])$$
$$\sigma[x] = \begin{Bmatrix} x, x > 0 \\ 0, x \le 0 \end{Bmatrix}$$

where the sigma function signifies that only *positive* input change contributes to reinforcement. Using Tesauro, one sees that learning occurs only when there is a positive change in input stimuli coupled with a substantial change in the neuronal output that is beyond what is normally expected. Even though only the positive change in input stimuli is used, the change in output may be positive or negative. Hence, synaptic weights may be made to increase or to decease.

The *Sutton-Barto* model (1981) uses a form of expectation closely related to that of the Rescorla-Wagner one, but whereas in the Rescorla-Wagner model associative strengths are changed based on the difference between received and expected US levels, in the Sutton-Barto one the weights are changed based on the difference between actual activity level and the time averaged level. The resulting law is

$$w_i(t+1) = w_i(t) + c * \overline{x_i}(t) * [y(t) - \overline{y}(t)]$$
$$\overline{x_i}(t) = \alpha \overline{x_i}(t-1) + x_i(t-1)$$
$$\overline{y}(t) = \beta \overline{y}(t-1) + (1 - \beta)y(t-1)$$

where an overbar on a parameter represents the corresponding trace value; α and β are constants that control the forgetting factor for the trace on the input stimulus and output response.

The Sutton-Barto adaptation law incorporates the traces of input and output signals as a means to account for the history of input and output stimuli. The input trace provides an intense input value only when the input is consistently present in the immediate past, hence placing more emphasis on more recent past events and less on the distant past. On the other hand, the output is "meaningful" and contains reinforcement value only when the actual output is different than the anticipated (or estimated) results. The change in output may be positive or negative, that is, reinforcement may be positive or negative. This allows the ability to learn as well as unlearn through association of deviation of action or inaction from anticipated results.

When the parameter α is not equal to zero, the trace of the input stimulus represents an exponentially decaying weighing function on all past inputs. In most of the Sutton-Barto simulations, the value of β is taken to be 0; then the adaptation law can be rewritten as $w_i(t + 1) = w_i(t) + c * x_i(t) * [z - y(t)]$. When β equals 0, the output trace simply reduces to the previous past value. Examination of the simplified Sutton-Barto rule shows that the learning is then related to the change in output values, that is, the temporal difference of the output, with an exponentially weighted input. If α is close to 1, forgetting is slow and learning depends on a large number of past input stimuli. If α is very close to 0, the weighing of past inputs diminishes quickly and learning primarily depends on the most recent input stimuli.

Sutton and Barto's model was a temporal refinement of the Rescorla-Wagner model. As Klopf (1978) noted, a signal trace of an input to a synapse may persist for some period of time, and that this trace might allow for increases in synaptic efficiency even if the input event and the firing event are separated in time. The Sutton-Barto model does effectively demonstrate the formation of positive associations for the experimental frameworks in which overlap of CS and US does not occur. However, it does not give positive associations when CS offset is extended beyond US onset. The Sutton-Barto model replicates all of the stimulus context behavior of the Rescorla-Wagner model, including blocking, overshadowing, conditioned inhibition, discrimination and pseudodiscrimination effects, as well as acquisition, extinction, and interstimulus interval (ISI) effects in trace conditioning.

The Sutton-Barto model, however, deviates from some other evidence found in animal experiments. For example, the model does not account for the initial positive acceleration in the S-shaped acquisition curves observed in classical conditioning. Also, it yields inaccurate predictions for several CS-US configurations involving significant overlap between CS and US durations. Some of these problems were later solved using reinforcement learning models (Barto, Sutton, & Anderson, 1983; Sutton, 1984, 1988). In the area of unsupervised learning, some of the difficulties of the Rescorla-Wagner and Sutton-Barto models were not overcome until the work of Klopf (1986) and his drive-reinforcement model (1988), to be discussed shortly.

After their original work on classical conditioning, Sutton and Barto (1982) extended their model to study how the CR usually anticipates the US in the simulation of conditioned inhibition and chaining associations. From this study, they moved into reinforcement learning models, which are outside the scope of this chapter. Such models are nevertheless important because they take temporal differences into account and move closer to an understanding of instrumental-operant learning (as opposed to classical-respondent learning models). For excellent treatments of reinforcement learning, see Kehoe (1989); Kehoe, Schreurs, and Grahm (1987); and Werbos (1989, 1990).

Klopf (1988) proposed what he called a *drive-reinforcement* (D-R) model of single neuron function, in which drives are viewed as signal levels in the nervous system, and reinforcers as changes in signal levels. The result was a model that was characterized by Caudill (1989) as the most biologically rigorous neural network model of classical conditioning in use at this time. The D-R model not only predicts the formation of associations when stimulus overlap occurs, but accurately predicts that weaker (but still positive) associations will form when CS offset and US onset are simultaneous or even temporally separated.

Klopf (1986) presented the essence of his drive-reinforcement model as an extension of Sutton-Barto (1981); it was to appear as a detailed model in Klopf (1988). He defined his learning rule and gave it the name of Differential Hebbian Learning (independently discovered by Kosko, 1986). Klopf (1979) suggested the use of a stimulus trace variable completely separate from the major signaling variable. Thus when activity at a synapse is eligible for modification, it remains eligible for a period of several seconds. A synapse's modifiability depends on the reinforcement level during this period of eligibility. Each synapse therefore can be viewed as having its own local trace mechanism. This trace mechanism mediates synaptic modification but does not directly alter other aspects of the neuron's behavior. Klopf further suggested that a trace could last for the relatively long times found in classical conditioning studies without interfering with continuing signal transmission.

The drive-reinforcement learning law of Klopf (1988) derives from the idea that it is *changes* in both input stimulus and output response that serve as reinforcers. But in addition, the present synaptic change is dependent on a weighted sum of past changes: The connection between two neurons is doubled, with one connection carrying an excitatory weight ($w_i^+(t)$) and a second connection carrying an inhibitory weight ($w_i^-(t)$). Klopf (1988, p. 87) further posited separate excitatory and inhibitory connections at each synapse, with total neuronal output equal to the sum of the activation from both excitatory and inhibitory responses. Reformulating Klopf's equation in a form consistent with other equations in this chapter, we have:

$$w_i^+(t+1) = w_i^+(t) + \sum_{k=1}^{\tau} c_j * |w_i^+(t-k)| * \sigma[x_i(t-k) - x_i(t-k-1)] * (y(t) - y(t-1))$$

$$w_i^-(t+1) = w_i^-(t) + \sum_{k=1}^{\tau} c_j * |w_i^-(t-k)| * \sigma[x_i(t-k) - x_i(t-k-1)] * (y(t) - y(t-1))$$

$$|w^+|, |w^-| \geq 0.1.$$

(8)

The drive-reinforcement model, unlike any of the previously mentioned models, has separate paths for excitatory and inhibitory reinforcement. In other words, every input stimulus has a positive excitatory weight and a negative inhibitory weight, with each weight being reinforced separately.

The form of the learning equation shows us that learning is based on the history of input stimuli and output responses. This historical information is more than just a trace as in the Sutton-Barto model. The actual change in input stimuli and the actual output values are available for adaptation with a decaying factor. In other words, learning can occur even though the CS and the US do not overlap in time, that is, there is a finite nonzero *interstimulus interval* (*ISI*).

The decaying or forgetting factor is designed to achieve diminishing effects similar to the results observed in ISI conditioning. Inclusion of weights in the reinforcement term produces an S-shaped acquisition curve. It provides quicker convergence, and at the same time serves as a bound to the learning. The separate weights for excitatory and inhibitory effects allow a particular input trace to produce different responses related to the role played by each of the two weights. Because both weights are updated with different strategy at different times, this allows for a very useful interaction, and is particularly evident during extinction and reacquisition paradigms. Klopf claimed that this model predicted a wide range of classical conditioning effects, including delay and trace conditioning, conditioned and unconditioned stimulus duration and amplitude effects, partial reinforcement effects, interstimulus interval effects, second-order conditioning, conditioned inhibition, extinction, reacquisition effects, backward conditioning, blocking, overshadowing, compound conditioning, and discriminatory stimulus effects (some of these conclusions are questioned in the next section).

In summary, we can logically organize the adaptation laws. All of them can be traced back to Amari's law, which basically says that weight change is cumulative and incremental with respect to the reinforcement. Hebb's learning rule specifies that the reinforcement is the multiplicative association between the input stimuli and the output response. The Widrow-Hoff law differs in that the association is not with the instantaneous output value, but rather with the difference between the output from the external target value. The GHT rule specifies that the output difference should not be dependent on an external target value, but rather on the previous output value. Tesauro's specifies that the association is not only with the change in output, but also with the change in input at the same time.

Rescorla-Wagner's learning rule, based on the Hullian approach of probability, specifies that the learning constant could be different at various stages of the paradigm and the "target" value is the maximum synaptic strength. Sutton-Barto's rule shows that the association should be dependent on the input trace rather than the present input value, and the output difference between the present output and the output trace. The drive-reinforcement rule uses the synaptic weights in the reinforcement term which is also dependent on the positive input change and output change. Furthermore, the D-R model specifies that there are two weights to each input stimulus, an excitatory weight and an inhibitory weight, each with separate updates.

These models can also be distinguished by their treatment of time. On a gross time scale, one speaks of successive trials with each trial consisting of presenting the US and CS and observing the response in total. On a finer time scale, one speaks of successive time steps with each time step showing how the US and CS are presented, that is, the temporal relationship of the US and the CS within a trial. Those models that do not account for this finer time scale, such as those of Hebb and Widrow-Hoff, are termed *non-real-time models*. Those models that account for this finer time scale, such as those by Tesauro, GHT, Sutton-Barto, and Klopf, are termed *real-time models*.

3. PSYCHOLOGICAL ASSESSMENT

In the preceding sections we have discussed various mathematical adaptation (learning) rules of several models of how a single neuron's activity might be simulated. The remainder of this chapter is devoted to further analyses of the neuronal models by testing them with various classical conditioning paradigms, followed by some of our conclusions.

The simulation environment used in this study was an interactive set of programs written in Turbo C++ for the IBM PC and compatibles. The main module provides a set of pulldown menu screens that allow the user to choose from several different learning strategies, conditioning paradigms, viewing modes, and setup options. When a learning rule had been selected, the user could change the operative constants for the learning model or continue with the default values.

A choice of input files was also given, each representing a particular conditioning paradigm, or experimental procedure. Nineteen procedures, representing all the conditioning paradigms simulated by Klopf (1988) for the D-R model, were selected. After choosing a procedure (file), the user could run one or more models on the input configuration it represents, producing a set of weight and output files. The user could also change the conditions of the experiment or configure new experimental procedures by altering an input file or creating new files with the editing facility provided. Specifically, the user could define the number of trials, the number of time steps, the onset and offset step and duration of the US, and select up to six CSs. After running a simulation, the View option reads the weight and output files and displays a screen of 2-dimensional input, output and weight graphs or a 3-dimensional view of the model's output (CR and UR) over the entire span of the experiment. The 2-D view screen may be altered to show the output of the learning element at different times across all trials and within each trial. The assumption for all models is that one time step equals 1 sec.

In the simulations presented, default values for all learning and trace constants were used. These defaults are chosen to coincide as closely as possible with those specified by the authors of these models. For example, the Rescorla-Wagner model learning constant c was set to 0.5, and all weights began at 0.0. For the Sutton-Barto model, the value for the learning constant was again 0.5, and all weights were initialized to 0.0. The value of the input trace control constant α was 0.9, trace values $\bar{x}(0)$ and $\bar{y}(0)$ were initialized to 0.0, and β was set equal to 0.0. The effect of this value for β in the learning equation is to make $\bar{y}(t)$ equivalent to $y(t-1)$. For the drive-reinforcement model, with $\gamma = 5$ and j initially equal to 1, the constant vector c was set equal to ($c_1 = 5.0$, $c_2 = 3.0$, $c_3 = 1.5$, $c_4 = 0.75$, $c_5 = 0.25$), and the absolute values of all weights have a lower bound equal to 0.01. For all experiments shown, the input nodes that received a US (unconditioned stimulus) had fixed weights of 1.0 and input nodes receiving CS (conditioned stimuli) had variable weights. The value for US input was always 0.5, and CS values were 0.2.

4. ANALYSES

In Pavlovian conditioning procedures, time is assumed to play a very important part. A conditioning stimulus, such as a bell, is typically presented to a hungry organism just *before* an unconditioned stimulus (US) is presented. The conditioning stimulus (CS) thus becomes associated with an unconditioned stimulus (food), and a conditioned excitation develops such that after a number of pairings, when the bell is rung, a conditioned response (CR) such as salivation occurs. When the amount of salivation in response to the bell is plotted relative to the number of trials, there is an initially positive acceleration of the curve changing to negative acceleration, which continues up to the point that an asymptotic level is reached, forming a somewhat S-shaped curve. It is assumed that the level of response (CR) to the CS represents the level of association that has developed between the CS and the US. This procedure, sometimes called delay conditioning, is defined so that CS onset precedes US onset and CS offset occurs at about the same time, or later than US onset (cf. Fig. 13.1).

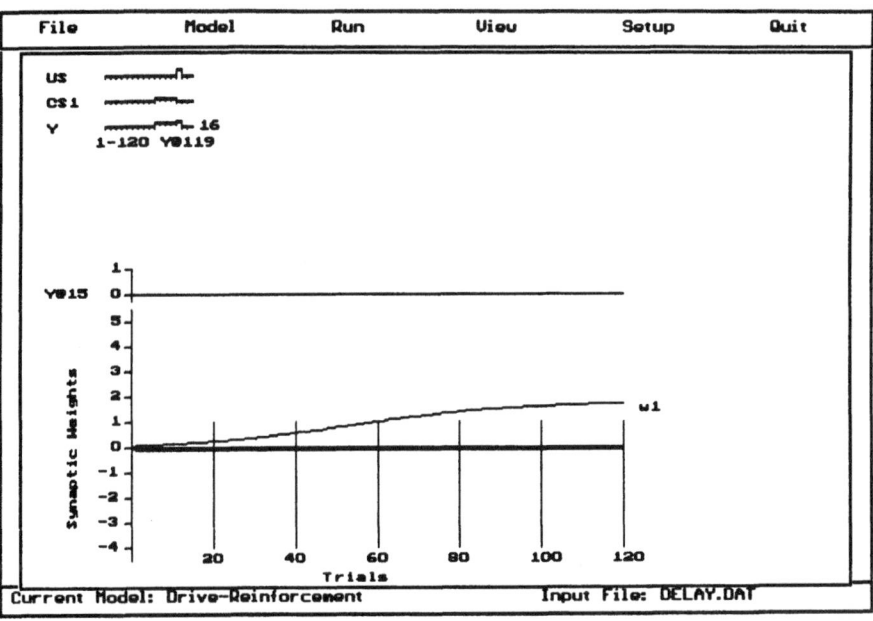

Fig. 13.1. Time course of a typical delay conditioning paradigm using the drive-reinforcement model.

In the computer simulations, we are dealing with discrete time. This means that when the CS goes off at the end of the time step preceding the onset of the time step when the US comes on, there is an assumption that the CS has gone off within one half of a time step before the US is presented. Following Klopf (1988), *"Onset* of a stimulus at time step t means that the stimulus was on during time step t and was not on during the preceding time step. *Offset* of a stimulus at time step t means that the stimulus was off during time step t and was not off during the preceding time step." On the other hand, if a CS goes off *during* the same time step *within* which the US is presented (comes on), then the two are said to be overlapping, that is, on *during* the same time step. In general, overlapping occurs with CS and US when the CS is on during any or all of the times steps when the US is also on. The CS can, of course, be adjusted to be on for as many time steps as the experimenter wishes.

When designing the simulations, we used Klopf (1988) as our guide because his drive-reinforcement model correctly simulates more classical conditioning paradigms than any of the other models. Thus we built all the models and files to work together from a common platform. Computer screens were designed to accommodate all seven of the above models and any we created for additional comparisons. Our data files were also centered around Klopf's onset/offset times and number of trial presentations for all 19 of the classical conditioning procedures (paradigms). When the simulator was finished, and we began the analyses, we found that using Klopf's onset/offset times for the other models sometimes limited their results, because inherent factors of a given learning rule were possibly not considered. In fact, some of the models failed to show learning when using his onset/offsets but do show learning when the onset/offsets are brought into line with learning equations in the other models.

Simulation	CS and US Timing			
Model	CS$_1$	CS$_2$	CS$_3$	US
Stimulus Duration				
Hebb	11/14/1-100	20/24/1-100	29/34/1-100	13/13/1-100
				23/24/1-100
				33/34/1-100
SB	10/13/1-100	19/23/1-100	28/33/1-100	13/14/1-100
				23/24/1-100
				33/34/1-100
DR	10/13/1-100	20/24/1-100	28/33/1-100	13/14/1-100
				23/24/1-100
				33/34/1-100
Blocking				
Hebb	11/14/1-160	11-14/101-160	--	13/14/1-160
SB and DR	10/13/1-160	10/13/101-160	--	13/14/1-160
Reacquisition				
Hebb	11/14/1-200	--	--	13/16/1-70
				13/16/141-200
SB and DR	10/13/1-200	--	--	13/16/1-70
				13/16/141-200
Second Order				
Hebb, SB, and DR	10/15/1-200	7/12/61-200	--	13/15/1-60

Table 13.1. Timing of the CS-US Configurations.

For example, the Hebb rule is based on simultaneous pre- and postsynaptic activity and must have overlapping CS-US presentations. In a delay conditioning paradigm, the CS must be presented for at least one time step after the US presentation begins. The Sutton-Barto and Gelperin-Hopfield-Tank models, on the other hand, cannot handle overlapping CS-US presentations because of the presynaptic trace element in these models. In the case of delay, the CS must terminate before or at the same time step when the US presentation begins. The comparisons made for our analyses were thus based on separate data files for these three models, and our results therefore differ from those of Klopf's. Onset and offset times for all of the simulations presented are listed in Table 13.1. As an example, CS_1 is on for time steps 11-14 for trials 1-100; CS_2 comes on at time step 20 and goes off at time step 24 for trials 1-100; CS_3 comes on at time step 29 and goes off at the end of time step 34 for trials 1-100; and the US for CS_1 is on for the duration of time step 13 only, for trials 1-100; the US is on for CS_2 for time steps 23 and 24 (for trials 1-100) and for CS_3 the US comes on again for time steps 33 and 34 (for trials 1-100). Note that pairings of US with all three CSs is shown in the upper left corner of each figure.

4.1. CS Duration

Figs. 13.2-13.9 give some of the simulator's output for an experiment varying the duration of the CS in delay conditioning. The length of delay for CS_1, CS_2, and CS_3 is 3, 4, and 5 time steps respectively. The experimental literature is in general agreement (Ayers, Haddad, & Albert, 1987) that extending the duration of the CS tends to slightly weaken the level of conditioning. Thus, we would expect an animal in a similar experimental procedure to have approximately the same S-shaped learning curve for CS_1, CS_2 and CS_3, with progressively lower asymptotic levels demonstrating weaker associative strengths as CS duration increases.

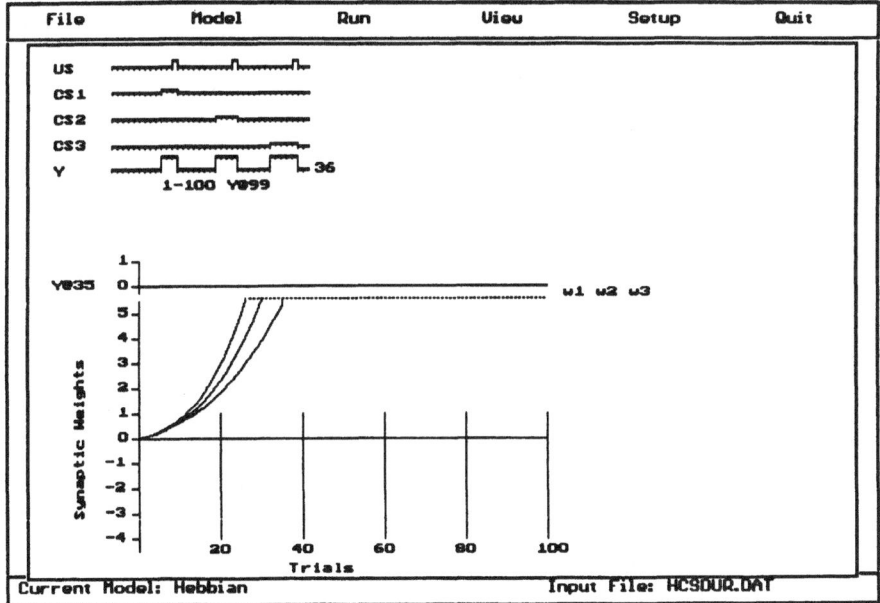

Fig. 13.2. Delay conditioning using the Hebbian model.

It should be noted that the model's output, as shown in the following graphs, represents changes in the synaptic weights of the single neuron which represents the conditioned response (CR). In our discussion of the results we are comparing the single neuron output to the known experimental results of the conditioned response found in whole animal behavior. Our findings show that the simulation output for Hebbian (HE) learning matches the known experimental results (Fig. 13.2).

The shortest duration (CS_1) forms the strongest association, as represented by the value of the synaptic weight (w_1), and the longest duration (CS_3) the weakest association (w_3). The results, however, show one of the major weaknesses of Hebbian learning; it provides no means of reducing the synaptic efficiency of the connection and fails to reach an asymptotic level. If not clamped artificially by the simulation environment at 5.5, the synaptic weight would continue to increase indefinitely.

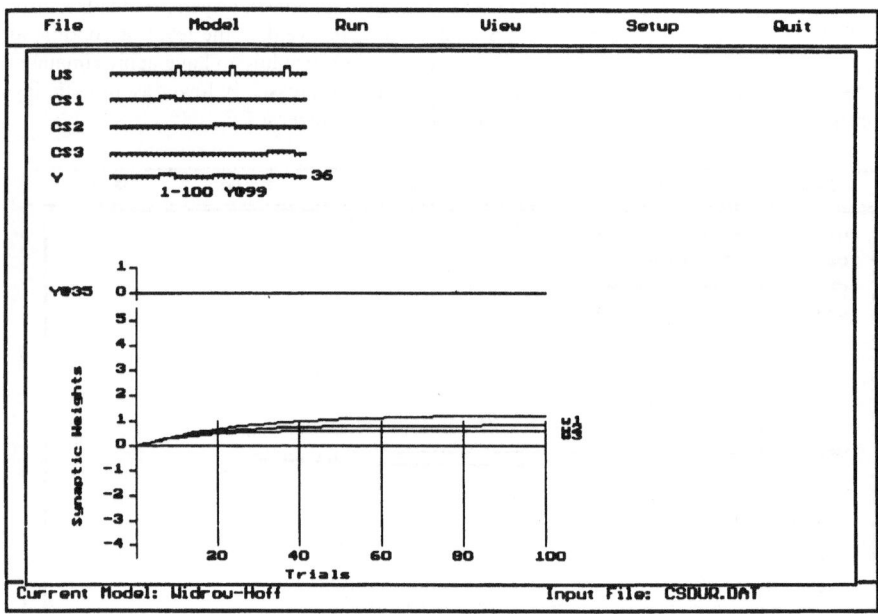

Fig. 13.3. Delay conditioning using the Hebbian model.

The Widrow-Hoff (WH), Rescorla-Wagner (RW), and Sutton-Barto (SB) models all show a learning curve for the weights of CS_1, CS_2, and CS_3 with the appropriate positive association, although the negative acceleration of the weight graph is not consistent with the experimental evidence. These models also give the correct ordering of associative strengths (Figs. 13.3-13.5). The Gelperin-Hopfield-Tank (GHT) model (not shown here for space reasons) shows that the CR for CS_1 conditions, but not for CS_2 or CS_3. The GHT model conditions for CS_1 because there is no overlap with the US. For the other CSs, there is overlap, and no conditioning occurs for this model. Duration of the CS appears to be a major factor in whether or not learning occurs.

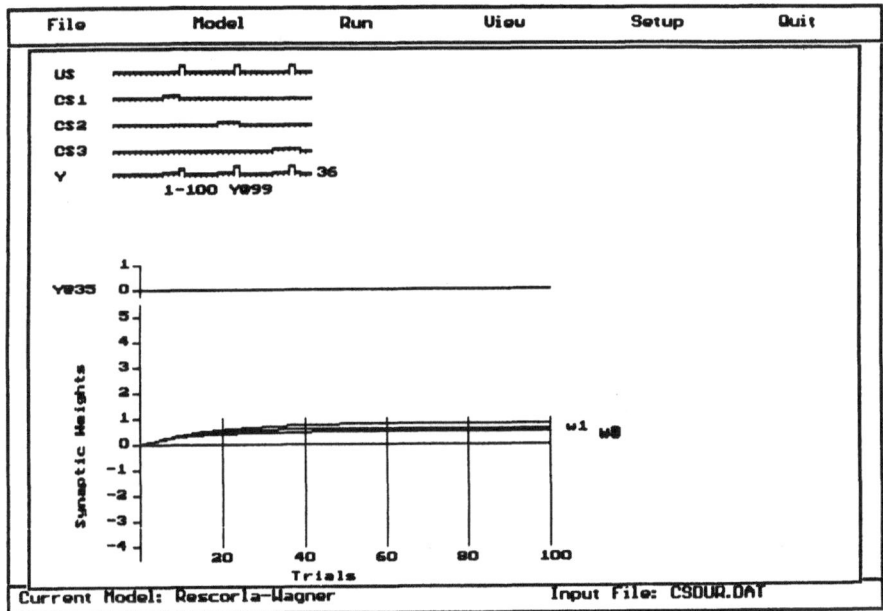

Fig. 13.4. Delay conditioning using the Rescorla-Wagner model.

The Tesauro (TE) model (Fig. 13.6), like the GHT, shows a negatively accelerated learning curve with higher associative strengths than the RW, WH, and SB models. The order effects of CS_1, CS_2 and CS_3 are also in accord with the experimental results for animal learning, in that less CS-US overlap results in an increase in the level of conditioning.

The drive-reinforcement (D-R) model in the same experimental framework is shown in Fig. 13.7. The formation of associations in the weight graphs produced is consistent with the S-shaped learning curves found in animal learning. However, the initial acceleration rate and asymptotic synaptic weight values are opposite to those found in animal research. Klopf (1988) stated, "whole-animal data may be insufficient to test these predictions, in that higher level attentional mechanisms may play a significant role when CS durations are extended beyond the US" (p. 95). The implication is that the D-R model findings are correct, and that further research at the single neuron might explain these results at a more molecular level.

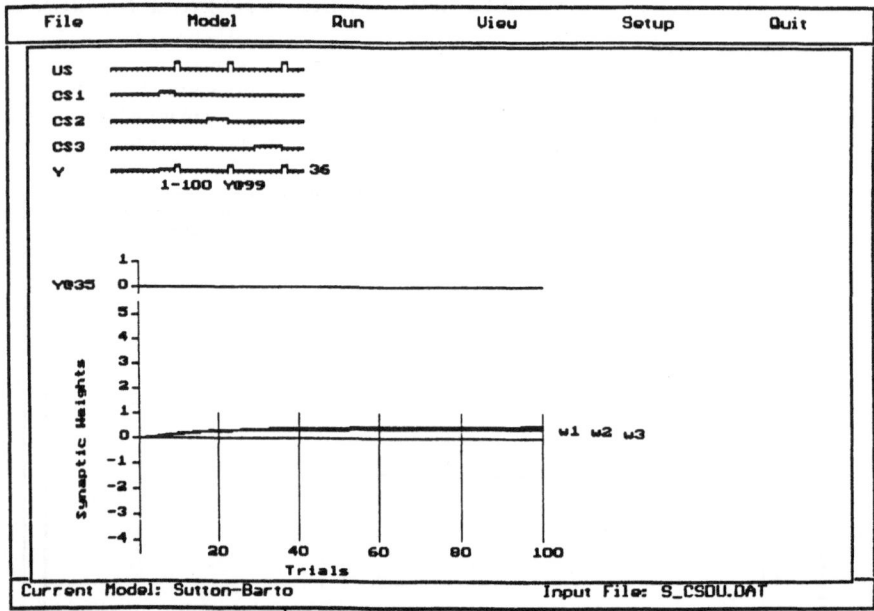

Fig. 13.5. Delay conditioning using the Sutton-Barto model.

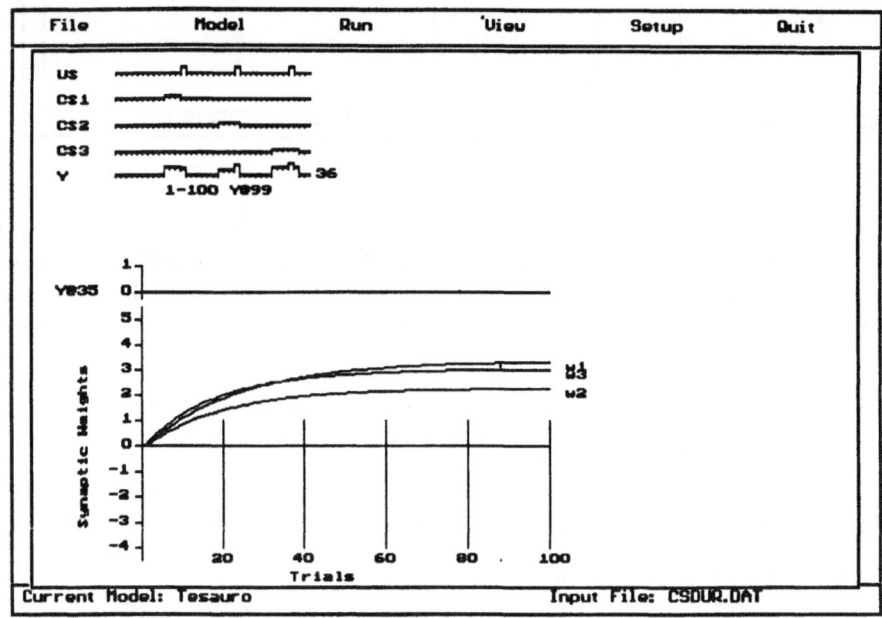

Fig. 13.6. Delay conditioning using the Tesauro model.

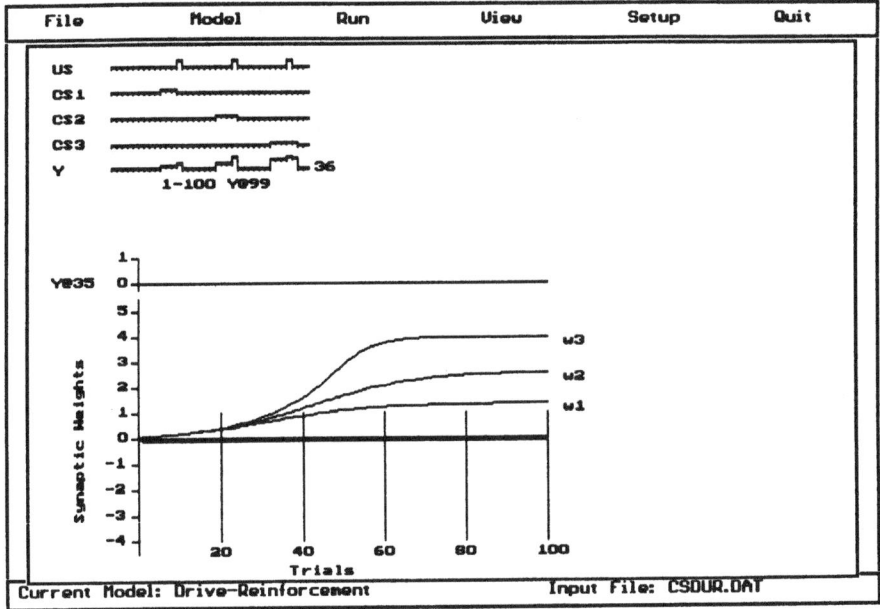

Fig. 13.7. Delay conditioning using the drive-reinforcement model.

From a whole-animal perspective, Klopf's original CS-US onsets and offsets contain a confounding factor. Both the CS duration and the absence or presence of a CS-US overlap are changed. The CS duration is increased from 3 for CS_1 to 5 for CS_3; however, there is an overlap for only the second two CS durations. Simulation of the D-R model without this confounding factor showed unexpected results. Figure 13.8 shows the D-R model run using the Hebb data files, which include an overlap for all three CSs. The results on weight acceleration are inconsistent with the known animal results. Also the asymptotic synaptic weight values do not correspond either to known animal results or to Klopf's predictions based on the original onset/offset times. When the D-R model was run using the Sutton-Barto files without CS-US overlap (Fig. 13.9) the results were different from the previous two simulation results, both in the acceleration curves and the asymptotes.

4.2. Blocking

Blocking is an example of a compound stimulus because it contains separable stimulus components. Kamin (1968) first gave an experimental group of rats a series of training trials pairing a light (CS) with shock (US) until the light was maximally effective in evoking the elicited response. Then, the same animals received further training in which a second stimulus, a tone, was also paired with shock and the light was also presented at the same time as the tone. No conditioning occurred to the tone. Kamin reasoned this was because the tone provided no new information to the rats, indicating that for something to be learned, or associated, there needs to be a "surprise" factor present. Bower (1970) noted that "the learning mechanism seems to become 'switched on' mainly when environmental events do not confirm expectations. These findings show that although temporal contiguity is necessary, the CS must also have predictive value." In the 2-D simulation of blocking (Fig. 13.10), it can be seen that the drive-reinforcement model best predicts the blocking phenomenon.

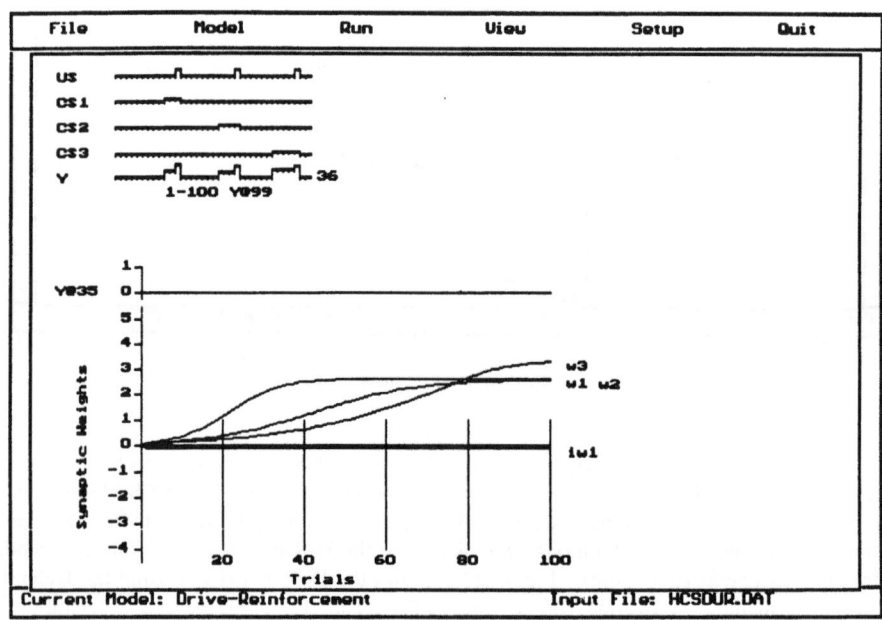

Figs. 13.8. Delay conditioning using the drive-reinforcement model with Hebb data file.

During trials 1-100, CS_1 is reinforced by the US in the first stage of conditioning, until the CS_1 excitatory weight approaches asymptote. Then, in trials 101-160, CS_1 and CS_2 are presented at the same time and reinforced by the US. Weight 2 (w_2), which represents CR_2 stays unchanged during the second stage, and is thus consistent with the experimental evidence.

The Tesauro model (not shown here) shows a nonsigmoidal acquisition curve for CS_1 and shows blocking when the CS overlaps the US onset. The Hebbian model shows the weights for both CS_1 and CS_2 as conditioned excitation, but does not show blocking (Fig. 13.14). The models that also show blocking to various degrees are Rescorla-Wagner, Sutton-Barto, Gelperin-Hopfield-Tank (Figs. 13.11-13.13), and Widrow-Hoff (not shown here).

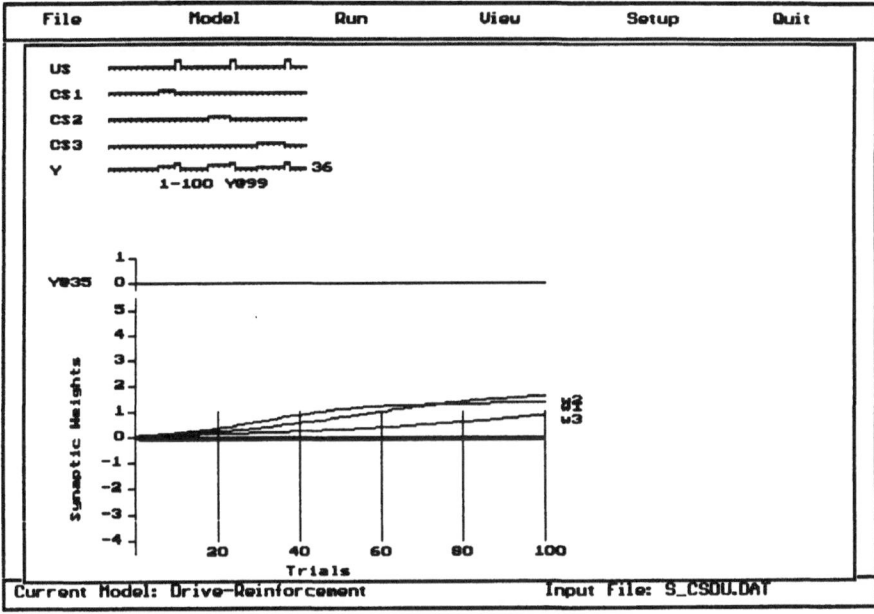

Fig. 13.9. Delay conditioning using the drive-reinforcement model with Sutton-Barto data file.

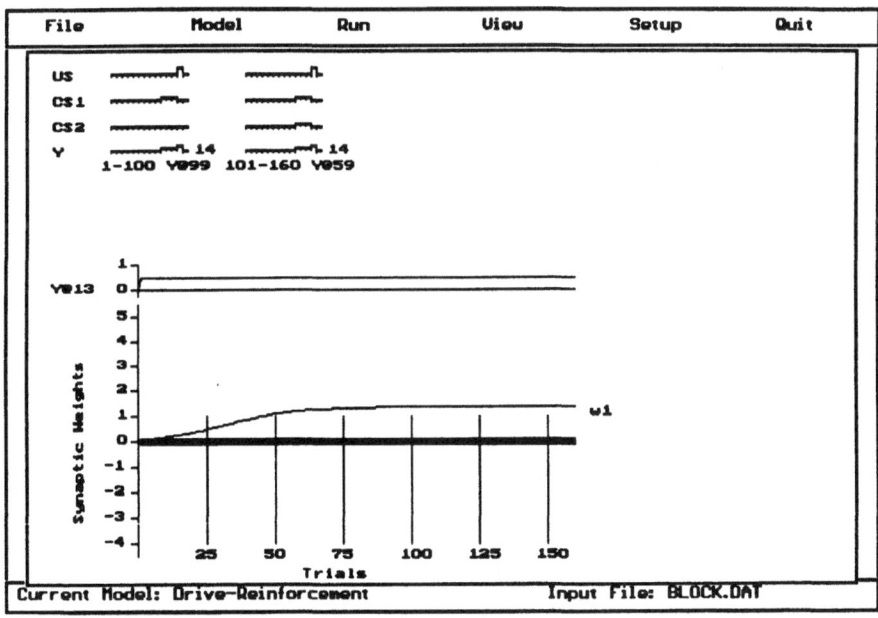

Fig. 13.10. Blocking simulation using the drive-reinforcement model.

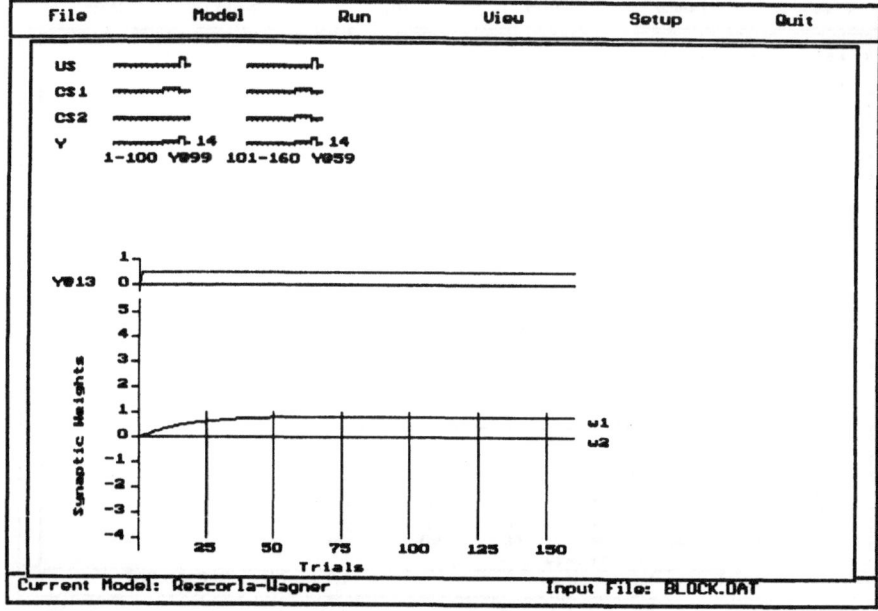

Fig. 13.11. Blocking simulation using the Rescorla-Wagner model.

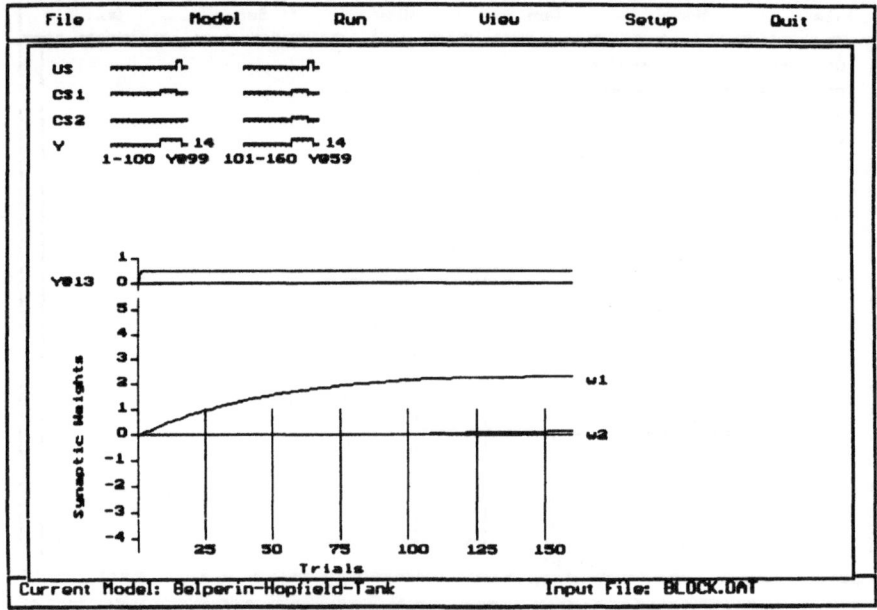

Fig. 13.12. Blocking using the Sutton-Barto method.

4.3. Reacquisition Analysis

The behavior of the D-R model in experiments examining the reacquisition of a previously extinguished association is completely in accord with the Pavlovian findings. The two-dimensional view of the input, output and weight graphs for this experiment is shown in Fig. 13.15.

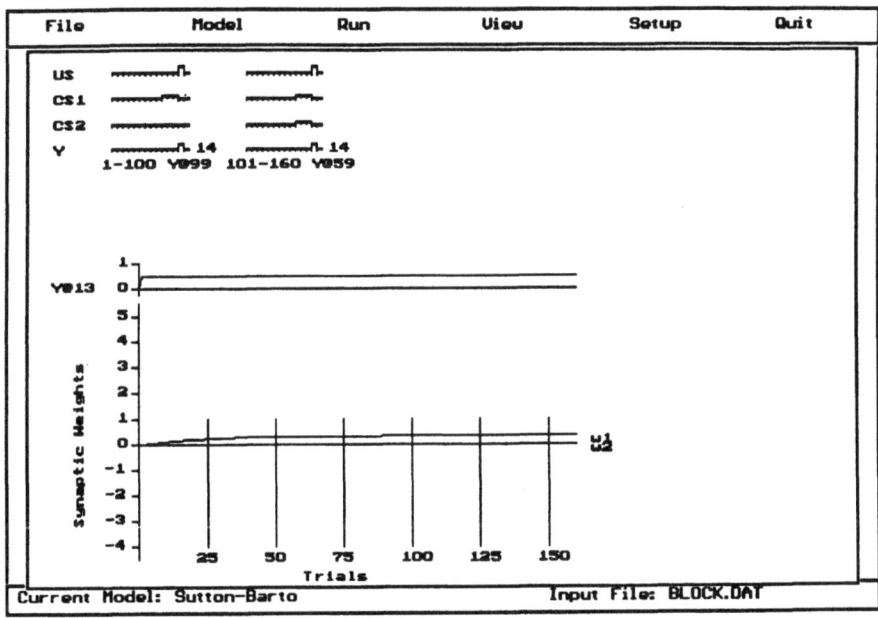

Figs. 13.13. Blocking simulations using the Gelperin-Hopfield-Tank model.

To demonstrate reacquisition requires three stages of conditioning. In Stage 1 (trials 1-70), a delay conditioning procedure is used to produce conditioned excitation. Note again the appropriate S-shaped acquisition curve this model forms for the excitatory synaptic weight (w_1) associated with CS_1.

In Stage 2 (trials 71-140), the inputs are configured so that the CS is no longer reinforced by the US. The resulting behavior is that extinction occurs, which for the D-R model means that the positive association of CS_1 to the US (w_1) decreases while the CS_1 inhibitory synaptic weight (iw_1) shows a small increase in absolute value. The effect of these changes is that the sum of w_1 and iw_1 approaches zero, causing the output of the neuron, $y(t)$, to approach zero. These findings are consistent with Pavlov's observations that response (salivation) to the CS (the bell) decreased from its previous asymptotic level when it was repeatedly given but not reinforced.

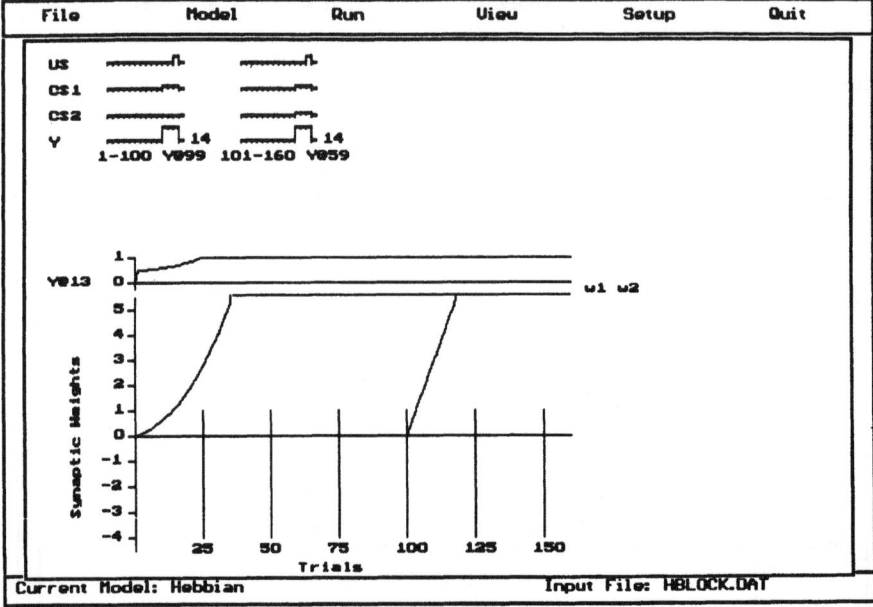

Fig. 13.14. Lack of blocking with the Hebbian model.

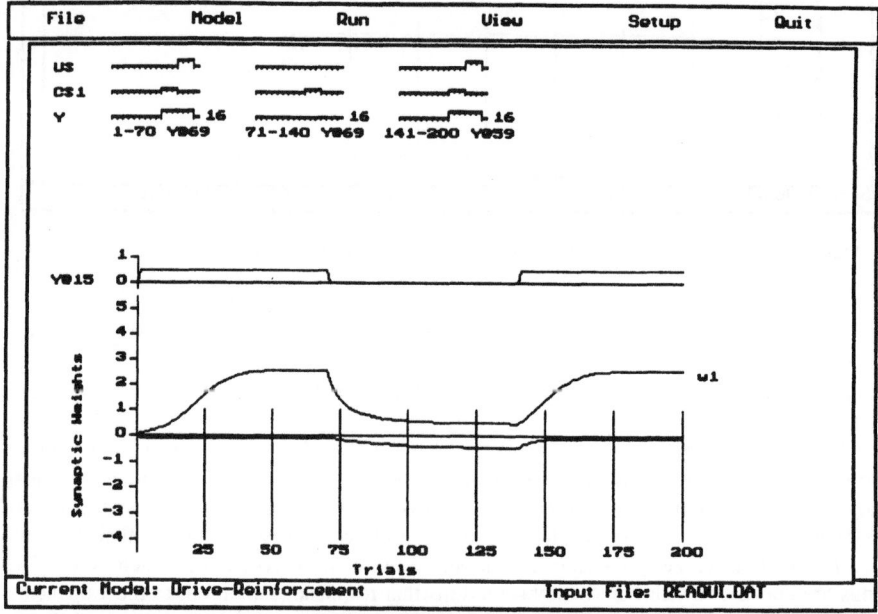

Fig. 13.15. Reacquisition with the drive-reinforcement model.

The critical part of the reacquisition paradigm is Stage 3 (trials 141-200). Pavlov found that when the CS was again reinforced by the US, reacquisition of the original positive association occurred more rapidly than in the initial stage. In Stage 1 of the D-R simulation, 60 trials occur before w_1 reaches its asymptotic level of 2.60. From the point where the US again begins to reinforce the CS (trial 141) to the point that w_1 returns to its previous level (trial 187) is just 46 time steps. Thus the D-R model correctly predicts the faster recovery found by Pavlov.

The SB and RW models (Figs. 13.16 and 13.17) both produce weight graphs that are negatively accelerating in the delay conditioning stages as noted in the discussion of CS duration effects. For both models, the association of CS to US formed in Stage 1 (w_1) extinguishes to zero in Stage 2. No explicitly inhibitory weights are used in either model, so the extinction of $y(t)$ can only be produced by reducing the level of association. The result of this total elimination of the previously formed association is that in Stage 3, reacquisition of association to asymptotic level takes precisely the same amount of time it did in Stage 1. These models thus do not correctly demonstrate the more rapid rate of reacquisition found in animal research studies.

The SB and RW (Figs. 13.16 and 13.17) and the GHT and WH models (not shown) all exhibit small amounts of acquisition for the first 71 trials, falling off gradually to about 0, then beginning to show reacquisition at trial 141. Note that the slope of the curves for w_1 (CR) does not occur more rapidly during reacquisition than during original acquisition. These models thus show support for reacquisition but do not predict the more rapid relearning of the CS as found in the experimental literature.

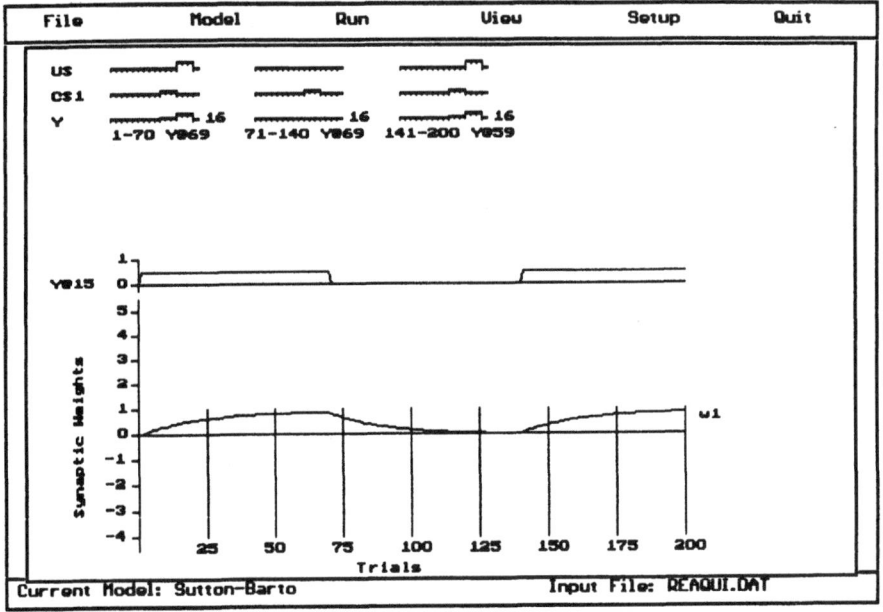

Figs. 13.16. Reacquisition using the Sutton-Barto model.

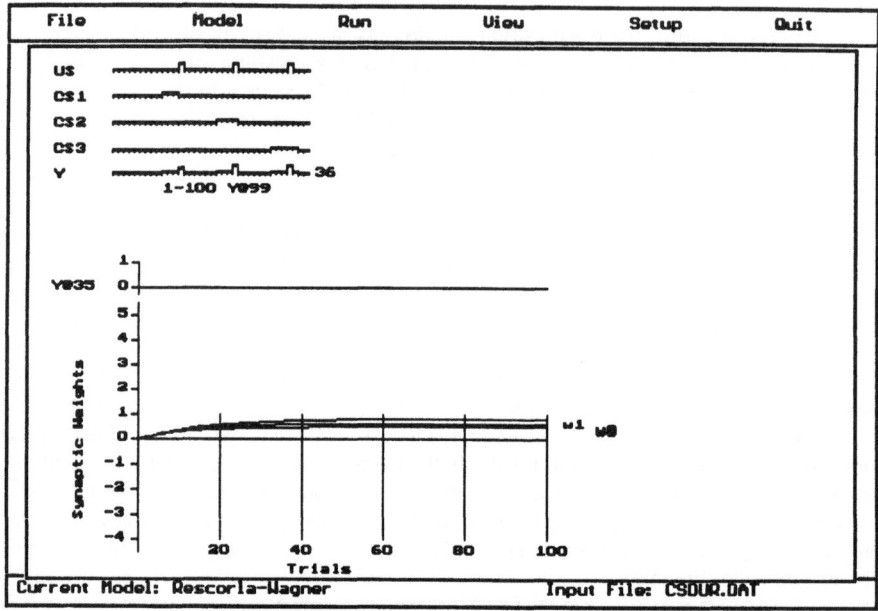

Figs. 13.17. Reacquisition using the Rescorla-Wagner model.

The Hebb model shows only an acquisition curve for Stage 1 (though clamped), as shown in Fig. 13.18. Actually, there are three graphs (inside Fig. 13.18) represented; at the top left, the onsets/offsets are shown for the US and CS_1, followed by the output y for the three stages of this procedure; the second graph is of the output y at the last time step minus 1, or; y at $t=15$ (in this case 16–1); and finally the third graph shows the growth of the synaptic weights (CR). The second graph, that of y at time 15, shows that the US goes off at trial 71 and remains off until the US is again paired with the CS at trial 141 and remains on to trial 199 (cross-referenced with the output files — not shown). It should be noted that at time step 15, only the US is presented, and therefore $y(t)$ only represents the UR (output).

An explanation might be appropriate, because all the models but one show the same results for $y(t)$: Whenever the US is not on, no learning takes place; but when the US is again paired with the CS, output results. What is reflected here is the combined strength of the US plus the CS in the form of a UR. With the Hebb model, CS_1 clamps thus preventing the second CS from affecting the CR in the third stage from appearing on the synaptic weight graph. The CR is not shown on the graph, yet the US elicits salivation, hence activity (UR), for trials 141-199.

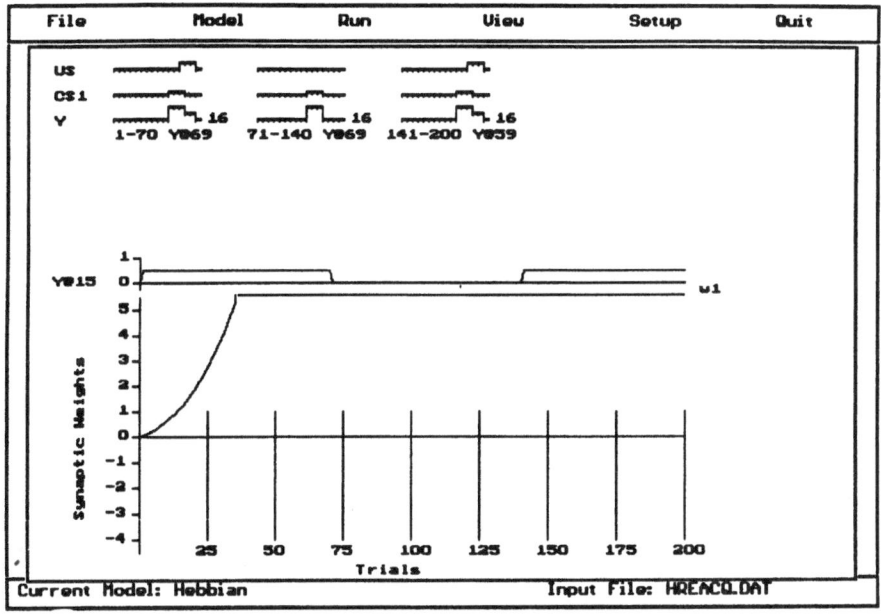

Fig. 13.18. Acquisition at Stage 1 using the Hebbian model.

What distinguishes the Tesauro model (not shown) is its rapid, strong rise in associative strength, (to a synaptic weight of 5.0, where the models above only achieved a strength between 1.0 and 2.0), beginning to fall off gradually at trial 71 and re-learning in Stage 3. However, the response to the second CS-US pairing did not occur more rapidly, as predicted by Pavlov.

4.4. Second-Order Conditioning

In second-order conditioning a weak association is formed between two CSs, one of which had previously been paired with the US. In Stage 1, CS_1 is paired with the US until a strong association has been formed. The CS_2 is presented with CS_1. Pavlov (1927) thought it was a transient response and found the second-order response to be weaker than the response to the first CS-US pairing. Pavlov claimed that the second CS would inhibit a CR that otherwise could have occurred. This suggests that CS_1 will serve to reinforce new learning (CS_2), much like a US does (Bower & Hilgard, 1981; Flaherty, 1985).

The D-R model clearly shows (Fig. 13.19) these processes in action. First, in Stage 1 (trials 1-60) CS_1 is reinforced by a US, giving a synaptic weight value slightly in excess of 4. In Stage 2 (trials 61-200) second-order conditioning occurs, but not as strongly as that produced in Stage 1, due to the effects of inhibition, as shown beginning at trial 71, labeled "iw_1 and iw_2"

on the graph (inhibitory weights 1 and 2). Although Klopf (1988) did not show the inhibitory weights in his graph (p. 100), his learning rule accounts for their presence. The results support Pavlov's (1927) contention that inhibition is partly explanatory for lower values of the second CR.

The Hebb model shows rapid acquisition for only CS_1 (Fig. 13.20), and also clamps the output. It is insensitive to second order conditioning. The RW and WH models (not shown) exhibit a monotonically rising curve, then extinction beginning at trial 71. Both also show a negative slope below the x axis for w_2 (CR_2). Neither shows learning during Stage 2, only inhibition. Note that in both models, weights 1 and 3 fall to the zero baseline.

The SB model (Fig. 13.21) shows minimal acquisition of the delay procedure in Stage 1 and only a slight association in Stage 2, as would be predicted. Tesauro's model (Fig. 13.22) shows acquisition and extinction for CS_1 as would be expected. However, CS_2 rapidly accelerates to the level where it must be clamped. The GHT (not shown here) shows some semblance of imitating the RW and WH models (above), and like them shows neither inhibition nor negative weights. Although the RW and WH models are similar mathematically, they do produce slightly different results.

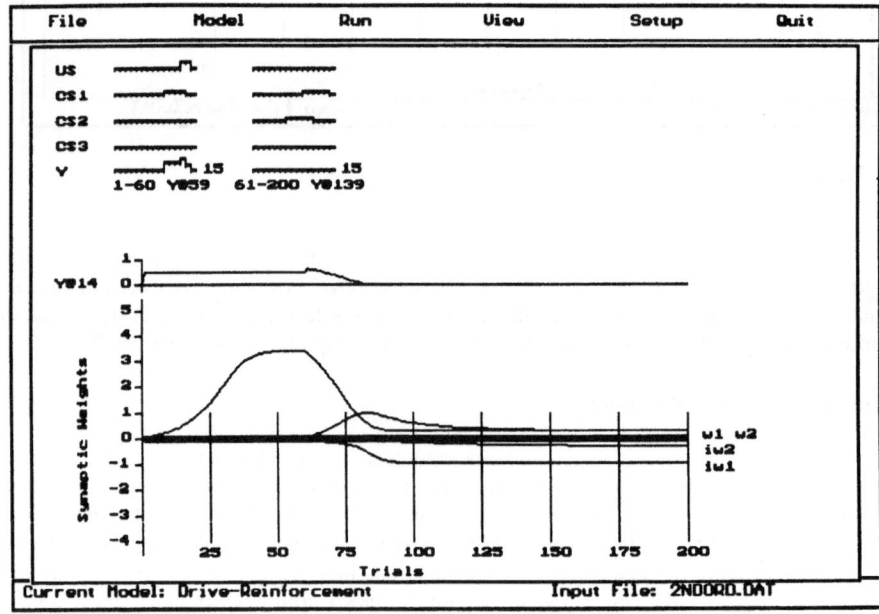

Figs. 13.19. Second-order conditioning with the drive-reinforcement model.

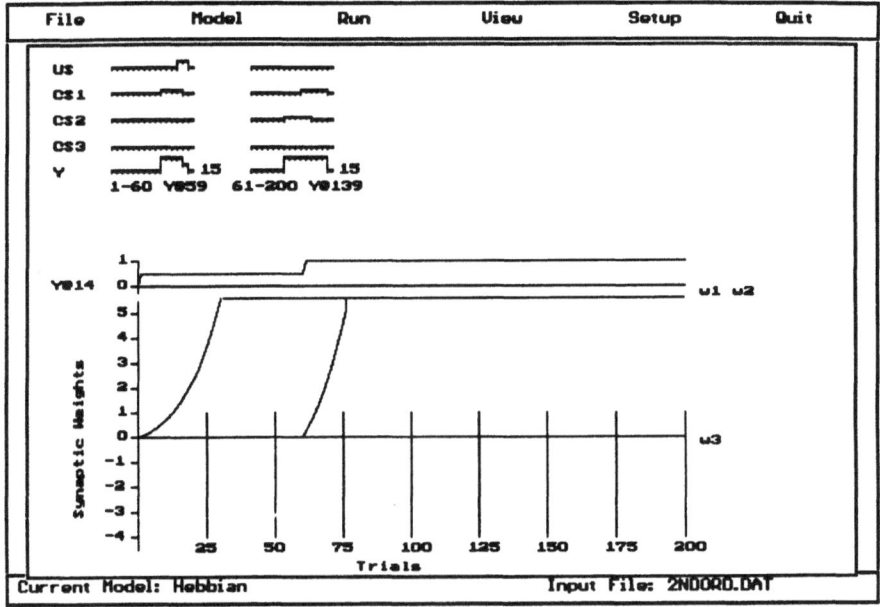

Figs. 13.20. Lack of second-order conditioning with the Hebbian model.

5. CONCLUSIONS

The computer simulator was developed to allow a uniform and consistent environment in which the different learning models could be duplicated and past research in the area of single-neuron learning replicated. This approach allows for a greater understanding of the mathematics involved in single-neuron computation, as well as the capabilities to test the models through the simulation of known classical conditioning results.

The various mathematical learning models have shown how diverse assumptions concerning the models appear to result in vastly different models. However, although there are many different forms of adaptation laws, there are several basic mathematical operations involved. First, weight adaptation is incremental; this is represented by derivatives (or first differences) of the weights. Second, weight adaptation is by means of associations; this is represented by multiplication. Third, parameter changes are more important than actual parameter values; this is represented by the derivatives (or first difference) of the parameters. Fourth, many models account for the temporal difference in event occurrences; this is represented by means of the memory feature. In other words, a history of past events is kept and is used to affect the present adaptation. Finally, the size of the memory can be asymptotically infinite, shortened to only the recent past, or even simplified to a single trace value.

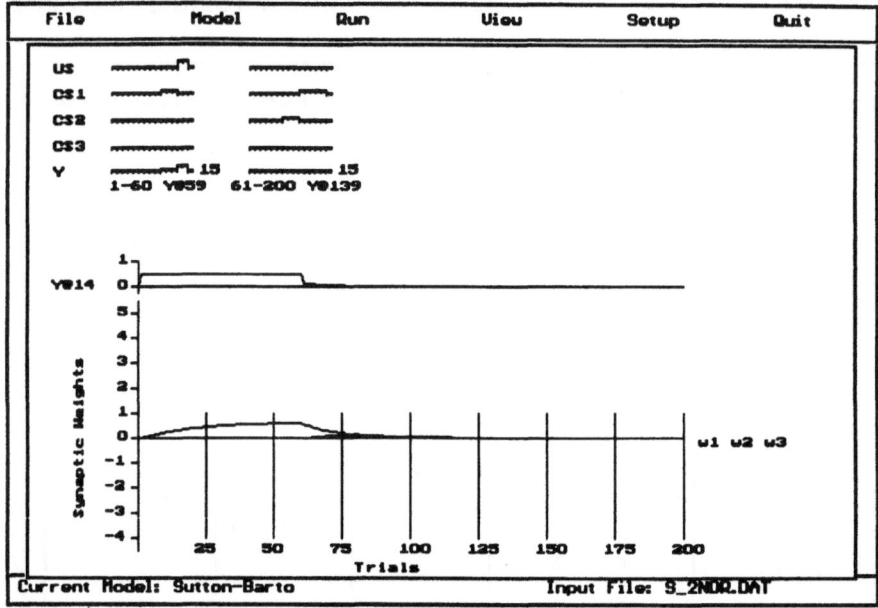

Fig. 13.21. Second-order conditioning with the Sutton-Barto model.

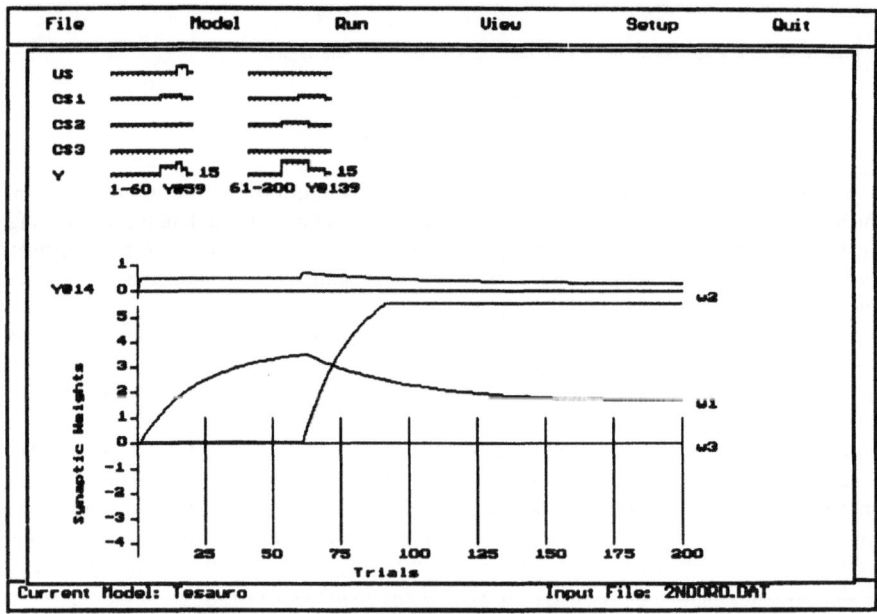

Fig. 13.22. Second-order conditioning with the Tesauro model.

The results of examination of real-time models on four different classical conditioning experimental designs have shown both the weaknesses and strengths of individual models. Manipulation of the experimental setup has allowed for a comprehensive comparison to establish the extent of their ability to model whole animal behavior. The designs chosen for comparison demonstrate differing elements in classical conditioning: (1) CS duration illustrates simple level conditioning with a single CS, (2) blocking incorporates a second, nonpredictive, CS stimulus, (3) reacquisition incorporates extinction and reacquisition, and (4) second-order incorporates higher order conditioning with a compound stimulus. Looking at the different learning models within the framework of the above elements of classical conditioning leads to differing conclusions about a model's overall performance. To demonstrate this approach, we make a comparison between two of the models: Hebb and drive-reinforcement.

Although the Hebb model is limited by its inability to clamp learning, its association between CS duration and US presentation matches known data when an overlap between the CS and US is allowed. This overlap represents the time factor inherent in the Hebbian model in that simultaneous pre- and postsynaptic activity is necessary for learning to occur. Grossberg and Levine (1987) called this the *synchronization problem*. A slight overlap between CS and US is known to be the best conditioning procedure and leads to the development of the strongest association (Bower & Hilgard, 1981). The Hebbian model consistently replicates known results for simple conditioning designs, including CS amplitude and US duration. The model cannot, however, replicate procedures that contain an interval of time between the CS and US (such as trace conditioning). In the CS duration simulation, when an overlap is present for all three stimuli, the D-R model is incapable of replicating this simple conditioning experiment. In order for the model to show three correct learning curves, shown in incorrect order, the experimental data must contain a confound between overlap and duration. The discrepancy in the order of association is explained, by Klopf (1988), at the neuronal level as opposed to whole-animal behavior comparisons made for other paradigms. The D-R model is inconsistent in replicating simple conditioning results (i.e., CS amplitude, interstimulus interval), which are either explained at the neuronal level or contain confounding experimental designs.

The D-R model is good at replicating conditioning designs that include more complex phenomena such as blocking, reacquisition, and second-order conditioning. This cannot be said for the Hebbian model, however, due to its mathematical limitations. Any CS that overlaps the US will form an association, as demonstrated by the two associations formed in blocking. It should be noted that Hebb could be set up to show blocking correctly by eliminating the overlap between the US and CS_2. Because the Hebbian model does not incorporate negative weights, extinction cannot occur, and therefore reacquisition is not possible. The Hebb model also cannot form an association between two conditioned stimuli as shown in second-order conditioning.

Since 1988, Klopf has extended his D-R model to include instrumental conditioning (Baird & Klopf, 1993; Klopf & Morgan, 1990; Klopf, Weaver, & Morgan, 1993) with a few D-R neurons in the form of "networks of trainable control systems, as models of nervous system function" (Klopf et al., 1993, p. 264). He referred to a control system that learns as an *associative control process* (ACP), which could lead to a neurobiologically and psychologically plausible process to explain the nervous system in the larger context of computational neuroethology.

The development of single-neuron mathematical models represents a bottom-up approach to understanding brain function. An understanding of how a single neuron operates is necessary to advance our ability to model behavior represented by networks of neurons. However, from a psychological perspective, several theoretical questions should be raised about the approach taken in modeling learning with a single neuron.

The basis of comparison between learning laws is their ability to model whole animal behavior. Although this gives a clear yardstick, supported by years of research, is it a valid measure? Research in the area of classical conditioning with single cell organisms has shown mixed results in their ability to form even simple associations (Abramson, 1994). This raises the question: Are we giving our mathematical neuron the ability to do more than is possible for a single biological neuron? If so, does it make a difference? Those who accept this perspective would argue that to empower the neuron with greater capabilities will change the functioning of the whole working together producing very different behavior.

It has been inferred that the model that most accurately duplicates whole-animal behavior is the best neuronal model. By this definition, Klopf's drive-reinforcement model most clearly, of those examined, demonstrates learning. However, the D-R model's inability to handle simple classical conditioning paradigms, such as CS duration and interstimulus intervals, casts doubt on its status as the best model. The Hebb model does replicate simple conditioning and neurophysiological research supports changes in neurons indicative of Hebbian learning (Kandel & Hawkins, 1993). However, the inability of Hebbian learning to explain more complex conditioning shows its limitations from a single-neuron perspective.

The problem may lie in trying to determine a single model as being the best overall. That is not to say that a comparison between models does not offer us interesting and important information. The fact that the Hebbian and D-R models have differing strengths in replicating classical conditioning may lead to the question: Does the complex brain system incorporate more than one adaptation law? A complex system may in fact deal with time discretely and simultaneously dependent on available information.

Recent results on the morphology, physiology, and synaptic architecture of single neurons (Segev, Rapp, Manor, & Yarom, 1992) give rise to one final theoretical question. They suggest that a single neuron should actually be modeled as a network, comprising input region (dendritic tree), cell body, and output region (axonal tree). Research has shown that the dendritic tree may not be isopotential (Claiborne et al., 1991, get this out of single neuron computation), which suggests synaptic modification cannot be represented by a single processing node. Although such modeling is beyond the scope of this chapter, it serves to add perspective to the study of single-neuron models that emulate whole-animal behavior.

The formal comparisons between the various mathematical models have raised some interesting questions from a psychological perspective. Although we can not offer concrete answers, we feel there is importance in the questions. Have we become conditioned to approach neuronal modelling from a single perspective? The following quotation may be of interest:

Pavlov thought dog saliva was extremely interesting. Therefore, he attended to it very closely. It was thus that a drop of dog saliva revolutionized our view of mind. One could argue that Pavlov did not discover anything, but that he himself was conditioned. Modern learning theorists emphasize that learning occurs when an organism discovers a discrepancy between the state of the world and the organism's representation of the world. ... Rescorla (1988) remarks that "organisms adjust their Pavlovian associations only when they are surprised." This statement suggests that conditioning and having an insight are more or less the same thing. To be surprised, one must be attending to or be expecting something ... Rather than say that Pavlov developed a theory, we could say that he became conditioned; rather than say that Pavlov's dogs were conditioned, we could say that they were surprised by a relationship between laboratory assistants and meat powder and that they developed a theory to account for this relationship. (Martindale, 1991, pp. 148-149)

In this study, we were conditioned to respond to single-neuron modeling from a mathematical perspective; however, we were "surprised" by the complexity of the psychological issues. Whether this will be insightful in furthering neuronal modeling research may depend on the conditioning of future theorists.

REFERENCES

Abrams, T. W., & Kandel, E. R. (1988). Is contiguity detection in classical conditioning a system or a cellular property? Learning in Aplysia suggests a possible molecular site. *Trends in NeuroSciences*, **11**, 128-135.

Abramson, C. I. (1994). *A Primer of Invertebrate Learning: The Behavioral Perspective.* Washington, DC: American Psychological Association.

Amari, S. (1977). Neural theory of association and concept-formation. *Biological Cybernetics*, **26**, 175-185.

Ayers, J. J. B., Haddad, C., & Albert, M. (1987). One-trial excitatory backward conditioning as assessed by conditioned suppression of licking in rats: Concurrent observation of lick suppression and defensive behaviors. *Animal Learning and Behavior*, **15**, 212-217.

Baird, L. C., & Klopf, A. H. (1993). A hierarchical network of provably optimal learning control systems: extensions of the associative control process (ACP) network. *Adaptive Behavior*, **1**, 321-352.

Barto, A. G., Sutton, R. S., & Anderson, C. W. (1983). Neuronlike adaptive elements that can solve difficult learning control problems. *IEEE Transactions on Systems, Man, and Cybernetics*, **SMC-13**, 835-846.

Bower, G. H. (1970). Analysis of a mnemonic device. *American Scientist*, **58**, 496-510.

Bower, G. H., & Hilgard, E. R. (1981). *Theories of Learning* (5th ed.) Englewood Cliffs, NJ: Prentice-Hall.

Carew, T. J., Hawkins, R. D., Abrams, T. W., & Kandel, E. R. (1984). A test of Hebb's postulate at identified synapses which mediate classical conditioning in Aplysia. *Journal of Neuroscience*, **4**, 1217-1224.

Caudill, M. (1989). *Neural Networks Primer* (Part VII, pp. 49-56). San Francisco: Miller Freeman.

Claiborne, B. J., Zador, A. M., Mainen, Z. F., & Brown, T. H. (1991). Computational models of hippocampal neurons. In T. McKenna, J. Davis, & S. F. Zornetzer (Eds.), *Single Neuron Computation* (pp. 81-116). New York: Academic Press.

Delgado, J. M. R. (1955). Evaluation of permanent implantation of electrodes within the brain. *Electroencephalography and Clinical Neurophysiology*, **7**, 637-644.

Flaherty, C. E. (1985). *Animal Learning and Cognition*. New York: McGraw-Hill.

Garcia, J., & Koelling, R. A. (1966). Relation of cue to consequence in avoidance learning. *Psychonomic Science*, **4**, 123-124.

Gelperin, A., Hopfield, J. J., & Tank, D. W. (1985). The logic of limax learning. In *Model Neural Networks and Behavior* (pp. 237-262). New York: Plenum Press.

Gould, S. J., & Lewontin, R. C. (1979). The spandrels of San Marcos and the Panglossian paradigm: A critique of the adaptionist programme. *Proceedings of the Royal Society of London*, **205**, 581-598.

Grossberg, S. (1967). Nonlinear difference-differential equations in prediction and learning theory. *Proceedings of the National Academy of Sciences*, **58**, 1329-1334.

Grossberg, S., & Levine, D. S. (1987). Neural dynamics of attentionally modulated Pavlovian conditioning: Blocking, interstimulus interval and secondary reinforcement. *Applied Optics*, **26**, 5015-5030.

Guthrie, E. R. (1952). *The Psychology of Learning* (rev. ed.). New York: Harper & Row.

Hebb, D. O. (1949). *The Organization of Behavior*. New York: Wiley.

Hull, C. L. (1947). Reactively heterogeneous compound trial-and-error learning with distributed trials and terminal reinforcement. *Journal of Experimental Psychology*, **37**, 118-135.

Kamin, L. J. (1968). Attention-like processes in classical conditioning. In M. R. Jones (Ed.), *Miami Symposium on the Prediction of Behavior: Aversive Stimulation* (pp. 9-31). Miami: University of Miami Press.

Kandel, E. R., & Hawkins, R. D. (1993). The biological basis of learning and individuality. In *Mind and Brain: Readings from Scientific American* (pp. 40-53). New York: W. H. Freeman.

Kehoe, E. J. (1989). Connectionist models of conditioning: a tutorial. *Journal of the Experimental Analysis of Behavior*, **52**, 427-440.

Kehoe, E. J., Schreurs, B. G., & Grahm, P. (1987). Temporal primacy overrides prior training in serial compound conditioning of the rabbit's nictitating membrane response. *Animal Learning and Behavior*, **15**, 455-464.

Klopf, A. H. (1972). *Brain function and adaptive systems: a heterostatic theory* (Rep. No. 133, AFCRL-72-64). L. G. Hanscom Field, Bedford, MA: Air Force Cambridge Research Laboratories.

Klopf, A. H. (1979). Goal-seeking systems from goal-seeking components; implications for A. I. *Cognition and Brain Theory Newsletter*, **3**, 54-62.

Klopf, A. H. (1982). *The Hedonistic Neuron: A Theory of Memory, Learning, and Intelligence.* Washington, DC: Hemisphere.

Klopf, A. H. (1986). A drive-reinforcement model of single neuron function: an alternative to the Hebbian neuronal model. In J. S. Denker (Ed.), *AIP Conference Proceedings 151: Neural Networks for Computing* (pp. 265-270). New York: American Institute of Physics.

Klopf, A. H. (1988). A neuronal model of classical conditioning. *Psychobiology, 16*, 85-125.

Klopf, A. H., & Morgan, J. S. (1990). The role of time in natural intelligence: implications of classical and instrumental conditioning for neuronal and neural network modeling. In M. Gabriel & J. Moore (Eds.), *Learning and Computational Neuroscience* (pp. 463-495). Cambridge, MA: MIT Press.

Klopf, A. H., Weaver, S. E., & Morgan, J. S. (1993). A hierarchical network of control systems that learn: modeling nervous system function during classical and instrumental conditioning. *Adaptive Behavior, 1*, 263-319.

Kosko, B. (1986). Differential Hebbian learning. In J. S. Denker (Ed.), *AIP Conference Proceedings 151: Neural Networks for Computing* (pp. 277-282). New York: American Institute of Physics.

Lashley, K. S. (1929). *Brain Mechanisms and Intelligence: A Quantitative Study of Injuries to the Brain.* Chicago: University of Chicago Press.

Le Cun, Y. (1985). A learning procedure for an asymmetric threshold network. *Proceedings of Cognitiva 85* (pp. 559-604). Paris, June, 1985.

Levine, D. S. (1991). *Introduction to Neural and Cognitive Modeling.* Hillsdale, NJ: Lawrence Erlbaum Associates.

Lieberman, D. A. (1990). *Learning: Behavior and Conditioning.* Belmont, CA: Wadsworth.

Malone, J. C. (1990). *Theories of Learning.* Belmont, CA. Wadsworth Publishing Co.

Martindale, C. (1991). *Cognitive Psychology: A Neural-network Approach.* Pacific Grove, CA: Brooks/Cole.

Olds, J., & Milner, P. (1954). Positive reinforcement produced by electrical stimulation of septal area and other regions of rat brain. *Journal of Comparative and Physiological Psychology, 47*, 419-427.

Parker, D. B. (1982). *Learning logic* (Invention Rep. No. 581-641, File 1). Stanford, CA: Stanford University, Office of Technology Licensing.

Parker, D. B. (1985). *Learning-logic* (Tech. Rep. No. 47). Cambridge, MA: MIT Center for Computational Research in Economics and Management Science.

Pavlov, I. P. (1927). *Conditioned Reflexes* (G. V. Anrep, Trans.). Oxford: Oxford University Press.

Pavlov, I. P. (1928). *Lectures on Conditioned Reflexes* (W. H. Gnatt, Trans). New York: International Publishers.

Rescorla, R. A. (1966). Predictability and number of parings in Pavlovian fear conditioning. *Psychonomic Science, 4*, 383-384.

Rescorla, R. A. (1973). Evidence for "unique stimulus" account of configural conditioning. *Journal of Comparative and Physiological Psychology, 85*, 331-338.

Rescorla, R. A. (1988). Pavlovian conditioning: it's not what you think it is. *American Psychologist, 43*, 151-160.

Rescorla, R. A., & Wagner, A. R. (1972). A theory of Pavlovian conditioning: Variations in the effectiveness of reinforcement and non-reinforcement. In A. H. Black & W. F. Prokasy (Eds.), *Classical Conditioning II: Current Research and Theory* (pp. 64-99). New York: Appleton-Century-Crofts.

Rumelhart, D. E., Hinton, G. E., & Williams, R. J. (1985). *Learning internal representations by error propagation* (ICS Rep. No. 8506). San Diego: University of California, Institute for Cognitive Science.

Segev, I., Rapp, M., Manor, Y., & Yarom, Y. (1992). Analog and digital processing in single nerve cells: dendritic integration and axonal propagation. In T. McKenna & S. F. Zornetzer (Ed.), *Single Neuron Computation* (pp. 173-198). New York: Academic Press.

Sutton, R. S. (1984). *Temporal credit assignment in reinforcement learning.* Unpublished doctoral dissertation, University of Massachusetts, Amherst.

Sutton, R. S. (1988). *Learning to predict by the methods of temporal differences* (Tech. Rep. No. 87-509.1). Waltham, MA: GTE Laboratories.

Sutton, R. S,. & Barto, A. G. (1981). Toward a modern theory of adaptive networks: Expectation and prediction. *Psychological Review, 88*, 135-170.

Sutton, R. S., & Barto, A. G. (1982). *Goal-seeking components for adaptive intelligence: an initial assessment* (Tech. Rep. No. AFWAL-TR-81-1070). Wright-Patterson Air Force Base, OH: Air Force Wright Aeronautical Laboratories.

Tesauro, G. (1986). Simple neural models of classical conditioning. *Biological Cybernetics, 55*, 187-200.

Wagner, A. R., & Brandon, S. E. (1989). Evolution of a structured connectionist model of Pavlovian conditioning; AESOP. In S. B. Klein & R. R. Mower (Eds.), *Contemporary Learning Theories: Pavlovian Conditioning and the Status of Traditional Learning Theory* (pp. 149-189). Hillsdale, NJ: Lawrence Erlbaum Associates.

Werbos, P. J. (1974). *Beyond regression: New tools for prediction and analysis in the behavioral sciences.* Unpublished doctoral dissertation, Harvard University.

Werbos, P. J. (1989). Neural networks for control and system identification. *Proceedings of the 28th IEEE Conference on Decision and Control, Tampa, FL, December, 1989.*

Werbos, P. J. (1990). Consistency of HDP applied to a simple reinforcement learning problem. *Neural Networks, 3*, 179-189.

Widrow, B., & Hoff, M. E. (1960). Adaptive switching circuits. *Institute of Radio Engineers, Western Electronic Show and Convention, Convention Record*, Part 4, 96-104.

14

On the Optimization of a Synaptic Learning Rule

Samy Bengio
INRS—Télécommunications, Québec

Yoshua Bengio, Jocelyn Cloutier, and Jan Gecsei
Université de Montréal

*The chapter by Samy Bengio, Yoshua Bengio, Jocelyn Cloutier, and Jan Gecsei, **On the Optimization of a Synaptic Learning Rule**, deals with different rules for synaptic modification — which are also considered in the chapters by Chance et al. and Carpenter. Bengio et al. deal partly with classical conditioning example, but unlike Chance et al., Bengio et al. do not work with specific conditioning models already in the literature. Rather, they work with a general mathematical structure for conditioning rules, inspired in part by biological data on **Aplysia**. These rules combine Hebb's rule with simulated effects of chemical modulation. They are combined with a variety of architectures that are also, in part, patterned after **Aplysia**.*

Bengio et al. apply statistical optimization methods (in a manner somewhat like Golden's chapter does) to figure out, within this structure, the best weights for performance of the functions they wish to perform. They train the network on a small sample of the tasks they wish it to learn and then try to minimize the generalization error; this issue of generalization is also dealt with in Tattersall's chapter in this book. The optimization methods used are three classes of widely used network methods: genetic algorithms (see also Dorsey and Johnson's chapter); gradient descent (see also Golden's chapter); and simulated annealing (see also Levine's chapter). In addition to conditioning, they applied their networks to classification problems and to calculating Boolean functions, the latter being similar to the thrust of Jagota's chapter.

This chapter explores a variety of parameter optimization problems and suggests that most of the methods have uses in different applications. Genetic algorithms seemed to work the best on classification problems. On Boolean function problems there was a trade-off between gradient descent and simulated annealing, with simulated annealing being more independent of initial conditions whereas gradient descent was faster. Experiments of the sort Bengio et al. are performing are of an exploratory character, and may contribute much to our knowledge of the utility of different optimization methods. The course of development or adult learning in human and animal nervous systems probably involves changes in weight parameters, so this work might eventually illuminate what optimization strategies are used in actual biological nervous systems.

ABSTRACT

This chapter presents a new approach to neural modeling based on the idea of using an automated method to optimize the parameters of a synaptic learning rule. The synaptic modification rule is considered as a parametric function. This function has local inputs and is the same in many neurons. We can use standard optimization methods to select appropriate parameters for a given type of task. We also present a theoretical analysis permitting to study the *generalization* property of such parametric learning rules. By generalization, we mean the possibility for the learning rule to learn to solve new tasks. Experiments were performed on three types of problems: a biologically inspired circuit (for conditioning in *Aplysia*), Boolean functions (linearly separable as well as non linearly separable) and classification tasks. The neural network architecture as well as the form and initial parameter values of the synaptic learning function can be designed using a priori knowledge.

1. INTRODUCTION

Many artificial neural network models have been recently proposed (see Hertz, Krogh, & Palmer, 1989, and Hinton, 1989, for detailed reviews), and each of them uses a different (but constant) synaptic update rule. We propose in this chapter to use optimization methods to search for new synaptic learning rules. Preliminary studies on this subject were reported in Bengio and Bengio (1991) and in Bengio, Bengio, and Cloutier (1991a, 1991b). Many biologically inclined researchers are trying to explain the behavior of the nervous system by considering experimentally acquired physiological and biological data for constructing their models (see, e.g., Byrne & Berry, 1989; Hawkins, 1989). These biologically plausible models constrain the learning rule to be a function of information locally available to a synapse. However, it has not yet been shown how such models could be efficiently applied to difficult engineering or artificial intelligence problems, such as image or speech recognition, diagnosis, prediction, and so forth.

Another approach, preferred by engineers, emphasizes problem solving, regardless of biological plausibility (e.g., error backpropagation; Rumelhart, Hinton, & Williams, 1986). The above two classes of models seem to be growing further and further apart. An objective of this chapter is to contribute to fill the gap between the two approaches by searching for new learning rules that are both biologically plausible and efficient compared to specialized techniques for the solution of difficult problems.

2. LEARNING RULE OPTIMIZATION

The most remarkable characteristic of a neural network is its capacity to adapt to its environment: it can learn from experience, and generalize when presented new stimuli. In both biologically motivated and artificial neural networks, this adaptation capacity is represented by a *learning rule*, describing how connection weights (synaptic efficacies) change. Although it is generally admitted that the learning rule has a crucial role, neural models commonly use ad hoc or heuristically designed rules; furthermore, these rules are independent of the learning problem to be solved. This may be one reason why most current models (some with sound mathematical foundation) cannot deal easily with hard problems. In this chapter we propose to improve a learning rule by adapting it to the problems to be solved. To do this, we consider the synaptic learning rule as a *parametric function*, and optimize its parameters using standard tools, such as gradient descent (Rumelhart, Hinton, & Williams, 1986), genetic algorithms (Goldberg, 1989) and simulated annealing (Kirkpatrick, Gelatt, & Vecchi, 1983). We make the following assumptions:

· The same rule is used in many neurons (this constraint may be relaxed to one rule for each type of neuron or synapse[1]). It is not plausible that each synapse or neuron in a network has its own rule. Actually, neural models described in the literature use a single rule (e.g., Hebb's, 1949), which dictates the behavior of every neuron and synapse.
· There exists a relation (possibly stochastic) between synaptic update and some information locally available to the synapse, that corresponds to the learning rule (i.e., synaptic update is not totally random). This relation may be approximated by a parametric function

$$f(x_1, x_2, ..., x_n; \theta_1, \theta_2, ..., \theta_m) \tag{1}$$

where x_i are variables of the function and θ_j are a set of parameters.

2.1. Variables and Parameters of the Learning Rule

Because the domain of possible learning algorithms is large, we propose to constrain it by using in Eq. (1) only already known, biologically plausible synaptic mechanisms. Hence, we consider only local variables, such as presynaptic activity, postsynaptic potential, synaptic strength, the activity of a facilitatory neuron, and the concentration of a diffusely acting neuromodulator. Figure 14.1 shows the interaction between those elements. Constraining the learning rule to be biologically plausible should not be seen as an artificial constraint but rather as a way to restrain the search space such that it is consistent with solutions that we believe to be used in the brain. This constraint might ease the search for new learning rules (Fig. 14.2).

2.2. General Form of the Learning Rule

From the above, and denoting $w(i, j)$ as the weight of the synapse from neuron i to neuron j, the general weight update function will have the form

---

[1] Biologists have found different types of neurons and synapses in the brain, but their characterization is far from complete (Gardner, 1987).

$$\Delta w(i,\ j) = \Delta w(x_1,\ x_2,\ ...,\ x_n;\ \theta_1,\ \theta_2,\ ...,\ \theta_m). \tag{2}$$

The synaptic update $\Delta w(i,\ j)$ of a synapse $i \rightarrow j$ is computed using Eq. (2), as a function of variables x_k, local to this synapse and a set of parameters θ_k. It is those parameters that we propose to optimize in order to improve the learning rule.

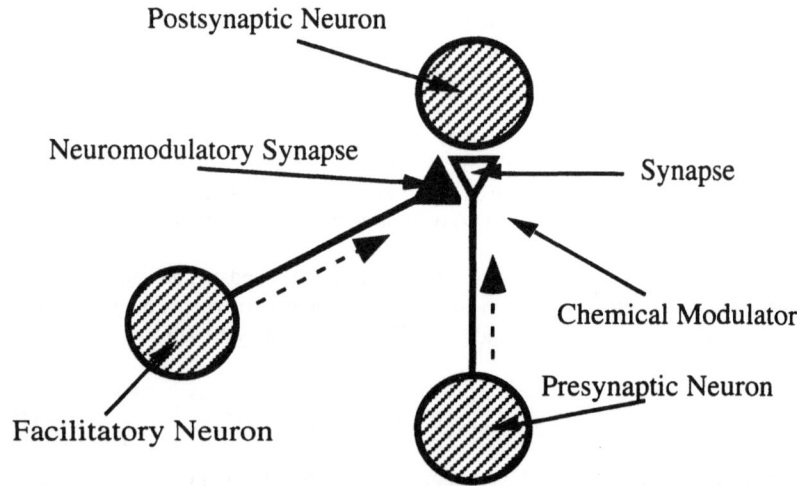

Fig. 14.1. Elements found in the vicinity of a synapse, that can influence its efficacy.

2.2.1. Example of a Parametric Learning Rule. Hebb's rule is probably the best known learning rule in connectionist models (Hebb, 1949). It suggests that a biologically plausible synaptic modification is to increase the weight of a connection when the presynaptic and postsynaptic neurons are active simultaneously. Under its most simple expression, Hebb's rule may be written as a parametric function as follows:

$$\Delta w(i,\ j) = \theta y(i)x(j) \tag{3}$$

where Δw is the weight update, $y(i)$ the activity of presynaptic neuron i, $x(j)$ the activity of postsynaptic neuron j, and θ a correlation constant. In this case, $y(i)$ and $x(j)$ are variables and θ a parameter.

2.3. Form of the Rule

Once the variables and parameters have been chosen, the learning rule must be given a precise form. In Eq. (3), it is simply the product of the variables and of the parameter. Currently available biological knowledge can help us design a more general form of the learning rule. For instance, it is now accepted that many mechanisms may interact in a synapse simultaneously but with different time constants, which suggests the inclusion of delays in the learning rule. We may also model a synapse with more details, considering for instance local interaction at the dendritic tree level (a synapse is then influenced by the neuron's local potential instead of its global activity).

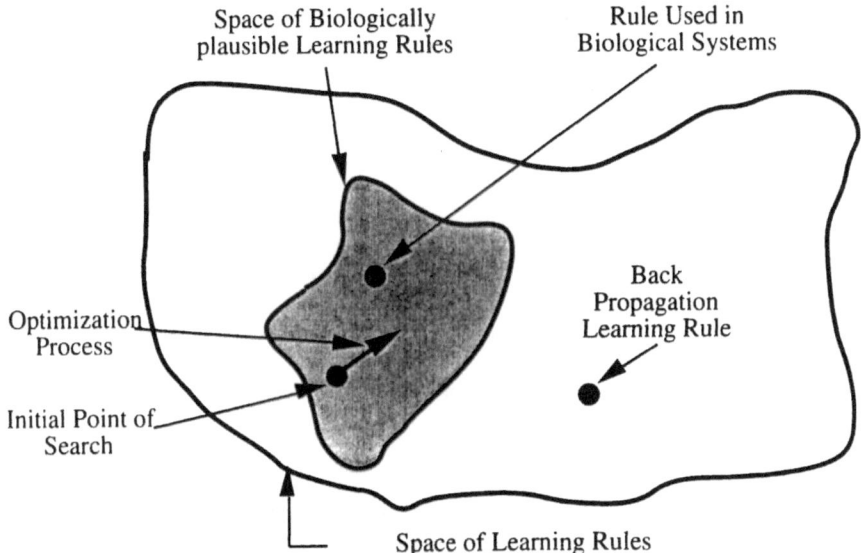

Fig. 14.2. Constraining the space of learning rules considered.

2.4. Task Diversification

Optimizing a learning rule to solve a given single task is an interesting and nontrivial problem in itself that is discussed in this chapter. We want this rule to be able to perform adequately on new instances of the task. However, it may be more interesting (and more difficult) to find a learning rule that can be used successfully on a number of different learning tasks. During the optimization process, the same rule (with the same parameters) must be used on all tasks considered. Thus, we have to optimize a learning rule with respect to its simultaneous performance on different neural networks learning different tasks. This constraint, suggested by Chalmers (1990), should yield rules of more general applicability. This raises the question of generalization over tasks. Let us suppose that we optimize a rule using a training set of tasks sampled from a certain function space. Generalization over tasks is the expected performance of the learning rule on a new function sampled from the same function space, that is, the same class of functions.

2.5. Parameter Optimization

Once the form of $\Delta w(\ )$ is determined, we have to search for values of parameters that optimize the learning rule's ability to solve different tasks. We now define the problem of optimizing a learning rule.

2.5.1. Learning and Optimization. Let X be a random variable with probability distribution P_X fixed but unknown, and let $\phi : X \rightarrow Y$ be an unknown function. For example, X may be a digital representation of a pixel matrix and Y the symbol (e.g., numbers, letters) corresponding to the image X. Or, X may be a measure of the economic activity on a given day, and Y the Dow Jones index on the following day (White, 1989).

Let J be a scalar cost function. For example, we can choose J to be the mean square criterion: $J(x, y) = (x - y)^2$. The goal of a learning system is then to produce a parametric function $\hat{\phi}(\theta): X \rightarrow Y$ which minimizes

$$C = \int J(\hat{\phi}(x;\theta),\phi(x)) P_X(x)\,dx \qquad (4)$$

by finding adequate parameters θ.[2] In order to minimize C with a supervised learning system, we usually proceed using a set of N examples $(x_i, \phi(x_i))$, for $i = 1, ..., N$, each chosen in (X,Y) independently using P_X. Training performance of such a system can be measured in term of the difference between ϕ and $\hat{\phi}$ for the N examples:

$$\hat{C} = \sum_{i=1}^{N} J(\hat{\phi}(x_i;\theta),\phi(x_i)) \qquad (5)$$

The generalization performance of such a system measures the difference between ϕ and $\hat{\phi}$ for points other than those used to calculate $\hat{C}$. To quantify the generalization property of a learning system, we now introduce the standard notion of *capacity*.

Let $z \in Z = (x, y) \in (X, Y)$ and let $G(z; \theta)$ be the set of parametric functions $g(z, \theta)$. For example in neural networks, g represents a network architecture and a cost function, θ is the set of weights, and z is the input and corresponding desired output vectors given to the network. More precisely, $g(z; \theta) = J(\hat{\phi}(x;\theta),\phi(x))$. When $J \in \{0, 1\}$, Vapnik (1982) defines the capacity of $G(z; \theta)$ as the maximum number of points $x_1, x_2, ..., x_h$ that can always be divided into two distinct classes with $G(z; \theta)$. Vapnik next describes an extension for the case where $J \in \mathbf{R}$. In this case, capacity is defined as the maximum number of input/output pairs $z_1, z_2, ..., z_h$ that can always be learned by $G(z; \theta)$ with a cost less than a threshold chosen to maximize h.

Theoretical studies such as Baum and Haussler (1989) and Vapnik and Chervonenkis (1971) give an upper bound on the number N of necessary instances of Z required to achieve

---

[2] For example, in a neural network, 0 is the set of weights.

generalization with given maximal error. More specifically, knowing that the capacity of a learning system is a measure of the number of functions it can approximate, which depends roughly on the number of weights in the network, we can relate the generalization error of a learning system with the number N of instances $z \in Z$ and the capacity h using the following formula (Vapnik, 1982):

$$\varepsilon \leq O\left(\sqrt{\frac{h}{N} \ln \frac{N}{h}}\right) \tag{6}$$

where ε is the difference between training error (using the N examples) and generalization error expected over all examples. This means that for a fixed number of examples N, starting from $h = 0$ and increasing it, one finds generalization improves until a critical capacity is reached. After this point, increasing h makes generalization worse. For fixed capacity, increasing the number of training examples N improves generalization (ε asymptotes to a value that depends on h). The specific results of Vapnik (1982) are obtained in the worst case, for any distribution P_X.

2.5.2. Optimization of the Parameters. Let L be a learning rule defined by its structure (fixed) and its parameters θ (considered to be variable). Optimizing the learning rule requires a search for the values of parameters minimizing a cost function.

Let $\{R_1, R_2, ..., R_n\}$ be a set of neural networks trained on n different tasks (and possibly having different structures), but using the same learning rule L. If C_i is the cost (as defined for instance by Eq. (6)) obtained with neural network R_i after being trained on its task, then the global cost function we wish to minimize is

$$C^* = \sum_i C_i \tag{7}.$$

Furthermore, if we want a learning rule to be able to solve any task, we should, in theory, optimize C^* with respect to the cost resulting from the learning of every possible instance of every one of the n tasks. Because this is usually impossible in practice, we can only find an approximation having the generalization error decrease with the increasing number of tasks used to optimize C^*. This follows from Vapnik and Chervonenkis' (1971) theorem, as long as those tasks are good representatives of the sets of all possible tasks, which we usually cannot verify.

More formally, we can define the capacity of a parametric learning rule as a measure of the number of rules it can approximate. Thus Eq. (7) holds for the generalization error of the learning rule where h is the capacity of the parametric learning rule (which is in fact a function of the number of parameters of the learning rule), N is the number of tasks used to optimize the parameters, and ε is the difference between training error (training tasks) and generalization error (on new tasks).

Thus, we can draw several conclusions from this extension. For example, it becomes clear that the expected error of a learning rule over new tasks should decrease when increasing the number of tasks (N) used for learning the parameters θ. However, it could increase with the number of parameters and the capacity of the learning rule class if an insufficient number or variety of training tasks are used in the optimization. This justifies the use of a priori knowledge in order to limit the capacity of the learning rule. It also appears more clearly that the learning rule will be more likely to generalize over tasks that are similar to those used for the optimization of the rule's parameters. In consequence, it is advantageous to use, for the optimization of the learning rule, tasks that are representative of those on which the learning rule will be ultimately applied.

2.6. Optimization Process

We use a two-step optimization process as shown in Fig. 14.3. The form of the learning rule is defined and the parameters are initialized either to random values within reasonable bounds, or to values corresponding to biological evidence (an example of this can be found in Section 3). We also define the set of tasks to be used in the process. Then we use the following algorithm:

1. We train n networks simultaneously on n tasks by using the current learning rule. Since $\Delta w(\ )$ is, with high probability, initially far from optimal, it is very unlikely that all tasks will be solved adequately. After a fixed number of learning steps, we compute the error obtained in each task from Eq. (6).
2. The objective function (such as Eq. (7)) is minimized by updating the parameters θ of $\Delta w(\ )$ according to the optimization method used.
3. We return to Step 1, using the new parameters of the learning rule, until we reach an acceptable learning rule. When the new learning rule is able to solve adequately all n tasks, we may want to test its capability to generalize on tasks that were not used in the optimization process. Many different optimization methods may be used to improve $\Delta w(\ )$. We can use local methods such as gradient descent, or global methods such as simulated annealing (Kirkpatrick et al., 1983) or genetic algorithms (Goldberg, 1989; Holland, 1975). Local methods are usually faster, but they can get trapped in local minima, whereas global methods are less sensitive to local minima but usually slower. Hybrid gradient descent/genetic algorithms were recently suggested (Davis, 1989; Whitley & Hanson, 1989). In our experiments we used gradient descent, simulated annealing, and genetic algorithms.

2.7. Problem Complexity

It is interesting to discuss briefly the complexity of the above optimization problem. We already have some knowledge of the complexity of neural network learning. Deciding if in the space of parameters of a neural network N there exists an adequate solution to an arbitrary task T is equivalent to the satisfiability problem, which is NP-complete (Judd, 1988). Consequently, the search for such solution in N must be NP-hard. However, experimentation shows that a complex task may be learned in polynomial time by a neural network if a sound approximation is acceptable (Hinton, 1989). General optimization methods such as gradient descent or simulated annealing can usually give good suboptimal solutions in polynomial time, even if they may need exponential time for the optimal solution. In our experiments we clearly cannot aim at exact optimization. Instead, we allocate a polynomial amount of time to each network using the current learning rule to solve its task. We are thus searching for a learning rule that can solve a set of tasks reasonably well in reasonable time.

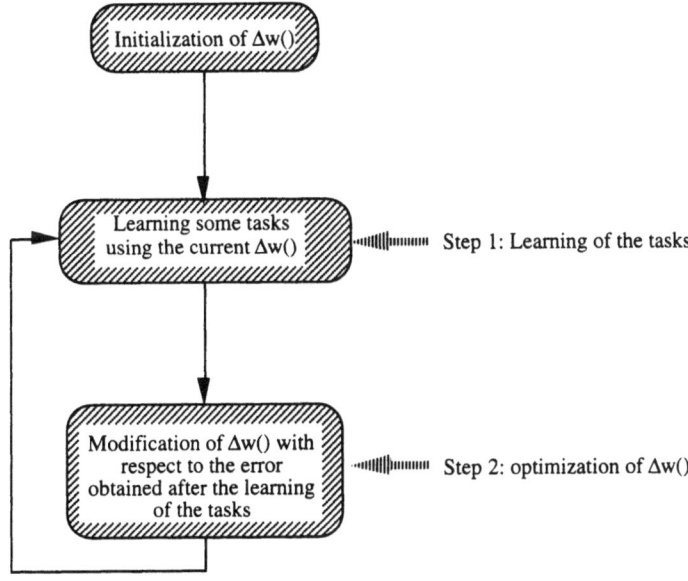

Fig. 14.3. Optimization steps.

3. PRELIMINARY EXPERIMENTS

To test the feasibility of our approach to the optimization of learning rules, we performed preliminary experiments with relatively simple problems (Bengio et al., 1991a, 1991b, 1993; Bengio & Bengio, 1990; Bengio, Bengio, Cloutier, & Gecsei, 1992). In this section we summarize the results. In these experiments, we used either gradient descent, simulated annealing, or genetic algorithms as optimization methods. The tasks were the following: conditioning, Boolean functions, and classification problems. Although preliminary results are positive, experimentation with more complex problems is needed in order to find useful synaptic learning rules.

3.1. Form of the Learning Rule

To facilitate the search for an optimal synaptic update rule $\Delta w()$, it is important to choose an adequate form of the learning rule. Here, by adequate we mean a form sufficiently rich to express a good solution (which is problem dependent), but sufficiently constrained to ease the search of the solution.[3] To do so, we can use some biological knowledge. Figure 14.4 shows the general form of the learning rules we use in the experiments. It consists of a certain number of a priori modules representing known or hypothesized biological synaptic characteristics. The resulting rule reflects a combination of a priori and free modules. The parameters determine the relative influence of each module on $\Delta w(\Delta)$.

The following equation is a concrete example of a learning rule used in the experiments that are described in this section:

$$\Delta w(i,j) = \theta_0 + \theta_1 y(i) + \theta_2 x(j) + \theta_3 y(\mathrm{mod}(j)) + \\ \theta_4 y(i) y(\mathrm{mod}(j)) + \theta_5 y(i) x(j) + \theta_6 y(i) w(i,j) \tag{8}$$

This is an instance of the general form described in Fig. 14.4. The function computed by this equation has sevem parameters, and integrates the following a priori modules:

· $y(i) x(j)$, Hebb's rule.
· $y(i) y(\mathrm{mod}(j))$, where $y(\mathrm{mod}(j))$ is a modulatory activity (chemical or neural). Hawkins (1989) described a conditioning model for *Aplysia* using such a mechanism.
· $y(i) w(i, j)$, where $w(i, j)$ is the synaptic weight at a the previous time frame. This term, suggested in Gluck's conditioning models (Gluck & Thompson, 1987), permits gradual forgetting.

3.2. Conditioning Experiments

The goal of our first experiment is to discover a learning rule that is able to reproduce some classical conditioning phenomena in animals. Conditioning experiments, first described by Pavlov (1932), are well known through experimental studies. For our experiments, we used Hawkins' model (Hawkins, 1989). We studied the following phenomena:

Habituation. Initially, a conditional stimulus CS_1 (e.g., a red light presented to an animal) pro duces a small response (e.g., the animal salivates slightly). By presenting CS_1 repetitively, the response gradually vanishes (i.e., the animal gets used to the stimulus, and reacts to it less and less).

Conditioning. A conditional stimulus CS_1 is followed by an unconditional stimulus U S (e.g., a red light followed by food). The response to CS_1 grows gradually (the animal salivates before seeing the food, and so on as the red light is turned on).

Blocking. After CS_1 has been conditioned, a second conditional stimulus CS_2 (e.g., a green light) is presented to the organism simultaneously with CS_1, both followed by an unconditional stimulus US. In that case, CS_2 is not conditioned (the animal will not salivate on green light only).

---

[3] Moreover, as we have seen in Section 2.5.2, if the form of the learning rule is too rich, capacity maybe too high to reach good generalization performance.

Second order conditioning. After CS_1 has been conditioned, CS_2 may be conditioned by presenting CS_2 followed by CS_1 (the animal will begin to salivate when it sees the green light, knowing that the red light follows, and that it is usually followed by food).

Extinction. After CS_1 has been conditioned, repetitive presentation of CS_1 not followed by US will reduce the animal's response to its original level (when no food follows the red light, the animal tends to lower its saliva response with time, eventually reaching its initial unconditioned level).

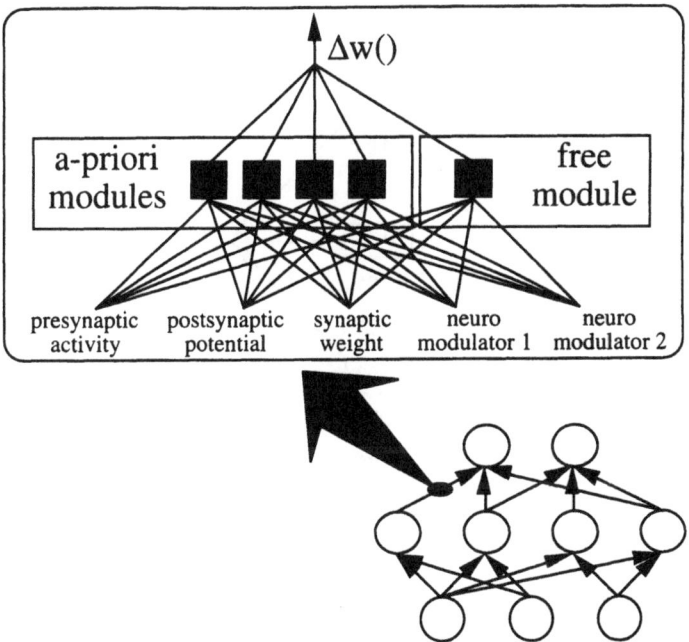

Fig. 14.4. A priori knowledge utilization.

In the conditioning experiments, the cost function to minimize (with gradient descent in this case) is defined following Eqs. (5) and (7) where $\hat{y}$ is the actual response of the network given an input sequence of stimuli, and y is the target behavior for the same input sequence. By fixing the initial values for some parameters (e.g., in Eq. (8), θ_3 is initialized to 1) and by initializing the other parameters to random values (in the range $[-1, 1]$), it was possible to find a set θ such that all five conditioning behaviors could be learned by the network in Fig. 14.6 with initial random weights. Figure 14.7 shows the evolution of the cost function during optimization. Figure 14.8 shows the results of all five conditioning tasks obtained from our new learning rule (an extensive analysis of the resulting rule will be done in a future paper). The results obtained by this learning rule are similar to Hawkins' (1989) experimental results.

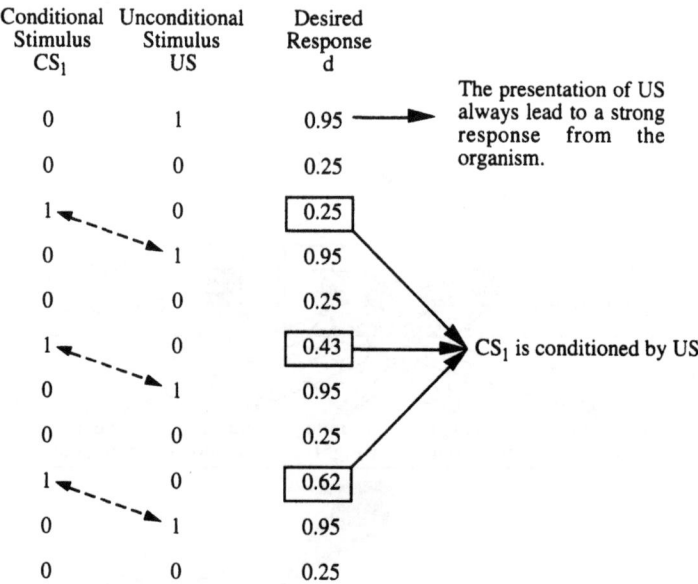

Conditional Stimulus CS_1	Unconditional Stimulus US	Desired Response d	
0	1	0.95	The presentation of US always lead to a strong response from the organism.
0	0	0.25	
1	0	0.25	
0	1	0.95	
0	0	0.25	
1	0	0.43	CS_1 is conditioned by US.
0	1	0.95	
0	0	0.25	
1	0	0.62	
0	1	0.95	
0	0	0.25	

Fig. 14.5. Conditioning by a sequence of stimuli and responses. The real sequence used for the experiments is much longer.

Figure 14.5 shows the way we modeled one of those behaviors (conditioning) with a sequence of stimuli and associated responses. The form of the learning rule we used is Eq. (8), and the network architecture is shown in Fig. 14.6. It is inspired by Hawkins' work (Hawkins, Abrams, Carew & Kandel, 1983; Hawkins, 1989). In this network, CS_1 and CS_2 are conditional stimuli, US is an unconditional stimulus, FN is a facilitatory neuron, and MN a motor neuron (it represents the animal's response to stimuli). CS_1 and CS_2 influence the motor neuron (through connections toward MN), and these connections are themselves modulated by a facilitatory neuron which takes into account two consecutive states of the system through connections with delay (e.g., when CS_1 is activated at time t and US at time $t + 1$).

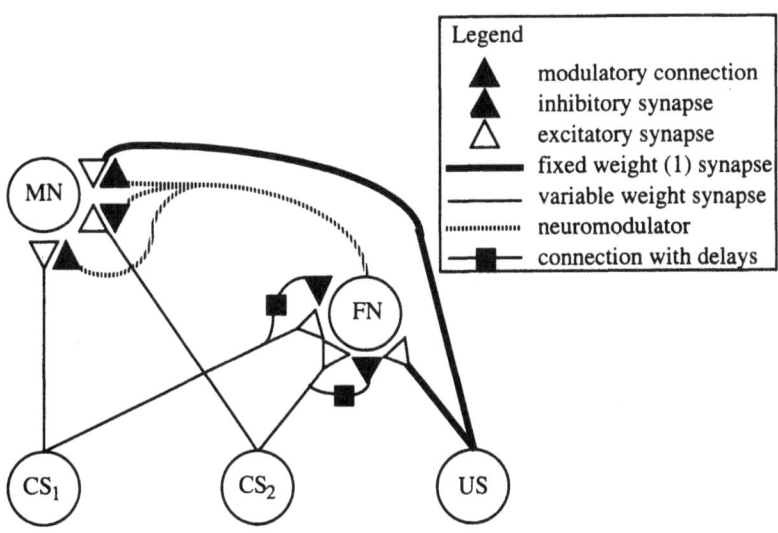

Fig. 14.6. Neural network used for the conditioning experiments.

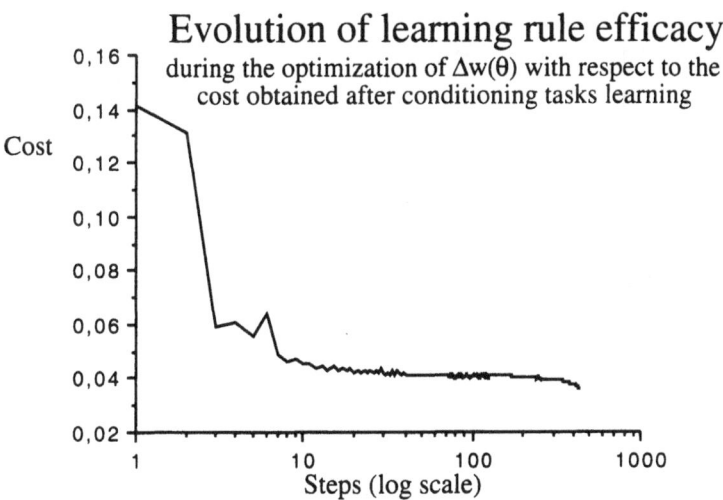

Fig. 14.7. Evolution of the learning rule efficacy.

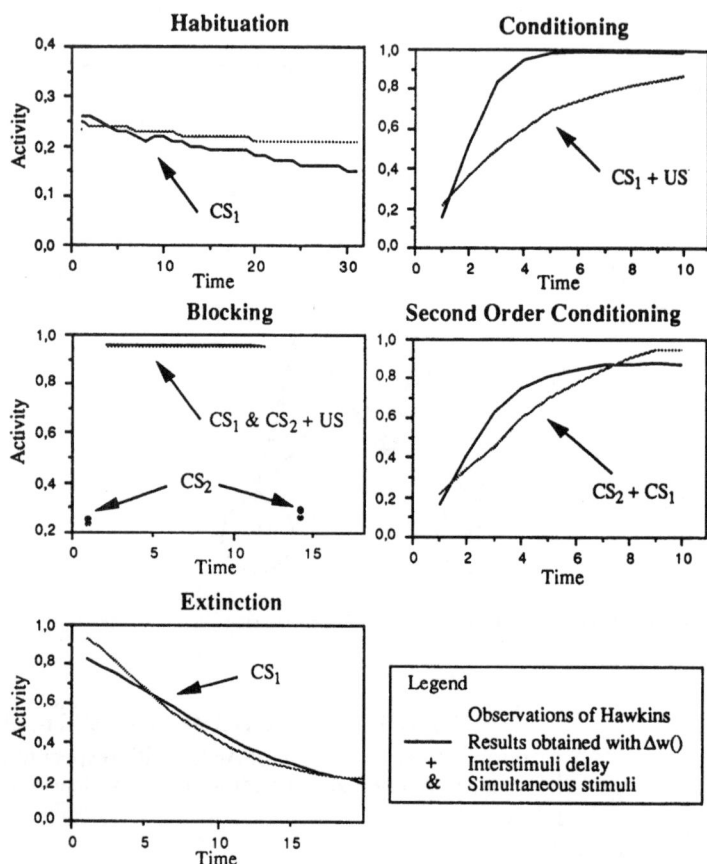

Fig. 14.8. Conditioning tasks computed by our new learning rule.

3.3. Experiment With Boolean Functions

The goal of these experiments is to explore in a very simple setting the possibility of optimizing a learning rule that could be used to train a network with hidden units. They allowed us to evaluate the applicability of our method to a simple computational problem. We used again the same learning rule from Eq. (8). Fully connected networks with two inputs, a single output, and one hidden unit were trained to perform linearly separable functions (such as AND, OR) and nonlinearly separable functions (such as XOR, EQ). Information provided to hidden units about their contribution to errors was fed through backward paths, with neurons that might modulate synaptic change on corresponding forward paths (Fig. 14.9).

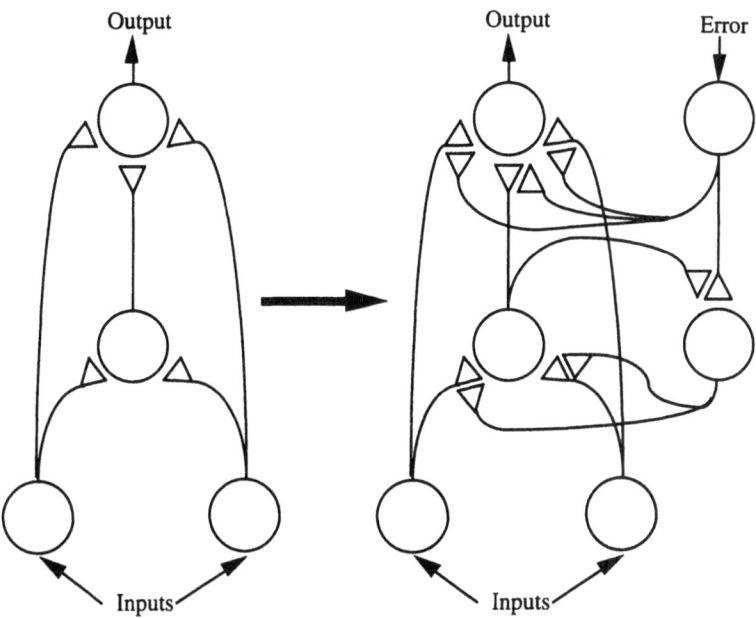

Fig. 14.9. Architecture transformation to enable local evaluation of the network error.

As for the conditioning experiments, a cost function was defined in terms of the difference between real and expected learning behavior of the network. This target behavior consists in learning some Boolean function within a cycle of 800 presentations of randomly selected noisy Boolean input patterns. A Gaussian noise was added to each input pattern for better generalization of the learning rule. Before each cycle, the network weights were initialized randomly, in order to allow the resulting rule to be as insensitive as possible to the network's initial conditions. Rule parameters θ were updated after each cycle. The results are summarized in Table 14.1.

Two optimization methods were used: gradient descent and simulated annealing. Gradient descent proved to be faster but sensitive to initial values of θ, whereas simulated annealing was slower (around 500 times slower) but insensitive to parameter initialization. In order to verify that gradient descent and simulated annealing were more efficient than random search, we also tried the following approach. For the random search, 50,000 vectors of θ, each with seven parameters within $[-1, 1]$, were chosen and those corresponding to the best learning performance on the training tasks were kept. The networks trained with that rule could not learn the nonlinearly separable functions completely (25% error at best), whereas the networks trained with a rule obtained with gradient descent or simulated annealing were able to learn the training tasks perfectly. Furthermore, this rule could also generalize to new (but similar) tasks including nonlinearly separable functions, as shown in Table 14.1. Another interesting observation is that, as expected with the capacity theory extended in Section 2.5.2 and with results of Chalmers (1990), generalization to new tasks is improved if more tasks are used for optimizing the learning rule.

3.4. Classification Experiments

The general problem of classification (Duda & Hart, 1973) is to establish a correspondence between a set of vectors and a set of classes. A classifier is a function $f: B \rightarrow C$, where $B \in \mathbf{R}^n$ is a set of n-dimensional vectors we want to classify and C is the set of classes. Many problems may be formulated in this way. An example is the optical character recognition problem, which is to associate an image to a character. Here, using a simple neural network architecture with two input units, one hidden unit and one output unit, we will search for a synaptic learning rule able to solve two-dimensional classification problems with two classes.

Number of Tasks	Type of Tasks	Number of Steps	Generalization (to New Tasks)		Optimization Method	Sensitivity to Initialization
			L	NL		
1	L	3	yes	no		
1	NL	15	yes	no	Gradient	yes
4	L	5	yes	no	Descent	
5	4L, 1NL	100	yes	yes		
1	L	100	yes	no	Simulated	
1	NL	1000	yes	no	Annealing	no
5	4L, 1NL	24000	yes	yes		

Table 14.1. Summary of Boolean Experiments. L stands for linearly separable task whereas NL stands for nonlinearly separable tasks.

Let the two classes be C_1 and C_2, and let $V_1 = \{v \in \mathbf{R}^2 | v$ belongs to $C_1\}$ and $V_2 = \{v \in \mathbf{R}^2 | v$ belongs to $C_2\}$ be the sets of vectors belonging respectively to C_1 and C_2. The task consists in learn whether each vector $v \in \mathbf{R}^2$ belongs to C_1 or C_2. To do this, we randomly select vectors $v \in \mathbf{R}^2$ belonging to C_1 or C_2. The network predicts the class C^* to which an input vector v belongs. The goal is to minimize (by modifying the connection weights) the difference between C^* and the correct class associated to a vector, for every training vector.

We performed experiments to verify the theory of capacity and generalization applied to parametric learning rules. In particular, we wanted to study the variation of the number of tasks N, the capacity h, and the complexity of the tasks, over the learning rule's generalization property (ε). Moreover, we did these experiments using three different optimization methods, namely, gradient descent, genetic algorithms, and simulated annealing. Experiments were conducted in the following conditions:

· Some tasks were linearly separable (L) and others were nonlinearly separable (NL).

· Each task was learned with 800 training examples and tested with 200 examples. A task was said to be successfully learned when there were no classification error over the test set.
· We used once again the network described in Fig. 14.9, with backward neurons that may provide error information to hidden connections.
· We tried two different parametric learning rules. Rule A was defined using biological a priori knowledge to constrain the number of parameters to seven, as in Eq. (8):

$$\Delta w(i,j) = \theta_0 + \theta_1 y(i) + \theta_2 x(j) + \theta_3 y(\mathrm{mod}(j)) + \\ \theta_4 y(i) y(\mathrm{mod}(j)) + \theta_5 y(i) x(j) + \theta_6 y(i) w(i,j)$$

where $w(i, j)$ is the synaptic efficacy between neurons i and j, $x(j)$ is the activation potential of neuron j (postsynaptic potential), $y(i)$ is the output of neuron i (presynaptic activity), and $y(\mathrm{mod}(j))$ is the output of a modulatory neuron influencing neuron j. Rule B had 16 parameters and w as defined as follows:

$$\Delta w(i,j) = \theta_0 + \theta_1 y(i) + \theta_2 x(j) + \theta_3 y(\mathrm{mod}(j)) + \theta_4 w(i,j) + \theta_5 y(i) x(j) \\ + \theta_6 y(i) y(\mathrm{mod}(j)) + \theta_7 y(i) w(i,j) + \theta_8 x(j) y(\mathrm{mod}(j)) + \theta_9 x(i) w(i,j) \\ + \theta_{10} y(\mathrm{mod}(j)) w(i,j) + \theta_{11} y(i) x(j) y(\mathrm{mod}(j)) w(i,j) + \theta_{12} y(i) x(j) w(i,j) \\ + \theta_{13} y(i) y(\mathrm{mod}(j)) w(i,j) + \theta_{14} x(j) y(\mathrm{mod}(j)) w(i,j) + \theta_{15} y(i) x(j) y(\mathrm{mod}(j)) w(i,j)$$ (11)

· A typical experiment was conducted as follows: We chose a parametric learning rule (A or B), an optimization method (genetic algorithms, gradient descent, or simulated annealing), a number of tasks to optimize the rule (1 to 9), and a complexity for the tasks (linearly separable, L, or nonlinearly separable, NL). Then we optimized the rule for a fixed number of iterations, and finally, we tested the new rule over other tasks (i.e., we tried to learn new tasks with their 800 training patterns and evaluate performance with a test over the remaining 200).

The first experiment was to verify that the number of tasks N used for the optimization had an influence on the rule's generalization performance. Figure 14.10 shows that for a given and fixed optimization method and capacity h, generalization error tends to decrease when N increases, as theory predicts.

The second experiment was to verify if the type of tasks used during optimization influences the rule's generalization performance. Figure 14.11 illustrates the results. We can see that when the rule is optimized using linearly separable tasks, generalization error on both linearly and nonlinearly separable tasks stays high, whereas if we use non linearly separable tasks during rule optimization, generalization error decreases when the number of tasks increases.

In the third experiment (Fig. 14.12), we verified if the capacity of a parametric learning rule influences its generalization performance. Here, we compared rules A and B (respectively with 7 and 16 parameters). As shown, if the number of tasks used for optimization is too small, the rule with the smallest capacity (A) is better, but the advantage tends to vanish when the number of tasks increases.

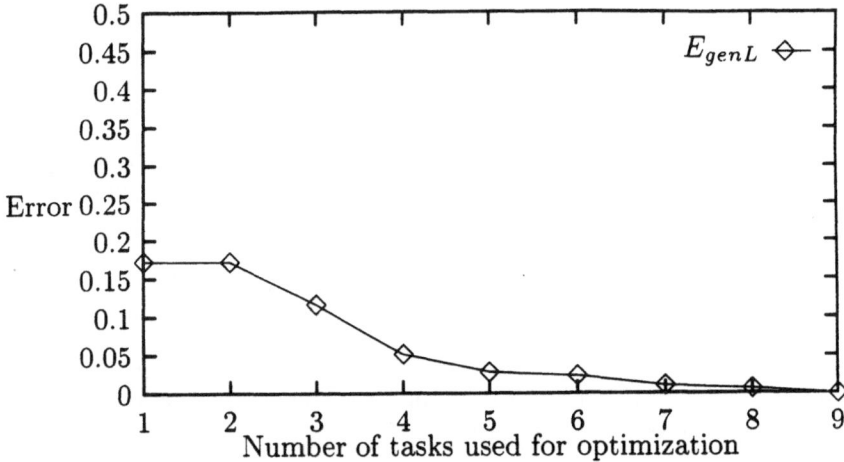

Fig. 14.10. Evolution of generalization error (E_{genL}) with respect to the number of tasks used during optimization. In this example, we used genetic algorithms and a rule with 7 parameters. Tasks were linearly separable.

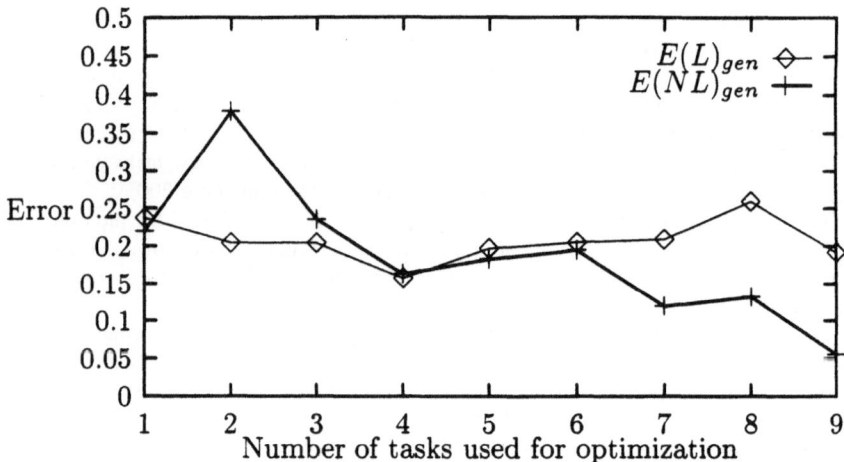

Fig. 14.11. Evolution of generalization error with respect to the task difficulty used during optimization. Here, we used genetic algorithms and a rule with seven parameters. $E(L)_{gen}$ represents generalization error when the rule is optimized on linearly separable tasks, whereas $E(NL)_{gen}$ represents generalization error when the rule is optimized on nonlinearly separable tasks.

In Fig. 14.13, we compare the use of different optimization methods to find the parameters of a learning rule. We compared two methods: genetic algorithms and simulated annealing.[4] Genetic algorithms seem generally better, especially when the number of tasks used for optimization is small.

The last figure (Fig. 14.14) shows how optimization error varies during optimization of the learning rule. At the beginning of the optimization process, training error on selected tasks is very high, but it decreases rapidly during the optimization process.

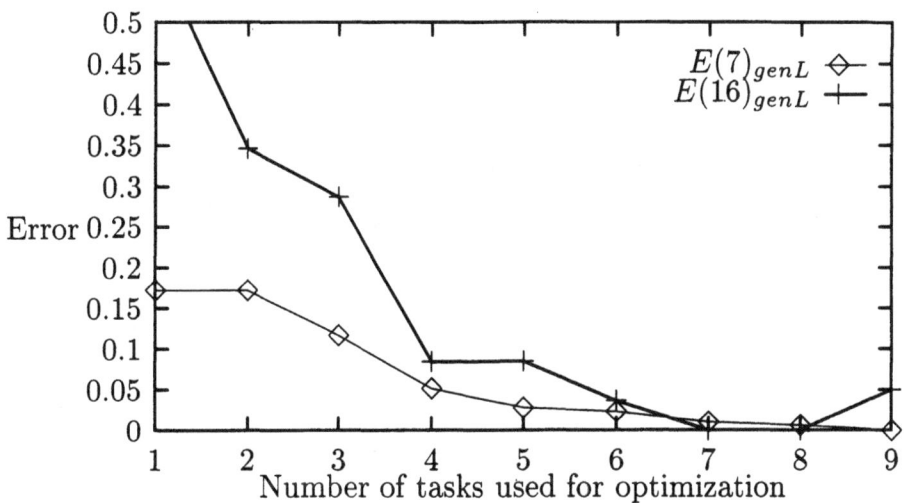

Fig. 14.12. Evolution of generalization error with respect to capacity of the parametric learning rule. Here, we used genetic algorithms and tasks were linearly separable. $E(7)_{genL}$ is the generalization error of rule A (with 7 parameters) and $E(16)_{genL}$ is the generalization error of rule B (with 16 parameters).

4. CONCLUSION

This chapter explored methods to optimize learning rules in neural networks. Preliminary results show that it is possible to optimize a synaptic learning rule for different tasks, while constraining the rule to be biologically plausible. Furthermore, we have established the conceptual basis permitting to study the generalization properties of a learning rule whose parameters are trained on a certain number of tasks. To do so, we have introduced the notion of capacity of parametric learning rules. The experimental results described here qualitatively agree with learning theory applied to parametric learning rules.

---

[4] Gradient descent always fell into local minima and thus was not able to give interesting results.

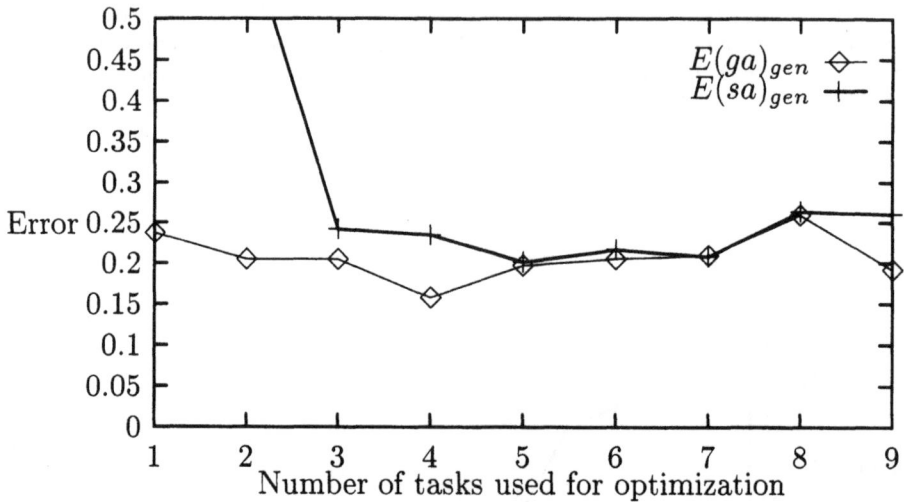

Fig. 14.13. Evolution of generalization error with respect to the optimization method. Here, tasks are linearly separable and the rule has 7 parameters. $E(ga)_{gen}$ represents generalization error with genetic algorithms, and $E(sa)_{gen}$ generalization error with simulated annealing.

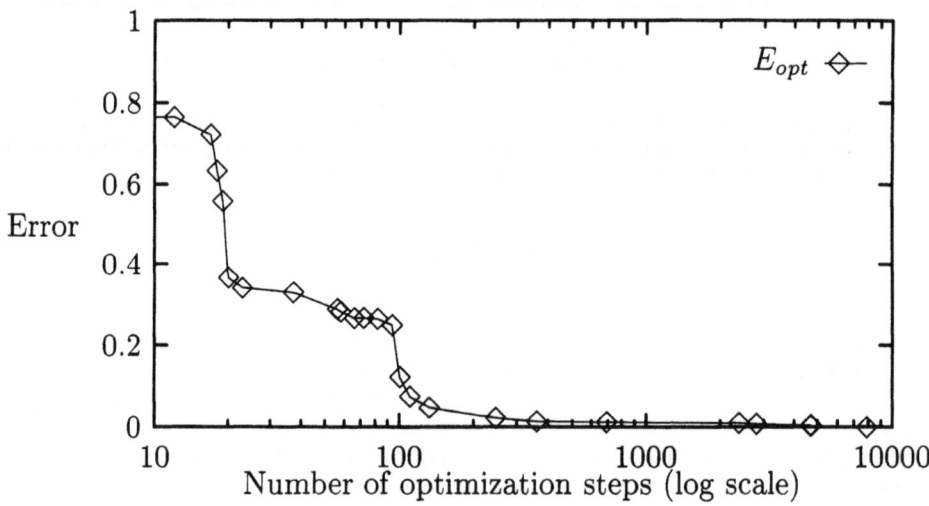

Fig. 14.14. Evolution of optimization error with respect to the number of steps during optimization of a parametric learning rule. Here, the optimization method is simulated annealing and the rule has 7 parameters.

The problems studied so far were quite simple, and it is important to improve the form of the learning rule and the optimization process in order to find rules that can efficiently solve more complex problems. In this section we discuss some improvements that should be considered for further research.

Optimization. There are two immediate ways to improve the optimization process. One is to refine the cost function by adding terms reflecting the quality of network generalization. Other optimization methods such as genetic programming (Koza, 1992) or second-order methods should be tested for their efficiency in handling diverse and more complex learning problems. Experiments show that a rule optimized to solve a simple task cannot solve a more difficult task, whereas the converse is often true (given that the number of tasks used to optimize the rule is sufficient).

Form of $\Delta w(\ )$. Optimization can yield good results only if the learning rule's constraints are soft enough to yield one or more acceptable solutions, but hard enough in order to yield good generalization performance and speed up optimization so that such solutions can be found in reasonable time. One possibility is to perform a preliminary analysis of the tasks to be learned and to "customize" the rule accordingly. Another one is to take in to account known (and probably useful) biological synaptic mechanisms such as:

· Temporal processing in a synapse (different input factors influence synaptic efficacy with different delays).
· Limiting the neurons to be excitatory or inhibitory (but not both). This is unlike most existing artificial neural network models in which a neuron may have both behaviors.
· More detailed modeling of operation of the neuron, taking in to account physical distance between synaptic sites and local influence of the local potential on the neural membrane on the dendritic tree.

Analysis of Resulting Learning Rules. It is important to analyze the obtained learning 26 rules. We should systematically compare our resulting rules with other learning techniques such as back-propagation and attempt to discover the underlying reasons for differences in their behavior.

REFERENCES

Baum, E. R., & Haussler, D. (1989). What size network give valid generalization? *Neural Computation*, 1, 151-160.
Bengio, S., Bengio, Y., Cloutier, J., & Gecsei, J. (1992). Aspects théoriques de l'optimisation d'une règle d'apprentissage. In *Actes de la Conference Neuro-Nimes 1992*, Nimes, France, November.
Bengio, S., Bengio, Y., Cloutier, J., & Gecsei, J. (1993). Generalization of a parametric learning rule. In *ICANN '93: Proceedings of the International Conference on Artificial Neural Networks, Amsterdam*, Netherlands.

Bengio, Y., & Bengio, S. (1990). *Learning a synaptic learning rule* (Tech. Rep. No. 751). Département d'Informatique et de Recherche Opérationnelle, Université de Montréal.

Bengio, Y., Bengio, S., & Cloutier, J. (1991). Learning a synaptic learning rule. *Proceedings of the International Joint Conference on Neural Networks 1991 — Seattle* (Vol. II, p. A-969). Piscataway, NJ: IEEE.

Bengio, Y., Bengio, S., & Cloutier, J. (1991). Learning synaptic learning rules. In *Neural Networks for Computing*, Snowbird, Utah, March.

Byrne, J., & Berry, W. (Eds.) (1989). *Neural Models of Plasticity.* Reading, MA: Addison-Wesley.

Chalmers, D. (1990). The evolution of learning: An experiment in genetic connectionism. In D. Touretzky, J. Elman, T. Sejnowski, & G. Hinton (Eds.), *Proceedings of the 1990 Connectionist Models Summer School.* San Mateo, CA: Morgan Kaufmann.

Davis, L. (1989). Mapping neural networks into classifier systems. *Proceedings of the Third International Conference on Genetic Algorithms* (pp. 375-378). San Mateo, CA: Morgan Kaufmann.

Duda, R. O., & Hart, P. E. (1973). *Pattern Classification and Scene Analysis.* New York: Wiley-Interscience.

Gardner, D. (1987). Synaptic diversity characterizes biological neural networks. In M. Caudill & C. Butler (Eds.), *Proceedings of the IEEE First International Conference on Neural Networks* (Vol. IV, pp. 17-22). San Diego: IEEE/ICNN.

Gluck, M. A., & Thompson, R. F. (1987). Modeling the neural substrate of associative learning and memory: A computational approach. *Psychological Review,* **94,** 176-191.

Goldberg, D. (1989). *Genetic Algorithms in Search, Optimization, and Machine Learning.* Reading, MA: Addison-Wesley.

Hawkins, R. D. (1989). A biologically based computational model for several simple forms of learning. In R. D. Hawkins & G. H. Bower (Eds.), *Computational Models of Learning in Simple Neural Systems* (pp. 65-100). San Diego: Academic Press.

Hawkins, R. D., Abrams, T. W., Carew, T. J., & Kandel, E. R. (1983). A cellular mechanism of classical conditioning in aplysia: Activity-dependent amplification of presynaptic facilitation. *Science,* **219,** 400-404.

Hebb, D. O. (1949). *The Organization of Behavior.* New York: Wiley.

Hertz, J., Krogh, A., & Palmer, R. (1991). *Introduction to the Theory of Neural Computation.* Reading, MA: Addison-Wesley.

Hinton, G. E. (1989). Connectionist learning procedures. *Artificial Intelligence,* **40,** 185-234.

Holland, J. (1975). *Adaptation in Natural and Artificial Systems.* Ann Arbor: University of Michigan Press.

Judd, S. (1988). On the complexity of loading shallow neural network. *Journal of Complexity,* **4,** 177-192.

Kirkpatrick, S., Gelatt, C. D., Jr., & Vecchi, M. P. (1983). Optimization by simulated annealing. *Science,* **220,** 671-680.

Koza, J. R. (1992). *Genetic Programming: On the Programming of Computers by Means of Natural Selection.* Cambridge, MA: MIT Press.

Pavlov, I. P. (1932). *Les Réflexes Conditionels.* Paris: Alcan.

Rumelhart, D. E., Hinton, G. E., & Williams, R. J. (1986). Learning internal representations by error propagation. In D. E. Rumelhart & J. L. McClelland (Eds.), *Parallel Distributed Processing: Explorations in the Microstructure of Cognition* (Vol. 1, pp. 318-362). Cambridge, MA: MIT Press.

Vapnik, V. N. (1982). *Estimation of Dependencies Based on Empirical Data*. New York: Springer-Verlag.

Vapnik, V. N., & Chervonenkis, A. Y. (1971). On the uniform convergence of relative frequencies of events to their probabilities. *Theory of Probability and Its Applications*, **16**, 264-280.

White, H. (1989). Learning in artificial neural networks: A statistical perspective. *Neural Computation*, **1**, 425-464.

Whitley, D., & Hanson, T. (1989). Optimizing neural networks using faster, more accurate genetic search. *Proceedings of the Third International Conference on Genetic Algorithms* (pp. 1888-1898). San Mateo, CA: Morgan Kaufmann.

15

Spatial Pattern Learning, Catastrophic Forgetting, and Optimal Rules of Synaptic Transmission

Gail A. Carpenter
Boston University

Gail Carpenter's chapter, **Spatial Pattern Learning, Catastrophic Forgetting, and Optimal Rules of Synaptic Transmission,** *introduces a neural network signal transduction rule. This synaptic transmission rule posits an adaptive threshold as the fundamental unit of long-term memory, rather than the traditional path weight, designed to multiply an axonal signal. The chapter shows how the adaptive threshold rule solves a type of catastrophic forgetting problem that can occur when synaptic transmission obeys the signal-times-weight product rule. The threshold rule, combined with a principle of atrophy due to disuse, leads to a new model, the* **distributed outstar,** *which achieves stable learning with codes that can be either winner-take-all or distributed. The distributed outstar generalizes the outstar (Grossberg, 1968a), in which a single source node projects to a field of target nodes. Outstar learning determines weight adaptation in the top-down adaptive filter of adaptive resonance theory (ART) models with winner-take-all code representations. The distributed outstar replaces the single outstar source node with a source field that may have arbitrarily many nodes and that can support arbitrarily distributed or compressed code representations.*

Carpenter's concern with optimal rules for synaptic transmission is shared with the chapters by Bengio et al. and Chance et al. Carpenter speculates that this type of investigation of different learning rules might help to guide the search by neuroscientists for learning rules in biological neurons. Also it can help guide the optimal design of memory units in optical and electronic neural networks. As Grossberg pointed out in his talk at the 1992 M.I.N.D conference (he could not write a chapter based on his talk due to other commitments), a system for sensory pattern classification has different requirements than a system for generating motor behaviors. A sensory system, Grossberg said, needs to encode impinging outside inputs in a relatively faithful manner, and therefore requires positive feedback such as occurs in outstars and in the many versions of ART. A motor system, on the other hand, needs to be able to perform preplanned target behaviors, and therefore requires error-correcting negative feedback such as occurs in the VAM model he developed with Daniel Bullock and Paolo Gaudiano. (Back propagation networks fall in the latter category, even though they are different from VAM in other ways.) Hence the work of both Carpenter and Grossberg reinforces the important point that optimal design of neural networks varies immensely with the tasks they are designed to perform.

ABSTRACT

It is a neural network truth universally acknowledged, that the signal transmitted to a target node must be equal to the product of the path signal times a weight. Analysis of catastrophic forgetting by distributed codes leads to the unexpected conclusion that this universal synaptic transmission rule may not be optimal in certain neural networks. The distributed outstar, a network designed to support stable codes with fast or slow learning, generalizes the outstar network for spatial pattern learning. In the outstar, signals from a source node cause weights to learn and recall arbitrary patterns across a target field of nodes. The distributed outstar replaces the outstar source node with a source field, of arbitrarily many nodes, where the activity pattern may be arbitrarily distributed or compressed. Learning proceeds according to a principle of atrophy due to disuse whereby a path weight decreases in joint proportion to the transmitted path signal and the degree of disuse of the target node. During learning, the total signal to a target node converges toward that node's activity level. Weight changes at a node are apportioned according to the distributed pattern of converging signals. Three types of synaptic transmission, a product rule, a capacity rule, and a threshold rule, are examined for this system. The three rules are computationally equivalent when source field activity is maximally compressed, or winner-take-all. When source field activity is distributed, catastrophic forgetting may occur. Only the threshold rule solves this problem. Analysis of spatial pattern learning by distributed codes thereby leads to the conjecture that the optimal unit of long-term memory in such a system is a subtractive threshold, rather than a multiplicative weight.

1. OPTIMAL RULES OF SYNAPTIC TRANSMISSION

When neural networks became popular in the 1980s, researchers struggled to define *neural network* with words that would include the diverse models in current use. As a step toward this definition, consider the question: What, if anything, do all the neural networks of the past fifty years have in common? The answer to this question is, most likely, nothing. However, the large majority of neural network models, from the McCulloch-Pitts (1943/1988) neuron to the many biological and engineering models at this year's conferences, have at least one thing in common, namely, the rule setting the net signal from a source node to a target node equal to a path signal times a synaptic weight (Fig. 15.1). This *product rule* of synaptic transmission is in such universal use that it is almost always treated as a nameless fact rather than a hypothesis, although neurophysiology so far neither confirms nor refutes this rule. Why, then, has this particular process found such widespread use? One answer is its computational power: the product rule sets the sum of weighted signals equal to the dot product of the signal vector and the weight vector. This dot product provides a useful measure of the similarity between the active path signal vector and the learned weight vector. However, utility and universality do not necessarily imply optimality.

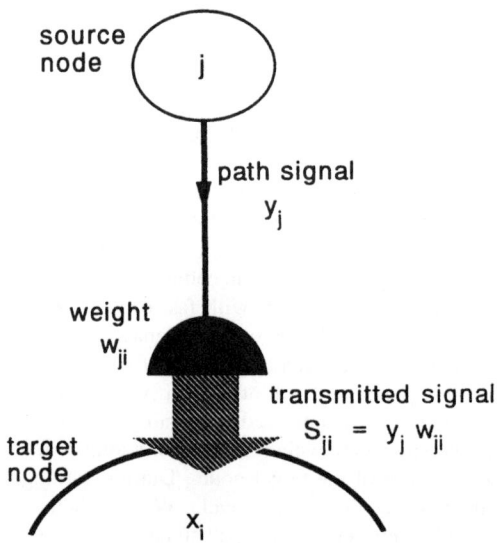

Fig. 15.1. The product rule postulates that the signal transmitted to a target node at a synapse is proportional to a path signal (y_j) times a weight (w_{ji}). This rule is a feature of nearly all neural network models.

This chapter describes a neural network learning problem for which the product rule is not computationally optimal. Solution of the learning problem requires a neural network design to support stable distributed codes. One such design is the *distributed outstar* (Carpenter, 1993, 1994), which solves the distributed code catastrophic forgetting problem when the product rule is replaced by an equally plausible synaptic transmission rule. This *threshold rule* postulates that the unit of long-term memory (LTM) is a subtractive threshold, rather than a multiplicative weight. In the process of solving a particular learning problem, therefore, computational analysis questions the optimality of a fundamental neural network design hypothesis.

2. OUTSTAR LEARNING AND DISTRIBUTED CODES

An *outstar* is a neural network that can learn and recall arbitrary spatial patterns (Grossberg, 1968a). Outstar learning and recall occur when a source node transmits a weighted signal to a target, or border, field of nodes. This network is a key component of various neural models of cognitive processing. For example, the outstar has been identified as a minimal neural network capable of classical conditioning (Grossberg, 1968b, 1974/1982c). In terms of stimulus sampling theory (Estes, 1955), the source node plays the role of a sampling cell. When the sampling cell is active, long-term memory traces, or adaptive weights, learn stimulus sampling probabilities of border field activity patterns. A sequence of outstars, called an *avalanche*, forms a minimal network for learning and ritualistic performance of an arbitrary space-time pattern (Grossberg, 1969). Within the adaptive resonance theory of self-organizing pattern classification, outstars learn the top-down expectations that are critical to code stabilization (Grossberg, 1976/1991). All neural network realizations of adaptive resonance theory (ART models) have so far used outstar learning in the top-down adaptive filter (Carpenter & Grossberg, 1987a/1991, 1987b/1991, 1990/1991; Carpenter, Grossberg, & Rosen, 1991a). The supervised ARTMAP system (Carpenter, Grossberg, & Reynolds, 1991) also employs outstar learning in the formation of its predictive maps. Outstars have thus played a central role in both the theoretical analysis of cognitive phenomena and in the neural models that realize the theories, as well as applications, of these systems.

An outstar is characterized by one source node sending weighted inputs to a target field. We here consider spatial pattern learning in a more general setting, in which an arbitrarily large source field replaces the single source node of the outstar. This *distributed outstar network* (Fig. 15.2a) is similar to the original outstar when the source field F_2 contains a single node. Then weights in the $F_2 \rightarrow F_1$ adaptive filter track the F_1 activity pattern when the one F_2 node is active.

At first, distributed outstar learning would appear to be modeled already in the ART top-down adaptive filter (Fig. 15.3a). However, to date, networks that explicitly realize adaptive resonance assume the special case in which F_2 is a *choice*, or *winner-take-all*, network. In this case, only one F_2 node is active during learning, so each F_2 node acts, in turn, as an outstar source node. We here consider how to design a spatial pattern learning network that allows the activity pattern at the coding field F_2 to be arbitrarily distributed (Section 3). That is, one, several, or all of the F_2 nodes may be active during learning. One possible design is simply to implement outstar learning in each active path. However, such a system is subject to catastrophic forgetting that can quickly render the network useless, unless learning rates are very slow (Section 4). In particular, if all F_2 nodes were active during learning, all $F_2 \rightarrow F_1$ weight vectors would converge toward a common pattern.

A learning principle of *atrophy due to disuse* leads toward a solution of the catastrophic forgetting problem (Section 5). By this principle, a weight in an active path atrophies, or decays, in joint proportion to the size of the transmitted synaptic signal and a suitably defined "degree of disuse" of the target cell. During learning, the total transmitted signal from F_2 converges toward the activity level of the target F_1 node. Atrophy due to disuse thereby dynamically substitutes total $F_2 \rightarrow F_1$ signal for the individual outstar weight. This seems a plausible step toward pattern learning by a coding source field instead of by a single source node. Unfortunately, this development is, by itself, insufficient. The network still suffers catastrophic forgetting if signal transmission obeys a *product rule*. This rule, now used in nearly all neural models, assumes that the transmitted synaptic signal from the jth F_2 node to the ith F_1 node is proportional to the product of the path signal y_j and the path weight w_{ji}. An alternative transmission process, used in a neural network realization of fuzzy ART (Carpenter, Grossberg, & Rosen, 1991b;

Carpenter & Grossberg, 1994), obeys a *capacity rule* (Section 6). However, catastrophic forgetting is even more serious a problem for the capacity rule than for the product rule.

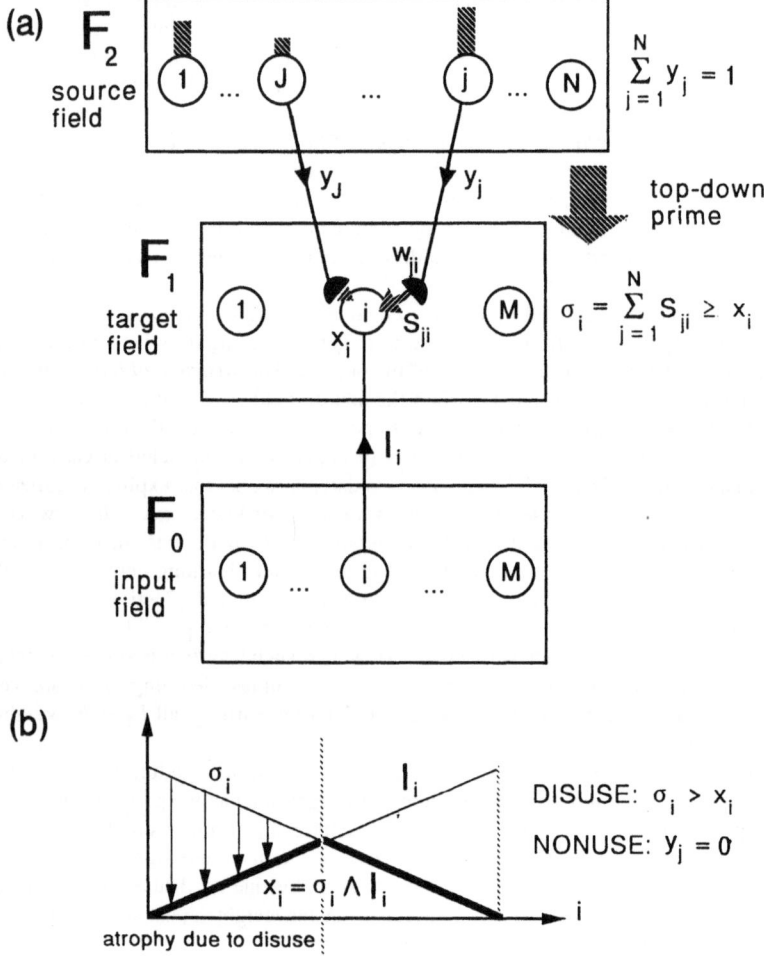

Fig. 15.2. Distributed outstar network for spatial pattern learning. During adaptation a top-down weight w_{ji}, from the jth node of the coding field F_2 to the ith node of the pattern registration field F_1, may decrease or remain constant. An atrophy-due-to-disuse learning law causes the total signal σ_i from F_2 to the ith F_1 node to decay toward that node's activity level x_i, if σ_i is initially greater than x_i. Within this context, three synaptic transmission rules are analyzed.

Fortunately, another plausible synaptic transmission rule solves the problem (Sections 7-9). This *threshold rule* postulates a transmitted signal equal to the amount by which the $F_2 \rightarrow F_1$ signal y_j exceeds an adaptive threshold τ_{ji}. Where weights decrease during atrophy-due-to-disuse learning, thresholds increase: formally, τ_{ji} is identified with $1 - w_{ji}$. When synaptic transmission is implemented by a threshold rule, weight/threshold changes are bounded and automatically apportioned according to the distribution of F_2 activity, with fast learning as well as slow learning. When F_2 makes a choice, the three synaptic transmission rules are computationally identical, and atrophy-due-to-disuse learning is essentially the same as outstar learning. Thus functional differences between the three types of transmission would be experimentally and computationally measurable only in situations where the F_2 code is distributed.

Computational analysis of distributed codes hereby leads unexpectedly to a hypothesis about the mechanism of synaptic transmission: the unit of long-term memory in these systems is conjectured to be an adaptive threshold, rather than a multiplicative path weight. Thresholds that determine a node's output signal have played an essential role in neural network models from the start (Hartline & Ratliff, 1957; McCulloch & Pitts, 1943/1988), and adaptive activation thresholds are a standard feature of back propagation models (Rumelhart, Hinton, & Williams, 1986). At the other end of the axon, however, these models all employ the standard signal-times-weight product rule to characterize synaptic transmission to a target node. Historically, early definitions of the perceptron specified a general class of synaptic transmission rules (Rosenblatt, 1958/1988, 1962). However, the electrical switching circuit model, which realizes multiplicative weights as adjustable gains, quickly became the dominant metaphor (Widrow & Hoff, 1960/1988). Over the ensuing decades, efficient integrated hardware realization of the linear adaptive filter has remained a challenge. In opto-electronic neural networks, the adaptive threshold synaptic transmission rule, realized as a rectified bias, may be easier to implement than on-line multiplication (T. Caudell, personal communication). Thus, even in networks where the product rule and the threshold rule are computationally equivalent, their diverging physical interpretations may prove significant in both the neural and the hardware domains. The adaptive threshold hypothesis completes the *distributed outstar learning law*, summarized in Section 10. Section 11 explicitly solves the distributed outstar equations, Section 12 illustrates distributed outstar dynamics with a network that has two nodes in the source field, and Section 13 concludes with a consideration of the physical unit of memory.

3. SPATIAL PATTERN LEARNING

The distributed outstar network (Fig. 15.2a) features an adaptive filter from a *coding*, or *source, field F_2* to a *pattern registration*, or *target, field F_1*. This filter carries out spatial pattern learning, whereby the adaptive path weights track the activity pattern of the target field, F_1. When F_2 consists of just one node ($N = 1$) the network is a type of outstar. During outstar learning, weights in the paths emanating from an F_2 node track F_1 activity. That is, when the jth F_2 node is active, the weight vector $\mathbf{w}_j$ converges toward the F_1 activity vector $\mathbf{x}$ of the target, or border, nodes at the outer fringe of the filter (Fig. 15.3). Although many variants of outstar learning have been analyzed (Grossberg, 1968a, 1972/1982), the essential outstar dynamics are described by the equation

$$\text{Basic outstar: } \frac{d}{dt} w_{ji} = y_j(x_i - w_{ji}). \tag{1}$$

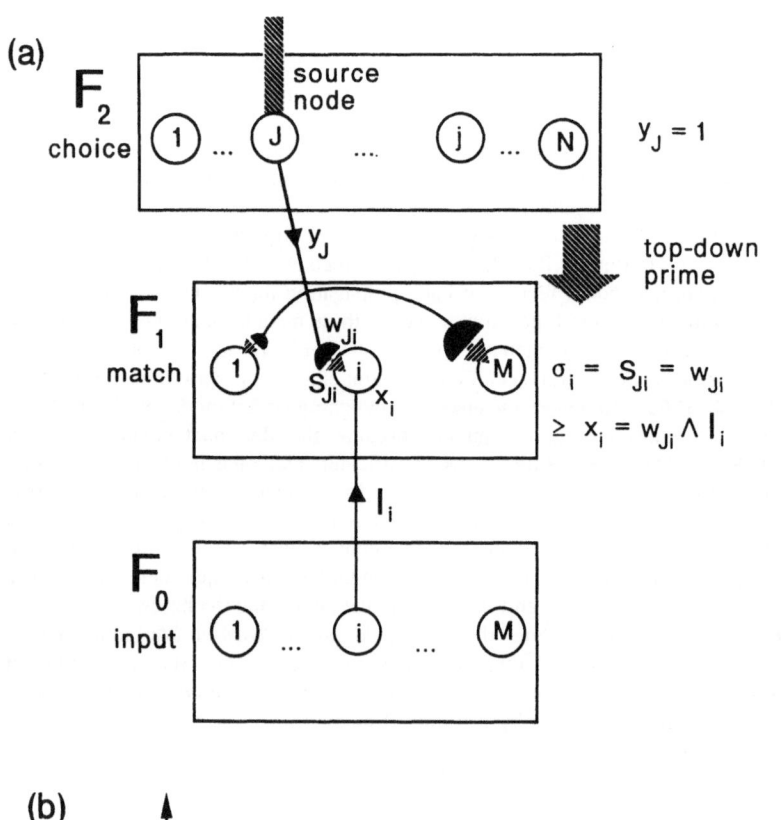

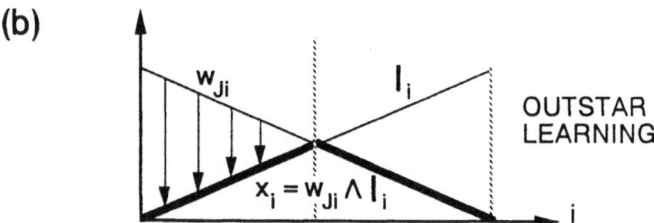

Fig. 15.3. ART 1/fuzzy ART.

This is the learning law used in the top-down adaptive filters of ART 1 (Carpenter & Grossberg, 1987a/1991), ART 2 (Carpenter & Grossberg, 1987b/1991), and fuzzy ART (Carpenter, Grossberg, & Rosen, 1991a). By (1), $w_{ji} \rightarrow x_i$ when $y_j > 0$. When $y_j = 0$, w_{ji} remains constant. The term $y_j x_i$ in Eq. (1) describes a Hebbian correlation whereby the weight tends to increase when both the presynaptic F_2 node j and the postsynaptic F_1 node i are active. The term $-y_j w_{ji}$ describes an anti-Hebbian process whereby the weight w_{ji} tends to decrease when the presynaptic node j is active but the postsynaptic node i is inactive ("pre- without post-"). The distributed outstar network does not constitute a stand-alone pattern recognition system. Like the outstar, this module would typically be embedded within a larger neural network architecture for supervised or unsupervised pattern learning and recognition. For example, in an ART system the top-down $F_2 \rightarrow F_1$ filter plays a crucial role in ART code stabilization. Additional network elements determine which F_2 code will be selected by an input $\mathbf{I}$ in the first place and implement search and other mechanisms of internal dynamic control (Carpenter & Grossberg, 1987a/1991). This chapter focuses only on design issues pertaining to the top-down adaptive filter.

4. CATASTROPHIC FORGETTING

The distributed outstar network for spatial pattern learning (Fig. 15.2a) needs to solve a potential catastrophic forgetting problem. Suppose, for example, that all F_2 nodes are active ($y_j > 0$) at some time when the ith F_1 node is inactive ($x_i = 0$) due, say, to the fact that there is no input to that node at that moment ($I_i = 0$). With fast learning, an outstar (Eq. (1)) would send all weights w_{ji} ($j = 1, ..., N$) to 0. Within an ART system, stability requirements imply that these weights then remain 0 forever. Moreover, no future input I_i to the ith F_1 node could even activate that node, once F_2 became active. If similar weight decays occurred at each F_1 node, all weights would decay to 0. The network would thus quickly become useless, quenching all F_1 activity as soon as any F_2 code was selected.

The special class of F_2 networks called choice, or winner-take-all, systems sidestep this catastrophic forgetting problem. A code representation field F_2 is a choice network when internal competitive dynamics concentrate all activity at one node (Grossberg, 1973). An F_2 code that chooses the Jth node is described by

$$F_2 \ choice: \ y_j = \begin{cases} 1 \ if \ j = J \\ 0 \ if \ j \neq J \end{cases}. \tag{2}$$

In this case, each F_2 node is identified with a class, or category, of inputs $\mathbf{I}$. Outstar learning (Eq. (1)) permits a weight w_{ji} to change only if the jth F_2 node is active. When F_2 chooses the node J, all other nodes $j \neq J$ are inactive. Only the weight w_{ji} tracks activity at the ith F_1 node, so

$$\mathbf{w}_J \rightarrow \mathbf{x}. \tag{3}$$

Even if w_{Ji} decays to 0, all other weights to the ith F_1 node remain unchanged when the Jth category is selected. These other weights w_{ji} ($j \neq J$) are thus reserved and can learn their own F_1 patterns when they later become active.

Choice represents an extreme form of short-term memory (STM) competition at F_2. By confining all weight changes to a single category, F_2 choice protects the learned codes of all the other categories during outstar learning. However, outstar learning poses a problem when F_2 category representations can be distributed. If a code y were highly distributed, with all $y_j > 0$, then the outstar learning law (Eq. (1)) would imply that all weight vectors w_j would converge toward the same F_1 activity vector x but not the asymptotic state of the weights. The severity of this problem can be reduced if learning intervals are extremely short. Then, because the rate at which w_j approaches x is proportional to y_j, little change will occur in weights w_{ji} with small y_j. If, however, many of the y_j values are nearly uniform or if learning is not always slow, catastrophic forgetting will occur as all weight vectors approach one common pattern that is independent of all prior learned differences.

An adaptation rule called the distributed outstar learning law solves this problem. Even with fast learning, where weights approach asymptote on each input presentation, the distributed outstar apportions weight changes across active paths without catastrophic forgetting. In the distributed outstar, the rate constant for an individual weight w_{ji} is an increasing function both of y_j, as in the outstar equation (1), and also of w_{ji} itself. When w_{ji} becomes too small, further change is disallowed. Weights, initially large, can only decrease during learning. Small weights can decrease further only when y_j is close to 1, which occurs when most of the F_2 STM activity is concentrated at node j. When F_2 activity is highly distributed only large weights, close to their initial values, are able to change. Moreover, for highly distributed codes, the maximum possible weight change in any single path is small.

The distributed outstar is derived from the notion that the sum of all $F_2 \rightarrow F_1$ individual path weights tracks target node activity during learning. A principle of atrophy due to disuse governs weight change, as described in the next section. Within this context, three signal transmission rules are examined (Section 6). An adaptive threshold rule for synaptic transmission is more computationally successful than either of the other two rules.

5. LEARNING BY ATROPHY DUE TO DISUSE

The principle of atrophy due to disuse postulates that the strength of an active path decays when the path is disused. Active "dis-use" is distinct from passive "non-use" (Fig. 15.2b), where the strength of an inactive path remains constant, as in outstar learning (Eq. (1)) (Fig. 15.3b). To define disuse, a specific class of target fields F_1 are considered. So far, no assumptions about the F_1 activity vector x have been made. The main hypothesis on F_1 is that, when F_2 is active, the total top-down input from F_2 to F_1 imposes an upper bound, or limit, on the maximum activity at an F_1 node. In addition to a bottom-up input I_i to the ith F_1 node, a top-down *priming* input σ_i from F_2 is assumed to be necessary for that node to remain active, once F_2 becomes active. This hypothesis is realized by the inequality

$$\text{Top-down prime: } 0 \leq x_i \leq \sigma_i, \tag{4}$$

where σ_i is the sum of all transmitted signals S_{ji} from F_2 to the ith F_1 node:

$$\sigma_i = \sum_{j=1}^{N} S_{ji} \tag{5}$$

(Fig. 15.2a). In particular, when F_2 is active but $\sigma_i = 0$, no activity can be registered at the ith F_1 node, for any bottom-up input $I_i \in [0, 1]$.

The top-down prime inequality (4) is closely related to the 2/3 Rule of ART (Carpenter & Grossberg, 1987a/1991), which implies that the ith F_1 node will be inactive ($x_i = 0$) if either the bottom-up input I_i is small or the total top-down input σ_i is small when F_2 is active. The 2/3 Rule was derived both from an analysis of system requirements for input registration, priming, and stable, self-organizing pattern learning and classification and from an analysis of the corresponding cognitive phenomena. In binary ART 1 systems with choice at F_2, the 2/3 Rule is realized by allowing the ith F_1 node to be active, when the Jth F_2 node is active, only if $I_i = 1$ and if σ_i exceeds a criterion threshold, where

$$\sigma_i = y_J w_{Ji}. \tag{6}$$

Fuzzy ART (Carpenter, Grossberg, & Rosen, 1991a), an analog extension of ART 1, realizes the 2/3 Rule by setting

$$x_i = I_i \wedge w_{Ji} \equiv \min(I_i, w_{Ji}) \tag{7}$$

when the Jth F_2 node is chosen (Fig. 15.3a). The symbol $\wedge$ in Eq. (7) denotes the fuzzy intersection (Zadeh, 1965). By Eqs. (2) and (6), when F_2 makes a choice,

$$\sigma_i = w_{Ji}. \tag{8}$$

Eqs. (7) and (8) suggest setting:

$$x_i = I_i \wedge \sigma_i \tag{9}$$

to define one class of F_1 systems that realize σ_i as a top-down prime, or upper bound, on target node activity x_i.

When F_2 primes F_1, by Eq. (4), the *degree of disuse* D_i of the ith F_1 node is defined to be

$$D_i = (\sigma_i - x_i) \geq 0. \tag{10}$$

When Eq. (9) defines F_1 activity,

$$D_i = (\sigma_i - I_i \wedge \sigma_i)$$
$$= \left\{ \begin{matrix} \sigma_i - I_i \ if \ \sigma_i \geq I_i \\ 0 \ if \ \sigma_i \leq I_i \end{matrix} \right\},$$
$$= [\sigma_i - I_i]^+$$

(11)

where $[...]^+$ denotes the rectification operator

$$[\theta]^+ \equiv \theta \vee 0 = \max(\theta, 0)$$

(12)

where $\vee$ denotes the fuzzy union (Zadeh, 1965). In this case, the degree of disuse at the ith F_1 node is the amount by which the top-down input σ_i exceeds the bottom-up input I_i at that node. A learning principle of atrophy due to disuse postulates that a path weight decays in proportion to the degree of disuse of its target node. We here consider a class of learning equations that realize this principle in the form

$$\frac{d}{dt} w_{ji} = -S_{ji} D_i.$$

(13)

Weights can then decay or stay constant, but never grow, when $S_{ji} \geq 0$ and $D_i \geq 0$. With the degree of disuse D_i defined by Eq. (10), the learning law (Eq. (13)) becomes:

$$Atrophy\ due\ to\ disuse: \frac{d}{dt} w_{ji} = -S_{ji}(\sigma_i - x_i)$$

(14)

(Fig. 15.2b). In Section 6, three synaptic transmission rules each define S_{ji} as a function of y_j and w_{ji}. In Sections 7 and 8 we will analyze atrophy-due-to-disuse learning and catastrophic forgetting for these three rules.
    Initially,

$$w_{ji}(0) = 1$$

(15)

for $i = 1, ..., M$ and $j = 1, ..., N$. The learning law (Eq. (14)) implies that a path weight w_{ji} can decay when the total top-down signal σ_i to the ith target F_1 node exceeds the node's activity x_i. The rate of decay is proportional to a path's contribution, S_{ji}, to the top-down signal. By Eq. (14), the sum of all weights converging on the ith node obeys the equation:

$$\frac{d}{dt}(\sum_{j=1}^{n} w_{ji}) = -\sigma_i(\sigma_i - x_i).$$

(16)

Thus if the F_1 pattern $\mathbf{x}$ and the F_2 pattern $\mathbf{y}$ are constant during a learning interval, and if $\sigma_i > x_i$ at the start of that interval, then one or more weights w_{ji} must continue to decay until σ_i converges to x_i.

When F_2 makes a choice, we see that

$$\sigma_i = S_{Ji} = w_{Ji},\qquad(17)$$

whereas $S_{ji} = 0$ $(j{\neq}J)$, for all three transmission rules. In this case the atrophy-due-to-disuse Eq. (14) reduces to

$$\frac{dw_{ji}}{dt} = -S_{ji}(w_{ji} - x_i)$$
$$= \begin{cases} -w_{Ji}(w_{Ji} - x_i) & \text{if } j = J \\ 0 & \text{if } j \neq J \end{cases}\qquad(18)$$

Comparing Eq. (18) with Eq. (16) illustrates the sense in which the total weighted signal σ_i in a distributed code replaces the weight w_{Ji} in a system where F_2 makes a choice. Note that w_{Ji} approaches x_i at a rate proportional to w_{Ji}. Equation (18) is thereby slightly different from the outstar equation (1), which reduces to

$$\frac{dw_{ji}}{dt} = \begin{cases} -(w_{Ji} - x_i) & \text{if } j = J \\ 0 & \text{if } j \neq j \end{cases}\qquad(19)$$

when F_2 makes a choice. Because $w_{Ji} = \sigma_i \geq x_i$, $x_i = 0$ if $w_{Ji} = 0$. Thus Eqs. (18) and (19) both imply that $\mathbf{w}_J \rightarrow \mathbf{x}$ while other $\mathbf{w}_j$ remain constant, as long as the Jth F_2 node remains active (Fig. 15.3b). With fast learning and F_2 choice the atrophy-due-to-disuse and outstar learning laws are equivalent. In this case, neither computational nor experimental analysis can differentiate outstar learning from atrophy due to disuse. The three synaptic transmission rules are similarly indistinguishable. However, when F_2 activity $\mathbf{y}$ is distributed, qualitative properties of learned patterns depend critically on both the learning law and the signal transmission rule, as follows.

6. SYNAPTIC TRANSMISSION FUNCTIONS

We now define three synaptic transmission rules. The F_2 path signal vector $\mathbf{y} = (y_1, ..., y_j, ..., y_N)$ is assumed to be normalized:

$$\sum_{j=1}^{N} y_j = 1\qquad(20)$$

but is otherwise arbitrary. Given a signal y_j from the jth F_2 node to the ith F_1 node, via a path with an adaptive weight w_{ji}, the net signal S_{ji} received by the ith F_1 node is assumed to be a function of y_j and w_{ji}:

$$S_{ji} = f(y_j, w_{ji}), \tag{21}$$

Each of the three rules corresponds to a physical theory of synaptic signal transmission in neural pathways. The present analysis uses computation alone to select one of these three rules over the others in a neural system for spatial pattern learning. The first synaptic transmission rule postulates that the $F_2 \rightarrow F_1$ signal is jointly proportional to the path signal y_j and the weight w_{ji}:

$$Product\ rule: S_{ji} = y_j w_{ji} \tag{22}$$

(Fig. 15.1). Synaptic transmission by the product rule is an implied hypothesis of most neural network models. The rule implies that σ_i, the sum of all transmitted signals to the ith F_1 node, equals the dot product between the $F_2 \rightarrow F_1$ path vector ($y_1, ..., y_j, ..., y_N$) and the converging weight vector ($w_{1i}, ..., w_{ji}, ..., w_{Ni}$). That is, the total signal from F_2 to the ith F_1 node is a linear combination of the path signals y_j:

$$\sigma_i = \sum_{j=1}^{N} y_j w_{ji} \tag{23}$$

with the coefficients w_{ji} fixed (McCulloch & Pitts, 1943/1988) or determined by some learning law. The total transmitted signal σ_i thereby computes the correlation between the $F_2 \rightarrow F_1$ path vector and the converging weight vector. Rosenblatt (1962/1988) considered synaptic transmission rules in the general form (Eq. (21)) when defining the perceptron. However, the product rule (Eq. (22)) and its linear matched filter (Eq. (23)) have since come into almost universal use.

A second synaptic transmission rule assumes that the path signal y_j is itself transmitted directly to the ith F_1 node until an upper bound on the path's capacity is reached. With this upper bound equal to the path weight w_{ji}, the net signal obeys the:

$$Capacity\ rule: S_{ji} = y_j \wedge w_{ji} \equiv \min(y_j, w_{ji}). \tag{24}$$

A capacity rule is suggested by the computational requirements of neural network realizations of fuzzy set theory, as in fuzzy ART (Carpenter, Grossberg, & Rosen, 1991b; Carpenter & Grossberg, 1994). Figure 15.4a illustrates how the product rule compares to the capacity rule. For each, the signal S_{ji} grows linearly when y_j is small. However, a product rule signal increases with y_j for all $y_j \in [0, 1]$, whereas a capacity rule signal ceases to grow when y_j reaches the upper bound w_{ji}. The geometry of the graph in Fig. 15.4a suggests a third signal function, to complete a transmission rule parallelogram. The third signal function describes a

$$Threshold\ rule: S_{ji} = [y_j - (1 - w_{ji})]^+. \tag{25}$$

It is awkward to interpret the transmission rule (Eq. (25)) in terms of the weight w_{ji}. However, a natural interpretation takes the unit of long-term memory to be a signal threshold τ_{ji} rather than the path weight w_{ji}. By setting

$$\tau_{ji} \equiv 1 - w_{ji}, \tag{26}$$

the threshold rule (Eq. (25)) becomes:

$$S_{ji} = [y_j - \tau_{ji}]^+. \tag{27}$$

In Eq. (27), the transmitted signal from the jth F_2 node to the ith F_1 node is the amount by which the path signal y_j exceeds an adaptive synaptic threshold τ_{ji}.

The three rules, Eqs. (22), (24), and (25), are identical if F_2 activity is binary, because for each rule,

$$S_{ji} = \begin{cases} w_{ji} & if \ y_j = 1 \\ 0 & if \ y_j = 0 \end{cases}. \tag{28}$$

In particular, the three synaptic transmission rules are computationally indistinguishable if F_2 makes a choice, by Eq. (2). However, when a normalized F_2 code is distributed, an adaptive system that uses either the product rule or the capacity rule can suffer catastrophic forgetting. The threshold rule solves this problem.

7. TRANSMISSION RULE COMPUTATIONS

When an F_2 code $\mathbf{y}$ is maximally compressed, the three synaptic transmission rules (Table 15.1) are computationally identical. Computations in this section demonstrate how the three rules diverge when the F_2 code is maximally distributed. Note that the weight adaptation equation (14) also learns spatial patterns in a system where x_i may sometimes be greater than σ_i. Then, the top-down signal vector σ would still track the F_1 spatial pattern vector $\mathbf{x}$. However, the top-down prime hypothesis (Eq. (4)) implies that weights can only decrease, and hence are guaranteed to converge to some limit in the interval [0, 1] for arbitrary learning and input regimes.

Product rule:	$S_{ji} = y_j \, w_{ji}$	(22)
Capacity rule:	$S_{ji} = y_j \wedge w_{ji}$	(24)
Threshold rule:	$S_{ji} = [y_j - (1 - w_{ji})]^+$	(25)

Table 15.1. Synaptic Transmission Functions.

Initial values. Consider an atrophy-due-to-disuse system (Eq. (14)) in its initial state, when no learning has yet taken place. Then, all $w_{ji} = 1$, so:

$$S_{ji}(0) = y_j(0) \tag{29}$$

for each of the three synaptic transmission rules (Table 15.1). Therefore, since the F_2 activity vector y is normalized (Eq. (20)),

$$\sigma_i(0) = \sum_{j=1}^{N} S_{ji}(0) = 1. \tag{30}$$

The following computations trace an example in which $x_i = I_i \wedge \sigma_i$, as in Eq. (9). Then

$$x_i(0) = I_i \in [0, 1], \tag{31}$$

by Eq. (30). The atrophy-due-to-disuse equation (14) then implies that x_i will remain equal to I_i for as long as I remains constant. During that time, as some or all weights w_{ji} decrease, the total top-down input σ_i will decay toward the bottom-up input I_i, no matter which transmission rule is selected. For each rule,

$$\frac{d}{dt} w_{ji} = -S_{ji}(\sigma_i - I_i) \tag{32}$$

Choice at F_2. When F_2 makes a choice, as in Eq. (2), $\sigma_i = w_{Ji}$, which converges toward I_i, by Eq. (32). All other weights w_{ji} ($j \neq J$) remain constant. Competition at F_2 hereby limits the maximum total weight change at each F_1 node. In fact, when F_2 makes a choice,

$$\begin{aligned}
\Delta(\sum_{j=1}^{N} w_{ji}) &= \sum_{j=1}^{N} [w_{ji}(0) - w_{ji}(\infty)] \\
&= [w_{Ji}(0) - w_{Ji}(\infty)] \\
&= (1 - I_i)
\end{aligned} \tag{33}$$

for all three signal transmission rules.

Distributed code at F_2. An F_2 code is maximally compressed when the system makes a choice. Consider now the opposite extreme, when an F_2 code is maximally distributed. That is, let:

$$y_j = \frac{1}{N} \tag{34}$$

for $j = 1, ..., N$. All weights $w_{1i}, ..., w_{Ni}$ obey equation (32) and all are initially equal, by Eq. (15). Therefore the weights w_{ji} ($j = 1, ..., N$) to a given F_1 node will remain equal to one another during learning, for any transmission function S_{ji}. However, these individual weight changes under the three transmission rules show significant qualitative differences, despite the fact that the total $F_2 \rightarrow F_1$ signal vector σ correctly learns the F_1 activity vector $\mathbf{x} = \mathbf{I}$ for all three. In particular, the nature of the pattern encoded by a given weight vector and the size of the total weight change at each F_1 node clearly distinguish the three rules, as follows.

Product rule. With the product rule (Eq. (22)),

$$S_{ji} = \frac{1}{N} w_{ji} \tag{35}$$

Therefore:

$$\sigma_i = \sum_{j=1}^{N} \frac{1}{N} w_{ji} = \frac{1}{N} \sum_{j=1}^{N} w_{ji} \tag{36}$$

and

$$\frac{d}{dt} w_{ji} = -\frac{1}{N} w_{ji} \left(\frac{1}{N} \sum_{k=1}^{N} w_{ki} - I_i \right). \tag{37}$$

Because all weights w_{ji} to the ith F_1 node remain equal during learning,

$$w_{ji} \rightarrow I_i \tag{38}$$

for $j = 1, ..., N$. Thus the maximum total weight change at an F_1 node i is

$$\Delta \left(\sum_{j=1}^{N} w_{ji} \right) = N(1 - I_i), \tag{39}$$

which could be anywhere from 0 (when $I_i = 1$) to N (when $I_i = 0$).

Capacity rule. With the capacity rule (Eq. (24)),

$$S_{ji} = \frac{1}{N} \wedge w_{ji} = \begin{cases} \frac{1}{N} & if \ \frac{1}{N} \leq w_{ji} \leq 1 \\ w_{ji} & if \ 0 \leq w_{ji} \leq \frac{1}{N} \end{cases} \tag{40}$$

Therefore

$$
\sigma_i = \begin{cases} 1 & if \ \dfrac{1}{N} \le w_{ji} \le 1 \ for \ all \ j \\[4mm] \displaystyle\sum_{j=1}^{N} w_{ji} & if \ 0 \le w_{ji} \le \dfrac{1}{N} \ for \ all \ j \end{cases} \tag{41}
$$

Eq. (41) accounts for all cases since $w_{1i} = ... = w_{Ni}$ during learning. Weights adapt according to

$$
\frac{d}{dt} w_{ji} = \begin{cases} -\dfrac{1}{N}(1 - I_i) & if \ \dfrac{1}{N} \le w_{ji} \le 1 \\[4mm] -w_{ji}(\displaystyle\sum_{k=1}^{N} w_{ki} - I_i) & if \ 0 \le w_{ji} \le \dfrac{1}{N} \end{cases}. \tag{42}
$$

By Eq. (42), unless $I_i = 1$, all weights w_{ji} shrink until they enter the interval $[0, \dfrac{1}{N}]$. Thus

$$
w_{ji} \rightarrow \begin{cases} \dfrac{I_i}{N} & if \ 0 \le I_i < 1 \\[3mm] 1 & if \ I_i = 1 \end{cases} \tag{43}
$$

for each $j = 1, ..., N$. The maximum total weight change at the ith F_1 node is

$$
\Delta(\sum_{j=1}^{N} w_{ji}) = \begin{cases} (N - I_i) & if \ 0 \le I_i < 1 \\ 0 & if \ I_i = 1 \end{cases} \tag{44}
$$

which lies between $N - 1$ and N, unless $I_i = 1$.

Threshold rule. With the threshold rule (Eq. (25)),

$$
S_{ji} = \begin{cases} (\dfrac{1}{N} - (1 - w_{ji})) & if \ (1 - \dfrac{1}{N}) \le w_{ji} \le 1 \\[4mm] 0 & if \ 0 \le w_{ji} \le (1 - \dfrac{1}{N}) \end{cases}. \tag{45}
$$

By Eqs. (14) and (45), weight w_{ji} ceases to change as it falls toward $(1 - \dfrac{1}{N})$. Thus, because all $w_{ji}(0) = 1$,

$$\sigma_i = 1 - \sum_{j=1}^{N} (1 - w_{ji}). \tag{46}$$

During learning,

$$\frac{d}{dt} w_{ji} = -(\frac{1}{N} - (1 - w_{ji}))(1 - \sum_{k=1}^{N} 1 - w_{ki} - I_i) \tag{47}$$

so

$$\sum_{j=1}^{N} w_{ji} \rightarrow N - (1 - I_i). \tag{48}$$

Therefore, because weights to the ith node remain equal as they decay:

$$w_{ji} \rightarrow 1 - \left(\frac{1 - I_i}{N} \right) \tag{49}$$

In other words, the threshold $\tau_{ji} \equiv 1 - w_{ji}$ rises from 0 until:

$$\tau_{ji} \rightarrow \left(\frac{1 - I_i}{N} \right) \tag{50}$$

Thus $\tau_{ji} \in [0, \frac{1}{N}]$ after learning. The total weight change at the ith node is

$$\Delta \sum_{j=1}^{N} w_{ji} = 1 - I_i \tag{51}$$

Like the weights, the sum of all threshold changes at the ith node is less than or equal to $1 - I_i$.

8. TRANSMISSION RULES, CATASTROPHIC FORGETTING, AND STABLE CODING

Compare now the different asymptotic weights learned under the maximally distributed F_2 code (Eq. (34)) using the three synaptic transmission rules. For all three rules the total top-down signal σ_i converges to the bottom-up signal I_i at each F_1 node i. However, the total weight changes vary dramatically (Fig. 15.4b), in contrast to the F_2 choice case, where the maximum total weight change at a given node equals $(1 - I_i) \in [0, 1]$ for all three rules.

Product Rule —Catastrophic Forgetting. With distributed F_2 activity and a product rule, all weights w_{ji} converge to I_i and the maximum total weight change is $N(1 - I_i) \in [0,N]$. The full range of all weight values is thus spanned upon presentation of the very first input. In particular, all weights w_{ji} ($j = 1, ..., N$) to the ith F_1 node decay to 0 if $I_i = 0$. Because weight values can only decrease during learning, these weights would remain equal to zero for all time. Moreover, the top-down prime hypothesis (Eq. (4)) implies that F_1 activity x_i would then always be zero for any future input I and any F_2 code y. Thus, the fact that a given component was zero on just one input interval would render that component useless for all future input presentations, unable to be registered in LTM or even in STM. Similarly, each $I_i = I_i^{(1)}$ value of the first input would set an upper bound on all future x_i values, because

$$
\begin{aligned}
x_i \le \sigma_i &= \sum_{j=1}^{N} y_j w_{ji} \\
&\le I_i^{(1)} \sum_{j=1}^{N} y_j = I_i^{(1)}
\end{aligned}
\tag{52}
$$

for any F_2 code y. If a sequence of inputs $I^{(1)}, I^{(2)}, ...$ were to activate the fully distributed code (Eq. (34)), each weight w_{ji} would converge toward the minimum of $I_i^{(1)}, I_i^{(2)}, ...$. Within a few input presentations, all weights w_{ji} would in, all likelihood, decay toward zero. This problem occurs for any distributed code y. In this sense, the product rule leads to catastrophic forgetting.

Capacity rule —Even-More-Catastrophic Forgetting. The situation with the capacity rule is even worse (Fig. 15.4b). When the F_2 code is fully distributed, all weights w_{ji} decay to $\dfrac{I_i}{N} \in [0, \dfrac{1}{N})$ unless $I_i = 1$; and the maximum total weight change at the ith node is $N(1 - I_i)$. Thus, unless I is a binary vector, the entire dynamic range of weight values is nearly exhausted upon the first input presentation.

Threshold Rule — Stable Coding. It is the adaptive threshold rule alone that limits the total weight change to $(1 - I_i) \in [0, 1]$ for maximally distributed as well as maximally compressed codes y. In fact, if y is any F_2 code that becomes active when all w_{ji} are initially equal to 1, then

$$
w_{ji} \rightarrow 1 - y_j(1 - I_i)
\tag{53}
$$

as in Eq. (49). Equivalently,

$$
\tau_{ji} \rightarrow y_j (1 - I_i)
\tag{54}
$$

by Eq. (26). Thus the total weight/threshold change at each F_1 node i is bounded by $1 - I_i$ for any code, provided only that y is normalized. An F_2 code y would typically be highly distributed, with all y_j close to $1/N$, when a system has no strong evidence to choose one category j over another. In this case, the change of each threshold τ_{ji} is automatically limited to the narrow interval $[0, y_j]$, reserving most of the dynamic range for subsequent encoding. Only when evidence strongly supports selection of the F_2 category node J over all others, with y_J therefore close to 1, would weights be allowed to vary across most of their dynamic range. In particular, it is only when y_j is close to 1 that a weight w_{ji} is able to drop, irreversibly, toward 0, if I_i is small. Even with fast learning, other weights w_{ji} to the ith node then remain large, even if all $y_j > 0$. This is because, by Eqs. (14) and (25), weight changes cease altogether when

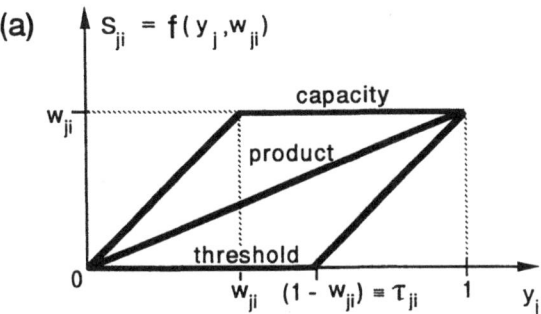

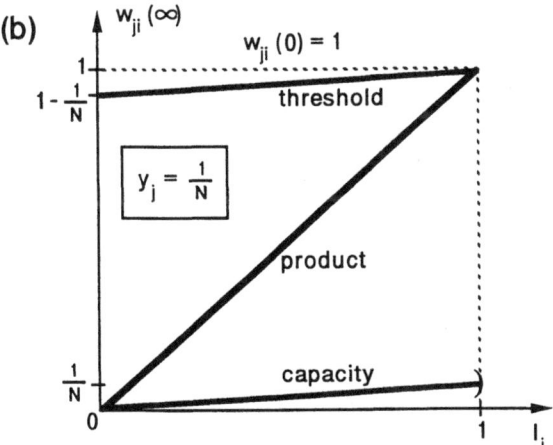

Fig. 15.4. (a) A synaptic transmission parallelogram. S_{ji} is the transmitted signal from the jth F_2 node to the ith F_1 node. By the product rule, $S_{ji} = y_j w_{ji}$. By the capacity rule, $S_{ji} = y_j \wedge w_{ji}$. By the threshold rule, $S_{ji} = [y_j - (1 - w_{ji})]^+ = [y_j - \tau_{ji}]^+$. The three rules agree when **y** is a binary code. (b) Asymptotic weight values for a fully distributed code, where $y_j = 1/N$. As a function of I_i, the dynamic range of $w_{ji}(\infty)$ depends critically upon the choice of synaptic transmission rule. During learning, weights decrease, from an initial value of $w_{ji}(0) = 1$, except when $I_i = 1$.

$$y_j \leq 1 - w_{ji} \equiv \tau_{ji} \tag{55}$$

The adaptive threshold τ_{ji} thereby replaces strong F_2 competition as the guardian, or stabilizer, of previously learned codes.

9. CONFIDENCE-PLASTICITY TRADEOFF

Figure 15.5 illustrates why the product rule and the capacity rule cause catastrophic forgetting and how the threshold rule solves this problem. During atrophy-due-to-disuse learning, if the ith F_1 target node is disused ($\sigma_i > x_i$), then the weight w_{ji} will decay in any path that sends a signal to the ith node ($S_{ji} > 0$) (Fig. 15.5a). When F_2 makes a choice, each of the three synaptic transmission rules allows weight change in only one path to each target node. However, if y_j is even slightly positive, both the product rule (Fig. 15.5b) and the capacity rule (Fig. 15.5c) allow weights w_{ji} to decay without limit, unless learning rates are very slow. In contrast, the threshold rule (Fig. 15.5d) implies that, even if the Jth F_2 node is active, the signal S_{ji} is still zero if the path threshold is large ($\tau_{ji} \geq y_j$); or, equivalently, if the path weight is small ($w_{ji} \leq 1 - y_j$). Only the positive signals S_{ji} sum to σ_i and only these signals can atrophy due to disuse. Threshold τ_{ji} remains small, and therefore plastic, if y_j is always small when $\sigma_i > x_i$. If y_J is large, τ_{ji} may increase toward 1. Once this occurs, however, $S_{ji} = 0$ for all F_2 codes $\mathbf{y}$ except those that compress most activity at the Jth node. Thus in a recognition system that allows an F_2 node to become highly active only when it is highly confident of its choice, the threshold rule automatically links confidence to stability. Conversely, when category selection is uncertain, distributed codes retain plasticity.

10. DISTRIBUTED OUTSTAR LEARNING

Computational analysis of distributed spatial pattern learning leads to selection of a synaptic transmission rule with an adaptive threshold. In terms of the threshold τ_{ji} in the path from the jth F_2 node to the ith F_1 node, a stable learning law for distributed codes is defined as the

$$\text{Distributed outstar:} \quad \frac{d\tau_{ji}}{dt} = S_{ji}(\sigma_i - x_i) \tag{56}$$

where S_{ji} is the thresholded path signal $[y_j - \tau_{ji}]^+$ transmitted from the jth F_2 node to the ith F_1 node and σ_i is the sum

$$\sigma_i = \sum_{j=1}^{N} S_{ji} = \sum_{j=1}^{N} [y_j - \tau_{ji}]^+. \tag{57}$$

Initially,

$$\tau_{ji}(0) = 0. \tag{58}$$

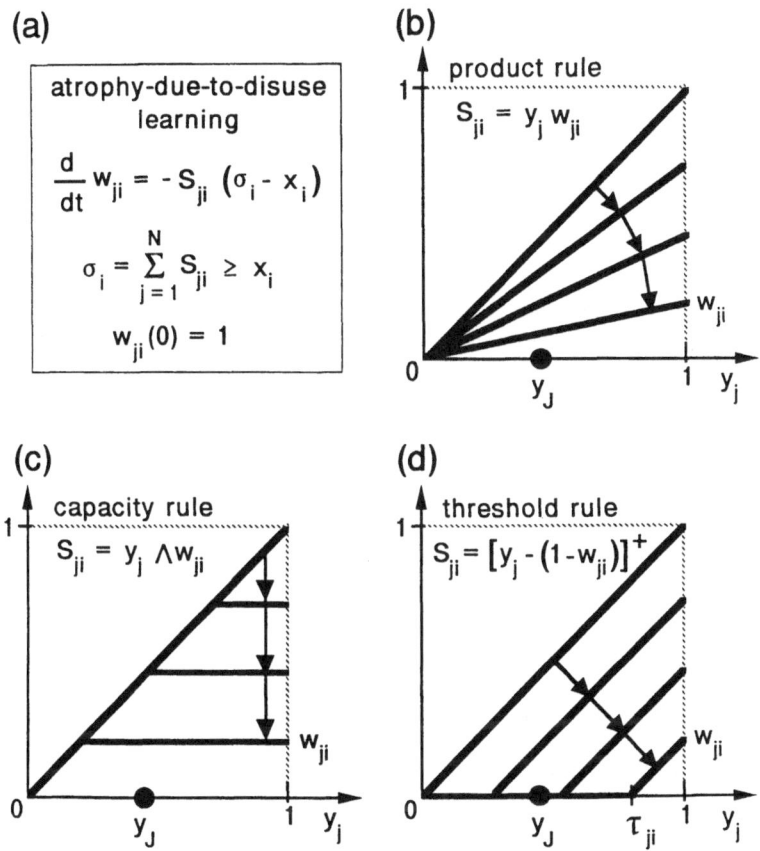

Fig. 15.5. (a) Atrophy-due-to-disuse learning causes a weight w_{ji} to decay at a rate proportional to (i) the signal from the jth F_2 node to the ith F_1 node and (ii) the degree of disuse, which equals the difference between total $F_2 \to F_1$ signal to the ith node and activity of that node. (b) When the Jth F_2 node is active, the product rule implies that the signal S_{Ji} to the ith F_1 node is positive. All weights w_{ji} therefore decay when $\sigma_i > x_i$, even if those weights are already small, causing catastrophic forgetting. (c) The capacity rule leads to catastrophic forgetting for the same reason as the product rule. (d) The threshold rule buffers learned codes against catastrophic forgetting by allowing only paths with sufficiently large weights (small thresholds) to contribute to the recognition code and hence to be subject to change during learning.

In a system such as ART 1 or fuzzy ART, the total top-down signal primes F_1. That is, σ_i is always greater than or equal to x_i. In the example computed in Section 12, $x_i = I_i \wedge \sigma_i$ (see Eq. (9)), so this hypothesis is satisfied. When $\sigma_i \geq x_i$, the distributed outstar allows thresholds τ_{ji} to grow but never shrink. The principle of atrophy due to disuse implies that a threshold τ_{ji} is unable to change at all unless (a) the path signal y_j exceeds the previously learned value of τ_{ji}; and (b) the total top-down signal σ_i to the ith node exceeds that node's activity x_i. In particular, if τ_{ji} grows large when the node j represents part of a compressed F_2 code, then τ_{ji} cannot be changed at all when node j is later part of a more distributed code, because threshold changes are disabled if $y_j \le \tau_{ji}$ (Fig. 15.5d).

11. DISTRIBUTED OUTSTAR SOLUTION

The form of the distributed outstar system (Eqs. (56)-(58)) is so simple that the equations can be solved in closed form. The formulas below give an explicit solution for an arbitrary input sequence with either slow or fast learning. Section 12 illustrates the geometry of this solution. Assume that an input $\mathbf{I}$ activates a distributed outstar field F_1 at some time $t = t_0$ and that $\mathbf{I}$ is held fixed for some ensuing interval. If $\sigma_i \le x_i$ at $t = t_0$, then τ_{ji} will remain constant during that interval, for all $j = 1, ..., N$. Similarly, τ_{ji} will remain constant if $y_j \le \tau_{ji}$ at $t = t_0$. Consider now a fixed F_1 index i such that $\sigma > x_i$ at $t = t_0$. Let

$$\Phi_i = \{j: y_j(t_0) > \tau_{ji}(t_0)\}. \tag{59}$$

For $j \in \Phi_i$,

$$\frac{d}{dt}\tau_{ji} = (y_j - \tau_{ji})(\sigma_i - x_i) \tag{60}$$

until y_j and x_i change. Geometrically, by Eq. (60), the projected vector of τ_{ji} values with $j \in \Phi_i$ follows a straight line toward the corresponding projected vector of y_j values. If all such τ_{ji} were to approach y_j then σ_i would converge to 0, by Eq. (57). Progress halts, however, as the τ_{ji} vector approaches the set of points where $\sigma_i = x_i$, by Eq. (60). Explicitly, for $t \ge t_0$, with y_j and x_i constant

$$\tau_{ji}(t) = \tau_{ji}(t_0) + \alpha(t)\frac{[\sigma_i(t_0) - x_i]^+}{\sigma_i(t_0)}[y_j - \tau_{ji}(t_0)]^+ \tag{61}$$

where $\alpha(t)$ is an exponential that goes from 0 to 1 as t goes from t_0 to ∞.

By Eq. (61), $\tau_{ji}(t)$ remains constant if $\sigma_i(t_0) \le x_i$ or if $y_j \le \tau_{ji}(t_0)$. If $\sigma_i(t_0) > x_i$ and if $j \in \Phi_i$, $\tau_{ji}(t)$ moves from $\tau_{ji}(t_0)$ toward

$$\tau_{ji}(\infty) = \tau_{ji}(t_0) + \frac{(\sigma_i(t_0) - x_i)}{\sigma_i(t_0)}(y_j - \tau_{ji}(t_0)) \tag{62}$$

as t goes from t_0 to ∞. In particular

$$\sigma_i(\infty) = \sum_{j \in \Phi_i} (y_j - \tau_{ji}(\infty))$$

$$= \sum_{j \in \Phi_i} (y_j - \tau_{ji}(t_0)) - \frac{(\sigma_i(t_0) - x_i)}{\sigma_i(t_0)} \sum_{j \in \Phi_i} (y_j - \tau_{ji}(t_0)) \tag{63}$$

$$= \sigma_i(t_0) - \frac{(\sigma_i(t_0) - x_i)}{\sigma_i(t_0)} \sigma_i(t_0)$$

$$= x_i$$

For the unbiased case where $t_0 = 0$, so all $\tau_{ji}(0) = 0$,

$$S_{ji}(0) \equiv y_j - \tau_{ji}(0) = y_j \tag{64}$$

and

$$\sigma_i(0) \equiv \Sigma_j S_{ji}(0) = \Sigma_j y_j = 1. \tag{65}$$

Thus

$$\tau_{ji}(t) = \tau_{ji}(0) + \alpha(t) \frac{[\sigma_i(0) - x_i]^+}{\sigma_i(0)} [y_j - \tau_{ji}(0)]^+, \tag{66}$$

$$= \alpha(t)(1 - x_i)y_j$$

$$S_{ji}(t) \equiv y_j - \tau_{ji}(t) = y_j - \alpha(t)(1 - x_i)y_j \tag{67}$$

$$= y_j(1 - \alpha(t)(1 - x_i))$$

and

$$S_{ji}(t) \rightarrow y_j x_i \tag{68}$$

as $t \rightarrow \infty$. By Eq. (68), when the system begins with no initial bias, the signal S_{ji} from the jth F_2 node to the ith F_1 node begins as y_j and converges toward the Hebbian pre- and post-synaptic correlation term $y_j x_i$.

12. DISTRIBUTED OUTSTAR DYNAMICS

The dynamics of distributed outstar learning are now illustrated by means of a low-dimensional example. Consider a coding network with just two F_2 nodes (Fig. 15.6(a)). Two top-down paths, with thresholds τ_{1i} and τ_{2i}, converge on each F_1 node. Assume that $x_i = I_i \wedge \sigma_i$, as in Eq. (9), and fix an F_2 code $y = (y_1, y_2)$, with

$$0 \le y_2 \le y_1 \le 1. \tag{69}$$

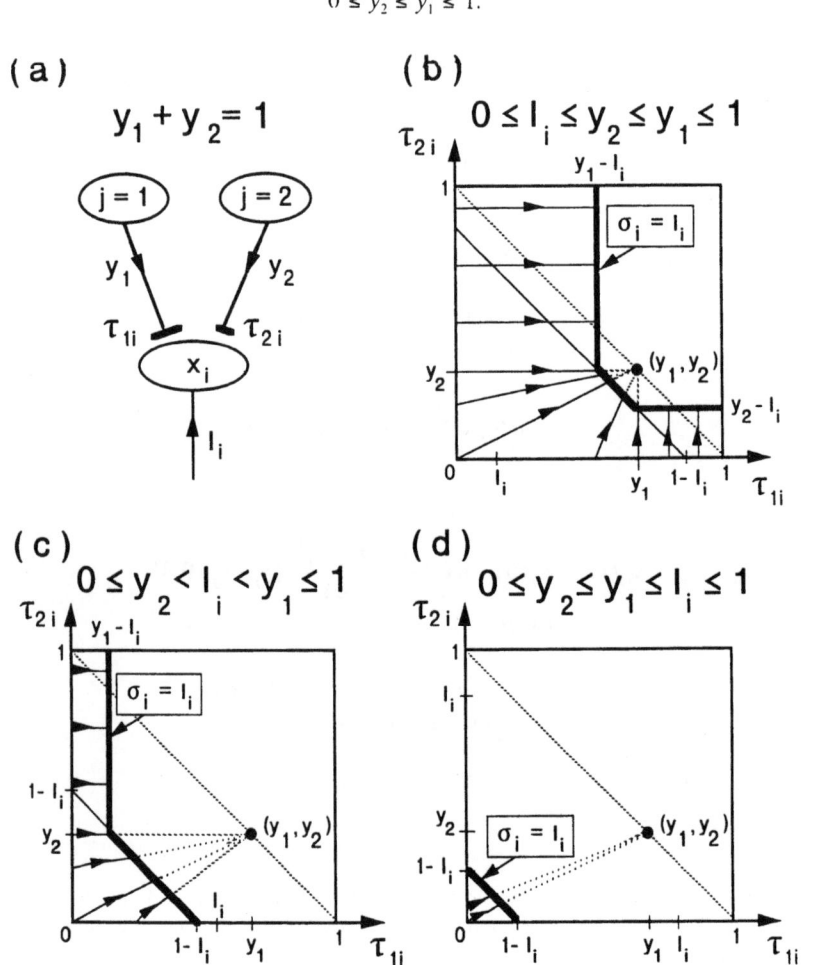

Fig. 15.6. (a) A distributed outstar whose coding field F_2 has two nodes. For each code y, $y_1 + y_2 = 1$ and $x_i = I_i \wedge \sigma_i$. When thresholds start out small enough, τ_{1i} and/or τ_{2i} increases toward $\{(\tau_{1i}, \tau_{2i}): \sigma_i = I_i\}$. (b) Threshold changes are greatest for small I_i. (c) When $I_i > y_j$, the jth node cannot dominate learning. Here, $I_i > y_2$, so τ_{2i} can change only when τ_{1i} also changes. (d) When I_i is large, only small thresholds can change at all.

By the F_2 normalization hypothesis (Eq. (20)), $y_1 + y_2 = 1$. By Eqs. (10), (11), (27), and (56), for $j = 1, 2$

$$\frac{d}{dt} \tau_{ji} = [y_j - \tau_{ji}]^+ [\sigma_i - I_i]^+ \tag{70}$$

where, by Eq. (57),

$$\sigma_i = [y_1 - \tau_{1i}]^+ + [y_2 - \tau_{2i}]^+. \tag{71}$$

Figure 15.6(b-d) shows the two-dimensional phase plane dynamics of the threshold vector (τ_{1i}, τ_{2i}) for a fixed input I_i. In each plot, trajectories that begin in the set of points where $\sigma_i > I_i$ approach the set where $\sigma_i = I_i$. Where $\tau_{1i}(t_0) < y_1$ and $\tau_{2i}(t_0) < y_2$, the point $(\tau_{1i}(t), \tau_{2i}(t))$ moves along a straight line from $(\tau_{1i}(t_0), \tau_{2i}(t_0))$ toward (y_1, y_2), slowing down asymptotically as

$$\begin{aligned}\sigma_i &= [y_1 - \tau_{1i}(t)]^+ + [y_2 - \tau_{2i}(t)]^+ \\ &= 1 - (\tau_{1i}(t) + \tau_{2i}(t)) \rightarrow I_i\end{aligned} \tag{72}$$

Only if $I_i = 0$ does (τ_{1i}, τ_{2i}) approach (y_1, y_2). Larger thresholds τ_{ji}, which make $\sigma_i \leq I_i$, are unchanged during learning. Small I_i values allow the greatest threshold changes (Fig. 15.6b). If $I_i = 0$,

$$\tau_{ji} \rightarrow y_j \tag{73}$$

as σ_i decreases to 0. Both thresholds grow if both are initially small. However, if one threshold is so large as to prevent $F_2 \rightarrow F_1$ signal transmission in the corresponding path, the other F_2 node "takes over" the code. For example, if $\tau_{2i}(t_0) \geq y_2$ there is no signal from the F_2 node $j = 2$ to the ith F_1 node, hence no threshold change in that path. If, then, $\tau_{1i}(t_0) < y_1 - I_i$, τ_{1i} increases until

$$\sigma_i = y_1 - \tau_{1i} \rightarrow x_i = I_i. \tag{74}$$

Larger I_i values permit threshold changes only for smaller initial threshold values. In Fig. 15.6c, τ_{2i} can change only if τ_{1i} changes as well, when both are initially small. In contrast, because y_1 is greater than I_i, τ_{1i} may increase, by itself, toward $y_1 - I_i$. Finally, for I_i close to 1 (Fig. 15.6d), adaptive changes can occur only if both τ_{1i} and τ_{2i} are initially small, as they are before any learning has taken place.

13. THE UNIT OF MEMORY

The distributed outstar network derives from a computational analysis of stable pattern learning by distributed codes. In the distributed outstar, the adaptive threshold rule of synaptic transmission solves a catastrophic forgetting problem caused by other rules. Because each formal transmission rule corresponds to a physical theory of synaptic transmission, computational analysis implies physiological prediction. Each transmission rule assumes a physical memory unit: a multiplicative weight (Fig. 15.7a), a fuzzy capacity or sieve (Fig. 15.7b), or a subtractive threshold (Fig. 15.7c). Experiments that probe distributed coding in a living organism may be able to test for the three types of memory unit. Similarly, distributed outstar computations imply distinct physical realizations of optical and electronic neural networks.

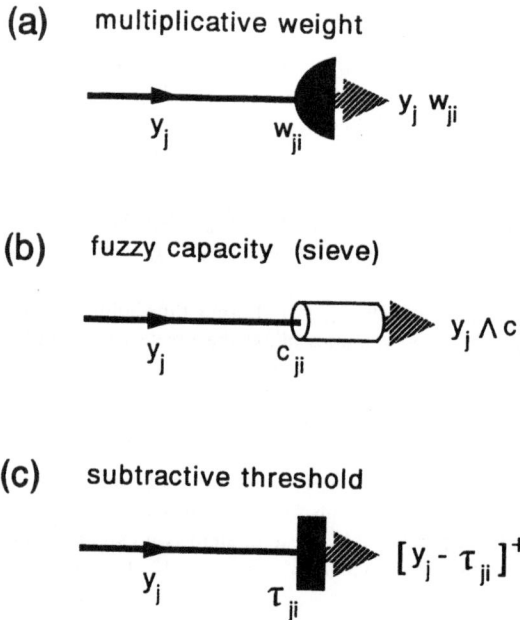

(a) multiplicative weight

y_j w_{ji} $y_j \; w_{ji}$

(b) fuzzy capacity (sieve)

y_j c_{ji} $y_j \wedge c_{ji}$

(c) subtractive threshold

y_j τ_{ji} $[y_j - \tau_{ji}]^+$

Fig. 15.7. (a) The product rule implies a physical substrate of memory that is a multiplicative weight (McCulloch & Pitts, 1943/1988). (b) The capacity rule implies a memory unit that is a fuzzy sieve (Zadeh, 1965). (c) The distributed outstar implies a memory unit that is a subtractive threshold.

ACKNOWLEDGMENTS

This research was supported in part by the National Science Foundation (NSF IRI-94-01659) and the Office of Naval Research (ONR N00014-95-1-0409 and ONR N00014-95-0657).

REFERENCES

Carpenter, G. A. (1993). Distributed outstar learning and the rules of synaptic transmission. *Proceedings of the World Congress on Neural Networks* (Vol. II, pp. 397-404). Hillsdale, NJ: Lawrence Erlbaum Associates.

Carpenter, G. A. (1994). A distributed outstar network for spatial pattern learning. *Neural Networks*, 7, 159-168.

Carpenter, G. A., & Grossberg, S. (1987a). A massively parallel architecture for a self-organizing neural pattern recognition machine. *Computer Vision, Graphics, and Image Processing*, 37, 54-115. Reprinted in G. A. Carpenter & S. Grossberg (Eds.) (1991), *Pattern Recognition by Self-Organizing Neural Networks* (pp. 316-382). Cambridge, MA: MIT Press.

Carpenter, G. A., & Grossberg, S. (1987b). ART 2: Stable self-organization of pattern recognition codes for analog input patterns. *Applied Optics*, 26, 4919-4930. Reprinted in G. A. Carpenter & S. Grossberg (Eds.) (1991), *Pattern Recognition by Self-Organizing Neural Networks* (pp. 398-423). Cambridge, MA: MIT Press.

Carpenter, G. A., & Grossberg, S. (1990). ART~3: Hierarchical search using chemical transmitters in self-organizing pattern recognition architectures. *Neural Networks*, 3, 129-152. Reprinted in G. A. Carpenter & S. Grossberg (Eds.) (1991), *Pattern Recognition by Self-Organizing Neural Networks* (pp. 451-499). Cambridge, MA: MIT Press pp. 451-499.

Carpenter, G. A., & Grossberg, S. (1994). Fuzzy ARTMAP: A synthesis of neural networks and fuzzy logic for supervised categorization and nonstationary prediction. In R. R. Yager & L. A. Zadeh (Eds.), *Fuzzy Sets, Neural Networks, and Soft Computing* (pp. 126-165). New York: Van Nostrand Reinhold.

Carpenter, G. A., Grossberg, S., & Reynolds, J. H. (1991). ARTMAP: Supervised real-time learning and classification of nonstationary data by a self-organizing neural network. *Neural Networks*, 4, 565-588. Reprinted in G. A. Carpenter & S. Grossberg (Eds.) (1991), *Pattern Recognition by Self-Organizing Neural Networks* (pp. 503-546). Cambridge, MA: MIT Press.

Carpenter, G. A., Grossberg, S., & Rosen, D. B. (1991a). Fuzzy ART: Fast stable learning and categorization of analog patterns by an adaptive resonance system. *Neural Networks*, 4, 759-771.

Carpenter, G. A., Grossberg, S., & Rosen, D. B. (1991b). *A neural network realization of fuzzy ART* (Tech. Rep. No. CAS/CNS TR-91-021). Boston University.

Estes, W. K. (1955). Statistical theory of spontaneous recovery and regression. *Psychological Review*, 62, 145-154.

Grossberg, S. (1968a). Some nonlinear networks capable of learning a spatial pattern of arbitrary complexity. *Proceedings of the National Academy of Sciences*, 59, 368-372.

Grossberg, S. (1968b). A prediction theory for some nonlinear functional-differential equations, I. Learning of lists. *Journal of Mathematical Analysis and Applications*, 21, 643-694.

Grossberg, S. (1969). Some networks that can learn, remember, and reproduce any number of complicated space-time patterns, I. *Journal of Mathematics and Mechanics*, 19, 53-91.

Grossberg, S. (1972). Pattern learning by functional-differential neural networks with arbitrary path weights. In K. Schmitt (Ed.), *Delay and Functional-Differential Equations and Their Applications* (pp. 121-160). New York: Academic Press. Reprinted in S. Grossberg (Ed.) (1982), *Studies of Mind and Brain* (pp. 159-193). Dordrecht, Holland: Reidel.

Grossberg, S. (1973). Contour enhancement, short term memory, and constancies in reverberating neural networks. *Studies in Applied Mathematics*, **52**, 217-257. Reprinted in S. Grossberg (Ed.) (1982), *Studies of Mind and Brain* (pp. 334-378). Dordrecht, Netherlands: Reidel.

Grossberg, S. (1974). Classical and instrumental learning by neural networks. In R. Rosen & F. Snell (Eds.), *Progress in Theoretical Biology*, Volume 3, pp. 51-141. New York: Academic Press. Reprinted in S. Grossberg (Ed.) (1982), *Studies of Mind and Brain* (pp. 68-156). Dordrecht, Netherlands: Reidel.

Grossberg, S. (1976). Adaptive pattern classification and universal recoding, II: Feedback, expectation, olfaction, illusions. *Biological Cybernetics*, **23**, 187-202. Reprinted in G. A. Carpenter & S. Grossberg (Eds.) (1991), *Pattern Recognition by Self-Organizing Neural Networks* (pp. 283-315). Cambridge, MA: MIT Press.

Hartline, H. K., & Ratliff, F. (1957). Inhibitory interactions of receptor units in the eye of *Limulus*. *Journal of General Physiology*, **40**, 351-376.

McCulloch, W. S., & Pitts, W. (1943). A logical calculus of the ideas immanent in nervous activity. *Bulletin of Mathematical Biophysics*, **5**, 115-133. Reprinted in J. A. Anderson & E. Rosenfeld (Eds.) (1988), *Neurocomputing: Foundations of Research* (pp. 18-27). Cambridge, MA: MIT Press.

Rosenblatt, F. (1958). The perceptron: A probabilistic model for information storage and organization in the brain. *Psychological Review*, **65**, 386-408. Reprinted in J. A. Anderson & E. Rosenfeld (Eds.) (1988), *Neurocomputing: Foundations of Research* (pp. 92-114). Cambridge, MA: MIT Press.

Rosenblatt, F. (1962). *Principles of Neurodynamics*. Washington, DC: Spartan Books.

Rumelhart, D. E., Hinton, G. E., & Williams, R. J. (1986). Learning internal representations by error propagation. In D. E. Rumelhart & J. L. McClelland (Eds.), *Parallel Distributed Processing. Volume 1: Foundations* (pp. 318-362). Cambridge, MA: MIT Press.

Widrow, B., & Hoff, M. E. (1960). Adaptive switching circuits. *1960 IRE WESCON Convention Record* (pp. 96-104). New York: IRE. Reprinted in J. A. Anderson & E. Rosenfeld (Eds.) (1988), *Neurocomputing: Foundations of Research* (pp. 126-134). Cambridge, MA: MIT Press.

Zadeh, L. (1965). Fuzzy sets. *Information Control*, **8**, 338-353.

16

A Generalized Autoassociator Model for Face Processing and Sex Categorization: From Principal Components to Multivariate Analysis

Hervé Abdi
University of Texas at Dallas and Université de Bourgogne à Dijon

Dominique Valentin and Alice J. O'Toole
University of Texas at Dallas

The chapter by Hervé Abdi, Dominique Valentin, and Alice O'Toole, *A Generalized Autoassociator Model for Face Processing and Sex Categorization: From Principal Components to Multivariate Analysis*, deals with a generalization of the autoassociator model due to James Anderson and his colleagues. The generalization Abdi et al. perform is essentially one of selective attention, that is, choosing which features to emphasize in a particular classification problem — in this case, classifying faces (represented by visual pixel patterns) as male or female. Thus the ultimate problem can be posed mathematically as a combination of **two** optimization problems. One is the optimal classification problem for the autoassociator itself, which is a subcase of a general problem discussed in Golden's chapter. The other is the optimal choice of coefficients, corresponding to features, in the autoassociation matrix. The problem of bias between features is dealt with by other chapters in this book, notably those of Leven and of Öğmen and Prakash.

Selective attention is used by Abdi et al. to determine which features are most distinctive in deciding the right categories for faces. The criterion they employ is that those features occuring **least frequently** are attended to the most. The rationale is that these are the most distinctive, and therefore decisive, features. The result of feature selection is to improve not the *autoassociator's accuracy but its speed of convergence. This is one of many possible ways to use selective attention in neural network models. In the chapters of Öğmen and Prakash, and of Levine, decisions on which features or attributes to emphasize are made on the basis of exploration mediated by outside reinforcement. In the work on consumer preference described in Leven's chapter, decisions on which attributes to emphasize are heavily influenced by the mood of the decision maker. The dynamics of selective attention within a feature module, however, might well be similar across all these cases, the differences lying in how this attention is influenced by connections from other modules of the network.*

ABSTRACT

In this chapter we propose a generalized version of the classical linear autoassociator that can be shown to implement a generalized least-squares approximation under linear constraints. The standard linear autoassociator is known to implement principal component analysis, whereas the generalized model implements the general linear model (e.g., canonical correlation). In practical terms, this generalization allows for the imposition of *a priori* constraints that enable differential weighting of both individual units of the input code and individual stimuli. As an illustration of the utility of the generalized model, we present simulations comparing the accuracy and learning speed of the standard and generalized versions of the autoassociator for the problem of categorizing faces by sex. We show that while the two models are equally accurate, the generalized model learns the task considerably faster than does the standard model.

1. INTRODUCTION

Recent years have witnessed a strong resurgence of interest in the field of neural networks. Some of the earliest models characterizing this resurgence were simple linear associative memory models (Anderson, Silverstein, Ritz, & Jones, 1977; Kohonen, 1977). Associative memories are capable of learning associations between input-output pairs such that the memory produces the appropriate output in response to a learned or "associated" input. The inner workings of these associative models and other related "neural" network models are reminiscent of the computational character of the brain. Specifically, the computations required to implement the storage and retrieval of information in the network can be carried out in parallel and the representation of individual learned associations is not localized in the memory, but rather, is "distributed" throughout the entire network.

The purpose of the present chapter is to propose a generalization of a particular case of a linear associator model known as an *auto*associator. The autoassociator can act as a content-addressable memory in the sense that it learns to associate inputs to themselves. As such, the model can operate also as a powerful pattern completion device, capable of reconstructing learned input stimuli with memory keys that have been degraded either by adding noise or by ablating parts of the code. Kohonen (1977), for example, showed that an autoassociative memory can be used to store images of human faces and reconstruct the original faces when features have been omitted or degraded.

When a linear autoassociative memory is viewed as a "neural network," the values in the weight matrix correspond to the connection strengths between the cells or units of the memory. Learning, in this framework, amounts to finding a set of connections between input units that minimizes the error in reconstructing the input stimuli. In the "standard" or "classical" autoassociator, all the units composing the memory are equivalent and independent, and all the stimuli to be stored in the memory are of equal importance. Although this kind of model is capable of solving many pattern recognition problems, other problems require additional constraints. Specifically, many real pattern recognition problems operate under well-established a priori constraints that can function either at the level of differentiating parts of the code and/or at the level of differentiating individual stimuli as a function of their "importance" in building the model representation.

One example of the way in which different parts of a code can be differentially important for solving a problem can be seen in the representational constraints that are implemented in many biological vision systems to enhance luminance contrast. Such constraints are required in order to make optimal use of the strongly limited bandwidth of the optic nerve, which transmits information from the retina to the cortex. The neural scheme operating in these visual systems capitalizes on the fact that individual parts of the retinal code are not equally informative. For example, areas of the retina that contain information about luminance contrast are more important than areas of the retina that are uniformly illuminated. The generalization of the autoassociator that we propose allows for a mechanism by which individual *units* of the memory can be assigned differential importance.

In addition to implementing representational constraints, it can be useful to implement a mechanism that allows for a differential weighting of individual stimuli. For example, to model human memory, it is often necessary to take into account factors that differentiate the importance of individual stimuli in building a representation of the problem. In the learning of lists, temporal interference between successive items is one such factor. This phenomenon can be easily simulated by implementing a differential weighting of the stimuli as a function of their position on the list. For example, in the case of retroactive interference, a more recent stimulus interferes with the memory of previously learned stimuli. More precisely, the importance of any given stimulus is inversely proportional to its position in the learning sequence. In addition to the differential weighting of parts of the code, the generalized autoassociator allows for the implementation of a priori biases in the stimulus set. The differential importance (and nonindependence) of both the units and the stimuli are defined as a set of constraints expressed via positive definite matrices operating on the autoassociator.

The classical linear autoassociator has been often analyzed in terms of the eigendecomposition or singular value decomposition of a matrix. Specifically, it has been shown that storing stimuli in an autoassociative memory amounts to creating the cross-product matrix of the stimuli and computing its eigendecomposition (Abdi, 1988, 1993, 1994b; Anderson et al., 1977; Kohonen, 1977). This is equivalent to computing the principal component analysis of the set of features used to describe the stimuli.

One advantage of this type of analysis is that it makes it clear that classical autoassociators implement least-squares approximation (or Wiener filtering; cf. Abdi, 1994a). In terms of optimization problems, the interest of the generalized autoassociator described in this chapter is that it implements a generalized least-squares approximation or a least-squares approximation under (linear) constraints. This technique is used in various settings. In multivariate statistical analysis, for example, canonical analysis (and hence the complete set of generalized linear models) can be easily derived within this framework (e.g., Mardia, Kent, & Bibby, 1979; Greenacre, 1984). As a consequence, neural networks can be easily shown to be equivalent to traditional statistical and optimization techniques.

This chapter is organized as follows. First, the basic features of the classical autoassociative model are briefly presented along with their relationship to the linear model of multivariate analysis. Then a generalization of this model is described and analyzed in terms of statistical and optimization problems. Specifically, we demonstrate that a generalized linear autoassociator implements the general linear model of multivariate statistics. Finally, we show that a linear generalized autoassociator implementing correspondence analysis (i.e., a specific case of the general linear model) can be used successfully to categorize a set of faces according to their sex. We show, in this specific application, that the generalized version of the autoassociator can learn the task as accurately as the standard version, but does so more quickly.

2. CLASSICAL MODEL

Objects to be stored in an autoassociative memory are represented by $I \times 1$ column vectors $\mathbf{x}_k$ whose I components code the values of the I features used to describe the objects. In a neural network implementation, these components represent the activation of the input units (i.e., cells). For convenience, the vectors $\mathbf{x}_k$ are assumed to be normalized so that $\mathbf{x}_k^T \mathbf{x}_k = 1$ (with $\mathbf{x}_k^T$ denoting the transpose of $\mathbf{x}_k$). The set of K stimuli to be stored in the memory is represented by an $I \times K$ matrix $\mathbf{X}$ in which the kth column is equal to $\mathbf{x}_k$. The autoassociative memory (or weight matrix) is represented by an $I \times I$ matrix $\mathbf{W}$. The values in the weight matrix correspond to the connection strengths between the units of the memory.

The stimuli are stored in the memory by changing the strength of the connections between units. This can be done using a simple Hebbian learning rule:

$$\mathbf{W} = \sum_{k=1}^{K} \mathbf{x}_k \mathbf{x}_k^T = \mathbf{X}\mathbf{X}^T. \tag{1}$$

Recall of a given stimulus $\mathbf{x}_k$ is given by $\hat{\mathbf{x}}_k = \mathbf{W}\mathbf{x}_k$, where $\hat{\mathbf{x}}_k$ represents the response of the memory. The quality of the response of the system can be measured by computing the cosine of the angle between $\mathbf{x}_k$ and $\hat{\mathbf{x}}_k$:

$$\cos(\mathbf{x}_k, \hat{\mathbf{x}}_k) = \frac{\mathbf{x}_k^T \hat{\mathbf{x}}_k}{\|\mathbf{x}_k\| \|\hat{\mathbf{x}}_k\|} \tag{2}$$

where $\|\mathbf{x_k}\|$ is the Euclidean norm of the vector $\mathbf{x_k}$ (i.e., $\|\mathbf{x_k}\| = \sqrt{\mathbf{x_k^T x_k}}$). A cosine of 1 indicates a perfect reconstruction of the stimulus.

When the stimulus set is composed of nonorthogonal stimuli, the associator does not perfectly reconstruct the stimuli that are stored. On the other hand, some new patterns are perfectly reconstructed, creating, in a way, the equivalent of a "false alarm" or "false recognition." These patterns are defined by the equation $\mathbf{Wu_r} = \lambda_r \mathbf{u_r}$, with $\mathbf{u_r^T u_r}$, where $\mathbf{u_r}$ denotes the rth eigenvector of $\mathbf{W}$, and λ_r the eigenvalue associated with that eigenvector.

From Eq. (1), it can be seen that the matrix $\mathbf{W}$ is equivalent to a cross-product matrix, and hence is positive semidefinite (i.e., all its eigenvalues are positive or zero). Consequently, $\mathbf{W}$ can be reconstructed as a weighted sum of its eigenvectors:

$$\mathbf{W} = \sum_{r=1}^{R} \lambda_r \mathbf{u_r u_r^T} = \mathbf{U \Lambda U^T} \text{ with } \mathbf{U^T U = I} \tag{3}$$

where $\mathbf{I}$ stands for the identity matrix, Λ represents the diagonal matrix of eigenvalues and R is the rank of the matrix $\mathbf{W}$. The eigenvectors in $\mathbf{U}$ are usually ordered according to their eigenvalues. This formulation makes clear the close relationship between the classical linear autoassociator and some techniques used in multivariate statistical analysis. Specifically, using an autoassociative memory to store and recall a set of objects is equivalent to performing a principal component analysis on the cross-product matrix of the feature set describing these objects (Anderson et al., 1977).

Associated with the technique of principal component analysis, is the notion of a distance. One way of looking at the eigendecomposition of the matrix $\mathbf{W}$ is to note that the Euclidean distance between stimuli, as well as the Euclidean distance between any stimulus and the average stimulus (i.e., the barycenter, or centroid, of the set of stimuli), is now decomposed orthogonally along the eigenvectors. Specifically, the Euclidean distance between stimuli k and k' is computed as:

$$d^2(\mathbf{x_k}, \mathbf{x_{k'}}) = (\mathbf{x_k} - \mathbf{x_{k'}})^T (\mathbf{x_k} - \mathbf{x_{k'}}). \tag{4}$$

The distance can be expressed also, through the eigendecomposition as

$$d^2(\mathbf{x_k}, \mathbf{x_{k'}}) = d^2(\mathbf{g_k}, \mathbf{g_{k'}}) = (\mathbf{g_k} - \mathbf{g_{k'}})^T (\mathbf{g_k} - \mathbf{g_{k'}}) \tag{5}$$

where $\mathbf{g_k}$ (respectively $\mathbf{g_{k'}}$) is the vector of the projections of stimulus k (respectively k') onto the eigenvectors. This suggests the use of principal component analysis to display the stimuli as they are "perceived" by the autoassociative memory.

A final point worth noting is that the eigenvectors and eigenvalues of the weight matrix $\mathbf{W}$ can be obtained directly using the singular value decomposition of the original matrix of stimuli $\mathbf{X}$. Formally,

$$X = U\Delta V^T \text{ with } V^T V = U^T U = I \tag{6}$$

where U represents the matrix of eigenvectors of XX^T, V represents the matrix of eigenvectors of $X^T X$, and Δ is the matrix of singular values that are equal to the square root of the eigenvalues of XX^T or $X^T X$ (they are the same). The projections, G, of the K stimuli of the training set on the R eigenvectors of the weight matrix can be found as

$$G = X^T U = V\Delta. \tag{7}$$

Within the framework of principal component analysis, G is the matrix of the projections of the stimuli on the principal components. From Eq. (7), it is easy to derive that the variance of the projections on a given eigenvector is equal to the eigenvalue associated with this eigenvector, that is, $G^T G = \Delta V^T V \Delta = \Lambda$. Likewise, the projections of a set of K' new stimuli (i.e., not learned by the memory), X_{new}, on the eigenvectors of W can be computed as $G_{new} = X_{new}^T U$. Within the framework of principal component analysis, G_{new} contains the projections of the supplementary elements (i.e., stimuli) on the principal components.

In order to improve the storage capacity of an autoassociative memory, most applications use the Widrow-Hoff learning rule. The Widrow-Hoff learning rule corrects the difference between the response of the system and the expected response by changing iteratively the weights in matrix W as follows:

$$W_{[t+1]} = W_{[t]} + \eta(X - W_{[t]}X)X^T \tag{8}$$

where η is a constant learning rate. The Widrow-Hoff learning rule can also be analyzed in terms of eigenvectors and eigenvalues (Abdi, 1994a). Hence, W at time t can be expressed as

$$W_{[t]} = U\Phi_{[t]}U^T \text{ with } \Phi_{[t]} = [1 - (1 - \eta\Lambda)^t]. \tag{9}$$

With a learning constant η smaller than $2\lambda_{max}^{-1}$ (λ_{max} being the largest eigenvalue), this procedure converges toward $W_{[\infty]} = UU^T$, which indicates that using the Widrow-Hoff error correction learning rule amounts to equalizing all the eigenvalues of W (i.e., to sphericizing the weight matrix).

3. GENERALIZED AUTOASSOCIATOR

3.1. Notation and Definition

First, the differential importance and nonindependence of both the stimuli and the cells of the memory allowed by the generalization are formalized as two sets of weights that correspond to the importance of individual stimuli and individual units (i.e., features describing the stimuli or equivalently memory cells), respectively. Specifically, the set of constraints imposed on the units is represented by a positive-definite matrix of order $I \times I$ denoted B. For example, if we want the importance of a unit to be inversely proportional to its use, B will be defined as the diagonal matrix of the inverse column margin of matrix $\mathbf{X}$ (i.e., $b_{i,i} = x_{i+}^{-1}$ with x_{i+} representing the total of the ith row of $\mathbf{X}$, or, equivalently, $x_{i+} = \Sigma_k \, x_{i,k}$). The set of constraints imposed on the stimuli is represented by a positive-definite matrix of order $K{\times}K$ denoted $\mathbf{M}$. For example, if we want to give a differential importance to each stimulus according to the value of its general activation, $\mathbf{M}$ will be defined as the diagonal matrix of the row margin of matrix $\mathbf{X}$ (i.e., $m_{k,k} = x_{+k}$ with x_{+k} representing the total of the kth column of $\mathbf{X}$, or, equivalently, $x_{+k} = \Sigma_i \, x_{i,k}$). Note that choices other than a diagonal matrix are possible for $\mathbf{B}$ and $\mathbf{M}$.

Second, in order to analyze the properties of the generalized autoassociator, we need to generalize some basic notions of Euclidean geometry. The generalized norm of vector $\mathbf{x}$, denoted $\mathbf{B}$-norm, is given by

$$\|\mathbf{x_k}\|_{\mathbf{B}} = \sqrt{\mathbf{x_k}^T \mathbf{B} \mathbf{x_k}}. \tag{10}$$

The biased orthogonality of the pair of vectors $\mathbf{x_k}$ and $\mathbf{x_{k'}}$, denoted $\mathbf{B}$-orthogonality, is given by:

$$\mathbf{x_k} \perp_{\mathbf{B}} \mathbf{x_{k'}} - \mathbf{x_k}^T \mathbf{B} \mathbf{x_{k'}} = 0. \tag{11}$$

The generalized cosine, denoted $\mathbf{B}$-cosine, is given by:

$$\cos_{\mathbf{B}} = \frac{\mathbf{x_k}^T \mathbf{B} \mathbf{x_{k'}}}{\|\mathbf{x_k}\|_{\mathbf{B}} \|\mathbf{x_{k'}}\|_{\mathbf{B}}}. \tag{12}$$

Finally, for convenience, the stimuli, x, are normalized in the metric defined by $\mathbf{B}$ (i.e., $\mathbf{x_k}^T \mathbf{B} \mathbf{x_k} = 1$).

3.2. Model Description

As in the classical model, the stimuli are stored in the memory by modifying the intensity of the connections between units, with the exception that during learning a differential importance is given to each stimulus. Formally, $\mathbf{W} = \mathbf{XMX}^T$. The effect of the constraints on the units (i.e., the bias matrix $\mathbf{B}$) can be interpreted as a filtering or recoding scheme for the original stimuli prior to storage in the memory. This effect can be modeled during the recall phase as a premultiplication of the stimuli by the matrix $\mathbf{B}$ before recall by multiplication through $\mathbf{W}$. Specifically, recall of a given stimulus $\mathbf{x}_l$ is obtained as

$$\hat{\mathbf{x}}_l = \mathbf{W} \mathbf{B} \mathbf{x}_l. \tag{13}$$

If the stimuli stored in the memory do not form a $\mathbf{B}$-orthogonal set, recall will not be perfect. The memory will add some noise (or cross-talk) to the original stimulus:

$$\hat{x}_1 = WBx_1 = \sum_k m_k x_k x_k^T Bx_1$$
$$= m_1 x_1 x_1^T Bx_1 + \sum_{k \neq l} m_k x_k x_k^T Bx_l \ . \tag{14}$$
$$= m_1 x_1 x_1^T Bx_1 + m_k \cos_B(x_k x_1) x_k$$

The quality of reconstruction of the stimulus can be evaluated using the generalized cosine between $\hat{x}_1$ and x_1 (cf. Eq. (12)).

If every pair of stimuli in the learning set is B-orthogonal, then the output of the memory will be proportional to the original stimulus:

$$\hat{x}_1 = m_1 \gamma_l x_1 + \sum_{l \neq k} m_k \cos_B(x_k x_1) x_k$$
$$= m_1 \gamma_l x_1 \tag{15}$$

with γ_l being a scalar equal to $x_l^T Bx_l$. When the stimuli stored in the memory are not B-orthogonal, some patterns will be perfectly reconstructed by the memory:

$$W\tilde{u}_r = \tilde{\lambda}_k \tilde{u}_r \text{ with } \tilde{u}_r^T B\tilde{u}_r = I. \tag{16}$$

The vectors $\tilde{u}_r$ are the "generalized eigenvectors" of W (these generalized eigenvectors can be computed using a standard eigendecomposition routine, cf. Wilkinson, 1965, and Appendix). Because the eigenvectors are B-orthogonal, and the eigenvalues are non-negative, the matrix W can be reconstructed as

$$W = \sum_r^R \tilde{\lambda}_r \tilde{u}_r \tilde{u}_r^T = \tilde{U}\tilde{\Lambda}\tilde{U}^T \text{ with } \tilde{U}^T B\tilde{U} = I. \tag{17}$$

Similarly, the Widrow-Hoff error correction learning rule (cf. Eq. (9)) can be generalized and W at time $t + 1$ can be expressed as

$$W_{[t+1]} = W_{[t]} + \eta(X - W_{[t]}BX)X^T$$
$$= \tilde{U}[I - (I - \eta\tilde{\Lambda})^{t+1}]\tilde{U}^T \ . \tag{18}$$

Abdi, Valentine, Edelman, and O'Toole (in press) showed that with a learning constant η smaller than $2\lambda^{-1}_{max}$ this procedure converges toward $W_{[\infty]} = \tilde{U}\tilde{U}^T$ where $\tilde{U}$ are the generalized eigenvectors of W.

The generalized eigenvectors and eigenvalues of the weight matrix $\mathbf{W}$ can be obtained directly using a generalization of the singular value decomposition of the matrix of stimuli $\mathbf{X}$. Formally,

$$\mathbf{X} = \tilde{\mathbf{U}}\tilde{\Delta}\tilde{\mathbf{V}}^T \text{ with } \tilde{\mathbf{V}}^T\mathbf{M}\tilde{\mathbf{V}} = \tilde{\mathbf{U}}^T\mathbf{B}\tilde{\mathbf{U}} = \mathbf{I} \qquad (19)$$

where $\tilde{\mathbf{U}}$ represents the matrix of generalized eigenvectors of $\mathbf{XX}^T$, $\tilde{\mathbf{V}}$ represents the matrix of generalized eigenvectors of $\mathbf{X}^T\mathbf{X}$, and $\tilde{\Delta}$ is the matrix of generalized singular values. The projections, $\tilde{\mathbf{G}}$, of the K stimuli of the training set on the R eigenvectors of the weight matrix can be found as

$$\tilde{\mathbf{G}} = \mathbf{X}^T\mathbf{B}\tilde{\mathbf{U}} = \tilde{\mathbf{V}}\tilde{\Delta}. \qquad (20)$$

From Eq. (20), it is easy to derive that the generalized variance of the projections on one eigenvector is equal to the eigenvalue associated with this eigenvector:

$$\tilde{\mathbf{G}}^T\mathbf{M}\tilde{\mathbf{G}} = \tilde{\Delta}\tilde{\mathbf{V}}^T\mathbf{M}\tilde{\mathbf{V}}\Delta = \tilde{\Lambda}$$

Likewise, the projections of a set of K' new stimuli (i.e., the test set), $\mathbf{X}_{new}$, on the eigenvectors of $\mathbf{W}$ can be computed as:

$$\tilde{\mathbf{G}}_{new} = \mathbf{X}^T_{new}\mathbf{B}\tilde{\mathbf{U}}. \qquad (21)$$

In terms of distances, the generalized autoassociator represents the stimuli using their generalized Euclidean distance. The generalized Euclidean distance between stimuli k and k' is computed as

$$d_\mathbf{B}^2(\mathbf{x}_k, \mathbf{x}_{k'}) = (\mathbf{x}_k - \mathbf{x}_{k'})^T\mathbf{B}(\mathbf{x}_k - \mathbf{x}_{k'}). \qquad (22)$$

The distance between stimuli k and k' can be expressed also, through the eigendecomposition of the generalized weight matrix as

$$d_\mathbf{B}^2(\mathbf{x}_k, \mathbf{x}_{k'}) = d_\mathbf{B}^2(\tilde{\mathbf{g}}_k, \tilde{\mathbf{g}}_{k'}) = (\tilde{\mathbf{g}}_k - \tilde{\mathbf{g}}_{k'})^T(\tilde{\mathbf{g}}_k - \tilde{\mathbf{g}}_{k'}) \qquad (23)$$

where $\tilde{\mathbf{g}}_k$ (respectively $\tilde{\mathbf{g}}_{k'}$) is the vector of the $\mathbf{B}$-projections of stimulus k (respectively k') onto the generalized eigenvectors.

Generalized Euclidean distances are widely used in a variety of applications. For example, Nosofsky (1992, p. 365 ff., Eqs. (1) and (3); see also Ashby, 1992), represents stimuli in his generalized context model (GCM) with a parameter standing for the strength of a stimulus, and with features weighted by an attentional parameter. Adapting his notation to the present chapter, Nosofsky's model can be seen as equivalent to representing the strength of a stimulus by the diagonal terms $m_{k,k}$ of $\mathbf{M}$ ($\mathbf{M}$ being a diagonal matrix in this case), and the attentional weights by the diagonal terms $b_{i,i}$ of $\mathbf{B}$ ($\mathbf{B}$ being diagonal also). Categorization can be then considered to be a function of the generalized distance to the centers of the categories. Another relatively well-known example of a generalized Euclidean distance is the "Mahalanobis" distance used in conjunction with discriminant analysis. In this case, the matrix $\mathbf{B}$ is the inverse of the between-features (or dimensions) correlation matrix.

4. CATEGORIZING FACES BY SEX

In recent years, a number of connectionist models have been applied to the problems of face recognition and categorization (for a review see Valentin, Abdi, O'Toole, & Cottrell, 1994). These models represent faces explicitly (Sirovich & Kirby, 1987; Turk & Pentland, 1991) or via a neural network architecture (Cottrell & Fleming, 1990; O'Toole & Abdi, 1989) in terms of the eigendecomposition of a matrix storing pixel-based descriptions of faces. The eigenvectors, in this framework, can be thought of as a set of features from which the faces are built. Likewise, the projections of the faces onto the eigenvectors can be interpreted as an indication of the extent to which each eigenvector characterizes individual faces. This type of approach suggests that faces can be efficiently represented using tools derived from multivariate statistical analysis. Specifically, previous work showed that complex perceptual discrimination such as the categorization of faces along visually derived dimensions (e.g., sex, race, age, etc.) can be achieved by a simple linear autoassociator (O'Toole, Abdi, Deffenbacher, & Bartlett, 1991; O'Toole, Abdi, Deffenbacher, & Valentin, 1993). Among these perceptual categorization problems, sex classification is one of the most biologically important and probably one of the easiest and fastest categorizations made by human beings. For example, Bruce, Ellis, Gibling, and Young (1987) reported an average sex categorization time of 613 ms for unfamiliar faces and 620 ms for familiar faces. In a more recent study, Burton, Bruce, and Dench (1993) reported that human subjects were able to classify photographs of 179 adults with respect to sex with 96% accuracy, even though the hair was concealed by a swimming cap. In this section, a generalized linear autoassociator is applied to the problem of categorizing faces according to their sex. To evaluate the usefulness of the generalized model, the performance of the generalized autoassociator is compared with the performance of a standard autoassociator on the same task.

In this specific application, the main idea was to have the cells of the memory take on differential importance as a function of their use. Specifically, each cell responds as the inverse of its use during the learning period. The rationale behind this coding scheme is to make the cells more discriminative. So, for example, if a cell is active for all the faces, it does not provide any information about a subset of specific faces. On the other hand, a cell that is active relatively rarely should be important for the detection of the subset of faces that triggers its activity. Formally, this is equivalent to defining $\mathbf{B}$ as being an $I \times I$ diagonal matrix with

$$b_{i,i} = \frac{x_{i,+}}{x_{+,+}} \tag{24}$$

where $x_{i,+}$ represents the total of the ith row of the face matrix $\mathbf{X}$ and x_{++} represents the grand total of $\mathbf{X}$.

Because there was no a priori reason to give more importance to some faces than to others, the face vectors were normalized (i.e., preprocessed) so that the sum of the pixels representing each face was equal to 1 (i.e., $\Sigma_i x_{i,k} = 1$, and $\mathbf{M} = \mathbf{I}$). In addition, to giving an identical importance to each stimulus in the learning set, this particular preprocessing has the advantage of transforming the matrix $\mathbf{X}$ into a "profile" matrix (i.e., each column of $\mathbf{X}$ adds up to 1). With this specific choice for $\mathbf{B}$ and $\mathbf{M}$, the generalized autoassociator implements the multivariate statistical analysis known as "correspondence analysis" (Benzécri, 1973; Greenacre, 1984; Weller & Romney, 1990) or as "dual scaling" (Nishisato, 1994). Strictly speaking, in correspondence analysis $m_{k,k}$ would be equal to $x_{+,k}/x_{++}$. However, because $x_{+,k} = 1$, our model is a particular case of correspondence analysis (i.e., when all the columns sum to a constant).

The generalized Euclidean distance associated with this technique is the so-called *chi-square* distance. It is essentially an informational distance. When the sum of squared distances of each point to the barycenter or centroid is computed, it produces the usual chi-square statistic used to analyze a contingency table in elementary statistics. Specifically,

$$\chi^2 = x_{+,+} \sum_k m_{k,k} d^2(x_k, c) = x_{+,+} \sum_k m_{k,k}(x_k - c)^T \mathbf{B}(x_k - c) \qquad (25)$$

where c gives the coordinates of the centroid or average face (cf. Benzécri, 1973; or Greenacre, 1984, for a proof).

To compare the classical and generalized autoassociators, two series of simulations were performed. For each simulation, faces were used as input for both a classical and a generalized autoassociator. The two autoassociators were then used to predict the sex of the faces via a perceptron approach. The first series of simulations evaluates the accuracy of the sex classification achieved by both models when full Widrow-Hoff learning is used. Previous work showed that the ability of the linear autoassociator to predict the sex of faces varied as a function of the number of faces in the training set and the number of eigenvectors used to reconstruct the faces (Abdi, Valentin, Edelman, & O'Toole, 1995; Valentin, Abdi, & O'Toole, 1996). Thus, the comparison between classical and generalized models was carried out using training sets of different sizes, and faces reconstructed with different numbers of eigenvectors to cover the performance range associated with these variations. The second series of simulations evaluates the number of iterations (i.e., speed of learning) necessary to reconstruct the faces in a way that allows a perfect sex categorization for faces in the training sets.

In the next two subsections, we show that although there is no difference in the accuracy with which faces can be categorized in the standard and generalized autoassociative model (see Simulation 1), there is a vast difference in the speed with which the learning takes place in the two models (see Simulation 2). The generalized model learns to classify faces by sex much more quickly than does the standard autoassociator.

4.1. Accuracy of Sex Classification

Stimuli: A set of 160 full-face pictures of young Caucasian adults, 80 females, and 80 males, was used in the following simulation. Each face was first digitized from slide using 16 gray levels to give a 151×225 = 33975 pixel image. The images were roughly aligned along the axis of the eyes so that the eyes of all faces were about the same height. None of the pictured faces had major distinguishing characteristics such as beards, glasses, or jewelry. To save processing time, each face was then compressed to a 46 × 31 = 1426 pixel image and coded as a 1426×1 vector x_k concatenated from the rows of the face image. The compression was done by local averaging using a 5×5 window. This compression technique reduces the number of pixels in an image and as a consequence filters out high frequency information. However, this is not a problem because we demonstrated earlier that there is enough information in 46 × 31 face images to classify them accurately according to their sex (Abdi et al., 1995; Valentin et al., 1996).

Procedure. Different samples of *N* (ranging from 2 to 110) faces were randomly selected (under the constraint that half of the faces were male and the other half female) from the original set of 160 faces and used as input for both a classical and a generalized autoassociator. The remaining faces were used to test the ability of the two models to generalize to new faces. The estimation of the sex of the faces was done by using a perceptron as a categorization network. The perceptron is a very simple neural network, and is equivalent to discriminant analysis (see Levine, 1991; Minsky & Papert, 1969). In the specific case of two face categories, the optimal classification procedure is equivalent to computing the coordinates of the barycenter (or center of gravity) of each class (i.e., sex), and then computing the distance to both barycenters for the face to be classified. The face is then classified as belonging to the sex with the closest barycenter (see Fig. 16.1).

Figure 16.1 illustrates the different steps of the categorization algorithm used in the following simulation:

· Step 1. For each training set, a classical and a generalized autoassociative memory were created from the face images using complete Widrow-Hoff learning and decomposed into eigenvectors. The weight matrix **W** at infinity was computed as

$$W_{[\infty]} = UU^T \text{ with } U^T U = I \tag{26}$$

for the classical model, and as

$$\tilde{W}_{[\infty]} = \tilde{U}\tilde{U}^T \text{ with } \tilde{U}^T B \tilde{U} = I \tag{27}$$

for the generalized model (cf. Eq. (9) and Eq. (18)).
· Step 2. The projections of all faces (learned and new) onto the eigenvectors of $W_{[\infty]}$ were computed as

$$G_{[\infty]} = X^T U \Delta^{-1} = V \text{ for the learned faces} \tag{28}$$

$$G_{[\infty]} = X_{new}^T U \Delta^{-1} \text{ for the new faces} \tag{29}$$

for the classical model (cf. Eq. (9)), and

$$\tilde{G}_{[\infty]} = X^T \tilde{U} \tilde{\Delta}^{-1} = \tilde{V} \text{ for the learned faces} \tag{30}$$

$$\tilde{G}_{[\infty]} = X_{new}^T \tilde{U} \tilde{\Delta}^{-1} \text{ for the new faces} \tag{31}$$

for the generalized model (cf. Eq. (27)). Recall that using a Widrow-Hoff learning rule amounts to sphericizing the weight matrix. As a consequence, after complete Widrow-Hoff learning, the variance of the projections onto each eigenvector is equal to 1 and hence Eqs. (7) and (20) reduce to Eqs. (29) and (31).

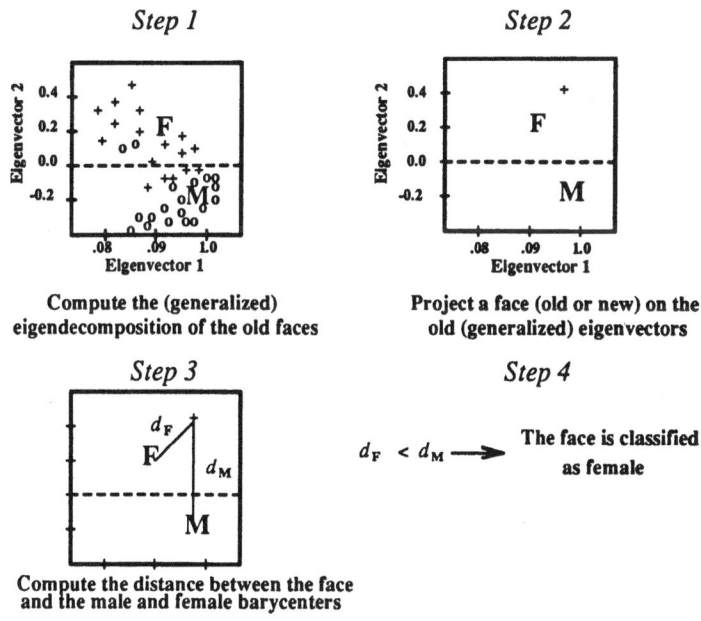

Fig. 16.1. Different steps used to classify the faces according to their sex.

· Step 3. For each model the coordinate vectors of the average male (**m**) and female (**f**) faces were computed by taking the mean of the projections of the male and female learned faces onto the L first eigenvectors (i.e., the ones with the largest eigenvalues), respectively:

$$\mathbf{m} = \frac{1}{J} \sum_{j \in (\text{male faces})}^{J} \mathbf{g}_j \text{ and } \mathbf{f} = \frac{1}{J'} \sum_{j' \in (\text{female faces})}^{J'} \mathbf{g}_{j'} \qquad (32)$$

where J represents the number of learned male faces, J' the number of learned female faces, $\mathbf{g}_j$ the vector of the projections of the jth male face on the first L eigenvectors, and $\mathbf{g}_j'$ the vector of the projections of the j'^{th} female face on the first L eigenvectors.

· Step 4. The categorization of a given face $\mathbf{x}_k$ was determined on the basis of the Euclidean distance between its projection $\hat{\mathbf{x}}_k$ onto the first L eigenvectors and the average faces:

$$\mathbf{d}(\hat{\mathbf{x}}_k, \mathbf{m}) = \|\hat{\mathbf{x}}_k - \mathbf{m}\| \text{ and } \mathbf{d}(\hat{\mathbf{x}}_k, \mathbf{f}) = \|\hat{\mathbf{x}}_k - \mathbf{f}\|. \qquad (33)$$

Faces closer to the average female face were classified as female, and faces closer to the average male face were classified as male.

The number of faces per training set (N) ranged from 20 to 110, and the eigenvectors used to reconstruct the faces (L) varied from 2 to $N - 2$ (where N is the rank of $\mathbf{W}$). To ensure that the model performance was not sample dependent, the categorization procedure was repeated 50 times for each condition.

Results. The average proportion of correct sex classification for both models varied as a function of the number of faces used to train the models ($N = 20$, 50, 80, and 110) and as a function of the number of eigenvectors used to reconstruct the faces ($L = 2$, 10, and $N - 2$). The data are summarized in Table 16.1, which shows that:

1. The performance of the classical autoassociator is similar to that reported in previous work. Specifically, the accuracy of categorization increases as a function of both the number of eigenvectors used to reconstruct the faces and the number of faces in the training set. The best performance (100% correct classification for the old faces and 85% for the new faces) was obtained with a training set of 110 faces and 108 eigenvectors. In previous work, using a similar categorization algorithm with a classical autoassociator, we found that with a training set of 158 faces 90% of the new faces were correctly classified as male or female (Abdi et al., 1995).

2. No substantial difference in performance accuracy can be seen between the two models.

In summary, the two models appear to be equally accurate on a sex classification task independently of the training set size and of the number of eigenvectors used to reconstruct the faces. This is not entirely surprising since the performance of the classical model is already impressive and probably difficult to improve. Comparable levels of performance were found using different models such as backpropagation networks (Cottrell & Metcalfe, 1991; Golomb, Lawrence, & Sejnowski, 1991) or HyperBF networks (Brunelli & Poggio, 1992). This general high level of sex categorization performance is probably due to the fact that sex discrimination is a linear problem.

# Eigenvectors	Classical Autoassociator		Generalized Autoassociator	
	Number of faces in the training set: 20			
2	.76	.73	.81	.74
10	.94	.77	.93	.76
18	1	.78	1	.76
# Eigenvectors	Number of faces in the training set: 50			
2	.76	.75	.79	.78
10	.85	.78	.87	.78
48	1	.80	1	.79
# Eigenvectors	Number of faces in the training set: 80			
2	.78	.80	.80	.79
10	.84	.80	.84	.80
78	1	.84	1	.84
# Eigenvectors	Number of faces in the training set: 110			
2	.78	.79	.81	.79
10	.84	.78	.84	.78
108	1	.85	1	.85
	Old	New	Old	New

Table 16.1. Proportion of Correct Classifications Obtained with a Classical Autoassociator Versus Proportion of Correct Sex Classification Obtained with a Generalized Autoassociator as a Function of the Number of Eigenvectors Used to Reconstruct the Faces and of the Number of Faces in the Training Sets. For each condition, the performance is averaged across 50 trials.

4.2. Learning Speed

The purpose of this second series of simulations was to compare the learning speed of the classical and the generalized model. The number of iterations of the Widrow-Hoff learning rule used to reconstruct the faces *prior* to sex classification was used as an indication of the learning speed. Specifically, the learning speed was defined as the minimum number of iterations necessary to achieve a perfect sex classification. Stimuli: The stimuli were the 160 face images used in the first simulation.

Procedure: For both models (classical and generalized), the complete set of faces (80 males and 80 females) was stored using a Widrow-Hoff learning rule (Eqs. (9) and (18), respectively) with different values of t. After each iteration, $W_{[t]}$ was decomposed into its eigenvectors:

$$W_{[t]} = U \Phi_{[t]} U^T \text{ for the classical associator}$$
$$W_{[t]} = \tilde{U} \Phi_{[t]} \tilde{U}^T \text{ for the generalized associator.} \qquad (34)$$

The effect of Widrow-Hoff learning is equivalent to projecting the columns of X onto the eigenvectors of W followed by an expansion or dilatation of the projections as indicated by the diagonal matrix $\Phi_{[t]}^{1/2}$. Specifically, the coordinates of the faces at time t were evaluated as

$$G_{[t]} = X^T U \Delta^{-1} \Phi_{[t]}^{\frac{1}{2}} = V \Phi_{[t]}^{\frac{1}{2}} \qquad (35)$$

for the classical autoassociator, and as

$$\tilde{G}_{[t]} = X^T \tilde{U} \tilde{\Delta}^{-1} \Phi_{[t]}^{\frac{1}{2}} = \tilde{V} \Phi_{[t]}^{\frac{1}{2}} \qquad (36)$$

for the generalized autoassociator.

The faces were then categorized according to their sex using Steps 3 and 4 of the categorization algorithm described in the previous section. This procedure was iteratively repeated until all the faces were perfectly classified. To make sure that the pattern of results we obtained was not due to a specific value of the learning constant η, three simulations were carried out using different values of η: $\frac{3}{2}\lambda^{-1}_{max}$, λ^{-1}_{max}, and $\frac{1}{2}\lambda^{-1}_{max}$, with λ_{max} representing the highest eigenvalue.

Results: The minimum numbers of iterations necessary to achieve a perfect sex classification with both the classical and the generalized model are presented in Table 16.2. Table 16.2 shows that for all three learning constants, the generalized autoassociator discriminated between male and female faces much faster than the classical one did. Although it took, on the average, 57 iterations to obtain perfect sex categorization with the generalized model, an average of 1933 iterations was necessary to obtain the same performance with the classical model. Note that, for both models, the male faces were categorized as "male" much faster than the female faces were categorized as "female." This bias can be explained by the fact that, for the specific set of faces used in these simulations, female faces are more widely scattered around their barycenter than are male faces. In other words, when not perfectly reconstructed by the memory, some female faces are closer to the male barycenter than to the female barycenter, and hence are categorized as "male". Additional iterations are necessary to reconstruct these specific faces to a level that enables differentiation from the male faces.

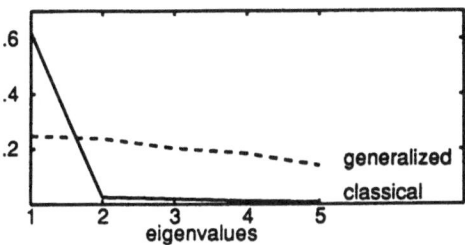

Fig. 16.2. Plot of first five eigenvalues obtained for the classical and the generalized models. Note that the difference between the first two eigenvalues is greater for the classical model than for the generalized model.

η	Classical		Generalized	
	Males	Females	Males	Females
$(3/2)\lambda_{max}^{-1}$	500	1200	15	30
λ_{max}^{-1}	1000	1500	20	50
$(1/2)\lambda_{max}^{-1}$	1500	3100	50	90

Table 16.2. Minimum Number of Iterations Necessary to Achieve a Perfect Categorization Using Widrow-Hoff Learning as a Function of the Type of Model and the Learning Constant.

Observation of the eigenvalues associated with the eigenvectors of $\mathbf{W}$ and $\tilde{\mathbf{W}}$ shows that the difference between the first and the second eigenvalues is much smaller for the generalized model than for the classical one (Fig. 16.2). Because, using a Widrow-Hoff learning rule amounts to equalizing iteratively the nonzero eigenvalues of the weight matrix (cf. Eqs. (9) and (18)) the faster learning rate of the generalized model might be due to this difference in the pattern of eigenvalues. In other words, the superiority of the generalized model in this application (i.e., with the specific set of constraints used here) might result from the fact that the generalized weight matrix, after simple Hebbian learning, is already almost sphericized. To check this hypothesis, we carried out a second series of simulations in which the first and the second highest eigenvalues were systematically equalized in both models. The results are presented in Table 16.3.

Table 16.3 shows that equalizing the first and second eigenvalues does reduce the difference in learning rate observed between the classical and generalized autoassociators. However, the generalized autoassociator is still learning faster than the classical one (56 vs. 83 iterations on the average). This result indicates that part of the superiority of the generalized model is indeed due to the difference in the eigenvalue ranges between the two models. However, it is not due to this difference alone.

5. DISCUSSION

In this chapter we have proposed a generalized version of the classical linear autoassociator. This model is of interest from both a theoretical perspective and a practical reason. First, the model makes interesting theoretical links between the neural network "learning" perspective and the statistical perspective of least-squares approximation. This analysis makes clear that, with a proper choice of constraints, generalized autoassociators can implement all of the techniques of the general linear model, including canonical correlation, discriminant analysis, and correspondence analysis. Second, numerous practical applications require the a priori imposition of constraints operating at the level of individual parts of the input code and at the level of individual stimuli.

	Classical		Generalized	
η	Males	Females	Males	Females
$(3/2)\lambda_{max}^{-1}$	20	50	15	30
λ_{max}^{-1}	40	70	20	50
$(1/2)\lambda_{max}^{-1}$	70	130	50	90

Table 16.3. Minimum Number of Iterations Necessary to Achieve a Perfect Categorization Using Widrow-Hoff Learning as a Function of the Type of Model and the Learning Constant, When the First and the Second Eigenvalues are Equalized.

In these simulations, we applied the generalized model to classifying faces by sex. Our method for imposing constraints made the generalized model equivalent to correspondence analysis, which differentially weights units of the input code as a function of their informational value. Although this manipulation did not change the accuracy of the model, it speeded up the learning considerably by biasing individual parts of the code to affect the structure of the feature space. Examples of this kind of prewiring constraints are common in many cognitive science applications. The simulations indicate that these generalized methods can yield valuable learning benefits when adequately utilized. The sex classification task presented here is but one example of the many possible schemes available for imposing linear constraints on a learning task. The many statistical "variations on a theme" that are commonly encountered in the literature are evidence for the ready applicability of other such schemes. The diversity of the neural network literature often makes it difficult to find a common statistical thread through the various models proposed to simulate human information processing. The generalized model we have proposed provides such a thread through the commonly used linear statistical and neural network models.

ACKNOWLEDGMENT

Thanks are due to June Chance and Al Goldstein for providing the faces used in the simulations and to Betty Edelman for helpful comments on a previous version of this chapter.

REFERENCES

Abdi, H. (1988). A generalized approach for connectionist auto-associative memories: interpretation, implications and illustration for face processing. In J. Demongeot (Ed.), *Artificial Intelligence and Cognitive Sciences*. Manchester: Manchester University Press.

Abdi, H. (1993). Précis de connexionisme. In J. F. Le Ny (Ed.), *Intelligence Artificielle et Intelligence Naturelle*. Paris: PUF.

Abdi, H. (1994a). *Les Réseaux de Neurones*. Grenoble: Presses Universitaires de Grenoble.

Abdi, H. (1994b). A neural network primer. *Journal of Biological Systems*, **2**, 247-282.

Abdi, H., Valentin, D., Edelman, B. G., & O'Toole, A. J. (1995). More about the difference between men and women: Evidence from linear neural networks and principal component approaches. *Perception*, **24**, 539-562.

Abdi, H., Valentin, D., Edelman, B. G., & O'Toole, A. J. (in press). A Widrow-Hoff rule for the generalization of the linear autoassociator. *Journal of Mathematical Psychology*.

Anderson, J. A., Silverstein, J. W., Ritz, S. A., & Jones, R. S. (1977). Distinctive features, categorical perception, and probability learning: Some applications of a neural model. *Psychological Review*, **84**, 413-451.

Ashby, F. G. (1992). Multidimensional models of categorization. In F. G. Ashby (Ed.), *Multidimensional Models of Perception and Cognition* (pp. 449-483). Hillsdale, NJ: Lawrence Erlbaum Associates.

Benzécri, J. P., (1973). *L'analyse des Données* (Vol. 2). Paris: Dunod.

Bruce, V., Ellis, H., Gibling, F., & Young, A. W. (1987). Parallel processing of the sex and familiarity of faces. *Canadian Journal of Psychology*, **41**, 510-520.

Brunelli, R., & Poggio, T. (1992, January). HyperBF Networks for sex classification. *Proceedings of the Image Understanding Workshop*, DARPA, San Diego.

Burton, A. M., Bruce, V., & Dench, N. (1993). What's the difference between men and women? Evidence from facial measurement. *Perception*, **22**, 153-176.

Cottrell, G. W., & Fleming, M. K. (1990). Face recognition using unsupervised feature extraction. *Proceedings of the International Neural Network Conference, Paris, France* (pp. 322-325). Dordrecht: Kluwer.

Cottrell, G. W., & Metcalfe, J. (1991). EMPATH: Face, sex and emotion recognition using holons. In R. P. Lippmann, J. Moody, & D. S. Touretzky (Eds.), *Advances in Neural Information Processing Systems 3* (pp. 564-571). San Mateo, CA: Morgan Kaufmann.

Golomb, B. A., Lawrence, D. T., & Sejnowski, T. J. (1991). Sexnet: A neural network identifies sex from human face. In R. P. Lippmann, J. Moody, & D. S. Touretzky (Eds.), *Advances in Neural Information Processing Systems 3* (pp. 572-577). San Mateo, CA: Morgan Kaufmann.

Greenacre, M. J. (1984). *Theory and Applications of Correspondence Analysis.* London: Academic Press.

Kohonen, T. (1977). *Associative Memory: A System Theoretic Approach.* Berlin: Springer-Verlag.

Levine, D. S. (1991). *Introduction to Neural and Cognitive Modeling.* Hillsdale, NJ: Lawrence Erlbaum Associates.

Mardia, K. V., Kent, J. T., & Bibby, J. M. (1979). *Multivariate Analysis.* London: Academic Press.

Minsky, M., & Papert, S. (1969). *Perceptrons: An Introduction to Computational Geometry.* Cambridge, MA: MIT Press.

Nishisato, S. (1994). *Dual Scaling: An Introduction to Practical Data Analysis.* Hillsdale, NJ: Lawrence Erlbaum Associates.

Nosofsky, R. M. (1992). Exemplar-based approach to relating categorization, identification, and recognition. In F. G. Ashby (Ed.), *Multidimensional Models of Perception and Cognition* (pp. 363-393). Hillsdale, NJ: Lawrence Erlbaum Associates.

O'Toole, A. J., & Abdi, H. (1989). Connectionist approaches to visually based feature extraction. In G. Tiberghien (Ed.), *Advances in Cognitive Psychology* (Vol. 2, pp. 124-140). London: Wiley.

O'Toole, A. J., Abdi, H., Deffenbacher, K. A., & Bartlett, J. C. (1991). Classifying faces by race and sex using an autoassociative memory trained for recognition. In K. J. Hammond & D. Gentner (Eds.), *Proceedings of the Thirteenth Annual Conference of the Cognitive Science Society.* Hillsdale, NJ: Lawrence Erlbaum Associates.

O'Toole, A. J., Abdi. H., Deffenbacher, K. A., & Valentin, D. (1993). A low-dimensional representation of faces in the higher dimensions of the space. *Journal of the Optical Society of America A,* **10**, 405-411.

Sirovich, L., & Kirby M. (1987). Low-dimensional procedure for the characterization of human faces. *Journal of the Optical Society of America A,* **4**, 519-524.

Turk, M., & Pentland, A. (1991). Eigenfaces for recognition. *Journal of Cognitive Neuroscience,* **3**, 71-86.

Valentin, D., Abdi, H., & O'Toole, A. J. (1996). Principal component and neural network analyses of face images: Exploration into the nature of the information available for classifying faces by sex. In C. Dowling, F. S. Roberts, & P. Thomas (Eds.), *Progress in Mathematical Psychology.* Hillsdale, NJ: Lawrence Erlbaum Associates.

Valentin, D., Abdi, H., O'Toole, A. J., & Cottrell, G. W. (1994). Connectionist models of face processing: A survey. *Pattern Recognition,* **27**, 1209-1230.

Weller, A. C., & Romney, A. K. (1990). *Metric Scaling: Correspondence Analysis.* Newbury Park, CA: Sage.

Wilkinson, J. H. (1965). *The Algebraic Eigenvalue Problem.* New York: Oxford University Press.

APPENDIX

Singular Value Decomposition

The generalized singular value decomposition can be computed from the standard singular value decomposition of a matrix. Let X be an $I \times K$ rectangular matrix, M be a $K \times K$ positive definite matrix, and B be an $I \times I$ positive definite matrix. The generalized singular value decomposition of X under the constraints of M and B is given as

$$X = \tilde{U} \tilde{\Delta} \tilde{V}^T \text{ with } \tilde{V}^T M \tilde{V} = \tilde{U}^T B \tilde{U} = I. \tag{37}$$

The first step is to compute the standard singular value decomposition of the matrix

$$Y = B^{\frac{1}{2}} X M^{\frac{1}{2}}$$

with

$$V^T V = U^T U = I.$$

The generalized singular value decomposition is then derived from the singular value decomposition of Y as

$$X = B^{-\frac{1}{2}} Y M^{-\frac{1}{2}} = B^{-\frac{1}{2}} U \Delta V^T M^{-\frac{1}{2}} = \tilde{U} \tilde{\Delta} \tilde{V}^T \tag{38}$$

with $\tilde{U} = B^{-\frac{1}{2}} U$, $\tilde{V} = M^{-\frac{1}{2}} V$, $\tilde{\Delta} = \Delta$. This satisfies the constraints of Eq. (37), namely

$$\tilde{U}^T B U = U^T B^{-\frac{1}{2}} B B^{-\frac{1}{2}} U = U^T U = I$$

and

$$\tilde{V}^T M V = V^T M^{-\frac{1}{2}} M M^{-\frac{1}{2}} V = V^T V = I.$$

17

A Neural Network for Determining Subjective Contours

Jayadeva and Basabi Bhaumik
Indian Institute of Technology, Delhi

*The chapter by Jayadeva and Basabi Bhaumik, **A Neural Network for Determining Subjective Contours**, poses a problem in visual pattern processing. Like the chapter by Abdi et al. that solved a different vision-related problem, it sees the solution to this problem in terms of optimality, using an energy function similar to the famous one used by Hopfield and Tank for the traveling salesman problem (TSP). Indeed the authors note the unity between the problem of finding subjective contours and the TSP, in that both involve connecting pre-specified points using a path of minimum length. The same approach has guided Jayadeva and Bhaumik in previous work on other problems in designing neural network for point connection tasks, such as the Steiner minimal tree and Steiner circuit problems.*

Jayadeva and Bhaumik do not make a case that their optimality criteria is what human perceivers actually use in constructing contours. But their solution to the network design problem, involving differential equations with multiple layers and lateral inhibition, is strikingly similar to designs by Grossberg and Mingolla (whom the authors cite) that are motivated by specific psychological data. Thus their work bears out the contention of Golden's chapter in this book, that many established neural network designs can be studied within an optimality framework, even if optimization was not explicitly used in the network's original development.

ABSTRACT

Subjective contours are amongst the most interesting of visual illusions, and involve a perceptual completion of an incomplete image. We first look at dot patterns which give the subjective impression of a single object. The contour is obtained by using Sequential Unconstrained Minimum Techniques to solve an optimization problem. The problem of finding subjective contours when the input image contains directionality information in the form of oriented edges is addressed next. The contours thus obtained are sharp or binary in nature, though subjective contours in practice have a diffuse appearance. We next propose a model where the pattern of activity of a grid of neurons represents the subjective contour, but the response of a neuron, which corresponds to the brightness at a point, is analog rather than binary.

1. INTRODUCTION

The human visual system is remarkably creative in the sense that it can easily extract features even from an incomplete image. Striking illustrations of this phenomenon can be found in visual illusions, involving subjective or illusory contours. A simple definition of an illusory contour is the boundary of a perceptually occluding surface that can be seen across a physically homogeneous region. It has been postulated that the generation of such contours is part of a filling-in process (Brigner & Gallagher, 1974; Kanizsa, 1976; Ullman, 1976) and plays an important role in vision. The discovery that specific cells in area 18 of the visual cortex respond to contours (von der Heydt, Peterhans, & Baumgartner, 1984) lends support to the view that they are an integral and important part of pre-attentive or low-level vision.

The modern era of illusory contour studies probably dates from Gaetano Kanizsa's work in the 1950s, which illustrated the brain's tendency to assemble disparate features into complete and simple forms. He also devised several figures that strikingly demonstrate this effect. Ullman (1976) suggested that the contour should be smooth and its curvature should be minimized. Grossberg and Mingolla (1987) considered it as a part of the boundary completion process; their model consists of a boundary contour system that fills in a boundary, and a feature contour system which fills attributes within a closed contour. Geman, Geman, Graffigne, and Dong (1990) used stochastic relaxation methods to partition a given image. Biologically oriented models of this process include those of Manjunath and Chellappa (1993), Ringer and Skrzypek (1993), and Sajda and Finkel (1993).

In this chapter, we address the problem from the optimality viewpoint. We consider two kinds of illusory contour: those arising when orientation information is present and those where it is unclear or ambiguous. We argue that an illusory contour is one that is optimal or minimal in the sense that it adds the least additional information to a given image. In our model the illusory contour is obtained from the pattern of activity on a layer of neurons. We illustrate the operation of the network on some typical examples.

The chapter is organized as follows: In Section 2, we show how an optimality criterion may be applied to the problem of determining subjective contours. The subjective contour is approximated by a piecewise-linear curve that is constrained to pass through given points of the image. The contour is formed by selecting the most activated neurons on a two-dimensional continuous grid. We first deal with image data where orientation information is absent. Such cases arise in images composed of dot patterns. We next address the case where edges, providing orientation information, are present at the given image points.

In Section 3, we present a model in which the subjective contour is formed by neurons on a two-dimensional grid but the activity level, which corresponds to the brightness, is a continuous analog function, instead of a binary function as used in Section 2. The activity is concentrated along the "subjective contour" but has a definite spread. Section 4 contains concluding remarks.

2. AN OPTIMALITY APPROACH

Consider the sets of dots in Fig. 17.1. A human subject quickly discerns shapes from these dot patterns. The dot patterns contain no actual lines supplying direction or orientation for completion, but a connected contour does quickly and automatically appear when we see points like this in fairly close proximity. We ask the question: given limited or insufficient information in the form of a set of points, what is the contour for which this set of points is a minimal representation? We suggest that the shortest tour, or the curve with minimal length spanning the points, is such a "minimal" contour. In other words, the curve with *minimal length* and passing through the given set of points is the subjective contour. In general, the "contour" may not be closed, and may consist of several segments.

In this section, we assume that the image is binary. Binarity permits one contrast only, such as black and white. Let I be the set of points of the given image, where the ith point is denoted by $\mathbf{p}_i$ ($i = 1, 2, ..., N$). These points are assumed to lie in a plane, i.e., $\mathbf{p}_i = (p_{i,1}, p_{i,2})$. The contour is composed of a curve which passes through the given points. We propose that the curve is optimal in the sense that it minimizes some objective function. The points $\mathbf{p}_i$ can be connected through any of a number of possible interconnection networks.

The class of interconnection, or network design, problems deals with such applications. Well known cases in this class include the Euclidean and rectilinear Steiner minimal tree, the Steiner circuit, and the traveling salesman (TSP) problems. In general, each of these interconnection problems is specified by a set of N points $\mathbf{p}_i$ ($i = 1, 2, ..., N$) with coordinates $(p_{i,1}, p_{i,2}, p_{i,3}, ..., p_{i,D})$ (in D dimensions) that have to be connected in a specified manner.

In a particular case we might seek a circuit of minimal length, or the shortest tree spanning the points. Further, several choices of length measures are possible, including the L_p norms. The interconnection network is approximated by a chain of piecewise-linear segments, with connections from the fixed (image) points to the chain. Let J be the set of endpoints of these segments, each endpoint being denoted by $\mathbf{x}_j = (x_{j,1}, x_{j,2})$, with j ranging from 1 to M. A possible link between a fixed point and a point on the chain either gets connected (1) or does not get connected up (respectively 0) to from a link or the edge of the graph we are drawing. The decision to connect or not to connect from a specific image point $\mathbf{i} \in I$ to a point $\mathbf{j} \in J$ could, for instance, come from a choice function $h(d_{ij}^a)$, for example,

$$h(d_{i,j}^a) = \begin{cases} 1, \ d_{i,j}^a = Min_{k \in j} d_{ik}^a \\ 0, \ otherwise \end{cases} \tag{1}$$

where d_{ij}^a denotes the distance between points $\mathbf{p}_i$ and $\mathbf{x}_j$, raised to the power a.

It has been shown that a large class of such interconnection problems can be solved by a neural network with the energy function $Min \sum_{i=1}^{M} \sum_{j=1}^{M} A_{ij} \|x_i - x_j\|_b^a + \sum_{i=1}^{N} \sum_{j=1}^{M} B_{ij} h(d_{ij}^a) d_{ij}^a$ subject to a set of constraints defined by the particular interconnection problem one is solving. From here onwards we consider interconnection problems in two dimensions only. Let $\mathbf{x}_i$ be the activity *vector* of neuron i, $i = 1, 2, ..., M$. This corresponds to a point $(x_{i,1}, x_{i,2})$ lying on an approximation to the optimal interconnection network. The variables a and b are positive scalars; $\mathbf{A}$ and $\mathbf{B}$ are weight matrices:

$$d_{ij}^{a} = \|\mathbf{p}_i - \mathbf{x}_j\|_b^a \qquad (2)$$

where $\|\mathbf{z}\|_p$ denotes the L_p norm of vector $\mathbf{z}$. $\|\mathbf{z}\|_p^a$ thus describes the L_p norm of vector $\mathbf{z}$, raised to the power a.

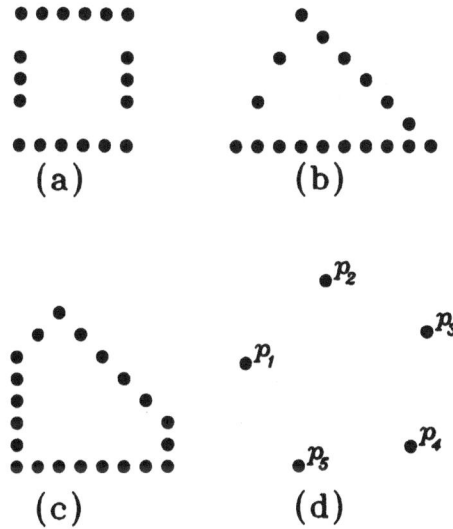

Fig. 17.1. A subjective contour can be seen with dot patterns.

Different choices of parameters in this formulation lead to different interconnection problems. For example, the Euclidean Steiner minimal tree problem (Jayadeva & Bhaumik, 1994a) is obtained with $a = 1$, $b = 2$, $A_{ij} = \delta_{i,j+1}$ and $B_{ij} = 1$, for all i and j; the rectilinear Steiner minimal tree problem (Jayadeva & Bhaumik, 1994b) is obtained by letting $a = 1$, $b = 1$, $A_{ij} = \delta_{i,j+1}$ and $B_{ij} = 1$, for all i and j.

The task of finding the subjective contour is also one from the same class. We propose that given a set of "points" $\mathbf{p}_i$ in a plane, we must find the curve C, which is the solution to the following optimization problem:

$$\text{Minimize } |C|$$
$$\text{subject to the constraints}$$
$$\text{dist}(\mathbf{p}_i, C) = 0 \tag{3}$$

where $|C|$ denotes the total length of C, and $dist(\mathbf{p}_i, C)$ is the minimum distance between a point $\mathbf{p}_i$ and the curve C.

As mentioned previously in this section, C is approximated in a piecewise-linear manner, i.e. we let C consist of M vertices $(x_{i,1}, x_{i,2})$ with straight lines joining $\mathbf{x}_i$ to $\mathbf{x}_{i-1}$ and $\mathbf{x}_{i+1}$. When C is closed, then the indices are read *modulo M*. The problem (3) can thus be rewritten as

$$\text{Min} \sum_{i=1}^{M} \|\mathbf{x}_{i+1} - \mathbf{x}_i\|_2^a$$
$$\text{subject to the constraints}$$
$$\text{Min}_{j=1, 2, ..., M} \|\mathbf{p}_i - \mathbf{x}_j\|_2^a = 0, \; i = 1, 2, ..., N \tag{4}$$

where $\|\mathbf{z}\|_2$ denotes the L_2 norm of vector $\mathbf{z}$.

2.1. Finding a Contour of Minimum Length

Equation (4) can be rewritten in the following form

$$\text{Min} \sum_{i=1}^{M} \|\mathbf{x}_{i+1} - \mathbf{x}_i\|_2^a \tag{5}$$
$$\text{subject to the constraints}$$
$$\sum_{j=1}^{M} h(d_{ij}^a) d_{ij}^a = 0, \; i=1, 2, ..., N \tag{6}$$

where $d_{ij}{}^a$ is given by Eq. (2) with $b = 2$ and $h(d_{ij}{}^a)$ is given by Eq. (1).

Equations (5) and (6) constitute a constrained nonlinear optimization problem, and can be solved in many ways. One class of methods that can be used are sequential unconstrained minimization techniques (SUMTs). In the next subsection we briefly introduce SUMTs and then use them to find the optimal contour.

2.2. Using SUMTs to Find the Optimal Contour

Sequential Unconstrained Minimization Techniques (SUMT), were originally proposed (McCormick, 1983; Fiacco & McCormick, 1968) as improved techniques for optimizing nonlinear constrained systems. Given the problem

$$\text{Min } f(\mathbf{x}), \ \mathbf{x} = (x_1, x_2, ..., x_M)^T$$
$$\text{subject to the constraints}$$
$$g_i(\mathbf{x}) \le 0, \ i = 1, ..., N \qquad (7),$$

one method from the class of exterior point or penalty SUMT methods solves a sequence ($p = 1$, 2, ...) of minimization problems of the form

$$Min \ P(x, \{r_{pi}\}) = f(x) + \sum_{i=1}^{N} (r_{pi})^{-1} \max[0, g_i(x)]^2, \quad p = 1,2,\cdots$$

where r_{pi} are perturbation parameters, which determine the relative contribution of cost and constraint terms in the objective function being minimized. The initial or starting point for each iteration in the sequence ($p = 1, 2, ...$) is the solution to the previous iteration. Solutions may lie on the boundary or in the interior of the feasible region. As this sequence of problems is minimized with an appropriate choice of decreasing r_{pi}, we obtain a sequence of local minima x_i^*. In the limit, as $r_{pi} \to 0$, the constraint terms become overwhelmingly large and the sequence of minima approaches the minimum of the original problem (Eq. (7)), which is our goal.

The problem of determining the optimal contour (Eqs. (5) and (6)) can therefore be solved through minimizing a sequence of expressions of the form

$$P(x) = \sum_{i=1}^{M} \|x_{i+1} - x_i\|_2^a + \sum_{i=1}^{N} (r_{pi})^{-1} \sum_{j=1}^{M} h(d_{ij}^a) d_{ij}^a$$

The proposed network, henceforth referred to as Layer 1, performs a gradient descent to minimize $P(\mathbf{x})$. It consists of $N + M$ neurons, each associated with an activity vector (x_{i1}, x_{i2}) corresponding to the coordinates of a curve vertex i. The activation values are therefore real numbers denoted by v_i, where $v_i = x_{i1}, \ i = 1, ..., M$; $v_{M+i} = x_{i,2}, \ i = 1, ..., M$; $v_{2M+i} = p_{i,1}, \ i = 1, ..., N$; $v_{2M+N+i} = p_{i,2}, \ i = 1, ..., N$. The differential equations describing how the state of each neuron changes are given by

$$\begin{aligned} \dot{v}_i &= \dot{x}_{i,1}, \ i = 1, \cdots, M \\ \dot{v}_{M+i} &= \dot{x}_{i,2}, \ i = 1, \cdots, M \\ \dot{v}_{2M+i} &= \dot{p}_{i,2} = 0, \ i = 1, \cdots, N \\ \dot{v}_{2M+N+i} &= \dot{p}_{i,2} = 0, \ i = 1, \cdots, N \end{aligned} \qquad (8)$$

where

$$\dot{x}_{jk} = -\frac{\partial P}{\partial x_{jk}} = (x_{j+1,k} - x_{j,k})/\|x_{j+1} - x_j\|_2^a + (x_{j-1,k} - x_{j,k})/\|x_{j-1} - x_j\|_2$$
$$+ \sum_{i=1}^{N} (r_{pi})^{-1} h(d_{ij}^a)(p_{i,k} - x_{j,k})/d_{ij}^a, \quad k = 1,2 \tag{9}$$

where the differential of $h(d_{ij})$ has been taken in the sense of generalized gradient (Clarke, 1983).
In practice, we approximate the discontinuous function $h(d_{ij})$ by a sequence of continuous

functions $\tilde{h}(d_{ij}^a) = \phi(d_{ij}^a, \beta_t^2)/\sum_{k=1}^{M} \phi(d_{ik}^a, \beta_t)$, parametrized by β_t, for example,

$$\tilde{h}(d_{ij}^a) = \exp(-d_{ij}^a/\beta_t^2)/\sum_{k=1}^{M} \exp(-d_{ik}^a/\beta_t^2).$$

The value of the parameter β_t is reduced with time, independent of the parameters r_{pi}. In the
limit, as $\beta_t \rightarrow 0$, $\tilde{h}$ tends to the nondifferentiable function h.

If a in Eq. (9) is assigned a value of 2, we obtain the following differential equations for
the neurons:

$$\dot{x}_{j,k} = -\frac{\partial P}{\partial x_{j,k}} = (x_{j+1,k} - x_{j,k}) + (x_{j-1,k} - x_{j,k}) + \sum_{i=1}^{N} (r_{pi})^{-1} h(d_{ij})(p_{i,k} - x_{j,k}),$$
$$k = 1,2 \tag{10}$$

which is similar to the equations used by Durbin and Willshaw (1987) in their elastic net
approach for the TSP.

However, reduction in values of the parameters r_{pi} in (9) and (10) is independent of β_t,
unlike in the elastic net of Durbin and Willshaw. A comparison between Eq. (9) and the elastic
net shows (Jayadeva, 1993) that the approach based on SUMTs converges faster, and more often
to a valid solution; it also yields better solutions (shorter tours) in general. Choosing $a = 1$
implies minimizing the Euclidean length rather than its square, and yields better solutions, as
shown in Table 17.1.

Figure 17.2 shows a network that can be used to calculate the subjective contour by using
Eq. (8). The circles in the figure denote activity components of the neurons. Each circle is thus
labeled with $x_{i,1}, x_{i,2}, p_{i,1}, p_{i,2}$.

The activities of the neurons labeled $p_{i,1}$ or $p_{i,2}$ do not change with time and correspond
to the given points of the image. Changes in the activities of the other neurons are governed by
Eq. (8). The equation indicates that to update $x_{i,1}$ (where i is any arbitrary neuron) requires the
values of $x_{i-1,1}, x_{i+1,1}$, and $p_{j,1}, j = 1, 2, ..., N$. The links in Fig. 17.2 depict this dependence for
one of the components of the activity vector of neuron i. The dependence of other activity values
on each other is similarly obtained from Eq. (8).

	$a = 1$	$a = 2$	$a = 3$
Convergence obtained in (number of examples)	99	100	88
Average number of iterations to the first solution	188.03	352.21	303.352
Average number of iterations to the best solution	304.757	378.32	412.147

Table 17.1. Choice of the Parameter a. Note: The table summarizes simulation results on a set of 100 randomly generated examples. Using $a = 1$ yielded better solutions than $a = 2$ (respectively, $a = 3$) in 43 (respectively, 23) examples; using $a = 3$ yielded better solutions than $a = 2$ in 40 cases.

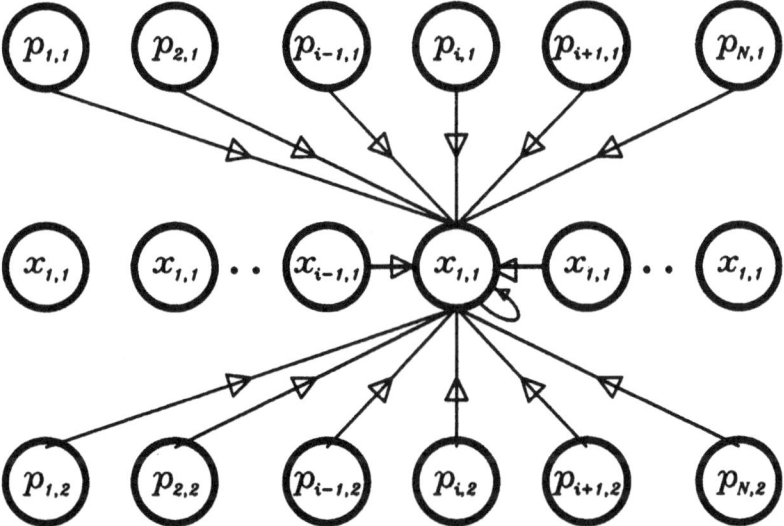

Fig. 17.2. Structure of Layer 1. Updating $x_{i,1}$ (where i is any arbitrary neuron) requires the values of $x_{i-1,1}$, $x_{i+1,1}$, and $p_{j,1}$, $j = 1, 2, ..., N$. The links in Fig. 17.2 depict this dependence of the activity values for the activity component $x_{i,1}$ of neuron i.

Figure 17.3 shows the minimal tours for some sample sets of points. Note that the completed contours correspond to a closed torus of the points, that is, to the solution of the traveling salesman problem for the point sets. Indeed, the TSP is the optimal interpolant for the given data in an information-theoretical sense. Observe that the perceived subjective contours connecting the dots in Fig. 17.1 match the tours (see Fig. 17.3) found with our neural network. In fact, the observation that human subjects are quickly able to determine very good TSP solutions with a map of the points (cities) visually was employed in a human-machine approach to the TSP by Krolak and Felts (1971).

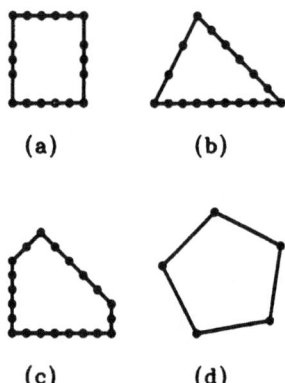

(a)   (b)

(c)   (d)

Fig. 17.3. Tours for some sets of points found in the process of testing the equations described in Section 2.2.

When a given image consists of several closely spaced points, the contour obtained in the manner described in this section consists of a large number of closely spaced line segments. If the image points are in close proximity, the grain of the contour becomes very fine, and the appearance of a curved contour may result, as in Fig. 17.4. Note that in comparison to Fig. 17.4(a), Fig. 17.4(b) gives the appearance of having a smooth curvature.

2.3. Subjective Contours with Orientation Information

We now extend the network to the case when orientation information is available in the form of directionality shown by line segments in the original pattern given. Figure 17.5 shows some examples of the subjective contour with orientation information. As may be seen in Fig. 17.5(c), directional restriction of contours to match even short lines in the initial pattern leads to curved shaping.

We suggest that the "optimal" curve in this case is one found by solving the problem subject to the constraints

$$Min \int_C dl + a_1 \int_C \left(\frac{\partial \theta}{\partial l}\right)^2 dl \qquad (11a)$$

$$dist(\mathbf{p}_i, C) = 0 \qquad (11b)$$

$$\theta(\mathbf{p}_i) = \lambda_i \qquad (11c)$$

where dl denotes an elemental length of the curve C, and θ denotes the angle between the tangent to the curve and the x-axis. Here, the two axes defining the Cartesian plane have been denoted by x_1 and x_2. The second term of Eq. (11a) corresponds to the rate of change of this angle, that is, it represents the elemental curvature of the curve C; a_1 is a scalar indicating its relative importance. The constraints imply that the curve must pass through a given set of points (corners, p_i terms), and further, that the first derivative (slope) of the curve at the given points must have a specific value at the given points.

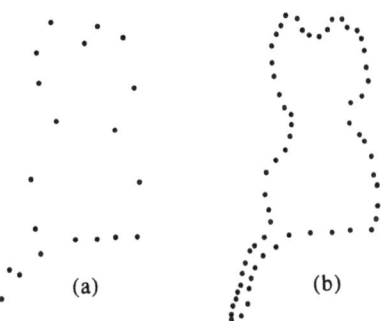

Fig. 17.4. The presence of a large number of closely spaced points can give rise to the appearance of a curved contour. The higher density of dots in (b) in comparison to (a) illustrates this effect.

Following the piecewise-linear curve obtained by the network discussed previously, we have a sequence of ordered points joining specified points. Because the curve already passes through the specified image points, the constraint (11b) is already satisfied. We therefore need only consider the chain of points on the curve that lie between successive points at which the orientation is specified.

Figure 17.6 shows two cases. Figure 17.6(a) shows a point $\mathbf{p}_i$ of the initial pattern/image, whose location is specified along with the orientation of the curve at that point, namely $\theta = \lambda_i$. Let the point $\mathbf{x}_j = (x_{j,1}, x_{j,2})$ be the point of the curve nearest to $\mathbf{p}_i$. In order to minimize the local curvature at point $\mathbf{p}_i$, $\mathbf{x}_j$ must lie on the line passing through $\mathbf{p}_i$ and making an angle λ_i with the x_1 axis. This will happen if it moves towards the point $\mathbf{q}_i$ shown in the figure. Note that $\mathbf{q}_i$ and $\mathbf{x}_j$ are equidistant from $\mathbf{p}_i$.

Figure 17.6(b) shows the other possible case, when the point $\mathbf{x}_k$ is not next to a given image point but is any arbitrary point on the curve, with interpolated points $\mathbf{x}_{k-1}$ and $\mathbf{x}_{k+1}$ being its neighbours on the curve. The local curvature at $\mathbf{x}_k$ is minimized if $\mathbf{x}_k$ lies on the line joining $\mathbf{x}_{k-2}$ and $\mathbf{x}_{k-1}$, as well as on the line joining $\mathbf{x}_{k+1}$ and $\mathbf{x}_{k+2}$. In other words, $\mathbf{x}_k$ must move toward $\mathbf{q}_{k-1}$ as well as toward $\mathbf{q}_{k+1}$. Again, note that $\mathbf{x}_k$ and $\mathbf{q}_{k+1}$ are equidistant from $\mathbf{x}_{k+1}$; $\mathbf{x}_k$ and $\mathbf{q}_{k-1}$ are equidistant from $\mathbf{x}_{k-1}$.

Consider, once again, a layer of $(N + M)$ neurons, whose activation values are real numbers denoted by v_i, where the v_i are as defined in Section 2.2. The equations of motion of these neurons are given in terms of slope angles by

$$\dot{v}_k = \dot{x}_{k,1} = (x_{k-1,1}-x_{k,1})/\|x_{k-1}-x_k\|_2 + (x_{k+1,1}-x_{k,1})/\|x_{k+1}-x_k\|_2 +$$
$$\alpha(d_1\cos(\theta_{k-2})-d_2\cos(\theta_{k+1})), \quad k=1,\cdots,M$$
$$\dot{v}_{M+k} = \dot{x}_{k,2} = (x_{k-1,2}-x_{k,2})/\|x_{k-1}-x_k\|_2 + (x_{k+1,2}-x_{k,2})/\|x_{k+1}-x_k\|_2 +$$
$$\alpha(d_1\sin(\theta_{k-2})-d_2\sin(\theta_{k+1}))$$

where $d_1 = \|x_{k-1} - x_k\|_2$ and $d_2 = \|x_{k+1} - x_k\|_2$ and α is a scalar to adjust the contribution from sloping to an appropriate scale compared to what appears in the distance metric terms.

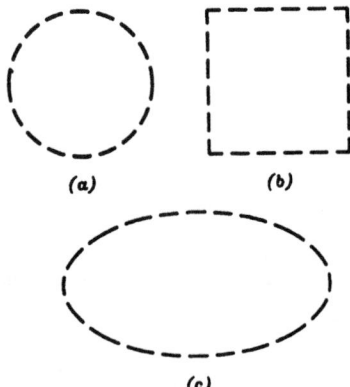

(a)   (b)

(c)

Fig. 17.5. Examples of patterns providing linear orientation information that affects the spread of subjective contours.

This layer of neurons, referred to as Layer 2, is also organized as a feedback network. Note that in contrast to Layer 1, $\dot{x}_{k,2}$ depends on x_k, x_{k-1}, x_{k-2}, x_{k+1}, and x_{k+2}. The equations of motion for points on the curve can now also be written as

$$\dot{v}_k^{(2)} = \dot{x}_{k,1}^{(2)} = \dot{x}_k^{(1)} + \alpha(d_1\cos(\theta_{k-2})-d_2\cos(\theta_{k+1})), \quad k=1,\cdots,M$$
$$\dot{v}_{N+k}^{(2)} = \dot{x}_{k,2}^{(2)} = \dot{x}_k^{(1)} + \alpha(d_1\sin(\theta_{k-2})-d_2\sin(\theta_{k+1})), \quad k=1,\cdots,M$$

where v_k^1 and v_k^2 denote the states of corresponding neurons of Layers 1 and 2 respectively. Figure 17.6 shows the structure of the network. Figure 17.7 illustrates the operation of the network on some examples. The orientation of the curve at each point $(x_{j,1}, x_{j,2})$ is shown by a small oriented segment through the point. Although the output of Layer 1 yields piecewise linear curves connecting the specified points, we have considerably perturbed the initial state in order to illustrate the convergence of the network from a random initial state. In practice, the curve obtained by Layer 1 is a very good initial state for Layer 2, and convergence is obtained rapidly.

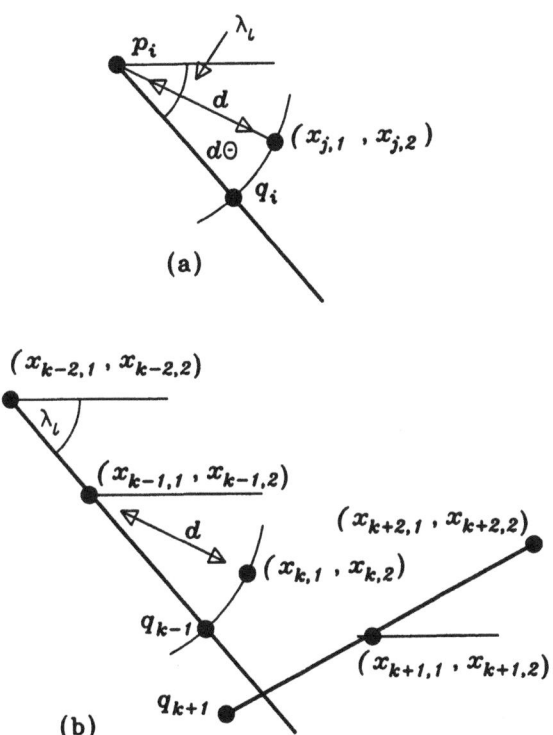

Fig. 17.6. How a curve point moves to minimize curvature.

Implementing this scheme appears cumbersome in practice. Furthermore, the contours obtained by the methods in Sections 2.2 and 2.3 are sharp, whereas subjective contours in reality appear to be diffuse, with a definite spread. We now turn from contour finding that selects only the most activated points to look at the associated "spread" of activation in the neighbourhood, a phenomenon occurring not only in vision, but also in semantic retrieval and elsewhere.

3. CONTOURS AS AN ACTIVITY PATTERN

In this section, we present a model in which the subjective contour is represented by the pattern of activity of a grid of neurons. However, the response of a neuron, which corresponds to the brightness at a point, is a continuous analog function. The pattern of activity thus obtained is concentrated along the "subjective contour," but has a diffuse appearance, with a definite spread. This is unlike the contours obtained in Section 2, which are sharp or binary (black and white) in nature. We first consider the case when orientation information is available at given "points" of the image.

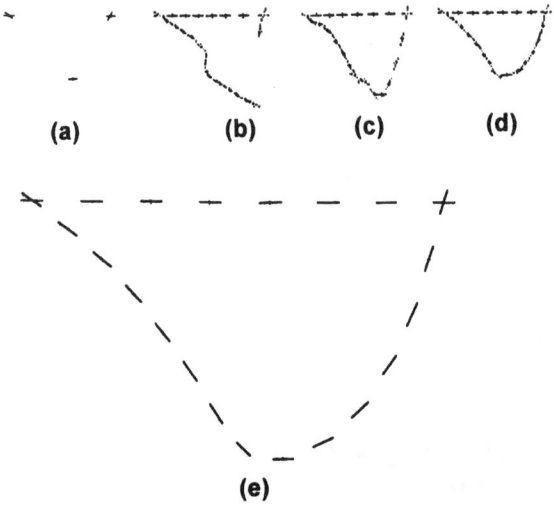

Fig. 17.7. (a) Specified corners of an image. The specified orientations are indicated by small line segments at the points. (b)-(d) Snapshots of the network illustrating the contour being generated. (e) Scaled image of the contour.

3.1. An Activity Based Model for Subjective Contours

We begin with well-known solutions for excitations in continuous space, using equations familiar from thermodynamics, electricity, and magnetism. Consider a thin slab in which the z axis is along the thickness, and the x and y axes lie along the plane of the slab. A point in the slab is specified by the coordinates (x, y, z). Each point corresponds to a set of neurons. Each neuron is maximally sensitive to a unique orientation. The presence of an oriented edge at (x, y, z) would activate a specific neuron sensitive to that orientation. Let $u(x, y, z, \Phi)$ denote the response of a neuron at the location (x, y, z) that is maximally sensitive to an orientation Φ.

At any given point, more than one neuron may be active, indicating that there is a spread of activation along different directions (orientations) at a point. This gives rise to the perception of a diffuse band of orientation at a point. For example, there may be a spread of activity with orientations ranging from $50°$ to $70°$ at a certain point, with a maximum of $60°$.

Let us consider the response of all the neurons in a plane of given thickness, say at z_j. If the responses of all these neurons were equal then we would perceive a region of uniform brightness. To perceive a contour, some neurons must, therefore, respond more strongly than others. When the input image is an incomplete contour, the neurons which receive oriented edges as input respond more strongly than others. We postulate that if some region in the neuronal plane is relatively more active than other regions, then a flow of activity takes place from the region of higher activity to those of lower activity. We denote the flow of activity by F_A, where $F_A = k \nabla u$. k is a constant of proportionality.

The total activity inside a volume V with surface S as its boundary is given as a function of Φ by

$$A(\phi) = \iiint_{x\,y\,z} u(x,y,z,\phi)\,dx\,dy\,dz \tag{12}$$

We assume there is no source or sink of activity inside the volume V. Therefore, the activity inside V will change only due to a flow of activity through the bounding surface S. The total flux of activity leaving V via S is given by

$$\iint_S F_A \cdot \vec{n}\,ds$$

where n is a unit vector normal to the surface, and d_s is a small element of the surface S. The change in activity inside volume V is

$$\frac{\partial A}{\partial \Phi} = \iint_S F_A \cdot \vec{n}\,ds. \tag{13}$$

From Eqs. (12) and (13), and following arguments similar to those used in the derivation of the heat flow equation (Pipes, 1958) we obtain

$$\frac{\partial u}{\partial \phi} = k\left[\frac{\partial^2 u}{\partial x^2} + \frac{\partial^2 u}{\partial y^2} + \frac{\partial^2 u}{\partial z^2}\right] \tag{14}$$

where k is a scalar constant. Because the contour is two-dimensional or planar, we can replace $u(x, y, z, \Phi)$ at a given z by $u(x, y, \Phi)$. In general, the flow may be anisotropic, and the constant k may be replaced by constants a and b for the x and y directions. Hence, we can rewrite Eq. (14) as

$$\frac{\partial u}{\partial \phi} = a\frac{\partial^2 u}{\partial x^2} + b\frac{\partial^2 u}{\partial y^2}. \tag{15}$$

The solution of this equation is obtained as

$$u(x,y,\phi) = (4\pi\phi\sqrt{ab})^{-1}\int_{-\infty}^{\infty}\int_{-\infty}^{\infty} f(v_1,v_1)\,exp(-[(x-v_1)^2/a+(y-v_2)^2)/b]/4\phi)\,dv_1\,dv_2 \tag{16}$$

for the initial condition

$$u(x, y, 0) = f(x, y). \tag{17}$$

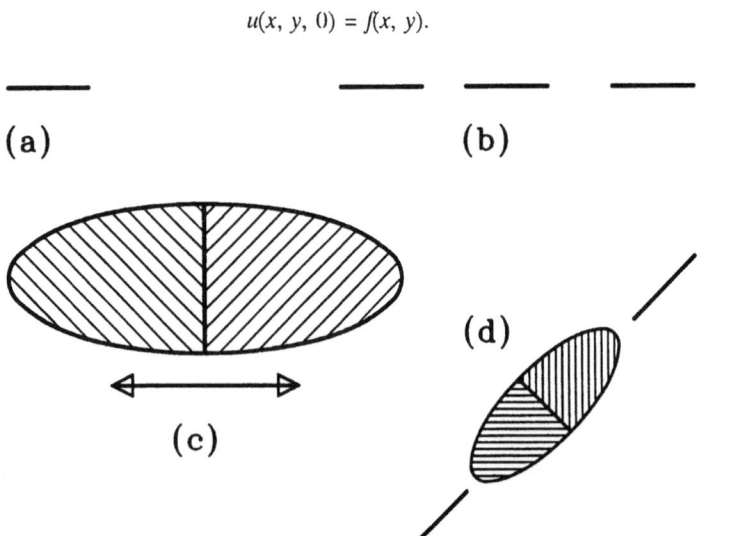

Fig. 17.8. (a) A single edge does not lead to the appearance of a subjective contour. (b) Additional edges are required in close proximity. (c) The receptive field of a neuron sensitive to subjective contour can be split into two halves. (d) Each half must receive input in order for the neuron to fire.

The initial condition given by Eq. (17) is an image point located at (x_0, y_0) and having an edge with orientation Φ_0. The variable Φ is measured with respect to the oriented edge at the input image point under consideration. Therefore, Eq. (16) reduces to

$$u(x,y,\phi) = (4\pi\phi\sqrt{ab})\,exp(-[(x-x_0)^2/a+(y-y_0)^2)/b]/4\phi) \tag{18}$$

When more than one input image point is given, the excitations from each input image point are computed and summed.

Consider an oriented edge as shown in Fig. 17.8(a), and note that no contour is possible when there is a single edge. In order for a boundary to be observed, there must be other oriented edges in reasonable proximity, as shown in Fig. 17.8(b). The activity will be at points that lie between two oriented edges, but not beyond the leftmost and rightmost ends.

Consider an oriented input with an orientation value of zero; that is, it lies along the x-axis and excites all neurons to its right. A neuron whose receptive field is aligned along the x axis and that lies to the right of this edge will receive maximal excitation. We assume that the receptive field of a neuron sensitive to subjective contours can be split into two halves, as shown in Fig. 17.8(c). The neuron will fire only if both halves of its receptive field receive excitation. For this to happen, the neuron must lie between at least two edges, whose orientations differ by

180°, as shown in Fig. 17.8(d). Alternatively, the output of a neuron sensitive to subjective contours may be the AND function of inputs from two orientation-sensitive neurons, whose selectivities differ by 180°. In the general case, the receptive field of a neuron will not lie exactly along the edge, and will receive less than maximum excitation, and its response will be weaker. The input to the neuron is then given by the dot product of a unit vector in the direction of its receptive field with the "source" excitation.

From Eq. (18) we find that the net input to a neuron at location (x,y) is of the form

$$u(x,y,\phi) = K\phi^{-1} exp(-[(x-x_0)^2/a+(y-y_0)^2)/b]/4\phi) cos(\theta)$$

where K is a constant and θ is the angular position of the point (x, y) with respect to the "source" at (x_0, y_0), that is, $\theta = arctan ((y - y_0)/(x - x_0))$. Note that the excitation is confined to points at locations such that $cos(\theta)$ is positive. The excitation to a neuron is obtained by a linear superposition of the excitations from different image points.

The output of the neuron located at position (x, y) and sensitive to an orientation λ, that is, $V_{x,y,\lambda}$ is given by $V_{x,y,\lambda} = F(U_{x,y,\lambda})$ where

$$F(x) = \begin{cases} 1, x>1/\beta \\ 0, x<0 \\ \beta x, 0<x<1/\beta \end{cases}$$

is the activation function, and β is a gain term. In the proposed neural network, neurons lie on a uniform grid in the xy-plane, that is, $V_{x,y,\lambda}$ is defined only at the grid points. Figures 17.9 through 17.15 show simulation results for some examples where subjective contours are perceived. In each figure, the oriented inputs (image points) are shown in part (a). The excitation to each half of the receptive field of a neuron is shown in part (b) by a line of proportional length with angle Φ. The excitations have been scaled in order to make them more prominent.

Observe that the activity is confined to a narrow region around the given image points. Also note that the excitations are in a narrow range of orientations for neurons close to the image point locations. The largest excitations correspond to the preferred direction of the contour. In Fig. 17.9, where the input is a set of orientations at the corners of a square, the activity is concentrated along the sides of the square. Although the neuron activities are along the edges for locations near the corners, they are very diffuse in the middle. Figure 17.10 shows another example where the input contains additional oriented edges along two sides of the square. Observe that the activity is in a much narrower band along these sides, as compared to Fig. 17.9. Figures 17.11 through 17.13 show other examples of a similar nature. Note the effect of additional oriented image points near the sides of the square as shown in Fig. 17.11. Increased activity can be seen near the top and bottom sides of the square in Fig. 17.12. An activity "bulge" is strongly visible along the top and bottom sides of the square in Fig. 17.13. Figure 17.14 shows an example where the contour forms the outline of a triangle. In the example in Fig. 17.15, the activities lie on the outline of a semicircle, and orientations change in a smooth manner.

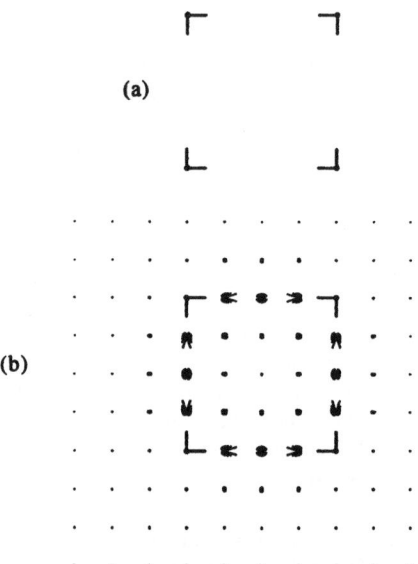

Fig. 17.9. (a) Input and (b) neuron excitations for an example where a subjective square is perceived.

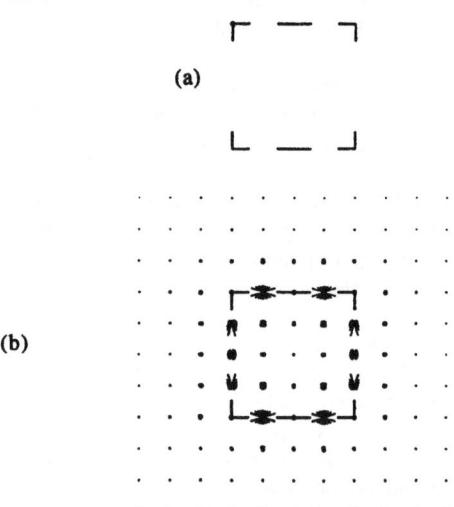

Fig. 17.10. (a) Input and (b) neuron excitations for an example where a subjective square is visible. The activity pattern is sharper along two sides of the square in this case as compared to the other two sides.

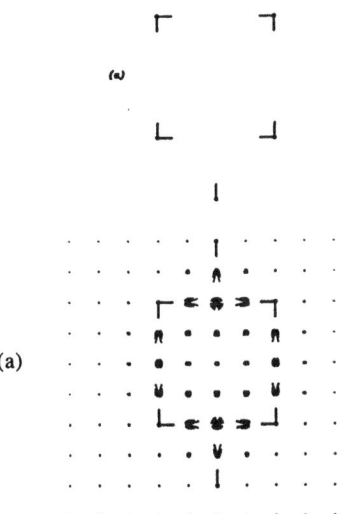

(a)

Fig. 17.11. (a) Input and (b) neuron excitations for an example where a subjective square is perceived. Note the effect of additional oriented image points near the sides of the square: the whole region becomes active.

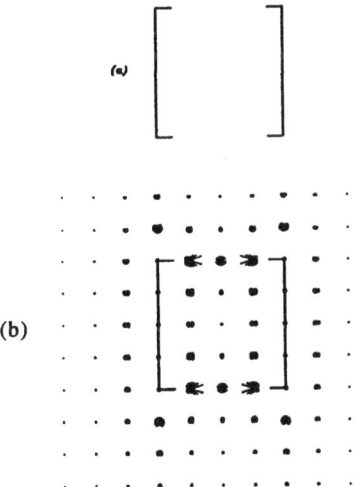

(b)

Fig. 17.12. (a) Input and (b) neuron excitations for an example where a subjective square is observed. In this example, the bulge near the bottom and top sides of the square is strongly visible.

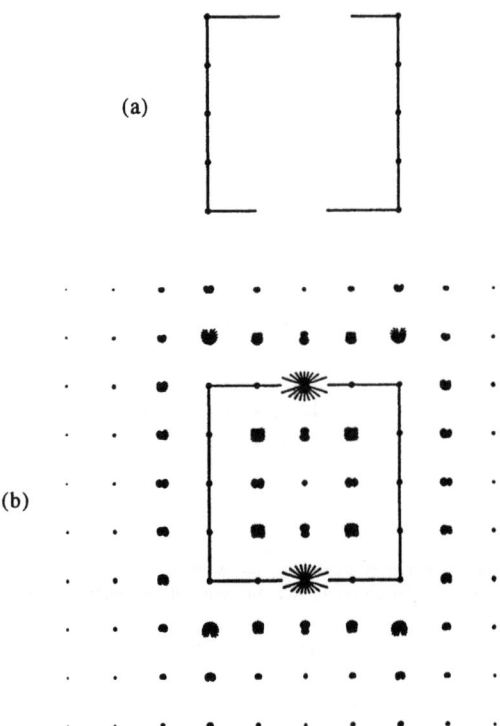

Fig. 17.13. (a) Input and (b) neuron excitations for an example where a subjective square is seen. Observe the increased activity along the top and bottom sides of the square.

The model discussed in this section does not include any mechanism for sharpening the orientation response. The inclusion of such mechanisms would make it possible to obtain much sharper contours. One way to incorporate such effects is to include a winner-take-all mechanism between neurons that lie at the same position but that are sensitive to different orientations.

4. CONCLUSION

Subjective contours are among the most striking of visual illusions and provide an insight into the human visual system. This chapter addresses the problem of determining subjective contours by using an optimality criterion. We posit that a subjective contour is one that is optimal or minimal in some sense.

Based on simple examples of dot patterns that give the impression of a single figure, we propose that the subjective contour in such cases is the curve of shortest length connecting the points. Such a curve adds the least additional information to the image, and is uniquely defined by the location of the points in the image. We formulate an optimization problem whose solution yields a minimal length contour passing through the dots of the image. The optimization task is solved by using sequential unconstrained minimization techniques (SUMTs), and the solution is closely related to elastic methods of solving the traveling salesman problem.

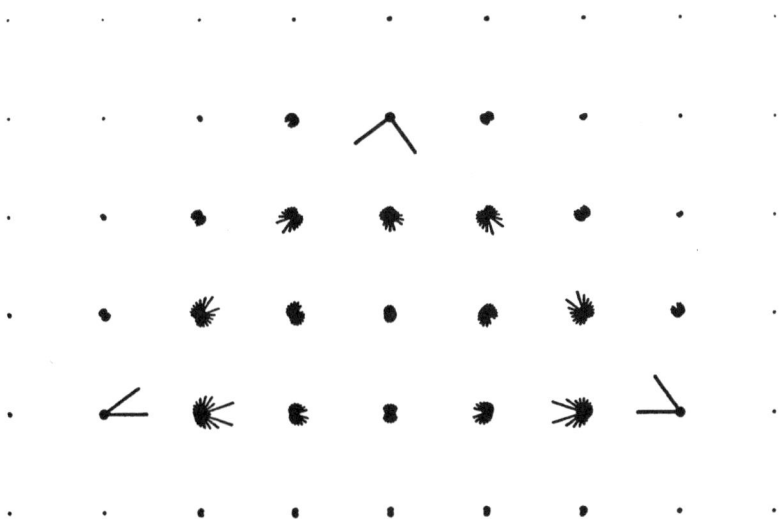

Fig. 17.14. Neuron excitations for an example where a subjective triangle is perceived.

We next look at images where orientation information is present in the form of directed edges at given corners or points of the image. The subjective contour in this case minimizes both length and curvature. We solve for the contour in a manner similar to the previous case. The contours obtained in this fashion are sharp ones found by selecting endpoints of a piecewise linear curve. In practice, subjective contours have a diffuse appearance. We develop a model where the contour is represented by the activity pattern of a plane grid of neurons, but the activity function, corresponding to brightness, is an analog function. The contours obtained in this case have a definite spread of activity. The activity of points in between is computed (filled in) in a feedforward manner, depending on relative locations of input points.

The filling-in process has been postulated as part of preattentive (low level) vision, and it has been argued that sequential approaches to early vision are biologically implausible (Marr,

1982). The criteria used can be extended to determination of three-dimensional subjective contours (Carman & Welch, 1992; Redies & Watanabe, 1993), that is, those where an impression of depth is perceived.

Fig. 17.15. Neuron excitations for an example where a subjective semicircle is seen.

REFERENCES

Brigner, W. I., & Gallagher, M. B. (1974). Subjective contour: Apparent depth or simultaneous brightness contrast? *Perceptual and Motor Skills*, **38**, 1047-1053.

Carman, G. J., & Welch, L. (1992). Three dimensional illusory contours and surfaces. *Nature*, **360**, 585-587.

Clarke, F. H. (1983). *Optimization and nonsmooth analysis.* New York: Wiley Interscience.

Durbin, R., & Willshaw, D. (1987). An analogue approach to the Travelling Salesman Problem using an elastic net method. *Nature*, **326**, 689-691.

Fiacco, A. V., & McCormick, G. P. (1968). *Nonlinear Programming Techniques: Sequential Unconstrained Minimization Techniques.* New York: Wiley.

Geman, D., Geman, S., Graffigne, C., & Dong, P. (1990). Boundary detection by constrained optimization. *IEEE Transactions on Pattern Analysis and Machine Intelligence*, **12**, 609-628.

Grossberg, S., & Mingolla, E. (1987). Neural dynamics of surface perception: Boundary webs, illuminants, and shape-from-shading. *Computer Vision, Graphics, and Image Processing*, **37**, 116-165.

Jayadeva (1993). *Optimization with neural networks.* Unpublished doctoral dissertation, Indian Institute of Technology, Delhi.

Jayadeva, & Bhaumik, B. (1994a). A neural network for the Steiner minimal tree problem. *Biological Cybernetics,* to appear.

Jayadeva, & Bhaumik, B. (1994b). A neural network for finding rectilinear Steiner trees. *Neural Network World,* communicated.

Kanizsa, G. (1976). Subjective contours. *Scientific American,* **234,** 48-52.

Krolak, P., & Felts, W. (1971). A man-machine approach toward solving the Travelling Salesman Problem. *Communications of the ACM,* **14,** 327-334.

Manjunath, B. S., & Chellappa, R. (1993). A unified approach to boundary perception: Edges, textures, and illusory contours. *IEEE Transactions on Neural Networks,* **4,** 96-108.

Marr, D. (1982). *Vision.* New York: W. H. Freeman.

McCormick, G. P. (1983). *Nonlinear Programming.* New York: Wiley.

Pipes, L. A. (1958). *Applied Mathematics for Engineers and Scientists.* New York: McGraw-Hill.

Redies, C., & Watanabe, T. (1993). Illusion and view stability. *Nature,* **363,** 119-120.

Ringer, B., & Skrzypek, J. (1993). A neural model of illusory contour perception. In F. H. Eeckman & J. M. Bower (Eds.), *Computation and Neural Systems* (pp. 184-187). Dordrecht: Kluwer Academic.

Sajda, P., & Finkel, L. H. (1993). Cortical mechanisms for surface segmentation. In F. H. Eeckman & J. M. Bower (Eds.), *Computation and Neural Systems* (pp. 195-199). Dordrecht: Kluwer Academic.

Ullman, S. (1976). Filling in the gaps: The shape of subjective contours and a model for Their Generation. *Biological Cybernetics,* **25,** 1-6.

von der Heydt, R., Peterhans, E., & Baumgartner, G. (1984). Illusory contours and cortical neuron responses. *Science,* **224,** 1260-1262.

IV

OPTIMALITY IN DECISION, COMMUNICATION, AND CONTROL

18

A Developmental Perspective to Neural Models of Intelligence and Learning

Haluk Öğmen and Ramkrishna V. Prakash
University of Houston

Haluk Öğmen and Ramkrishna Prakash's chapter, **A Developmental Perspective to Neural Models of Intelligence and Learning,** *describes a decision-making neural network that is applied to robotics. The choices that the network makes about what objects to approach are not, as Öğmen and Prakash point out, based on "minimization of a* **fixed, predetermined,** *global 'cost function.'" Instead, the choices are based on three interacting sets of criteria: reinforcement (which is close to a traditional utility function, as Rosenstein's chapter in this book points out); novelty (which is close to the "diversity generation" criterion of Prueitt's chapter); and habit.*

Öğmen and Prakash's network combines such design principles as adaptive resonance (see also Carpenter's chapter), opponent processing, and lateral inhibition. In this network, the interplay between reinforcement, novelty, and habit is dynamic; which of these factors is more important is heavily dependent on the environmental context in which the network (or robot) finds itself. Even the stimulus criteria that maximize reinforcement can change dynamically with the context. In fact, the network categorizes objects based on a variety of stimulus attributes, and which attribute or feature of the stimulus is most important changes with context. The network modifies previous work of Samuel Leven and Daniel Levine, who interpreted the effects of frontal lobe damage on a card sorting task in terms of inability to change a categorization criterion. A similar ability to change the relative weighting of criteria is manifested in the face recognition network from the chapter by Abdi, O'Toole, and Valentin, which uses a very different architecture (modified brain-state-in-a-box).

Hence this chapter goes far toward answering the question also raised in Levine's chapter: how does a neural network learn to use previous information about reward to develop high-order rules for what is rewarding? In addition to providing theories for cognitive functions of several brain areas, the capacity for such rule formation and learning will enable neural networks to incorporate some of the capabilities of traditional heuristic programs from artificial intelligence.

ABSTRACT

In the first part of the chapter, we introduce a general developmental model of intelligence and argue that the strict notion of optimization does not constitute a proper level of *analysis* for understanding human intelligence because a fixed, a priori construct (*objective function*) cannot account entirely for the interactive nature of intelligence. In the second part of the chapter we address the *synthesis* question by presenting simulations of a specialized architecture of the general model. The synthesis of this architecture was not based on the traditional optimization methodology in that the minimization of a fixed, predetermined global "cost function" was not the central design principle. No attempt was made to explicitly synthesize such a function. Rather we approached the problem in a more dynamical setting by integrating specialized architectures whose continuous-time interactions synthesize goals that change according to the interactions between the system and the environment. We present several simulations illustrating the interplay between system and environmental dynamics and show, for example, how the system can change its criterion (modify its "cost function") according to prevailing environmental conditions.

1. INTRODUCTION

For three decades, artificial intelligence (AI) research largely ignored the role of experience and learning in intelligent behavior. This neglect fueled the resurgence and popularity of neural networks during the last decade. However, the reaction has been so strong that most neural network models consider experience and learning as the *only* determinants of intelligent behavior. Typically, a homogeneous network ("blank slate") is trained by stimuli ("unsupervised learning") or by stimuli and responses ("supervised learning"). As opposed to traditional artificial intelligence models where a designer puts a priori constructs in a nonadaptive box (see Fig. 18.1a), most neural networks can be viewed as an adaptive box devoid of a priori constructs (Fig. 18.1b). A closer examination of these models reveals, however, that the diagram shown in Fig. 18.1c is a more appropriate representation because a designer is required to interface these networks to the environment. In the absence of a designer, a passive system operating in a high-dimensional complex environment will rapidly be overwhelmed by the combinatorial explosion of possible stimuli combinations in time. In fact, practice clearly shows the need for a designer in these networks to structure the environment (e.g., selection of appropriate features, desired response patterns, introducing some a priori bias to the network by selecting the number of hidden units, etc.) This problem has been called the *bias-variance dilemma* (Geman, Bienenstock, & Doursat, 1992) or *stability-plasticity dilemma* (Grossberg, 1980).

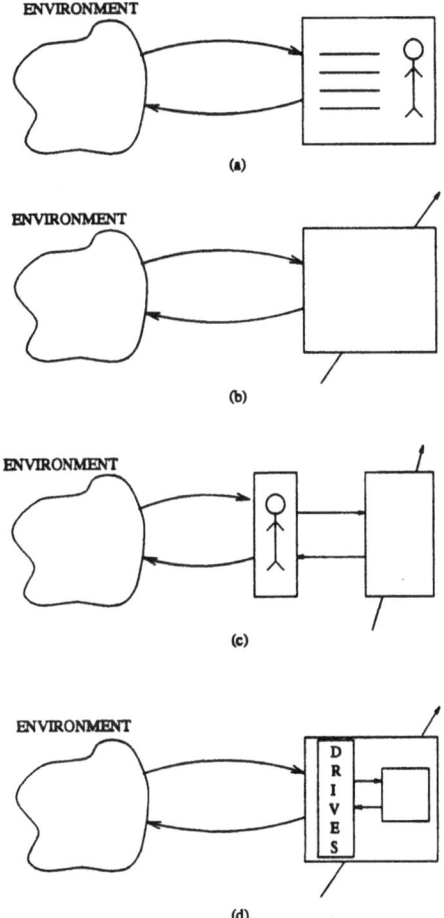

Fig. 18.1. Diagrammatic representation of (a) artificial intelligence, (b, c) several neural networks, and (d) some behavioristic neural models.

Once an external designer is introduced, the role of the network is reduced to a "continuous memory" capable of interpolating and extrapolating the structures created by the designer. Therefore, the main theoretical issue in these networks has been the synthesis of an *optimal memory* according to a criterion selected by the designer. Optimization may be a useful and sufficient tool for designing and analyzing such networks. However, the main intelligent agent, the designer, is not characterized by such an analysis. In order to link the environment to the system without a designer, several behaviorist models introduced "intervening variables" such as expectations and drives (reviewed in Zuriff, 1985; see Fig. 18.1d), a tradition that continues in some current neural network models (e.g., Sutton & Barto, 1981). The extent to which such constructs can solve the problem depends on their origins and nature. If they are synthesized purely from experience, then Fig. 18.1d becomes equivalent to the purely empirical model shown in Fig. 18.1c. If they are innate and do not change through experience, then we are faced with the problem of the construct shown in Fig. 18.1a, namely, the inability for the system to expand autonomously its "mental universe."

2. A DEVELOPMENTAL MODEL OF LEARNING

Our approach to this problem finds its roots in an invertebrate model of sensory-motor reflexes (Öğmen, 1992a) shown in Fig. 18.2. The figure illustrates the general form of the model specialized for the landing reflex of the fly (Öğmen & Moussa, 1993). The model consists of three components: sensory, motor, and sensory-motor gate layers. The necessity of sensory-motor gate layers arises from the observation that no matter how well the sensory part is tuned to the triggering features of the reflex, since many of these features can occur in a variety of contexts in the environment, a context-dependent interpretation of incoming stimuli is necessary. For example, an expanding pattern when the fly is in the landed position (e.g., approaching predator) triggers the initiation of flight. An expanding pattern during flight either is ignored (background) or triggers landing (approach to a site), depending on the internal state (e.g., flight velocity, habituation level) as well as on stimulus characteristics (Öğmen & Moussa, 1993). Context-dependent filtering of the input is implemented by lateral connections within a sensory-motor gate layer, indicating a competition to take control of the motor system between multiple reflexes. This input gating is genetically wired. Therefore, this model interacts with the environment through innate, a priori structures. However, as we mentioned earlier, although learning can fine-tune these reflexes, the system will never be able to go beyond its innate constructs. Therefore, although such a model can explain learning in lower species and in some simple human reflexes, we suggest that it is inadequate to account for human intelligence in general.

We generalized this model (Öğmen, 1992a, 1992b, 1995) by using the Piagetian concept of *scheme* with the functions of *adaptation* (assimilation and accommodation) and *organization* (Piaget, 1963). The augmented model, which we will interpret as a sensorimotor reflex for definiteness, is shown in Fig. 18.3. The filled circular symbols represent the sensory and motor elements involved in a reflex that plays an essential role in the development of cognitive functions. Sucking, grasping, and rooting are examples of such reflexes (Piaget, 1963). The scheme is a totality, a property imposed by units that we call *affective units*. The triangular symbol (a^+), a *primary affective unit*, is connected, directly or indirectly, to all elements of the scheme and unifies the scheme into a totality. As shown in the figure, the rectangular units, which we call *secondary affective units*, bring an organization to the scheme by imposing a cyclic order (in this case $s_1 \rightarrow m_1, m_1 \rightarrow s_2, s_2 \rightarrow m_2, m_2 \rightarrow s_3, s_3 \rightarrow m_3, m_3 \rightarrow s_1$). Comparing this model to its ancestor shown in Fig. 18.2, one can see that the secondary affective units are similar to the sensory-motor gate circuits. However, although the model of Fig. 18.2 is organized into a totality only by the external stimuli, the totality and closure of the scheme in Fig. 18.3 are its inherent organizational properties. When organization and adaptation go hand in hand, learning involves not only the modulation of existing local connectivity in a scheme, but also an active modulation of cyclic totalities (Öğmen, 1995).

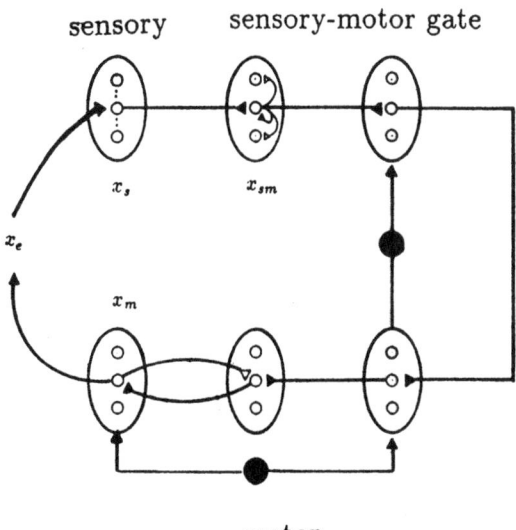

Fig. 18.2. General structure of the invertebrate model for sensory-motor reflexes. Variables x_e, x_s, x_{sm}, and x_m denote environmental, sensory, sensory-motor, and motor variables, respectively. (From Öğmen, 1992a; adapted with permission.)

According to this model, sensory, motor, and affective components are inseparable, and more generally *all cognitive acts involve affective components* which regulate the schemes of the subject. The subject is trying to assimilate all possible objects into his or her schemes. In order to do so in a complex environment, the subject adapts his or her schemes to the environmental data without destroying already existing structures. Each time such a reorganization succeeds the subject becomes better adapted to the object. The idea of comparing different equilibria of the subject is certainly a useful tool of analysis, and indeed intelligence can be seen as a progression from one equilibrium to a better equilibrium. Such a process can be seen as a form of optimization. However, we propose that the strict notion of optimization does not offer the proper level of analysis because an objective function rooted in an a priori environment and an a priori subject cannot cope with an interactively changing system. The need for a repetitive updating of the objective function necessitates an understanding of the changing schemes of the subject which in turn requires the study of interactions between subject's schemes and the environment, the proper level of analysis according to our approach.

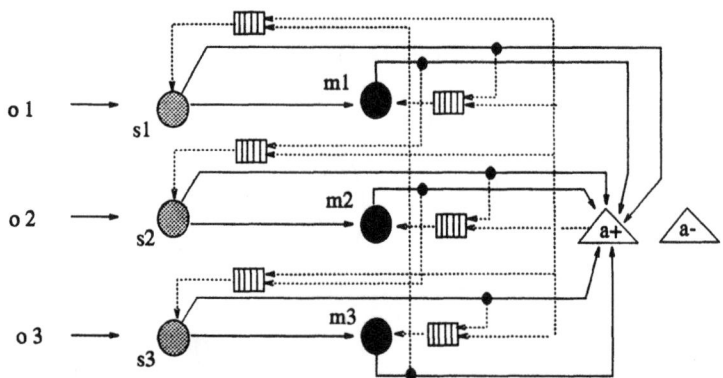

Fig. 18.3. General developmental model for a simple scheme. Variables *o*, *s*, and *m* denote environmental, sensory, and motor variables, respectively. Triangular and rectangular nodes represent primary and secondary affective units, respectively. Solid and dashed lines show connectivities between sensory, motor, and affective components.

In a specific version of this general model, we showed how generalizing and recognitory assimilation interfaces the environment in a context sensitive way to the system (Prakash & Öğmen, in press) through simulations that were mainly directed to reproducing some characteristics of the reflex stage. In the subsequent stages, active target seeking plays a crucial role in the modification of existing schemes (secondary circular reactions, Piaget, 1963) and the formation of new schemes. In the rest of this chapter, we present a specialized architecture for target selection.

3. TARGET SELECTION ARCHITECTURE

Having argued against optimization as an analysis tool we now consider whether optimization can be an adequate design tool. There is a tradition in modelling literature that views model design as a problem in optimization theory (reviewed in Poggio, Torre, & Koch, 1985; Yuille, 1989). We argued elsewhere that the strict notion of optimization does not constitute an adequate design tool for modeling continuous-time vision (Öğmen, 1993a, 1993b, in press). A major shortcoming of the optimization approach is its inability to cope with continuous-time interactive nature of biological nervous systems via its fixed a priori object function. We illustrate continuous-time interactions between the environment and our model by simulating a specialized part of the general model, the target selection architecture.

3.1. General Structure

This architecture is a component of the grasping reflex and fulfills the target selection function. The general structure of the network is shown in Fig. 18.4. The shaded pathway from visual inputs conveys features (determined in general by the assimilation scheme in which this module is inserted) of the visually attended target to categorization networks. The visual attention is controlled by the attentive scanning network, which receives inputs from the spatial novelty and categorization networks. The signal from the spatial novelty network biases the scanning process toward objects placed in novel locations (e.g., an object that is moved to a new location will attract the attention of the system), whereas the signal from the categorization network freezes the scanning process so that while the categorization is in process, the input is held constant. The first type of categorization ("good-bad") distinguishes harmful objects and prevents the system from grasping such objects. The second type of categorization ("object-type") analyzes the appearance of objects. The output of this categorization network is processed by a novelty detector. The target selection process combines signals from the categorization networks and determines the object that will be grasped.

In the following sections, we analyze this network by computer simulations. All neural architectures were simulated on an Amdahl supercomputer. The ordinary differential equations (ODE) defining the networks were solved using a numerical ODE solver (the Runge-Kutta-Fehlberg 4-5 method) developed by Sandia Laboratories, Albuquerque, NM. The user interface of the simulation enabled the modification of external signals (introduction or removal of objects and external reinforcement signals) by interrupting the program as and when needed. On interruption all the state variables of the network are pushed onto the stack of the computer and the interrupt is handled. On returning back from the interrupt, these state variables are reloaded back and the network equations are solved from the same internal state of the network before the interrupt occurred. This architecture has also been implemented and tested in a robotic setup composed of a camera, a DATACUBE real-time image processing board, and a PUMA 562 robotic arm with a CYBERNETICS controller (Öğmen & Prakash, 1993).

3.2. Spatial Novelty and Attentive Scanning

In a first series of simulations we show how the input stage of the architecture is interfaced with the environment. The network attends to novel stimuli while ignoring (filtering) others. The spatial novelty network consists of an array of gated dipoles (Grossberg, 1972), each sensing a discrete spatial position.[1] The outputs of the spatial novelty network project to the attentive scanning module. This module consists of (a) a "winner-take-all" network wherein each neuron excites itself and inhibit all others, and (b) a feedback pathway through additional layers (delay and inhibitory neurons) to regulate cyclic scanning of attention (see Fig. 18.5).

---

[1] Currently spatial locations have been restricted only to two dimensions by keeping the depth constant.

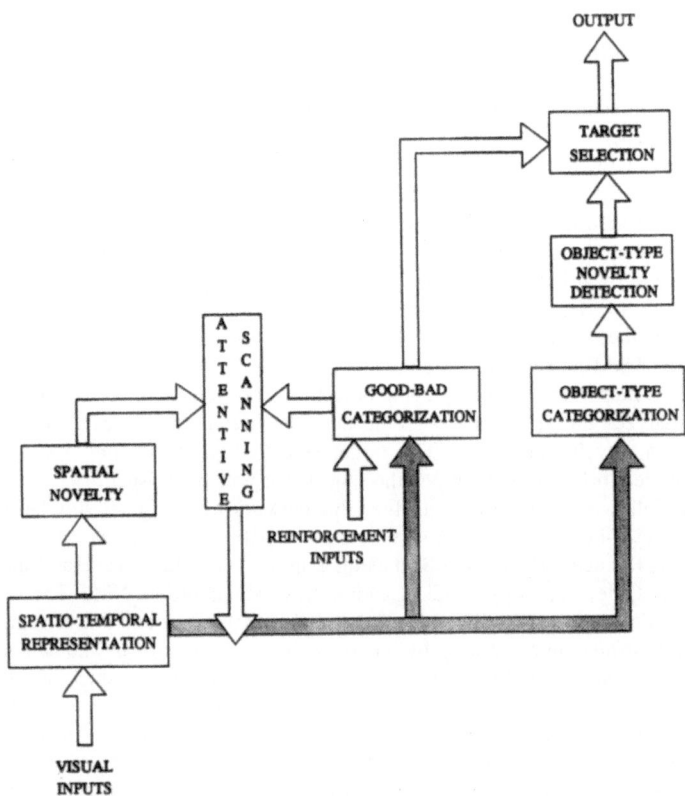

Fig. 18.4. Block diagram of the target selection architecture.

We first demonstrate novelty detection by simulating the gated dipole network along with the winner-take-all layer of the attentive scanning module. The combined network was first proposed by Levine and Prueitt (1989; see also Levine, 1991) and also forms the seed of the object-type novelty detection module (see Figs. 18.4 and 18.16). Equations presented in Appendix A.1 were solved numerically for inputs shown in Fig. 18.6. The four panels show the input signals at four distinct spatial locations as a function of time. A signal value of 1 (0) indicates the presence (absence) of an object at that spatial location. Each panel in Fig. 18.7 shows the activity of a neuron in the "winner-take-all network" (see Fig. 18.5) sampling the input shown in the corresponding panel of Fig. 18.6. A neuron wins the competition when its activity exceeds its threshold value indicated by a dashed line in Fig. 18.7. As one can see from Fig. 18.6, the temporal order according to which inputs are introduced proceeds from the upper to the lower panel. Objects are removed from the environment according to the reverse order. The suprathreshold activities of neurons in Fig. 18.7 show that the neuron that samples the most novel input wins the competition. Thus, other things being equal, novelty guides the attention of the network. Note also that during competition, brief transient signals may go above threshold in the winner-take-all network. In order to eliminate such transients an additional feedforward on-center-off-surround network is introduced as shown in Fig. 18.5.

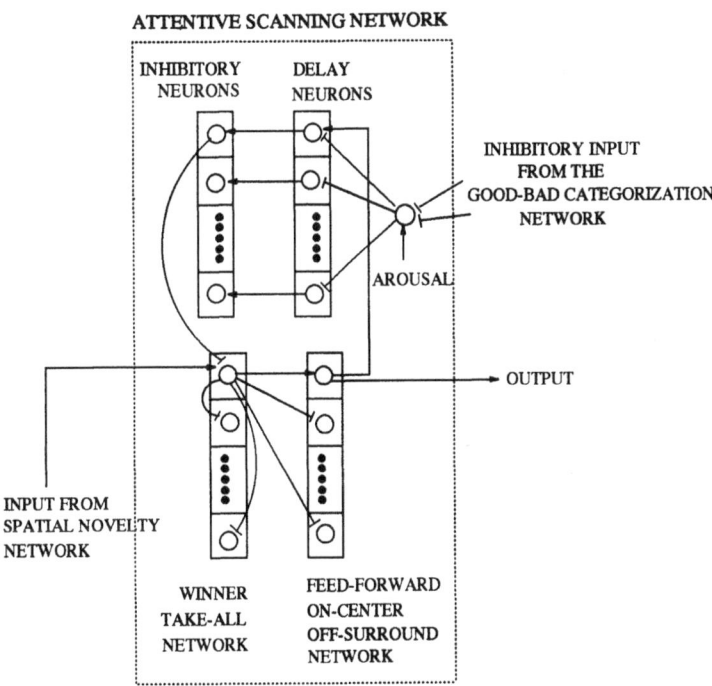

Fig. 18.5. Attentive scanning network.

Once a neuron wins the competition it will remain the winner as long as the inputs do not change. The upper two layers in Fig. 18.5 are added to prevent this and to generate a cyclic scanning of all objects in the environment by the attentional system. The cyclic scanning is achieved by a delayed inhibitory recurrent signal, which reduces the net input of the winning neuron thereby allowing other neurons to win the competition. Simulations illustrating this

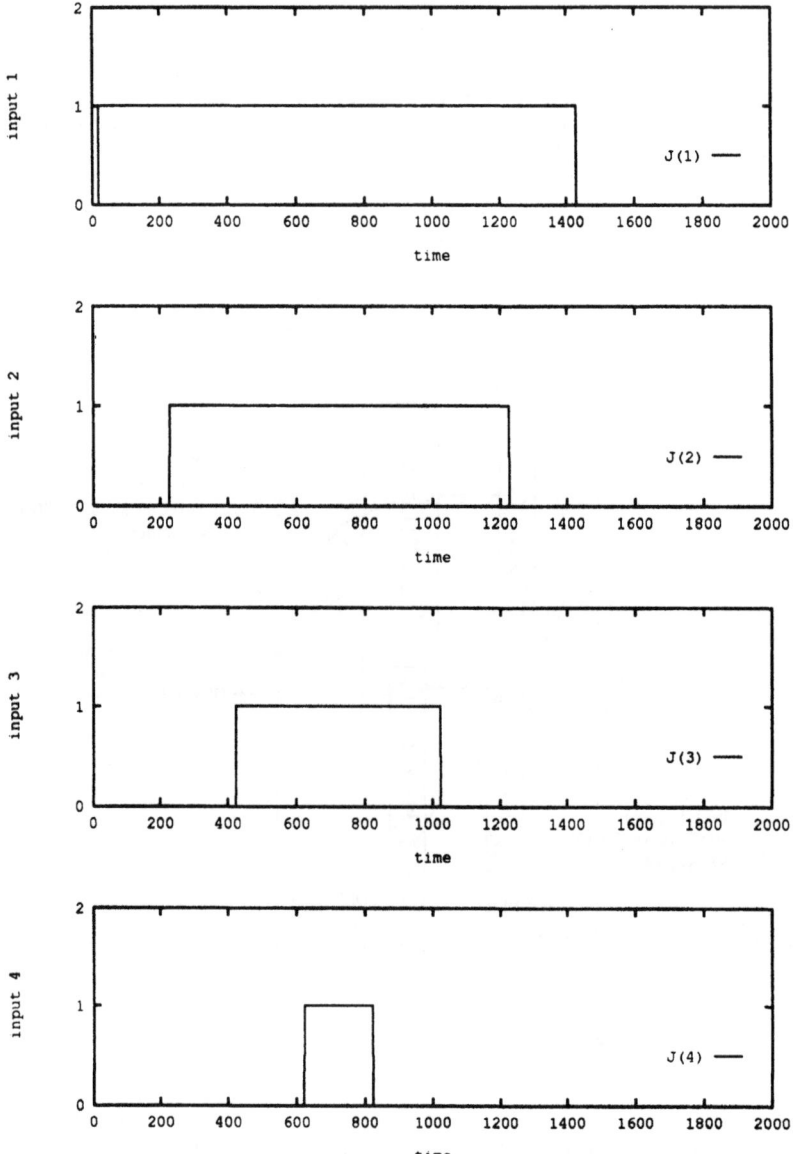

Fig. 18.6. Inputs to the novelty detection network. The four panels in this figure represent the sequence in which four inputs at four distinct spatial positions are introduced and removed. A high signal (1) implies the presence and a low signal (0) indicates the absence of an object at a particular spatial location. The response of the network is shown in Fig. 18.7.

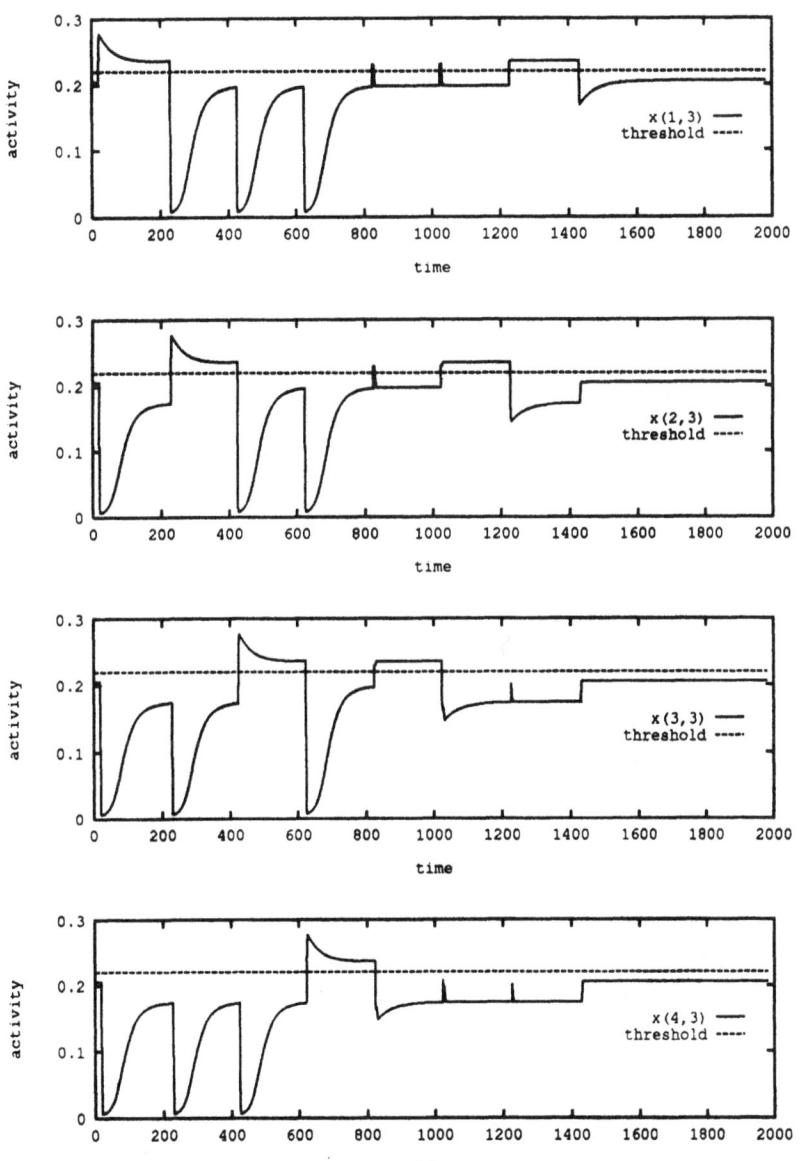

Fig. 18.7. Each panel shows the activity of a neuron in the "winner-take-all network" (see Fig. 18.5) sampling the input shown in the corresponding panel of Fig. 18.6. A neuron wins the competition when its activity exceeds the threshold value indicated in the figure.

scanning process are shown in Fig. 18.8. The lower panel in Fig. 18.8 shows the sequence of presentation of inputs to the network. Three inputs, placed at three different spatial locations,are presented successively to the network. The upper three panels show the activities of output neurons of the attentive scanning module sampling these three spatial locations. Following the introduction of the first input, the system starts to scan this input. When the second and third inputs are introduced, all three inputs are scanned serially. As one can see from the simulation results, after some time the novelty of inputs vanishes and the system stops scanning the inputs.

3.3. Good-Bad Categorization

An important aspect of grasping behavior concerns the avoidance of harmful objects such as hot objects. Basic information is provided through simple reflexive circuits and is used in the model as "reinforcement" signals. However, because reinforcement signals are unspecific, the internal criterion of the network in selecting the relevant features must be dynamic. To understand this, consider the situation shown in Fig. 18.9. Before deciding to reach for an object, the system has to categorize it as a "good" or a "bad" object. Bad objects are typically those that have been correlated with negative reinforcement signals and the system avoids such objects. Assume that the pattern shown at the bottom of the figure illustrates the features of the input object and that the two patterns shown at the top of the figure illustrate the templates for good and bad objects. As seen from the figure, this categorization problem is ambiguous in that if color is taken as the categorization criterion then the input will be categorized as a good object, whereas if shape is taken as the categorization criterion then the input will be categorized as a bad object. The choice of the categorization criterion is guided by habit and reinforcement signals.

The "good-bad" categorization network is a variant of an architecture proposed by Leven and Levine (1987; see also Levine, 1991) and is shown in Fig. 18.10. It consists of a cascade of two ART networks (Carpenter & Grossberg, 1988): layers F_1 and F_2 categorize inputs into object-types; layers F_2 and F_3 categorize inputs into good and bad categories. As the reinforcement signal is nonspecific, one must ensure that it is correctly assigned to the current "choice" of internal criterion. This is achieved by the circuitry shown in the left part of the figure. Match neurons encode which internal criterion is currently being used by the network to achieve categorization. Habit neurons memorize past experiences of the network. Bias neurons combine habit and reinforcement cues to select an appropriate internal criterion for categorization. This selection of the internal criterion, encoded in the activities of bias neurons, is used to gate the bottom-up weights of the F_1—F_2 ART network. Thus this neural network combines reinforcement signals with its past experiences to dynamically modify its internal criterion for categorizing input objects. The decision and ambiguity neurons were added to the Levine-Prueitt model to enable this network to function in a continuous nonalgorithmic fashion in a dynamic environment (Öğmen & Prakash, 1992, 1993; Öğmen, Prakash, & Moussa, 1992). The categorization layers (F_2 and F_3) of the ART network can generate transients due to competition amongst the category neurons during categorization. If this network were interfaced with other systems, these transients could introduce undesirable effects. To avoid this, a *decision layer* was added. The decision neurons, which form a feedforward on-center-off-surround network, filter transients and pass only the steady-state decisions of the ART network to other systems.

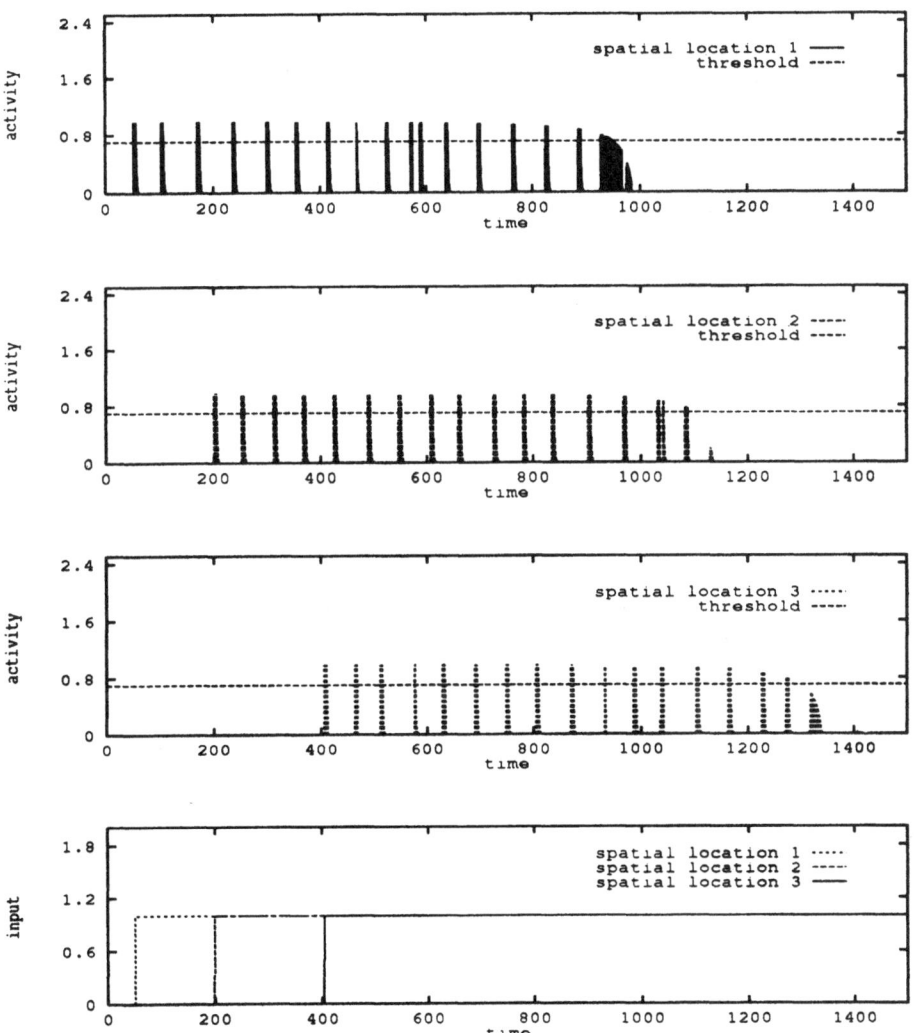

Fig. 18.8. The top three panels illustrate the activities of the output neurons of the attentive scanning network (see Fig. 18.6) sampling the three objects shown in the last panel. The network intermittently scans these three objects until their novelty wears out.

Template
for the
"good" category

Template
for the
"bad" category

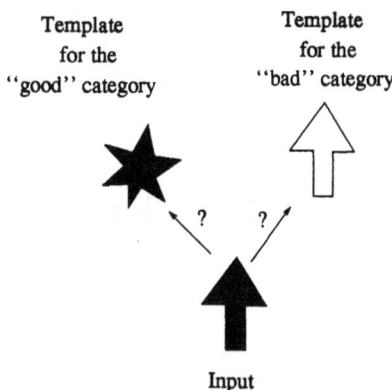

Input

Fig. 18.9. Input object shown at the bottom of the figure has to be categorized into one of two categories whose templates are shown at the top of the figure. The template at left may be for example for "good objects" that the system has learned to pick; the template at right may be for "bad objects" that the system has learned to avoid. The categorization here is ambiguous in that if color is taken as criterion then the input is a good object but if shape is taken as criterion then the input is a bad object.

When the prior history of reinforcement and habits does not allow to resolve ambiguities such as the one illustrated in Fig. 18.9, the ambiguity neuron resolves the conflict by biasing one of the categories. The ambiguity neuron monitors the decision neuron layer to determine if the input object has been categorized within a certain duration of time. This monitoring is achieved by integrating a constant input that indicates the presence of an input to be categorized. When a categorization is made, the integration is reset. However, if the network fails to reach categorization before the integration reaches a threshold value, the ambiguity neuron generates an output that biases one of the possible categories over the others.

In our simulations, objects presented to the network had three features and each feature had four possible types. For example, four possible colors could be blue, red, green, and yellow. This is illustrated graphically in Fig. 18.11, where for definiteness features are interpreted as size, color, and shape. Thus 4 × 4 × 4 = 64 distinct objects could be presented to the network. Each object is graphically represented by a three-dimensional vector as shown in Fig. 18.11b. Four different bar patterns (solid, thick stripe, thin stripe, and thick-thin stripes) are used to represent graphically the four possible types of each feature. The first object shown in Fig. 18.11b consists of thick-stripe, solid, and thin-stripe bar patterns. The first thick-stripe bar implies that the object is of medium size, the second solid bar implies that the object is blue, and finally the third thin-stripe bar implies that the object is a circle. Similarly, the graphical representation for the second object consisting of thick-thin-stripe, thick-stripe, and solid bars implies that the object is a huge red square. Another aspect of stimuli is the length of time they were presented to the network before a categorization was made. This duration is represented graphically by the width of the bars.

The lower part of Fig. 18.12 shows a case in which a medium blue circle was presented to the network between 5 and 10 time-unit followed by a huge red square that was presented between 40 and 50 time-units. The output graphs shown in Fig. 18.12 plot the activity of the four category neurons (in layer F_2 of the network in Fig. 18.10), each representing a particular

GOOD-BAD CATEGORIZATION NETWORK

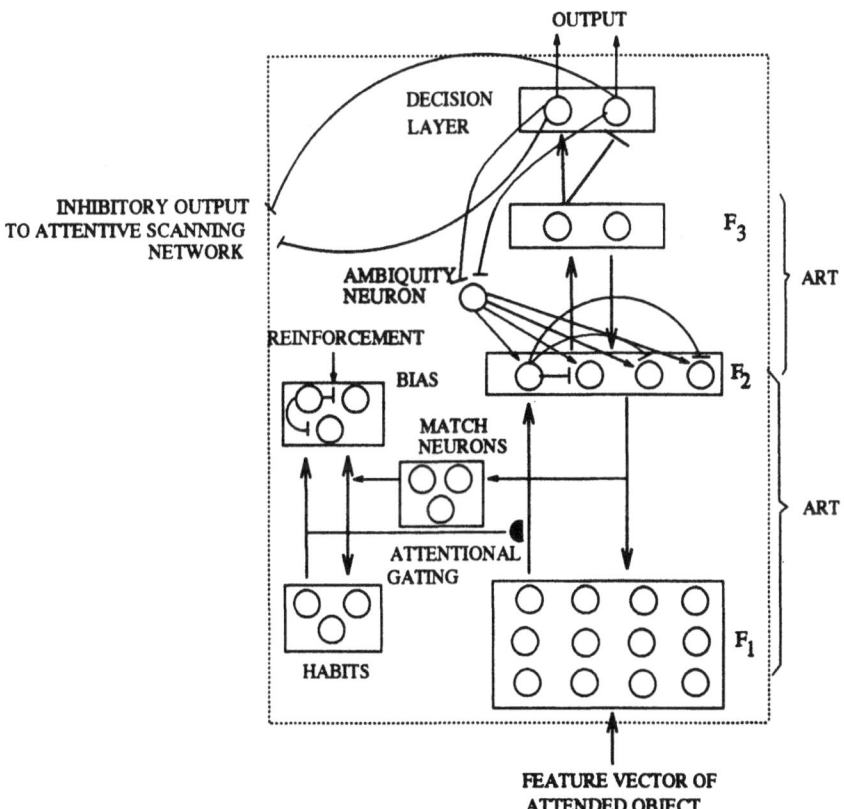

Fig. 18.10. Good-bad categorization network. (Adapted from Leven & Levine, copyright © 1987 IEEE; reprinted with permission of the publishers.)

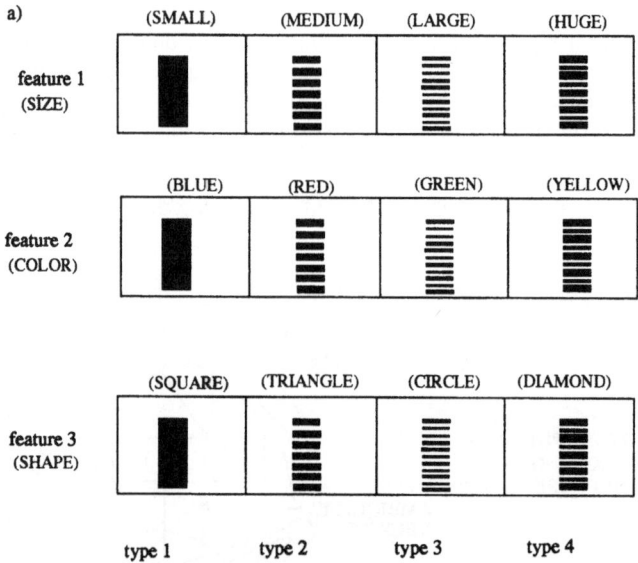

a)

type 1   type 2   type 3   type 4

b) Examples of a graphical representation of objects

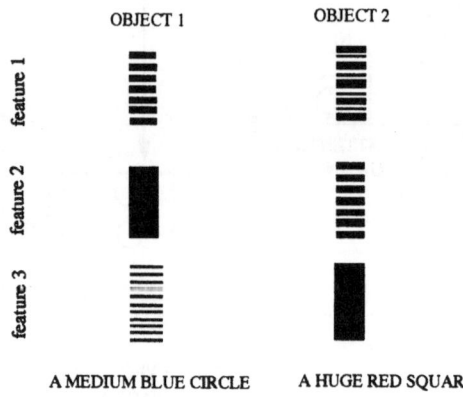

Fig. 18.11. Graphical representation of inputs.

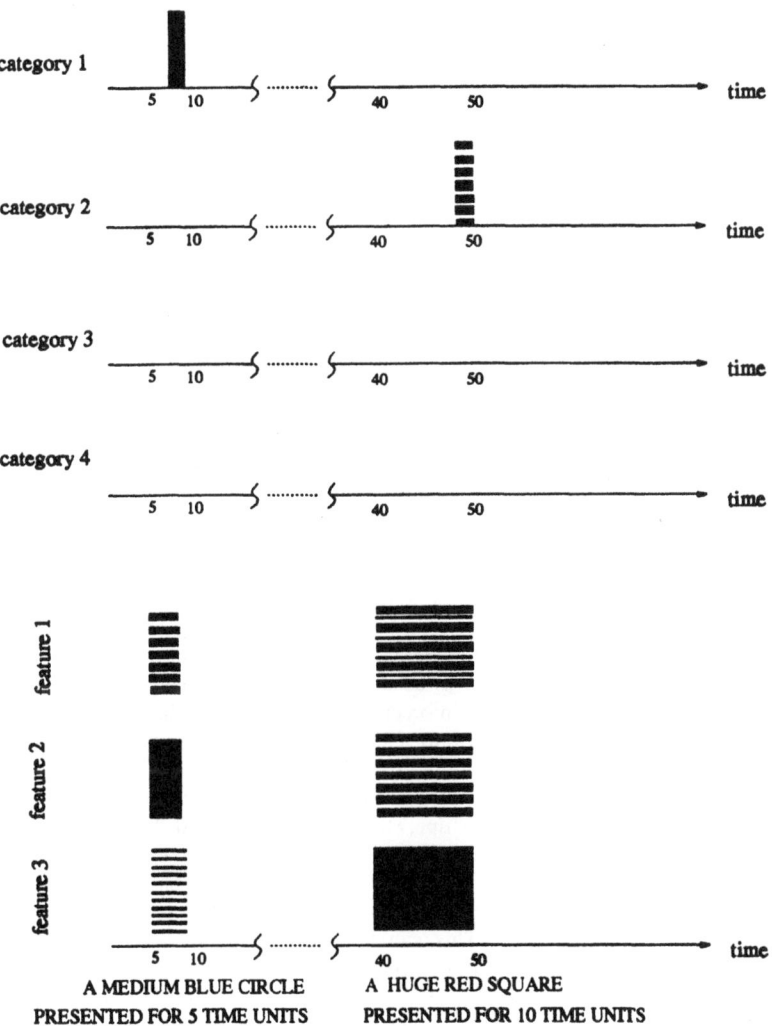

Fig. 18.12. Graphical representation of the F_2 category neuron activities of the good-bad categorization network and identification of the criterion used by the network.

type a feature can take. By comparing the feature types of the medium blue circle shown in Fig. 18.12, one can see that it can be categorized to category 2 if size is used as the criterion (medium is the second type of feature size as shown in Fig. 18.11), to category 1 if color is used as criterion (blue is the first type of feature color), and finally to category 3 if shape is used as criterion (circle is the third type of feature shape). The activities of the network category neurons are plotted by bars that have the same pattern as those used to represent particular types of features to allow an easy identification of the network criterion. That the medium blue circle is categorized into category 1 because the network used feature 2 (color) as criterion is made explicit by the bar pattern used in plotting the category neuron activity.

The upper three panels of Fig. 18.13 describe graphically the inputs presented to the network at different time instants, with each panel representing an input feature as previously described. These inputs are categorized into one of four possible categories. Figure 18.14 shows the category chosen by the network for a given input. Each panel represents the activity of a neuron in the category layer of F_2 the ART network (see Fig. 18.10). The suprathreshold activity implies that the input is categorized to that particular category. Category neuron activities have similar styling as the four possible types of each feature as mentioned before. At any instant of time, at most one category has a supra-threshold "bar" indicating that the network classified the input object to that category. The feature used by the network to categorize the input can be identified by comparing which of the first three panels of Fig. 18.13 has a bar pattern similar to the category panel at the given instant. The first input presented to the network is a medium, blue circle represented by the thick-stripe, solid, and thin-stripe bar patterns.

The network categorizes this input to category 3 represented by a thin-stripe bar as shown in Fig. 18.14. Thus the internal criterion used by the network was feature 3. The fourth panel in Fig. 18.13 represents the reinforcement signal given to the network for its behavior (categorization). Initially feature 3 was used by the network to categorize the objects, but on receiving two consecutive negative reinforcements at around 380 time-units and 430 time-units the network shifts its internal criterion to feature 2. Positively reinforcing the network causes it to continue the use of feature 2 as the criterion for categorization. This can be verified by considering the input presented to the object just prior to 600 time-units. The input is a huge, green circle represented by thick-stripe, thin-stripe, and solid bar patterns, respectively. Comparing the category neurons outputs at the same time interval, category 3 neuron's activity represented by thin-stripe bars is above threshold. Thus feature 2 has been used as the criterion to classify the input object. Later, a change in the internal criterion to feature 1 is achieved by issuing negative reinforcements. This can be verified by observing that the network has categorized the object presented just after 1000 time-units using feature 1 as the criterion.

The upper three panels of Fig. 18.15 display the activities of the bias neurons (see Fig. 18.10) for the three features used as categorization criteria. These neurons integrate the reinforcement and habit signals to modulate the internal criterion used by the network. Initially, because the criterion used by the network was feature 3, the bias neuron 3 is active (the dashed line represents threshold). After receiving negative reinforcement signals, the activity of bias neuron 3 falls below threshold. This results in a rise in the activity of bias neurons 1 and 2 toward the threshold. Positive reinforcement issued to the network for using feature 2 causes the

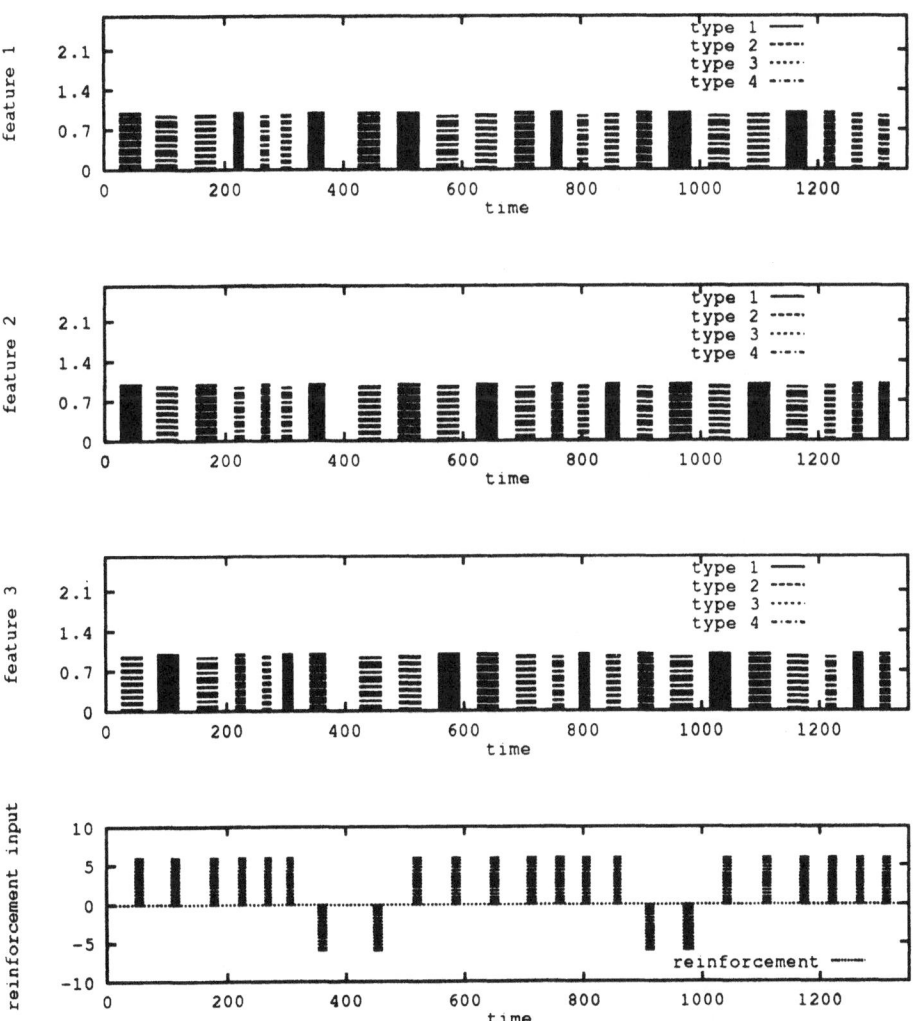

Fig. 18.13. The first three panels describe the features of inputs presented to the network at different time instants. Each input possesses three features (e.g., size, color, and shape) and each feature can take one of four possible values (types). Hence 64 different inputs can be presented to the network. Each of the first three panels represents a feature. The four different styles of bars in each panel represent the four different types of a given feature (e.g., for color they may correspond to blue, red, green, and yellow). At any instant of time, the bars represented by the three panels describe the properties of the input presented to the network, for example, the first input is of type 2 of feature 1, type 1 of feature 2 and type 3 of feature 3. The categorizations performed by the network are presented in Fig. 18.14. The last panel describes the reinforcement signals delivered to the network in response to its categorization of the object.

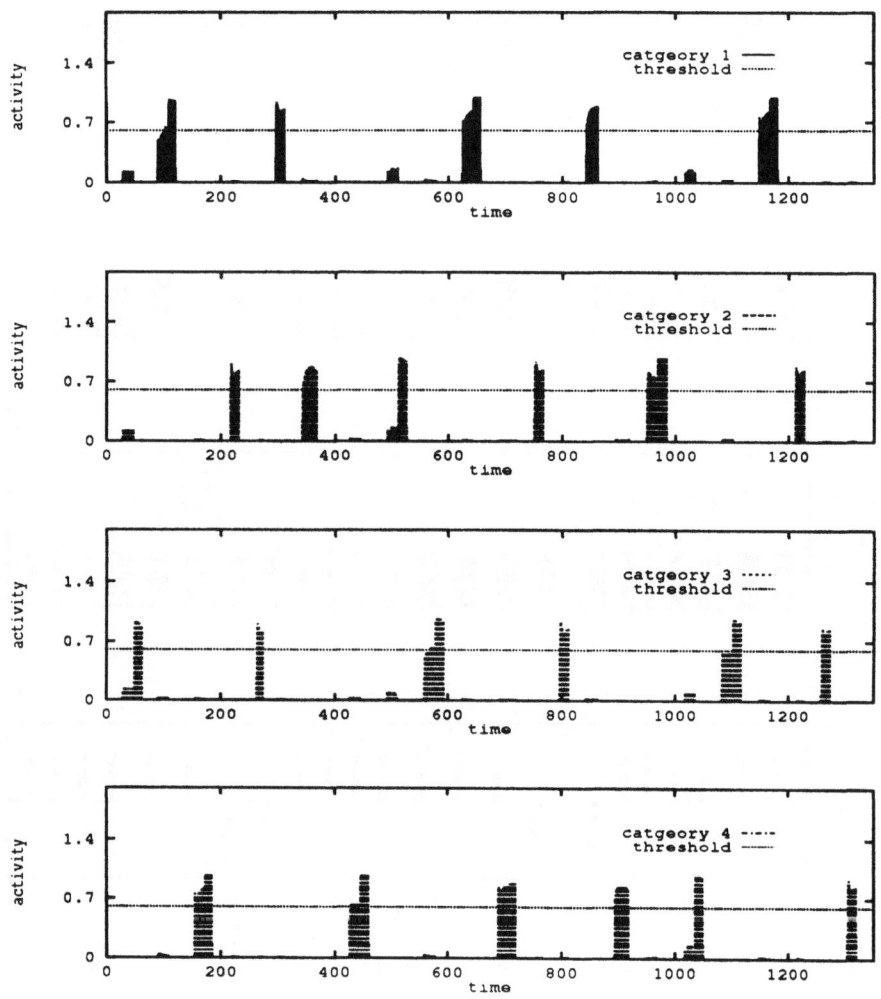

Fig. 18.14. The activities of categorization layer neurons in the F_2 layer of the good-bad categorization network. Each of the panels represents the activity of a single category neuron as a function of time.

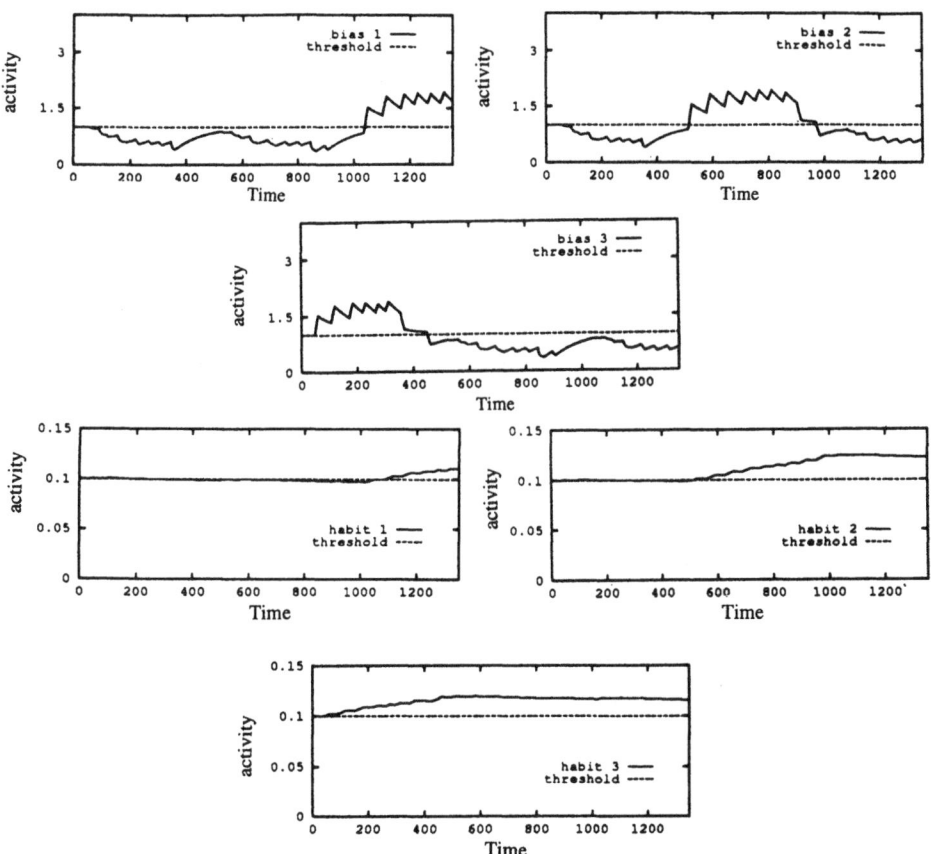

Fig. 18.15. The top and bottom three panels plot the activities of the bias and habits neurons, respectively, of the good-bad categorization network.

activity of bias neuron 2 to increase above threshold. Similarly, later when the network uses feature 1 and is rewarded, the activity of bias neuron 1 rises above threshold. The lower three panels display the activity of the habit neurons (see Fig. 18.10). These neurons encode how many times a given feature was used to categorize objects. Initially the activity of habit neuron 3 increases as feature 3 is used by the network to categorize the objects. However, as features 2 and 1 are later used for similar durations, the activities of the three habit neurons are nearly the same. Notice also the slow decay of habit neurons when the corresponding feature is not used, illustrating that previous experiences are not easily forgotten.

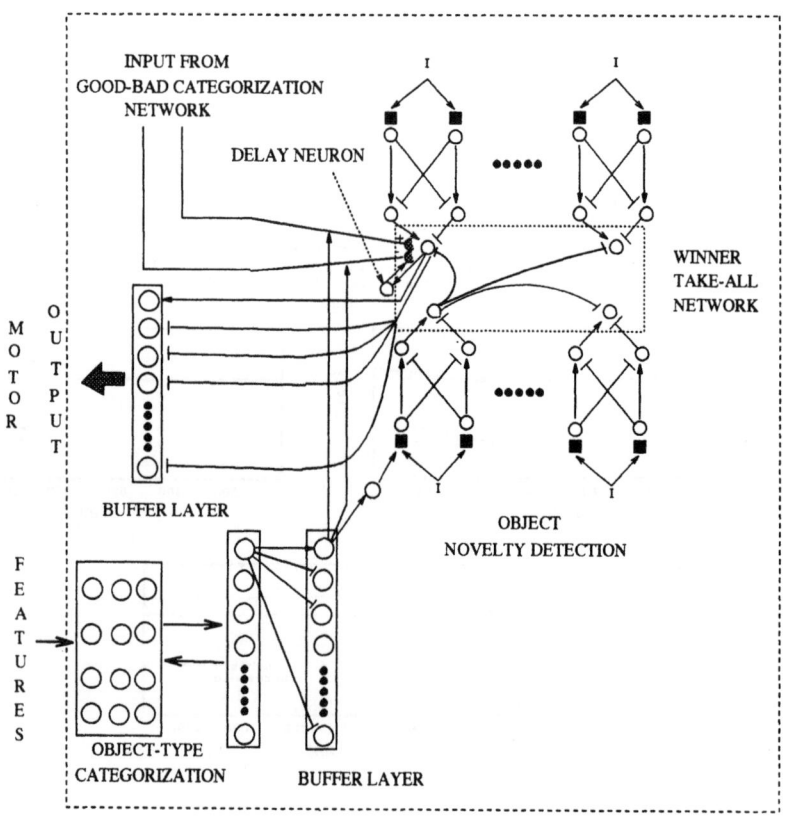

Fig. 18.16. Object-type categorization, object-type novelty, and target selection architectures. (From Levine & Prueitt, 1989. Adapted with permission.)

3.4. Object-Type Categorization, Object-Type Novelty, and Target Selection

Figure 18.16 shows the architecture of object-type categorization, object-type novelty, and target selection modules shown in Fig. 18.4. The first two layers in the lower left part of the figure correspond to a categorization network (ART) that categorizes inputs into different object types, and the third layer is a buffering layer similar to the layer shown in Fig. 18.10. The output of each buffered category neuron feeds a gated dipole to produce an object-type novelty detection. Finally, the outputs of these gated dipoles feed to a winner-take-all target selection network, which makes the decision regarding the object to be grasped. The decision is buffered and sent to motor control circuits. An additional input from the good-bad categorization network biases the competition in the target selection layer by inhibiting (exciting) negatively (positively) reinforced objects.

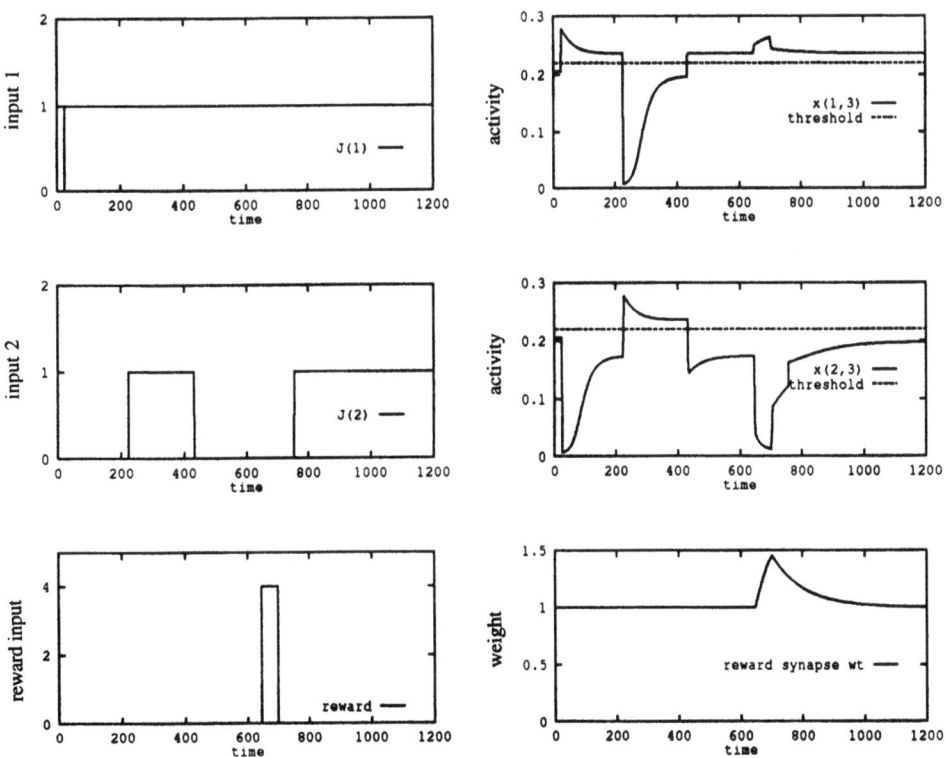

Fig. 18.17. Effects of positive reinforcement.

Simulation results shown in Fig. 18.17 illustrate how positive reinforcement can bias the target selection process. The left column shows the timing of visual (input 1 and input 2) and reward inputs that were delivered to the network. The upper two panels in the right column plot the activities of two neurons, $x(1,3)$ and $x(2,3)$, in the winner-take-all network of Fig. 18.16.

Because these two neurons compete, only one can be suprathreshold at a given time. When activity of $x(1,3)$ $(x(2,3))$ goes above threshold, the system selects input 1 (input 2). When input 1 is introduced, the network selects and grasps this input object. When input 2 is introduced, the network selects this second object. The novelty of the second input biases the competition. The activities $x(2,3)$ and $x(1,3)$ go above and below threshold, respectively. When input 2 is removed, the network switches back to input 1. Now, while the network has selected input 1, a positive reinforcement is delivered for about 20 time-units as shown in the bottom panel of the left column. After this reinforcement, when input 2 is introduced again, the network continues to select input 1 although input 2 is relatively more novel. This is the result of the previously delivered positive reinforcement that the network associated with input 1. The bottom right panel shows the synaptic weight by which the positive reinforcement is encoded into the long-term memory of the system.

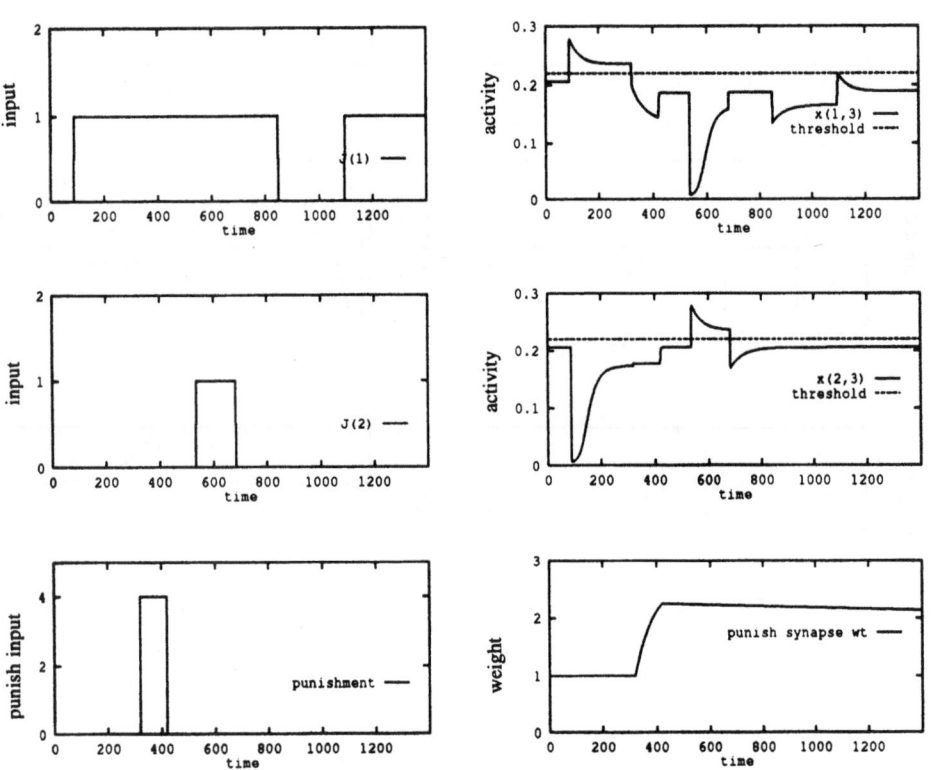

Fig. 18.18. Effects of negative reinforcement.

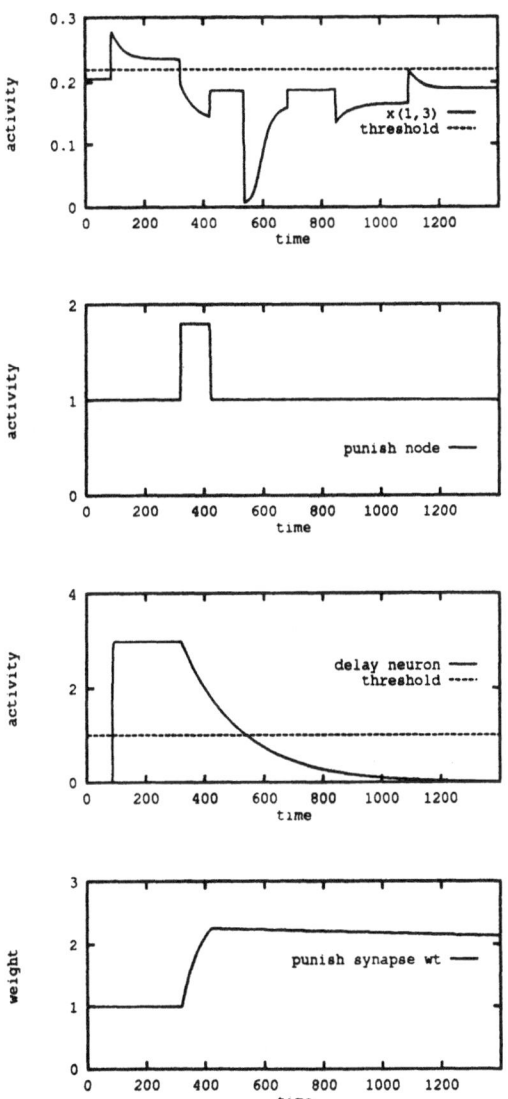

Fig. 18.19. Temporal dynamics of avoidance learning. While there is very little temporal overlap between post- and presynaptic activities ($x(1,3)$ and punish node, respectively), the overlap between the delay neuron and the punish node enables the coding of negative reinforcement into synaptic weights.

Negative reinforcement, on the other hand, yields opposite effects, as shown in Fig. 18.18. When an input is associated with punishment, the network learns to avoid that input. Even when the input reappears much later, its novelty is not strong enough to bias the network to select it. Thus the network learns to avoid punishing inputs even though they could be relatively more novel. The effects of both punishment and reward fade away with time if further reinforcements are not issued, and eventually novelty dominates. As one can see by comparing the decay of synaptic weights associated with positive and negative reinforcement, our parameter choice implied a longer retention of negative reinforcement. The connections from good-bad categorization network to the winner-take-all target selection layer are learned. Encoding of the short-term memory activity (STM) of the reward ("good") neuron into long-term memory (LTM) follows the classical Hebbian learning rule with decay. This is possible because the reward node is connected via excitatory connections to neurons in the target selection layer. As a result, when reward is delivered this reinforces the activity of the postsynaptic neuron that is suprathreshold (thereby crediting reward to the current choice). This creates a sustained temporal correlation of pre- and postsynaptic activities as required by Hebbian learning.

However, the "punish node" ("bad") has inhibitory connections with the neurons in the winner-take-all circuit to depress the activity of the neuron that is above threshold. As a result of this inhibitory effect, pre- and postsynaptic activities remain simultaneously active only for a brief period of time. This leads to an ineffective coding of the punish neuron activity into LTM via a Hebbian learning rule that requires a temporal correlation of pre- and postsynaptic activities. To avoid this problem, a delay neuron whose STM trace follows a delayed version of the suprathreshold activity of the postsynaptic neuron is introduced along with a modified learning rule which is given in the Appendix. Simulations illustrating this learning process are presented in Fig. 18.19. Although there is very little temporal overlap between post- and presynaptic activities ($x(1,3)$ and punish node, respectively), the overlap between the delay neuron and the punish node enables the coding of negative reinforcement into synaptic weights.

4. CONCLUSION

In summary, we argued in this chapter that the strict notion of optimization is neither a proper tool of analysis nor a proper tool of design for interactive systems evolving in a dynamic and complex environment. The major shortcoming of the optimization technique comes from its inability to capture in an a priori fixed function the "goals" of the system.

ACKNOWLEDGMENT

This material is based in part upon work supported by the Texas Advanced Technology Program under grant 003652023 and by a grant from NASA-JSC.

REFERENCES

Carpenter, G. A., & Grossberg, S. (1988). The ART of adaptive pattern recognition by a self-organizing neural network. *IEEE Computer*, **21**, 77-88.

Geman, S., Bienenstock, E., & Doursat, R. (1992). Neural networks and the bias /variance dilemma. *Neural Computation*, **4**, 1-58.

Grossberg, S. (1972). A neural theory of punishment and avoidance, II: Quantitative Theory. *Mathematical Biosciences*, **15**, 253-285.

Grossberg, S. (1980). How does a brain build a cognitive code? *Psychological Review*, **87**, 1-51

Leven, S. J., & Levine, D. S. (1987). Effects of reinforcement on knowledge retrieval and evaluation. In M. Caudill & C. Butler (Eds.), *Proceedings of the First International Conference on Neural Networks* (Vol. 2, pp. 269-279). San Diego: IEEE/ICNN.

Levine, D. S. (1991). *Introduction to Neural and Cognitive Modeling*. Hillsdale, NJ: Lawrence Erlbaum Associates.

Levine, D. S., & Prueitt, P. S. (1989). Modeling some effects of frontal lobe damage: Novelty and perseveration. *Neural Networks*, **2**, 103-116.

Öğmen, H. (1992a). On the neural substrates leading to the emergence of mental operational structures. *Proceedings of the Third International Workshop on Neural Networks and Fuzzy Logic*, NASA Conference Publication 10 111 (Vol. 1, pp. 30-39).\

Öğmen, H. (1992b). *On the developmental basis of self-organization and its active exploratory Components*. University of Houston Systems, Neural-Nets, and Computing Technical Reports, No. 92-05, 1992. (Also in *Proceedings of 2nd Behavioral and Computational Neuroscience Workshop*, Washington, DC, May 1992.)

Öğmen, H. (1993a). A neural theory of retino-cortical dynamics. *Neural Networks*, **6**, 245-273.

Öğmen, H. (1993b). Continuous-time global vision with analog, specialized, and interacting neural networks. *Information Sciences*, **70**, 5-25.

Öğmen, H. (1995). Dynamics of development. University of Houston Systems, Neural-nets, and Computing Technical Report, No 95-01.

Öğmen, H. (in press). Sensorial nonassociative learning and its implications for visual perception. In O. Omidvar (Ed.), *Progress in Neural Nets* (Vol. V). Norwood, NJ: Ablex.

Öğmen, H., & Moussa, M. (1993). A neural model for nonassociative learning in a prototypical sensory-motor scheme: The landing reaction in flies. *Biological Cybernetics*, **68**, 351-361.

Öğmen, H., & Prakash, R. V. (1992). *Self-organization via active exploration in robotic applications*. NASA Technical Report.

Öğmen, H., & Prakash, R. V. (1993). *Self-organization via active exploration in robotic applications: Hybrid implementation*. NASA Technical Report.

Öğmen, H., Prakash, R. V., & Moussa, M. (1992). Some neural correlates of sensorial and cognitive control of behavior. In D. W. Ruck (Ed.), Science of Artificial Neural Networks, *Proceedings of SPIE*, **1710**, 177-188.

Piaget, J. (1963). *The Origins of Intelligence in Children*. New York: Norton.

Poggio, T., Torre, V., & Koch, C. (1985). Computational vision and regulation theory. *Nature*, **317**, 314-319.

Prakash, R. V., & Öğmen, H. (1996). A neural model for the initial stages of the sensori-motor level in development. In D. Vincent & M. Witten (Eds), *World Scientific Series in Mathematical Biology and Medicine, Vol. 5: Computational Medicine, Public Health and Biotechnology: Building a Man in the Machine* (pp. 846-860). World Scientific.

Sutton, R. S., & Barto, A. G. (1981). Towards a modern theory of adaptive networks: Expectation and prediction. *Psychological Review*, **88**, 135-170.

Yuille, A. L. (1989). Energy functions for early vision and analog networks. *Biological Cybernetics*, **61**, 115-123.

Zuriff, G. E. (1985). *Behaviorism: A Conceptual Reconstruction.* New York: Columbia University Press.

APPENDIX: EQUATIONS AND PARAMETERS USED TO MODEL THE TARGET SELECTION NETWORK

A.1. Spatial Novelty and Attentive Scanning

The spatial novelty network is implemented using an array of gated dipole networks. The differential equations for the transmitter nodes $vz_{i,1}$ and $vz_{i,2}$ of the gated dipole are given by

$$\frac{dvz_{i,1}}{dt} = \alpha(\beta - vz_{i,1}) - \gamma(I+J_i)vz_{i,1} \quad i = 1,2,\cdots,n$$

$$\frac{dvz_{i,1}}{dt} = \alpha(\beta - vz_{i,2}) - \gamma I vz_{i,2} \quad i = 1,2,\cdots,n$$

where α is the transmitter replenishment rate, β is the maximum amount of transmitter, γ is the rate of transmitter depletion, I and J_i are the background and specific inputs, respectively, and $vx_{i,1}$ and $vx_{i,2}$ are respectively the "ON" and "OFF" channel neurons of the ith gated dipole.[2] Shunting equations have been used to model the "ON" and "OFF" channel neurons $vx_{i,1}$ and $vx_{i,2}$:

$$\frac{dvx_{i,1}}{dt} = -A vx_{i,1} + (B - vx_{i,1})(I+J_i)vz_{i,1} - vx_{i,1}I vz_{i,2} \quad i = 1,2,\cdots,n$$

$$\frac{dvx_{i,2}}{dt} = -A vx_{i,2} + (B - vx_{i,2})I vz_{i,2} - vx_{i,2}(I+J_i)vz_{i,1} \quad i = 1,2,\cdots,n$$

where A is the passive decay rate and B is the upper saturation level of their activities. The outputs of the spatial novelty network feed into the $vx_{i,3}$ neurons of the winner-take-all layer,

---

[2] Note that although we implemented a two-dimensional array of gated dipoles, for simplicity the equations are written using a single spatial index i.

which along with delay and inhibitory layer neurons s_i and l_i, respectively, regulate attentive scanning. The equations for the neurons in these three layers are

$$\frac{dvx_{i,3}}{dt} = -A_1 vx_{i,3} + (B - vx_{i,3})(C_1 vx_{i,1} + G_3 g(vx_{i,3} - \theta))$$
$$- vx_{i,3}(l_i + H\sum_{j\neq i} f(vx_{j,3} - \theta) + C_1 vx_{i,2}) \qquad i = 1,2,\cdots,n$$

$$\frac{ds_i}{dt} = -As_i + (B_1 - s_i) G_1 g(p_{i,1} - \theta_1) - s_i H_1 g(a - \theta_2) \qquad i = 1,2,\cdots,n$$

$$\frac{dl_i}{dt} = -Al_i + (B_1 - l_i) G_2 g(s_i - \theta_i) \qquad i = 1,2,\cdots,n$$

with $f(x) = \dfrac{(5.5x)^5}{(0.001 + (5.5x)^5)} u(x)$ and $g(x) = xu(x)$, where A_1 and A are passive decay rates; B_1 and B are the upper bounds for activity; H_1, C_1, G_1, G_2, and G_3 are positive gains; and θ, θ_1, and θ_2 are threshold constants. The term $u(x)$ is the unit step function. a is the arousal neuron which inhibits scanning until the currently attended object is classified. The neurons $p_{i,1}$ form the decision layer that buffers transients generated during competition. The equations for the arousal neuron and the neurons in the decision layer are presented below.

$$\frac{da}{dt} = -A_f a + (B_1 - a)Arousal - aG_4 \sum_{j=1}^{2} g_2(p_{j,2} - \theta_3)$$

$$\frac{dpx_{i,1}}{dt} = -A_p p_{i,1} + (B_2 - p_{i,1}) Wg(vx_{i,3} - \theta) - p_{i,1} W\sum_{j\neq i} g(vx_{j,3} - \theta) \qquad i,j = 1,2,\cdots,n$$

with $g_2(x) = 10.0u(x)$, where A_f and A_p are passive decays, B_1, and B_2 are upper bounds for activity, W and G_4 are gain constants, θ_3 is a threshold, and $Arousal$ is a constant positive signal. $p_{j,2}$ neurons are the decision neurons of the good-bad categorization network.

A.2. Good-Bad Categorization Network

This network, which categorizes objects based on experience and reinforcement, consists of a cascade of ART networks; the first ART network (F_1–F_2) ascertains which features should be used for categorizing the inputs, and the second ART network (F_2–F_3) categorizes the categories obtained from the first ART network into good and bad categories. The neurons in layers F_1, F_2, and F_3 are represented by bx_i, by_j, and bz_k, respectively. Previous experience is encoded by the habit neurons h_k which, together with external reinforcement signals R^+ and R^-, modulate the selection of the categorization criterion via bias neurons Ω_k. The match neurons Φ_k correlate the nonspecific reinforcement signals with the relevant decision criterion, thus ensuring an appropriate credit assignment. The activities of the F_3 layer neurons are buffered by the layer of $p_{i,2}$ neurons before they go to the other two networks. The ambiguity neuron a ensures that a decision is reached in case of uncertainty. The differential equations for this network are

$$\frac{dbx_i}{dt} = -Abx_i + (B - bx_i)[I_i + \sum_{j=1}^{3} f(by_j)bu_{j,i}^1] - bx_i[\sum_{j=1}^{3} f(by_j) + I], \qquad i = 1,2,\cdots,12$$

$$\frac{dby_j}{dt} = -Aby_j + (B - by_j)[f(by_j - \theta_4) + \sum_{i=1}^{12} g(\Omega_{\left[\frac{i+3}{4}\right]}bx_i)bw_{i,j}^1$$

$$+ W_1(a - \theta_1)^+ bv_j + \sum_{k=1}^{2} f(bz_k)bu_{k,j}^2]$$

$$- by_j[\sum_{r \ne j} f(by_r - \theta_4) + \sum_{k=1}^{2} f(bz_k) + I] \qquad j = 1,2,\cdots,4$$

$$\frac{dbz_k}{dt} = -Abz_k + (B - bz_k)[f(bz_k - \theta_4) + \sum_{j=1}^{4} g(by_j)bw_{j,k}^2]$$

$$- bz_k[\sum_{r \ne k} f(bz_r - \theta_4) + I] \qquad k = 1,2$$

$$\frac{d\Omega_k}{dt} = -E(\Omega_k - \theta_3) + \{(F - \Omega_k)[(h_k - \theta_2)^+ + \alpha R^+ + g(\Omega_k)]$$

$$- \Omega_k[\alpha R^- + G\sum_{r \ne k} g(\Omega_r)]\}f(\Phi_k) \qquad k = 1,2,3$$

$$\frac{dh_k}{dt} = Hh_k[(J - h_k)(\Phi_k - \theta_2)^+ - (\Phi_k - \theta_2)^-], \quad k = 1,2$$

$$\frac{d\Phi_k}{dt} = -A\Phi_k + (B - \Phi_k)\left[\sum_{i=4k-3}^{4k} \sum_{j=1}^{4} I_i g_1(by_j - \theta_5)bu_{j,i}^1\right] - \Phi_k I \qquad k = 1,2,3$$

$$\frac{dp_{i,2}}{dt} = -Ap_{i,2} + (B - p_{i,2})g(bz_i - \theta_6) - p_{i,2}\left[\sum_{j \ne i} g(bz_j - \theta_6) + I\right] \qquad i = 1,2$$

$$\frac{da}{dt} = -Aa + (B - a)\sum_{i} g_1(I_i) - a[Y\sum_{j} g(p_{j,2} - \theta_7) + I] \qquad j = 1,2; \ i = 1,2,\cdots,6$$

with $f(x) = 1/(1 + e^{-15(x-0.5)})$, $[x]^+ = xu(x)$, $[x]^- = -xu(-x)$, $g(x) = xu(x)$, and $g_1(x) = u(x-0.05)$, where A, A_1, and E are passive decay constants; B, F, and J are the upper bounds for activity; I is a reset signal; $w_{i,j}^1$ and $w_{j,k}^2$ are fixed bottom-up weights and $u_{j,i}^1$ and $u_{k,j}^2$ are fixed top-down weights for the two ART networks respectively; α, Y, G, H, and W_1 are positive gain constants; θ_1, θ_2, θ_3, θ_4, θ_5, θ_6, and θ_7 are positive threshold constants; I_i is the input feature-vector of the object; and bv_j are fixed weights from the ambiguity neuron a to the F_2 layer neurons. These weights filter the ambiguity neuron feedback that biases one of the category neurons in situations where the input object cannot be uniquely categorized to a particular category.

A.3. Object-Type Categorization, Object-Type Novelty, and Target Selection Network

The object-type categorization network is an ART network. The terms fx_i and fy_j denote the activities of the neurons in the input and category layers of this ART network, respectively. The various object-type categories are buffered by the layer of $p_{i,3}$ neurons to filter transients generated during categorization. The output of the buffered layer is fed to the object-novelty network via a layer of slowly decaying neurons q_i, which memorize previous occurrences of similar object type. The differential equations for the object-type categorization network are:

$$\frac{dfx_i}{dt} = -Afx_i + (B-fx_i)(I_i + \sum_{j=1}^{64} f(fy_j)fu_{j,i}) - x_i(\sum_{j=1}^{64} f(fy_j)+I) \qquad i=1,2,\cdots,12$$

$$\frac{dfy_j}{dt} = -Afy_j + (B-fy_j)(f(fy_j) + \sum_{i=1}^{12} g(fx_i)fw_{i,j} - fy_j(\sum_{r\neq j} f(fy_r - \theta)+I) \, j=1,2,\cdots,64$$

$$\frac{dp_{i,3}}{dt} = -Ap_{i,3} + (B-p_{i,3})W_1 g(by_i - \theta_2) - p_{i,3}(\sum_{j\neq i} g(by_j - \theta_2)+I), \qquad i=1,2,\cdots,64$$

$$\frac{dq_i}{dt} = -A_q q_i + (B-q_i)p_{i,3} \qquad i=1,2,\cdots,64$$

with $f(x) = 1/(1 + e^{-15(x-0.5)})$ and $g(x) = xu(x)$, where A and A_q are passive decay constants; B is the maximal activity of the neurons; I_i are the components of the input feature-vector and I is the reset signal; $w_{i,j}$ and $u_{j,i}$ are the bottom-up and top-down weights of the ART network, which are fixed; and θ_1 and θ_2 are threshold constants. The novelty network comprises an array of gated dipoles similar to the spatial novelty network where $cz_{i,1}$ and $cz_{i,2}$ are the "ON" and "OFF" channel transmitter nodes and $cx_{i,1}$ and $cx_{i,2}$ are the output neurons of the "ON" and "OFF" channels, respectively. The differential equations for the object-novelty network are

$$\frac{dcz_{i,1}}{dt} = \alpha(\beta - cz_{i,1}) - \gamma[I+g_1(q_i - \theta_3)]cz_{i,1} \qquad i=1,2,\cdots,64$$

$$\frac{dcz_{i,2}}{dt} = \alpha(\beta - cz_{i,2}) - \gamma Icz_{i,2} \qquad i=1,2,\cdots,64$$

$$\frac{dcx_{i,1}}{dt} = -Acx_{i,1} + (B-cx_{i,1})[I+g_1(q_i - \theta_3)]cz_{i,1} - cx_{i,1}Icz_{i,2} \qquad i=1,2,\cdots,64$$

$$\frac{dcx_{i,2}}{dt} = -Acx_{i,2} + (B-cx_{i,2})Icz_{i,2} - cx_{i,2}[I+g_1(q_i - \theta_3)]cz_{i,1} \qquad i=1,2,\cdots,64$$

with $g_1(x) = u(x)$, where α is the transmitter replenishment rate, β is the maximum amount of transmitter, γ is the rate of transmitter depletion, I and q_i are the background and specific inputs, respectively, A is the passive decay constant, B is the maximal activity of the output neurons, and θ_3 is a threshold constant.

The target selection network consists of a winner-take-all layer of neurons $x_{i,3}$, and weights the novelty of the object versus its good-bad qualities. The $p_{i,3}$ neurons ensure that the novelty and behavioral qualities of the same object are compared by the $cx_{i,3}$ neurons. A buffer layer of neurons $p_{i,4}$ filter transients that occur during target selection and outputs the appropriate motor command decision. The equations for these neurons are

$$\frac{dcx_{i,3}}{dt} = -Acx_{i,3} + (B - cx_{i,3})[cx_{i,1} + G_2g(cx_{i,3} - \theta_4) + G_1 p_{1,2} w_{i,r} p_{i,3}]$$
$$- cx_{i,3}[cx_{i,2} + H\sum_{j \neq i} f(cx_{j,3} - \theta_4) + G_2 p_{2,2} w_{i,p} p_{i,3}] \qquad i = 1,2,\cdots,64$$

$$\frac{dp_{i,4}}{dt} = -Ap_{i,4} + (B - p_{i,4}) W_1 g(cx_{i,3} - \theta_2) - p_{i,4} W_1 \sum_{j \neq i} g(cx_{j,3} - \theta_2) \qquad i,j = 1,2,\cdots,64$$

with $f(x) = \dfrac{(5.5x)^5}{(0.001 + (5.5x)^5)} \, u(x)$ and $g(x) = xu(x)$, where A is the passive decay rate, B is the upper bound of neuron activity, G_1, G_2, H, and W_1, are positive gain constants, θ_4 is the threshold constant, and $p_{1,2}$ and $p_{2,2}$ are the output of the good-bad categorization network.

The good-bad properties of a given object are encoded in the weights from the category neurons to the respective object-type novelty neurons. $w_{i,r}$ and $w_{i,p}$ encode these good-bad properties of a given object and y_i neurons trace the activity of the $x_{i,3}$ neurons of the object-type novelty network. The learning rule for the excitatory weights and the modified learning rule for the inhibitory weights are

$$\frac{dw_{i,r}}{dt} = -A_2(w_{i,r} - \theta_5) + (M - w_{i,r}) B_2 g(p_{1,2} - \theta_6) g(cx_{i,3} - \theta_4) + C_2$$

$$\frac{dw_{i,p}}{dt} = -A_2(w_{i,p} - \theta_5) + (M - w_{i,p}) B_2 g(p_{2,2} - \theta_6) g(y_1 - \theta_7) + C_2$$

$$\frac{dy_i}{dt} = -A_2 y_i + (B - y_i) g(cx_{i,3} - \theta_4)$$

where A_2 is the passive decay constant, M and B are the maximum values of the weights can assume and the maximum activity of the trace neurons, respectively, θ_5, θ_6, and θ_7 are positive threshold constants, By_2 is a positive gain, C_2 is a positive constant to ensure that the weights never go to zero, and $p_{1,2}$ and $p_{2,2}$ are the output neurons of the good-bad categorization network.

The parameters for spatial novelty and attentive scanning network were: $\alpha = 1.0 \times 10^{-2}$, $\gamma = 1.0 \times 10^{-2}$, $\beta = 10.0$, $I = 1.0$, $J_i = 1.0$, $Arousal = 3.0$, $A = 1.0$, $A_1 = 0.1$, $A_I = 10.0$, $A_p = 5.0$, $B = 1.0$, $B_1 = 5.0$, $B_2 = 2.0$, $C_1 = 0.1$, $G_1 = 3.0$, $G_2 = 500.0$, $G_3 = 2.0$, $G_4 = 10.0$, $H = 0.15$, $H_1 = 100.0$, $W = 100.0$, $\theta = 0.24$, $\theta_1 = 0.6$, $\theta_2 = 0.4$, $\theta_3 = 0.55$. The parameters for the good-bad categorization network were $A = 1.0$, $A_1 = 0.001$, $B = 1.0$, $E = 0.01$, $F = 3.0$, $G = 10.0$, $H = 0.001$, $I_i = 10.0$, $I = 10.0$, $J = 3.0$, $W_1 = 50.0$, $\alpha = 0.01$, $\theta_1 = 0.9$, $\theta_2 = 0.1$, $\theta_3 = 1.0$, $\theta_4 = 0.30$, $\theta_5 = 0.20$, $\theta_6 = 0.25$, $\theta_7 = 0.60$, $Y = 10.0$. The parameters for object-type categorization, object-type novelty and target selection network were $A = 1.0$, $A_1 = 0.0001$, $A_2 = 0.01$, $B = 1.0$, $B_2 = 0.5$, $C_2 = 0.01$, $G_1 = 0.05$, $G_2 = 0.5$, $H = 95.0$, $I = 1.0$, $I_i = 1.0$, $I = 10.0$, $M = 2.0$, $W_1 = 10.0$, $\alpha = 0.01$, $\gamma = 0.01$, $\beta = 10.0$, $\theta_1 = 0.3$, $\theta_2 = 0.25$, $\theta_3 = 0.3$, $\theta_4 = 0.22$, $\theta_5 = 1.0$, $\theta_6 = 0.55$, $\theta_7 = 0.1$.

19

The Income-Choice Approach and Some Unsolved Problems of Psychopathology — A "Bridge Over Time"

G.-Z. Rosenstein
Faculty of Medicine, Hebrew University, Jerusalem

Gershom-Zvi Rosenstein's chapter, **The Income-Choice Approach and Some Unsolved Problems of Psychopathology — A "Bridge Over Time,"**[1] *is an application of the author's general theory of the brain's affective (reward and punishment) systems that posits a biological variable called* **income** *that the organism tries to maximize. His income, defined by analogies between psychology and economics, is some "inverted-U" shaped function of the intensity of stimulation of the brain's affective systems. Hence, there is some optimal level of stimulation, and the organism tries to avoid being overstimulated or understimulated.*

In this chapter, Rosenstein attempts to apply the income-choice approach (ICA) previously developed in his 1991 book to understanding many aspects of schizophrenia: for example, positive and negative symptoms, stereotyped behavior, and spontaneous remissions. He sees most of the symptoms of schizophrenia as attributable to attempts by the organism to maximize income in the face of abnormally high affective stimulation. This is of particular interest to those of us studying the riddle of optimality, because mental illness is often seen as an exaggeration of suboptimal behavior in normals (see, e.g., the discussion of frontal lobe damage in Levine's

---

[1] For 15 years (1973-1988) G.-Z. Rosenstein was excluded from organized scientific life because he was a "refusenik" in the former Soviet Union — with no right to leave the country and with no scientific position because of desire to do so. That is why this article, in which we attempt to unite old findings with the modern state of the art, can be seen as a "bridge over time."

chapter). Rosenstein, by contrast, describes schizophrenic behaviors and ideations as an optimal response to an abnormal biochemical environment in the brain.

This chapter does not include a detailed neural network theory of the various subcortical areas involved in schizophrenia (mainly, those of the dopaminergic reward system). Nor does it indicate the network processes by which an expectation of income is calculated and influences the positive or negative associations that a person develops to particular sensory stimuli. Rosenstein does, however, connect the ICA tentatively with a version of the "dopamine hypothesis" of schizophrenia, by identifying one form of income with dopamine inputs to the nucleus accumbens. He states that there are other forms of income as well, and leaves a more definite biological identification of income to future work.

The main contribution of his book was specifying an abstract mathematical function that could serve as a useful organizing principle for some theories of neural control of behavior in normal and abnormal biochemical environments. This chapter extends this work by applying it to problems of schizophrenia. It also suggests an approach to therapy for schizophrenia, based on restoration of normal stimulation, or compensation of abnormal stimulation, by increase or decrease of stimulation, by means of increase or decrease of stimulation of selected subdivisions of the brain's affective system.

ABSTRACT

This chapter aims to outline a unified approach to several unsolved problems of behavioral regulation, mainly related to the puzzle of schizophrenia. The income-Choice approach (ICA), proposed originally in the 1970s, has recently been summarized in a book (Rosenstein, 1991). One of the main problems to which this approach has been applied is modeling behavior disturbances.

Biological income[2] is a basic variable in the control of biological systems hypothesized by the author. This idea is modeled after the notion of money in economics with its all-penetrating role in regulation of market activities (including financial market and market of information) by price of money variations (inflation-deflation mechanism). Biological income is suggested to play a similar role in biological control and choice of behavior on different levels of biological organization. The income can be presented in many forms, accumulated, and spent on all kinds of activities of the brain and organism. The goal of the model is to maximize the income function in the course of the model's lifetime. In Rosenstein (1991), the income is defined by assumption on intensities of streams of impulses directed to the reward system. In this chapter, besides this dynamic form of representation of "money" in the model, we discuss dopamine (and some other substances) as candidates to represent income in a form stable enough to be accumulated and distributed among brain subnetworks.

Specifically, the ICA is applied to modeling the following problems (see Table 19.1):

---

[2] This income is of the marginal utility function type used in economics to describe, for example, the value of utility of a product as dependent on its quantity.

· The causes of catecholamine distribution change in the schizophrenic brain
· The role of dopamine in information processing
· The fact that in schizophrenics, observations prevail over expectations
· The anhedonia hypothesis of neuroleptic action
· Explanation of the nature of stereotypic behavior in comparison with adjunctive-type behavior
· The origin of the so-called "positive" and "negative" symptoms in schizophrenia

No.	Problem	Question	Answer
1	Catecholamine distribution	What causes it?	CDC phenomena in the schizophrenic brain can be caused by augmented streams of impulses (ASI) directed to the reward system, in agreement with our model (Rosenstein, 1991, 1994; Vaisbord & Rosenstein, 1968a, 1968b).
2	Positive symptoms in schizophrenia	Why are they produced by the schizophrenic?	Positive symptoms can cause increase of income or prevent decrease of income when ASI reduce the value of external sources of income. This is why schizophrenics produce these symptoms when the brain's inner market needs them.
3	A main difference between the psychology of normal and schizophrenic subjects is that in schizophrenia "observations prevail over expectations."	Why is this property associated with schizophrenia?	"Observations prevail over expectations" can be seen be seen as a metaphorical description of a theorem proven earlier (Glazunov et al., 1971; Rosenstein, 1991), in which this property (called "flat mind effect") is obtained from our model.
4	Anhedonia hypothesis (AH)	Can it be supported by the income-choice approach (ICA)?	Yes. It seems that AH should be raised to the rank of "theory of anhedonia" because it was presented as a theory of the phenomena independent of the experimental results from which AH was deduced.
5	Adjunctive behavior (AB)	What is the nature of AB? Why does it occur?	AB is a subset of the set of behaviors whose common property is to be conferred directly with the regulation of the reward system. This idea can be proven by experiments with reward system activity registration at a time when the animal is producing adjunctive behavior.

Table 19.1. Summary of Basic Topics in This Chapter.

1. INTRODUCTION

During the years 1963-1972, the income-Choice approach (ICA) to the problems of biological regulation was developed in Moscow by the author and his colleagues (the complete bibliography is given in Rosenstein, 1991).[3] According to the ICA, a reward system calculates a certain function (income function of our model) defined on the set of intensities of streams of impulses directed to the nucleus of the reward system. The proposed income function is of a kind known in economics as a marginal utility curve (MUC) (Fig. 19.1). The input and output activities of the system are regulated to maximize the expected income. The streams of impulses directed to the reward network are seen as one form of "money" in our model. Other forms of "money," more stable and that can be stored, include dopamine and maybe other biochemical substances as endorphins. The bioeconomic metaphor as a whole has yet to be adjusted to the working brain. The types of currencies, storehouses, currency exchange structures, and so on, can be only preliminarily assigned today to definite substances and constructions in the brain. The marginal utility curve is of the same type (Fig. 19.1) for each form of currency, not only for streams of impulses directed to the reward system.[4,5]

This kind of change of optimal behavior (or of the set of optimal behaviors) under the influence of augmented (in comparison to the norm) streams of impulses (ASI) is interpreted as a "behavior disturbance" in our model. Note that the same MUC-type income function is foreseen for other forms of inner "currency" too (as dopamine), besides the streams of impulses directed to the reward network.

According to our assumption, income is being accumulated, distributed, and spent on different activities of the system. Some types of income functions were discussed earlier (Rosenstein, 1991; Vaisbord & Rosenstein, 1965, 1967). It was shown that ASI directed to the reward system can change substantially the set of optimal behaviors in our model (Glazunov, Rosenstein, & Jablonsky, 1971; Glazunov et al., 1972; Rosenstein, 1991, Chapter 6). This effect is recognized in the model as a behavioral disturbance.

Models of behavior disturbances (Glazunov et al., 1971, 1972; Rosenstein, 1972, 1991; Vaisbord & Rosenstein 1968a, 1968b) and of epileptic-type seizure control (Antik, Arshavskii, & Rosenstein, 1972; Arshavskii, Meshman, & Rosenstein, 1972; Meschersky & Rosenstein, 1969; Rosenstein, 1969, 1991) were proposed in the framework of ICA.

According to our approach, each one of the possible causes of schizophrenia proposed by existing different theories of the illness causes direction of ASI to the reward system or to reduction of existing stimulation, thus incurring the phenomenology of schizophrenia. This basic phenomenology, that is, the substantial deviation from the set of comprehensible to all ("normal")

---

[3] The reader is advised to look at Chapters 1, 4, 6 and 7, Introduction and Conclusion in the 1991 book in the course of reading the present chapter.

[4] The universal character of this U curve is connected to the optimal behavior of systems with limited resources. See discussion and examples in Rosenstein (1991), Chapter 1.

[5] After the first draft of this chapter was written, there appeared a chapter by Liddle (1994), where he wrote: "In rats, the level of locomotor activity produced by administration of dopamine agonists follows a U-shaped curve. As dose increases there is initially an increase in amount of activity but, at higher doses, total activity decreases as the animal become engaged in repetitive, stereotypic activity." See below our discussion of the cause of stereotypic behavior in schizophrenics. In the same chapter by Liddle, a discussion can be found about volition and dopamine in schizophrenics that is of interest from the ICA angle.

behaviors, can be "colored" by the specificity of the particular reason because of what caused the stimulation of the reward system to change (whether it is a stress-producing event reflected in the memory of a subject, a physiological reason, or a change genetically programmed at a certain age in the intensity of streams of impulses from a particular source directed to the reward system, etc.). In case of a stress-producing event, for example, the reflection of the event in the memory of the subject can become a source of additional stimulation switched on by associations connected to the contents of the event. Various types of therapy used in psychopathology (particularly in the treatment of schizophrenia) could be seen to correspond with possible methods of interruption or compensation of ASI in our model (Rosenstein, 1991; Vaisbord & Rosenstein, 1968a, 1968b).

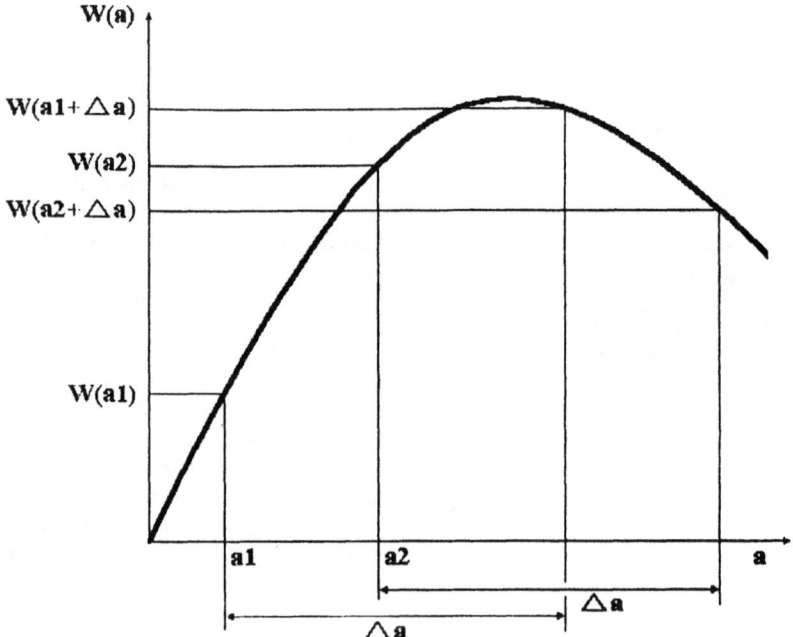

Fig. 19.1. Income function of marginal utility curve (MUC) type. Terminology: $W(a)$, MUC-like income function; $a1$: density of stream of impulses directed to reinforcement center if the system chooses behavior 1; $a2$, density of stream of impulses directed to reinforcement center if the system chooses behavior 2; Δa, density of augmented stream of impulses (ASI) directed to reinforcement center; $W(a1)$, income received by system if behavior 1 chosen; $W(a2)$, income received by system if behavior 2 chosen; $W(a1+\Delta a)$, income received by system with ASI directed to reinforcement center if behavior 1 chosen; $W(a2+\Delta a)$, income received by system with an additional center as if behavior 2 was chosen. Without ASI, $W(a_2) > W(a_1)$, hence behavior 2 is preferred. However, $W(a2+\Delta a) < W(a1+\Delta a)$, so with ASI, behavior 1 is preferred.

According to today's popular opinion, in schizophrenics observations prevail over expectations. This trait of schizophrenic phenomenology (as opposed to normal behavior) is seen in learning and conclusion-making tests, such as the Charpentier delusion formation. This trait was deduced formally from our model (Glazunov et al., 1971; 1972; Rosenstein, 1991), in which the following property was found: The a posteriori distribution of probabilities formed when a system with disturbed behavior learns is "closer" to the even distribution than a distribution formed when a "normal" system learns. This means that the "disturbed" network can perform better in dealing with novel information than the "normal" one (Glazunov et al., 1971, 1972).

It should be admitted that the modern knowledge of reward networks is yet far from being sufficient even in the case of animals, not to mention humans. The notion of income was proposed by generalization of limited information obtained from experiments with self-stimulation and on the basis of bioeconomic metaphor. One can hope that the applications of income presented here can stimulate further studies, and that new income-dependent phenomena will be discovered in biological control systems on different levels of their organization.

2. INCOME-CHOICE APPROACH (ICA) AND NEUROPHYSIOLOGY OF SCHIZOPHRENIA

Our model of behavior disorders (Glazunov et al., 1971, 1972; Rosenstein, 1972, 1991; Vaisbord & Rosenstein, 1965, 1968a, 1968b) contains the following basic axioms and theorems:

1. The model is a system (network) whose goal is to maximize the predicted income accumulated during the lifetime of the system. The duration of the life of the system can depend on its turn on the quantity of accumulated income.

Several examples of functionals that can be maximized by biological systems of control are discussed in Chapter 3 of Rosenstein (1991). For example, it can be the predicted income accumulated by the system in any moment during its lifetime. In humans these goal-functions can be to some extent reprogrammed consciously or subconsciously. In animals, the possible reprogramming is much more limited and can be done mostly under external pressure.

2. The income function is of the marginal utility curve (MUC) type (Fig. 19.1). As depicted in economic models, it reflects the fact that the derivative of the value of a commodity (including money) decreases with an increase in quantity of the commodity (for large enough quantities of the commodity possessed by the system). In the general case, it is connected to limitations on resources. Examples of optimal control with such limitations are discussed in Chapter 1 of Rosenstein (1991).

The impulses sent to the reward system are seen as one of the forms of "money" in our model.

3. The income of the model is calculated by its reward network. In the variant of the model we dealt with in Rosenstein (1991), income is dependent on the stream of impulses sent to the reward nuclei at a given moment and on the background intensity of stimulation of these nuclei. As stated above, the additional experimental information is needed to develop other variants of reward networks. In particular, the interaction of several reward nuclei should be taken into account. In the book (Rosenstein, 1991), it was done only in a preliminary way for

two — a "positive" and a "negative" nucleus (see Rosenstein, 1991, Chapter 1). In this chapter we discuss dopamine as another possible form of biological income.

4. Disturbed behavior is caused in the model by additional (augmented) streams of impulses (ASI) directed to the nucleus of reward system. These ASI can originate from many different causes. (A similar idea can be applied to other forms of inner currency of the organism, too, if the corresponding reward structures for these forms of currency is discovered).

It was shown for a particular example of income function where formal analysis has been done (Rosenstein, 1991, Chapter 6) that ASI can change the optimal behavior in a substantial part of all possible cases. It is an important hint from the side of the theory of optimal control that our hypothesized additional stream of impulses directed to the reward network can be the real cause for the disturbance of the behavior in the model.

5. A number of existing theories of schizophrenia refer to different causes of the illness. And according to the ICA, each of them should cause a substantial change in the set of impulses directed to the reward network. This change can be constant or conditional, that is, dependent on time or on some events in the inner physiological world of the sick person. (It is obvious that these physiological events can be in their turn produced by psychological, genetic, or in some cases even by social factors.)

6. The normalization of behavior in our model is related to restoration of the set of streams or compensation of ASI directed to the reward system. The compensation can be provided by a stream of impulses directed to the competing nuclei of the reward system. Existing modes of treatment for schizophrenia appear to correspond with various forms of interruption and/or compensation of ASI.

7. It was shown that a characteristic trait of the procedure of learning[6] is that the speed of formation of the a posteriori distribution during learning in a system with "disturbed" behavior is slower than in a "normal" one. We call this phenomenon the *flat mind effect* (FME) in schizophrenia.[7] FME is due to the fact that in our model the reward calculated by the reward system of the schizophrenic brain is usually smaller than in the normal brain. The effect is caused by ASI directed to the reward system of the schizophrenic brain and by nonlinear character of income function that is of the marginal utility curve (MUC) type (see Fig. 19.1, and also Fig. 19.3 below). It leads to a kind of phenomena that can be seen as *inflation* of the reinforcement. This inflation can be of two opposite signs: *hyperinflation* or *hypoinflation*. We will see later that it corresponds on the level of human behavior to so-called "positive" and "negative" symptoms in schizophrenic subjects.

---

[6] We refer to the statistical type of learning procedures.

[7] The difference between the prior and posterior distributions formed by a schizophrenic subject is less than that in a normal subject for the same number of learning steps. In particular, the posterior distributions formed by schizophrenics are "flatter" than in the norm for the same quantity of steps of learning if the prior distribution is uniform. This is called "flat mind effect" (FME). It includes the phenomenon known to clinicians as "flatness of affect" but is more general.

2.1. Biological Income, Calculation, and Representation

The first question we address is: What is the relationship, if any, between the ICA and known models of schizophrenia based on the well-established phenomenon of catecholamine distribution change (CDC) in the schizophrenic brain?[8]

I recently published an article (Rosenstein, 1994) that offered a preliminary answer to this question. More complete information follows.

2.1.1. Income-Choice Approach (ICA) and Catecholamine Distribution Change (CDC) Phenomena in Schizophrenia. The ICA approach is perfectly suited to the CDC, although it is not limited by it. The ICA presents explanations for a number of contradictions and limitations of the CDC hypothesis of schizophrenia and affords a preliminary answer to the basic question of why CDC occurs in the schizophrenic brain.

Involvement of the catecholamine (CA) system in reward mechanisms of the brain is already well established (Crow, 1972; Fibiger & Phillips, 1986; Olds & Yuwiler, 1972; Wise, 1978). The basic facts are the following:

1. All hypothalamic nuclei contain dopamine (DA). Sizable quantities of DA and noradrenalin were found (Axelrod, 1977) in more than 130 nuclei of the cat brain. Some single DA neurons are estimated to have over 0.5 million synaptic endings (Anden, Fuxe, Hamberger, & Kokfelt, 1966), suggesting that their function is to modulate the activity of large neuronal networks (German & Bowden, 1974).

2. Ascending catecholamine-containing fiber pathways travel through the lateral hypothalamic region, from which high rates of self-stimulation are most consistently obtained (Crow, 1972).

3. Modern studies provide in vivo evidence that increased CA release and turnover in the limbic system and frontal cortex is a correlate of intracranial self-stimulation at the origin and terminations of the mesolimbic and mesocortical CA pathways (Fibiger & Phillips, 1986; Gratton, Hoffer, & Gerhardt, 1988).

4. Drugs that impair central CA transmission diminish or abolish self-stimulation responses (Poschel & Ninteman, 1966; Fantie & Nakajima, 1987; Fibiger & Phillips, 1986).

5. Some drugs that enhance the release of CA by nerve activity — for example, amphetamines and cocaine — increase the self-stimulation responses (Crow, 1972).

---

[8] The most widely accepted pathophysiological explanation for the symptoms of schizophrenia is the "dopamine hypothesis," which suggests that schizophrenic symptoms are due primarily to hyperactivity in the dopamine system. Neuroleptic drugs have been shown to produce blockade of dopamine receptors in animals. The drugs that produce the strongest blockade in animals are also those most effective in reducing symptoms of schizophrenia in humans. Drugs that enhance dopamine transmission, such as amphetamines, tend to exacerbate the symptoms of schizophrenia. Functional hyperactivity of the *D2* receptor may be involved in producing some of the symptoms of schizophrenia. It may be that other neurotransmitter systems are involved in the neurochemistry of schizophrenia as well, such as norepinephrine, serotonin, and gamma-aminobutyric acid systems (Talbott, Hales, & Yudofsky, 1988). From the point of view presented in this chapter, this means that the above neurotransmitters may play a role of "money" along with dopamine in the whole "market" or in some "submarkets" of the brain.

6. Studies dealing with specific drive manipulations tend to support the general hypothesis that self-stimulation involves the activation of neuronal pathways ordinarily involved with the processes of natural reinforcement (Esposito & Kornetsky, 1978; Olds & Yuwiler, 1972).

To summarize points 1-6, it can be seen that secretion of CA is dependent on electrical stimulation of the reinforcement nuclei, and thus the reward impact of these streams of impulses directed to the reinforcement nuclei depends, to a large extent, on the speed of DA secretion. It means that additional (or augmented) streams of impulses directed to the reward nuclei of the brain can cause the CDC phenomenon (Rosenstein, 1994). These ASI directed to the reward nuclei were proposed earlier (Glazunov et al., 1971; Rosenstein, 1991; Vaisbord & Rosenstein, 1968a, 1968b) as the cause for behavioral disorders.

It has been shown in a number of cases that a change in quantity of CA (particularly DA) present in different subsystems of the brain causes substantial changes in the quantitative and qualitative character of their productivity. A few examples are found in the following studies. Brozoski, Brown, Rosvold, and Goldman (1979) provided direct evidence showing that DA may give selective support to a specific cortical function in monkeys; Ljungberg and Enquist (1987) showed that the low doses of D-amphetamine that cause a depletion of DA pools have a disruptive effect on the ability of rats to organize behavior into functional sequences; Joyce (1983) showed that DA depletion in basal ganglia induces a profound hypokinesia in rats, and elevated DA causes locomotor hyperkinesia with stereotyped behavior (Iversen, 1977); and Oke and Adams (1987) linked elevated thalamic DA with possible sensory dysfunctions in schizophrenia. Additional examples are found in McKenna (1989) and Reynolds (1989). Infrahuman studies have revealed that central DA-containing systems are the crucial element of a broad spectrum of goal-seeking behaviors (Fibiger & Phillips, 1986; Panksepp, 1981; Simon, Scatton, & LeMoal, 1980).

The process of mobilization of CA (and, perhaps of some of their predecessors and derivatives "from storage pools to the functional pools," Stinus, LeMoal, & Cardo, 1972) and their release from these pools by stimulation of the reward system, induced by nerve firing, is analogous to a process of exchange of two different "currencies" of the brain — the streams of impulses directed to the reward network and the amount of dopamine. The income function, $W(a)$, where a is density of streams of impulses directed to the reward system of the brain (see Fig. 19.1), was proposed in Vaisbord and Rosenstein (1968a, 1968b) and subsequently discussed in detail in Glazunov et al. (1971, 1972) and Rosenstein (1972, 1991, 1994).

The streams of impulses directed to the reward network seems to present the "inner currency" of the organism in dynamic form. Thus the dopamine is a more stable form of this currency that can be accumulated, stored and distributed among different subsystems of the brain (see Footnote 8). Other forms of presentation of the income can exist too. And they can be as distinct from the two forms mentioned above as these forms are distinct from each other, or as in economic systems the state's gold reserve is distinct from records on the hard disk in the bank computer. Investigation of the brain's "system of banks and currencies" can become one of the important topics in the neurophysiology of the near future.

Next we compare the basic premises of the model based of presentation of inner currency as streams of impulses directed to the reward network with some of those derived from conceptualizations of the role of CA (in particular DA) in the organization of activities in the brain, especially in the schizophrenic brain.

2.1.2. The Role of Dopamine in the Processing of Information. "It is generally thought that the dopaminergic input has a modulatory rather than an information-carrying role" (Early, Posner, Reiman, & Raichle, 1989).

Stein and Wise (1971) connected the etiology of schizophrenia with progressive damage of the noradrenergic reward system. In our model (Rosenstein, 1991; Vaisbord & Rosenstein, 1968a, 1968b), the etiology of schizophrenia is associated with ASI directed to the reward system. However, this is seen not as damage, but mainly as a cause for a functional (and, in many cases, reversible) change in the set of optimal behaviors of the organism. Our model allows for remissions, often seen in schizophrenic patients, and makes it possible to correlate between several existing theories of the origin of schizophrenia and causes of ASI directed into the reward system. Therapeutic approaches to schizophrenia can be associated with possible methods for restoration or compensation of ASI.

Miller (1984) argued that DA acts "to set the threshold for inductive inference." An excess of DA permits associations that would normally be rejected as coincidental to become established in memory. In other words, the plausible effects of a septo-hippocampal DA excess might be not only to induce erroneous attributions of significance — or delusional meaning — but also to promote the generation of abnormal inductive inferences or delusional beliefs (McKenna, 1987). This suggestion reflects known results of learning experiments in animals. However, it is completely unclear *why* the excess permits associations that would normally be rejected.

Our model clarifies the situation (Glazunov et al., 1971, 1972; Rosenstein, 1991; Vaisbord & Rosenstein, 1968a, 1968b). In the model, we can have an excess or a deficit in one of the types of currency (stimulation of reward system, dopamine or other possible forms of currency) in general or in one of local "storehouses" of the brain. The excess leads to "hyperinflation" on the brain market and it is connected to the special verbal, behavioral, and motor production that can be called, in the wording of Crow (1980), "positive symptoms" of schizophrenia. The long-lasting deficit (when the system fails to find ways to increase the income) is related to the "negative symptoms" of schizophrenia. The particular set and type of these symptoms is defined by the particular distribution of dopamine, or other possible forms of currency, in the local "storehouses" of the brain.

According to the model, an excess of a given type of currency causes a change in the reward received by the organism during learning (training) because of the nonlinear MUC-type income function (see Fig. 19.1) that defines the running value of this type of currency on the inner market of the brain. In models with "disturbed" behavior relative to models with "normal" behavior, this decrease in reward brings all effects enumerated by Miller (1984) and McKenna (1987). In particular, it causes a change of threshold for inductive inference, thereby permitting associations that would normally be rejected. These experimental phenomena are deducible from our theoretical model (Glazunov et al., 1971, 1972; Rosenstein, 1991; Vaisbord & Rosenstein, 1968a, 1968b). For example, well-known differences in the formation of Charpentier (and other) delusions in schizophrenics as compared to normal subjects were obtained by formal analysis of the model (Glazunov et al., 1971, 1972; Rosenstein, 1991).

The "involvement of DA in reinforcement processes" (Beninger, 1983; Crow, 1973; McKenna, 1987; Wise, 1978, 1982) is currently viewed as a well-established suggestion. The effect of change of DA should depend on the quantity of income (not only in the form of DA!) already stored in the "banks" of the brain and on the intensity of the streams of reinforcement impulses directed to the reward system of the brain (see Fig. 19.2) at the time. A marginal utility curve of the same type as seen in Fig. 19.1 describes the effect of change of the value of the "brain currency" in the form of dopamine.

2.1.3. DA, Nucleus Accumbens, and ICA. Several well known modern attempts to solve the puzzle of schizophrenia (see Early et al., 1989; Gray, Feldon, Rawlins, Hemsley, & Smith, 1991; Swerdlow & Koob, 1987) suggest excess DA in the nucleus accumbens. According to Neill (1982), "DA transmission in the nucleus accumbens regulates the amount of effort (energy) the animal is willing to expend to achieve a goal and that positive reinforcing stimuli transiently enhance DA transmission biasing the animal to expend more effort."

In Neill's scheme, "the animal continually monitors the effort (energy) which has been expended on a behavior and decides whether to maintain that behavior." It can be seen that the quantity of DA in the nucleus accumbens is related to such abstract notions as "effort" and "energy." However, to decide "whether to maintain that behavior," we need to be able to compare not only future expenditures but also the predicted profit under maintenance versus nonmaintenance of the behavior.

The notion of income is designed to make possible such comparisons, at least in principle. We think that DA is one of several forms of this abstract income used to attend to brain activities and to be attended by them (because the goal of the system is connected in our model with maximization of the income function during the system's lifetime). Several examples of particular income functions are discussed in Rosenstein (1991, Chapter 1).

What causes excess DA in the nucleus accumbens (NAC)? According to our assumption (Rosenstein, 1994), the CDC phenomenon in the schizophrenic brain is caused largely by ASI directed to the reward system. This ASI is seen as the main cause for the disturbance of behavior (Rosenstein, 1972, 1991; Vaisbord & Rosenstein, 1968a, 1968b), and the CDC phenomenon is a secondary effect produced by the ASI. The influence of ASI on the excess of CA (and DA in particular) in NAC (mainly in the self-stimulation paradigm) is a fairly well-established fact (see Blaha & Phillips, 1990; Nakahara, Ozaki, Niura, Miura, & Nagatsu, 1989; Nakahara, Ozaki, Kapoor, & Nagatsu, 1989; Stellar & Corbett, 1989).

It is interesting to note that, according to Colle and Wise (1988), injection of amphetamine into the nucleus accumbens facilitates the rewarding effects of brain stimulation. This is analogous to the effect shown in Fig. 19.2, where a decreased quantity of income in any form in the store (as exemplified by amphetamine-binding DA receptors in the NAC) causes an increase in the rewarding value of the streams of impulses directed to the reward system. Our view is also supported by data showing that parts of the ventral striatum, such as the nucleus accumbens, olfactory tubercle and the ventromedial caudate-putamen are in a unique position to mediate motivational and associative influences on behavioral output (Mogenson, Jones, & Yim, 1980). We view excess DA in the nucleus accumbens — seen in particular as one of the main operational pools of income in the brain — as a result of ASI directed to the reward system. It means that the basic assumption of several modern approaches to the schizophrenia problem regarding the excess of DA in the NAC of the schizophrenic brain (see Early et al., 1989; Gray et al., 1991; Swerdlow & Koob, 1987) can be easily deduced from our model.

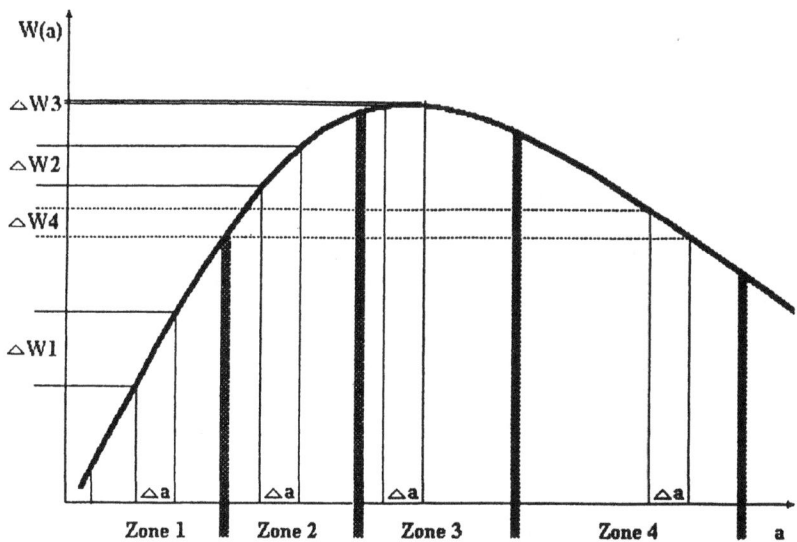

Fig. 19.2. Income function of marginal utility curve type and the hedonic feeling. a = quantity of currency accumulated by the brain's reward system. $W(a)$ = income received by the brain. Zone 1 (superhedonic) — $\Delta W_1 >> \Delta W_2$. Zone 2 (hedonic) — $\Delta W_1 >> \Delta W_2 >> \Delta W_3$. Zone 3 (anhedonic zone) — $\Delta W_2 >> \Delta W_3 \geq 0$. Zone 4 (antihedonic) — $\Delta W_4 < 0$.

Concurrently, we believe that some of the other theoretical approaches (such as the one of Gray et al., 1991, emphasizing damage to the subicular input into the nucleus accumbens) afford useful models of origination of some schizophrenia-like symptoms. However, to meet the challenge of the puzzle of schizophrenia as a whole, we should take into account the basic principles of control and information processing in the human brain, along with details of its neurophysiology. Optimal control approaches seem to be useful in this context. They "treat reinforcement and operant behavior in the context of the demand law which asserts that as the price of a commodity increases the consumption of the commodity decreases" (Sinnamon, 1982).

It is important to state that in biological systems of regulation, as in economics, one of the commodities should be the most "abstract" (the "money" of the system) that makes it possible to compare and interchange other commodities. As in economics, the demand law can be applied to each of them and the "money" can be presented in many different forms. The value of other commodities and of the income received by the system can be measured using this "money."

What plays the role of this "most abstract" commodity in biological systems seems to be a crucial question of tomorrow's neuroscience. Additional short discussion of the topic can be found in the Afterword to this chapter.

2.1.4. Various Forms of Currency in the Domain of Brain Activities. Dopamine appears to be an important member of the set of currencies representing income in brain activities. However, it is only a member of the set. For this reason, according to the ICA, it appears natural that "a monolithic DA hypothesis formulating that increased DA activity leads to schizophrenia appears to be too simplistic a notion" (van Kammen, 1979). For example, medial forebrain bundle self-stimulation is much more dependent on dopamine systems than prefrontal cortex self-stimulation (Corbett, 1990) and a major role is seen for accumbens DA in the medial forebrain bundle (MFB) reward, but not for the DA in caudate and medial frontal cortex (Stellar & Corbett, 1989). Unemoto, Takeichi, Kurumiya, and Olds (1984) viewed their experimental results as negative evidence for the hypothesis that DA innervations in the medial prefrontal cortex are critical neural substrates for self-stimulation. They suggested that activation of intrinsic neurons in the MFB is responsible for self-stimulation in the region. A number of researchers supported the hypothesis that stimulus-stimulus associative learning occurs even if dopamine function is blocked (Benninger, 1982). It is known, for example, that a thalamic rat will learn to obtain hypothalamic stimulation (Huston & Borbely, 1974). In this experiment, the cortical, hippocampal, striatal, and amygdaloid CA terminals are aspirated, yet hypothalamic stimulation can still be effectively used to reward simple responses (Wise, 1978).

According to Fibiger and Phillips (1986), DA neurons do not form an exclusive link in the neuronal systems mediating brain stimulation reward in the lateral hypothalamus and non-DA systems exist that can support intracranial self-stimulation. According to Gray (1982), noradrenalin and serotonin serve to label incoming stimuli as "important" or "important associated with punishment." Dopamine may also act to label incoming stimuli as "important" (perhaps as "important associated with reward"). The word "important" can be naturally translated into the language of the present article as "promising a large change in the income." The role of various CAs and opiates in relation to self-stimulation were discussed by many researchers (Crow, 1972; Esposito & Kornetsky, 1978; Fibiger & Phillips, 1986; Oades, 1985; Wise, 1978).

The stream of electrical impulses directed to the reinforcement nucleus occurs as one of the most widely used and most fundamental "dynamic" forms of the income representation in the brain because it causes the production of other (more stable) forms of income by stimulation of appropriate nuclei in the brain. In reinforcement experiments these streams represent, it seems, the basic form of income in the physiological realm of the brain that can be in use even when other forms (see Huston & Borbely, 1974; Wise, 1978) are almost excluded.

According to the ICA, the presence of many different forms of "currency" in biological systems can be expected. What we have termed "biological income" seems to be a notion of universal importance behind the various forms of "currency." In the Afterword to this chapter we briefly address how the income might be more precisely defined.

2.2. ICA and the Anhedonia Hypothesis of Neuroleptic Action

The anhedonia hypothesis (Gray & Wise, 1980; Wise, Spindler, de Wit, & Gerber, 1978) is explained briefly as follows. Tests under continuous reinforcement schedules suggest that neuroleptics blunt the ability of reinforcers to sustain response at doses that largely spare the ability of an animal to initiate response. Neuroleptics can also impair response (although not response *capacity*) that is normally sustained by environmental stimuli (and associated expectancies — G.R.) in the absence of a primary reinforcer. Neuroleptics also blunt the euphoric impact of amphetamine in humans.

These data suggest that the most subtle and interesting effect of neuroleptics is a selective attenuation of motivational arousal that is (a) critical for goal-directed behavior; (b) normally induced by reinforcers and associated environmental stimuli, and (c) normally accompanied by the subjective experience of pleasure (Wise, 1982). The modified anhedonia hypothesis states that the normal functioning of some non-identified dopaminergic substrate (which could be one or more of several dopaminergic projections in the brain) and its efferent connections are necessary for the motivational phenomena of reinforcement and incentive motivation and for the subjective experience of pleasure that usually accompanies these phenomena (Wise, 1982).

The ICA offers an assumption analogous to the anhedonia hypothesis, as follows. Let the feeling of pleasure produced by a stream of impulses to the reward system or by an injected dose of neuroleptics[9] be measured by the change in value of the income function produced by these agents (this proposition agrees with some known theories of emotions — see Simonov, 1965). This change is dependent on the quantity of the agent ("money") received by the organism at a given moment, and on the income accumulated (in currency of any form) up to this moment, in accordance with a marginal utility curve (see Fig. 19.2).

It can be seen from Fig. 19.2 that there are four established zones of different feelings: Zone 1, where the feeling is very acute; Zone 2, where the feeling is moderate (normal); Zone 3, where the feeling is substantially blunted, and Zone 4 where the feeling is reversed. This is related to functions $\Delta W_1 >> \Delta W_2 >> \Delta W_3 \geq 0 > \Delta W_4$.

Zones 2, 3, and 4 can be naturally marked as "hedonic" (Swerdlow & Koob, 1978), "anhedonic" (Wise, 1978, 1982; Wise et al., 1978) and "antihedonic" (Klemm, 1982). It seems reasonable to mark Zone 1 as "superhedonic." Evidently, the existence of anhedonic and antihedonic zones can be explained by the appearance of ASI directed to the reward zones.

Many clinicians, beginning with Bleuler, have empirically commented on the chronic defect in capacity for experiencing pleasure in schizophrenia (see Rado, 1969). Therefore, the relationship between ASI and schizophrenia could be approached from another angle.

2.2.1. United Reward System of the Brain. We hypothesize that the reward system calculates income as a function of streams of impulses directed to the network of interconnected

---

[9] As it was shown elsewhere, neuroleptics can be seen not only as agents making an impact on CA pools but as ASI producing agents (see Rosenstein, 1991, Chapter 6). A definite dose of neuroleptics could be seen as equal to a stream of impulses of given density and duration directed to the reward system if they produce the same change in the quantity of DA accumulated by the organism.

reinforcement nuclei of an organism. However, it is unclear how different modalities of reward and punishment are weighed and compared, even in the case that they are presented only in the form of streams of impulses.

Modern experimental data hint at the existence of unified reward systems in the brain. As Mora and Ferrer (1986) stated: "Some findings have given rise to the hypothesis that several single feed-back pathways or single circuits exist between points of self-stimulation in the medial prefrontal cortex and points of self-stimulation in other areas of the brain." The hypothalamus is seen as playing one of the central roles in the unified reward system.

"The lateral hypothalamus is involved in integrative functions related to emotion, reward, aversion and learning. It is, however, unclear whether the medial forebrain bundle (MFB) forms a substrate common to the anterior and posterior hypothalamic areas or whether information regarding rewarding and aversive stimuli converges on and is integrated by the same hypothalamic neuron" (Ono, Nakamura, Nishijo, & Fukuda, 1986). In this context, a subcortical control system with the reticular formation, hypothalamus, and thalamus as its principal components has been proposed (Pay, 1979, 1981).

It appears that the motivational or connective aspect is the primary contribution of the hypothalamus (Anderson & Haymaker, 1974; Pay, 1979). The limbic system is envisaged as sorting and synthesizing an assemblage of information to be fed into the hypothalamus. The information concerns thirst and hunger, cognition, and the whole realm of affect and sexual function in their somatic and autonomic accompaniments (Pay, 1981).

According to data of Pribram and his colleagues (Pribram, 1991), "hippocampal resections shift bias with which organism in general approaches the situations it is faced with toward caution and inferotemporal damage shifts bias toward risk." This means that these structures participate in evaluation of future rewards and can be seen in this respect as functional parts of the reward system of the organism. However, alternative points of view on the role of hippocampal formation were published too (see Ridley, Aitken, & Baker, 1989).

In fact we are only at the very beginning of understanding how the "banks" of the brain work and interact and what kind of "inner currencies" are involved in this interaction.

3. ICA AND THE PROBLEMS OF NEUROPSYCHOLOGY OF SCHIZOPHRENIA

3.1. In Schizophrenia "Perceptions Violate Expectations." Why?

According to the currently prevailing opinion of schizophrenia theorists, "there indeed exists a basic psychological dysfunction in schizophrenia; or, at least, only a limited number of such dysfunctions" (Gray et al., 1991). It was argued that it is first of all "a weakening of the influence of stored memories of regularities of previous input on current perception" (Hemsley, 1987), "when perceptions violate expectations" (Einhorn & Hogarth, 1980).

The differences in known behavioral phenomena, such as latent inhibition (Lubow, 1973, 1989), blocking effect (Kamin, 1968), and partial reinforcement extinction (Gray, 1975), between normal subjects and schizophrenics are seen by some authors as confirmation of Hemsley's suggestion. Very similar effects can be produced by dopamine-releasing drugs and modulated by neuroleptics in similar experimental paradigms with animals, which is seen as confirmation of the role of CDC phenomena in schizophrenia in humans, on one hand, and strong support for Hemsley's suggestion on the other (Gray et al., 1991). A fourth phenomenon (besides latent inhibition, blocking effect, and the partial reinforcement estimation) that could be added to the list is the change of the gating effect in schizophrenics as opposed to normal subjects (Braff & Geyer, 1990; Freedman et al., 1987).[10]

Hemsley's suggestion can be immediately deduced from our model of behavior disturbances with no additional hypotheses. The "flat mind effect" (FME) mentioned previously in the section on ICA and the Problems of Neurophysiology of Schizophrenia (see also Fig. 19.3) was deduced from the basic assumptions of our model (Glazunov et al., 1971). FME is the gradual elimination of previous experience in the schizophrenic mind, and therefore constitutes "a weakening of the influence of stored memories of regularities of previous input on current perception," as proposed by Hemsley (1987). Thus, "less and less the subject forms his own impressions, and more and more he is impinged upon by his environment" (Ascombe, 1987).

In our model, the schizophrenic does not intend to form his or her own impressions because he or she does not expect a substantial increase in income, having previously been consistently underpaid by his or her reward system due to the ASI directed to the reward network (Glazunov et al., 1971, 1972; Rosenstein, 1991; Vaisbord & Rosenstein, 1968a; also see Fig. 19.1), as explained above. By the general income-choice approach we come to the impairment in generation of willed intentions in schizophrenics noted also by Frith (1987) — however, without Frith's special suggestion that in schizophrenia willed intentions are not correctly monitored.

3.2. The "Searching for Income" Brain

According to our approach, the brain can be viewed in particular as a system "hunting for income." In this model, the new sources of substantial income for the schizophrenic brain are much more limited than for the normal brain. Being constantly underpaid (Rosenstein, 1991; Vaisbord & Rosenstein 1968a, 1968b) and, therefore, descending step by step to the state of "flat mind," the schizophrenic brain is more strongly motivated[11] to find new — even if nonhabitual or in some sense risky — "fields of hunting," while the milestones of his or her previous experience sink into the calm waters of equal probability of expected events (see Fig. 19.3; Glazunov et al., 1971; Rosenstein, 1991) that we call the "flat mind" effect.

In this situation, three main strategies for finding new sources of income should be expected.

---

[10] We plan to publish an analysis of how the gating effect in schizophrenia relates to these three phenomena, and to Hemsley's (1987) proposition, as a separate article now in preparation.

[11] The organism storage pools ("banks") contain less and less income in any form, and often the mobilization of DA and other forms of income used on the different levels of the organization is interrupted (Stinus et al., 1972). It brings the organism to a choice — stagnation of the "inner market" (with development of negative symptoms of schizophrenia) or intensive search for new income-producing strategies.

3.2.1. The Perception Strategy. Finding new perceptual space by variation of thresholds of channels of perception — and thereby new programs of stimulation of the reward system of the organism — is the *perception strategy.* It is a kind of "environment switching" discussed by us earlier in detail (Meschersky & Rosenstein, 1969; Rosenstein, 1991; Vaisbord & Rosenstein, 1965, 1967). This strategy could consist of switching on usually underused or nonused channels of conscious and subconscious perception.

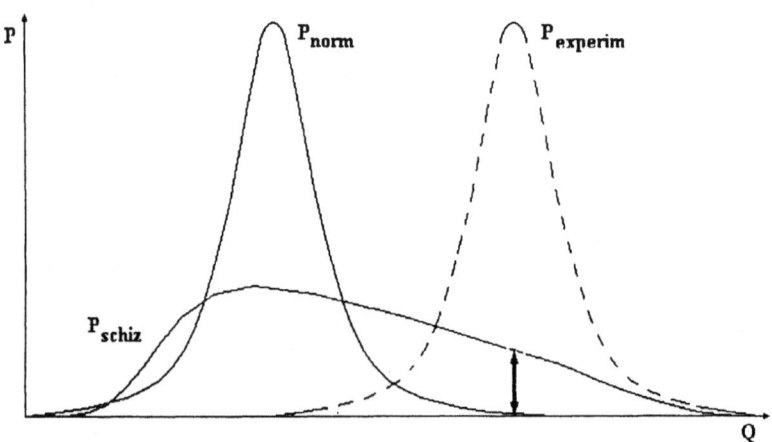

Fig. 19.3. Flat mind effect (FME) in the schizophrenic brain. P_{norm} denotes a typical a posteriori distribution of parameter Q formed by a "normal" brain (a brain without ASI directed to the reward system, i.e., $\Delta a = 0$) in the course of experiments in hypothesis formation. Q is any parameter evaluated by a normal or schizophrenic subject in the course of a learning procedure. P_{schiz} denotes a typical a posteriori distribution of Q formed by the schizophrenic brain according to our model (flat mind effect). $P_{experim}$ denoted the distribution of the value of Q of an "unexpected" pattern to the subject in the course of the experiment (see Rosenstein, 1991, Fig. 7.3). This figure suggests that schizophrenics with FME are "more ready" for unexpected experiences than "normal" subjects (as in Charpentier illusion formation — for other examples see Rosenstein, 1991).

Previous research (Antik et al., 1972; Arshavskii et al., 1972; Meschersky & Rosenstein, 1969; Rosenstein, 1969, 1991) poses the system of control of brain activity in epileptic seizures as an example of this strategy, which can also be viewed as a policy of attention control (see Meschersky & Rosenstein, 1969). Thus the known attention problems in schizophrenia can be understood from the same point of view of optimal control directed to maximization of the income.

It was shown (Antik et al., 1972; Arshavskii et al., 1972; Rosenstein, 1991) that the balance of activities in the lateral and ventromedial nuclei of the hypothalamus can be shifted towards the "positive" or the "negative" side by self-regulation of the perception thresholds in some channels of perception. In animal models of epilepsy, as predicted in Rosenstein (1969) and later shown in experiments, such a shift can substantially decrease or increase the occurrence of seizures (Antik et al., 1972; Arshavskii et al., 1972; Rosenstein, 1991). All this can be seen as a starting point for origination of a theory of attentional disturbances in schizophrenics.

3.2.2. The Production Strategy. The brain can develop new production (mental, verbal, motor, etc.) to stimulate its reward system. "Some schizophrenics, particularly those who are chronically isolated from others, will report that they 'enjoy' their hallucinatory experiences" (Hoffman, 1986). However, this is usually not the case.

Hallucinations are in most cases of an unpleasant and highly emotional character. They are often formed as messages capable of creating the expectation of great possible loss (according to the system of inner values of the subject). Therefore, hallucinations can be an effective inner resources mobilizing tool, if they raise readiness to fight and to win[12] in defence of the most cherished values of the subject. In this way, the expected income of the organism can be substantially increased. This approach is in agreement with the observation that self-stimulation behavior can be elicited from various "aversive" brain structures (Cazala, 1986). (In this connection see Paragraph 3.2.4.)

3.2.3. "Positive Symptoms" in Schizophrenia. In previous publications (Rosenstein, 1991; Vaisbord & Rosenstein, 1968a, 1968b), we speculated that schizophrenia-like behavioral disorders can be determined by an additional (augmented) inward input to the brain reinforcement zones. This idea is supported by the observation of parallel participation of brain's CA in self-stimulation (Crow, 1972; Gratton et al., 1988; Olds & Yuwiler, 1972) and in production of "positive" symptoms in schizophrenia (Early et al., 1989; Gray et al., 1991; McKenna, 1987; Swerdlow & Koob, 1987). Additional arguments for this point of view follow.

3.2.4. "Positive Symptoms" and Self-Stimulation. Positive symptoms in schizophrenia can be considered a peculiar variant of self-stimulation. The subject is so involved in his or her own hallucinations and paranoid ideas that he or she almost totally detaches from objective reality and pays no attention to the outside world, as do animals in the process of intensive self-stimulation. The brain, obsessed by hallucinations and paranoid ideas, as during classical self-stimulation, is a closed system: hallucinatory images and obsessive ideas, because they reflect the electrical impulses in the self-stimulation paradigm, have a tendency to self-determination. This means that an initial image or idea can produce successive related images and ideas, which can, in their turn, stimulate and determine their own reappearance. This clinical observation is supported on a physiological level by the fact that vivid and meaningful images (produced, for instance, by the review of a movie) cause increased CA secretion (probably not only in the suprarenal glands — Carruthers & Taggart, 1973 — but in the brain as well).

The increased brain CA is consequently able to stimulate the reappearance of the same kind of psychic activity that becomes "profitable" in the sense of ICA. The positive feedback between quantity of CA and reappearance of symptoms can be partly responsible for maintaining

---

[12] An example is rumors spread by newspapers about a would-be war that brings about an increase of patriotic feelings and the whole "energy" of the society that has to be ready to fight and to win.

CDC in the schizophrenic brain and can explain, to some extent, the increase of brain CA activity in schizophrenia. The ASI mechanism is very likely involved in this process. It means that the images of a strong emotional character cause the ASI in the reward network, and the increase of DA production follows as explained in Section 2.1.1.

   3.2.5. "Standard" and "Aversive" Symptoms. A special class of schizophrenia production is formed by standard images and verbal hallucinations of an "aversive" character. According to our model, this phenomenon appears due partly to the optimal strategy of income production of the brain that possesses a source of additional stimulation of the reward system. If the stimulation is directed to "negative" nucleus of the reward system, the constantly underpaid brain can try to synthesize hallucinatory inner messages (visual, auditory, etc.) of "high energy" to fill the empty "banks" from new sources of income.

   In some cases, standard blocks of information of "aversive" character are preferred when forming these inner messages. The messages of aversive character can be taken from the memory of the subject, newly synthesized or allowed to pass the mind filters if received from other sources. These standard aversive blocks have two advantages. Being aversive, they provoke such feelings as hate, fear, or anxiety and will switch on, at least at the beginning, powerful income producing mechanisms of defense. The expected income will increase due to mobilization of inner resources and positive mental pictures projected by the sick brain — the future victorious battles, overcoming of fatal dangers, ingenious enemies defeated by the bold and prudent principal hero, who is the same sick person in search of income for his starving reward system.

   Being "standard," these blocks of information do not need new expenditures of income to be produced and they can be included with time in new chains of associations and references in the subject's memory for long-lasting usage. These two properties together — aversive and standard — are able to guarantee the production of substantial quantities of income for the starved brain (see also Section 3.2.2 where the discussion of aversive symptoms begins).

   Through thousands of years, schizophrenics have described visual and audio messages ("visions" and "voices") mostly of aversive character and of amazing similarity despite huge differences in their cultures, social surroundings, and personal experiences. This fact cannot be avoided but has to find a place in the future unified theory of the illness. In short it is a strong hint that some objective sources of information are allowed to be switched out by informational filters of the schizophrenic brain.

3.3. Stereotyped Behavior

   Stereotyped behavior — one of the typical positive symptoms of schizophrenia and a concurrent hyperdopaminergic state — is clearly associated with reinforcement. As Katz (1982) said, "Certain repetitive acts, by virtue of their kinesthetic consequences, are intrinsically reinforcing."

   Repetitive mental, verbal, or motor production of reinforcing character can be used by the organism as a tool for producing positive stimulation of the reward network when the brain has a deficit of income in stock. The brain is therefore trying to replenish the stock by such means as self-produced stereotyped acts where the organism is in full command and its expenses are minimal. Stereotyped behavior can produce a stream of impulses in "positive" centers or restrain impulses from other sources directed to the "negative" centers in order to prevent a reduction of income. The same mechanism can be responsible for the typical schizophrenic phenomena of speaking aloud and continued inner speech (also see the previous paragraph).

In animals, according to Iversen (1977), "dopamine agonist drug stimulation ... induces a general locomotor hyperkinesia which quickly gives way to the phenomena of stereotypy, where there is an increasing repetition of one or a few responses from the normal repertoire of the animal." McKenna (1987) added: "the normal variability of behavior gradually diminishes and is replaced by the incessant performance of simple acts like sniffing, lifting and rearing. The conceptualization of dopamine as reinforcement is by no means incompatible with such findings."

The particular type of behavior may depend on special characteristics of ASI directed to the subject's reward system. If strong outbursts of ASI directed to "negative" nuclei appear, they result in rapid attacks of an acute deficit in income such that the lost income should be immediately compensated. The stereotyped behavior looks like the most suitable instrument in this situation. In the case of long-lasting and not too dramatic deficit in income, it can be compensated by self-organization in the network of the brain by an inner source of stimulation of the reward system. In some types of personalities it can be a stream of "creative" activities as is well-known in the psychology of schizophrenia. This is one of the topics in the framework of the income-choice approach that deserves more detailed treatment in the future.

3.4. "Adjunctive" Behavior

According to ICA, the same basic idea underlies so-called "adjunctive" behavior and the "displacement activities" associated with it (Fibiger & Phillips, 1986) when "irrelevant" behaviors are repeatedly produced during periods of tension and uncertainty. These irrelevant behaviors, such as "nonmotivated" acts of schizophrenics, cannot be explained in terms of physiological deficit. However, they can be seen again as a way to regulate the stream of impulses into the reward network. It is natural to expect from the ICA viewpoint that "a lesion of the brain dopamine system would disrupt displacement behavior as does a lesion of the lateral hypothalamus" (Robbins & Koob, 1980).

Experiments of the "adjunctive" type accompanied by direct recording of the activity of reinforcement nuclei in animals can easily prove or disprove this suggestion.

3.5. Self-Stimulation and Body Resistance

It has been established that self-stimulation increases body resistance and decreases the vulnerability of an organism to various forms of disease (Rotenberg & Arshavsky, 1979; Rotenberg, 1984).[13] This result can be expected from the ICA point of view: Self-stimulation

---

[13] This development originated in this author's model of income-dependent control of epileptic seizures (Rosenstein, 1969, 1991). It was predicted (Rosenstein, 1969) and later verified (Antik et al., 1972; Arshavskii et al., 1972) that the prevention or realization of epileptic attacks in animals can be controlled by proper stimulation of the reward system — by increasing or decreasing the income of the organism. Subsequently, it was shown by

produces a large reserve of income ("free money") in the storehouses of the brain, and this reserve is ready to be used to prevent the illness. In parallel, it was established in clinical studies that the acute phase of psychotic disorders with obvious positive symptoms is usually accompanied by increased somatic health (Anisman & Zacharka, 1982; Arshavsky & Rotenberg, 1976), whereas the suppression of these symptoms by neuroleptics is often accompanied by somatic disorders (Rotenberg, 1984).

These observations show that positive symptoms in schizophrenia can be income producing phenomena, and as such, can be provoked by the need of the brain to find or create new sources of income.

3.6. The United Perception-Production Strategy

The "perception" strategy can be used by the brain in its search for sources of income in combination with the "production" strategy. In this case, for example, the threshold of perception for inner speech can be decreased (perception strategy) and, at the same time, the informational system of the brain can be activated (production strategy) and the "personalities" whose conversations are heard by the schizophrenic as inner speech can be "allowed to come," that is, released by the filtration system of the brain, or newly synthesized.

The use of this perception-production strategy means, of course, more severe separation from everyday reality. This separation increases as more channels of perception of the inner world of the subject being brought into the game and as more elaborate "personages" are brought from the subject's inner world to the attention of his or her subconscious levels of information processing and raised afterwards to the level of consciousness.

Some of these channels of contact with inner personages, if organized into complex traces in the subject's memory, cannot always be switched off by the brain even if "undesirable" for the subject (contradicting his or her former points of view, beliefs or interests). This inability can be due to a deficit of income in stock to perform the switching-off process (it can be seen as a lack of "willpower" in the person), or it may occur that the subject has become dependent on sources of income (even "aversive" ones) and, according to an automatic inner evaluation, the risk of living without these sources of income is too big to accept. It is worth mentioning that the behavior of the subject depends, to a great extent, on the subject's evaluation of the future. According to our assumption, the reward system of the brain calculates *expected* income. This evaluation can be influenced by the information deduced from the hallucinatory pictures, and thus the subject can be literally caught by his or her own perception or production strategies. The person masters more and more his or her new inner sources of income and becomes less dependent on external events and more distant from his or her social surroundings.

---

other authors that dopaminergic supersensitivity follows ferric-induced limbic seizures (Csernansky, Bonnet, & Hollister, 1985) and dopamine agonists have an anticonvulsant effect (Loscher & Czuczwar, 1986). In addition, our model explains the mechanism of origination of schizophrenic-like symptoms in epileptics (see the Conclusion to Rosenstein, 1991).

4. AFTERWORD

We hope to come, step by step, to the understanding that "effort," "energy", "reward," "punishment," "stimulation of reward networks," "catecholamines," "endorphins," and so on, can serve as suitable, through inexact, descriptions of a more universal and subtle concept in control of biological systems that we call "biological income." This income is a theoretical abstraction of something that all living creatures strive for.

In another article in progress, I am applying the income theory to the often—discussed relationship between schizophrenia and genius. Space does not permit a detailed exposition here, but it was shown in the example of the Charpentier illusion (Glazunov et al., 1971, 1972; Rosenstein, 1991) why the performance of schizophrenics can be better than that of normals in unexpected and hardly predictable situations. This is due to the possibility of less restricted associations and ways of drawing conclusions arising from the flat mind effect. This quality of schizophrenics, which in our model is due to ASI directed to the reward system of the schizo- phrenic brain, is also typical of highly talented people. At the same time, as it was mentioned above, the deficit of "income" can serve as a "drive" for creative activities. This observation can support the type of interconnection between schizophrenia and talent (genius) suggested in our model.

Again, we come to the bioeconomic metaphor. Various forms of payment ("money") used in business life are not completely adequate forms of economic income, either. Processes of inflation and deflation that cause dramatic changes in the income in economics express the changes of the "value of money," which is explained by economists, in many cases, through such human factors as psychological preferences, mass expectations, prejudices, or even epidemic madness, leaving in this "value of money" a good deal of mystery.

Besides economics, there are two other sources of knowledge that lead to the income idea in the field of biological control. Both of them are even more mysterious than the "value of money." One source is the *psychic energy* suggested by Carl Jung, which can be seen as one of the sources of biological income idea in philosophy of biological control. Another and most fundamental among the sources of the income idea is the *Chaiut* ("life energy" in a rough translation from Hebrew). It is partly reflected in other cultures under different names. Streams and stocks of this "energy" can be regulated in humans by many means, including rules of ethical conduct, meditation, and prayer.

My feeling is that we are now close to a paradigm in neurophysiology that will be one of motive forces of research in the field of biological control and organization in the near future. Not only terms such as "hedonic feeling," "pleasure," "joy," but even "delight," "happiness" and "love" on their highest levels are much closer to what one may have in mind when thinking about income in the control systems of the brain (especially in the human brain) than all the particular substitutes of income mentioned in the first lines of this Afterword. It is, in fact, what we strive for in the course of our lives, and from whence originate to a large extent, "energy," "resources," and "efforts" that are invested to meet the challenges of life. It is well known now in practical medical science (see Siegel, 1986) that the old expression "the wounds of the winners are healed quicker" is far from being only a motto.

"Joy is the food of the soul"[14] was written by giants of the spirit long before our time. Tomorrow, if not today, this old suggestion has a good chance of becoming a regular part of brain and behavioral sciences.

ACKNOWLEDGMENTS

This chapter is written in memory of Mr. Abraham Harman, the former President and Chancellor of the Hebrew University, Jerusalem. There are souls that make this world sunny.

I thank Karl Pribram, Shlomo Samueloff, Vadim A. Rotenberg, Daniel Levine, Ehud Avissar, and Israel Elkind for valuable comments. In particular, Dr. Rotenberg provided comments about the "positive symptoms" in schizophrenia, and Professor Pribram introduced the term "inflation of reinforcement" that coins a basic feature of the ICA.

Important financial support was obtained from the S. H. and Helen R. Scheuer Family Foundation, from the Wolfson Family Charitable Trust, and from the Inner Fund of the Division of Research and Development of Hebrew University, with the recommendations of Professors Rami Rahamimoff and Moshe Abeles. I also thank Pavel Kaspler and Larisa Trembovler for assistance with the bibliography, Hanna Siegal-Rais and Mona Blumberg for typing the manuscript, and the Department of Physiology (Head, David Lichtstein) of the Hebrew University Medical School for an encouraging atmosphere.

REFERENCES

Anden, N. E., Fuxe, K., Hamberger, B., & Kokfelt, T. (1966). A quantitative study of the nigro-neostriatal dopamine neuron system in the rat. *Acta Physiologica Scandinavica*, **67**, 306-312.

Anderson, E., & Haymaker, W. (1974). Breakthrough in hypothalamic and pituitary research. In D. F. Swaab & J. P. Schade (Eds.), *Progress in Brain Research, Vol. 41: Integrative Hypothalamic Activity* (pp. 1-60). New York: Elsevier.

Anisman, H., & Zacharka, R. M. (1982). Depression: the predisposing influence of stress. *Behavioral and Brain Sciences*, **5**, 89-138.

Antik, A. P., Arshavskii, V. V., & Rosenstein, G. S. (1972). II. Role of motivation centers in the control of the convulsive activity of the brain. *Biofizika*, **17**, 4, 681-686.

Arshavskii, V. V., Meshman, V. F., & Rosenstein, G. S. (1972). I. Role of the motivation centers in the control of the convulsive activity of the brain. *Biofizika*, **17**, 515-520.

Arshavsky, V. V., & Rotenberg, V. S. (1976). Search activity and its influence on experimental and clinical pathology. *Zhurnal Nervnoy Degatelnost*, **26**, 424-428.

Ascombe, R. (1987). The disorder of consciousness in schizophrenia. *Psychopharmacology*, **72**, 17-19.

---

[14] R. Shneur Zalman Schneerson (1965, first English edition), *Likutie Tora*, Kehot Publication Society, Brooklyn, NY. This is the famous "Tania," the classic of Jewish Hasidic literature, published for the first time in Hebrew in the year 1797 with references to much, much older sources.

Axelrod, J. (1977). Catecholaminergic systems in the brain. *Acta Neurologica Scandinavica* (Suppl. **64**), 85-89.

Beninger, R. J. (1983). The role of dopamine in locomotor activity and learning. *Brain Research Reviews*, **6**, 173-196.

Blaha, C. D., & Phillips, A. G. (1990). Application of in vivo electrochemistry to the measurement of changes in DA release during intracranial self-stimulation (SS). *Journal of Neuroscience Methods*, **34**(1-3), 125-133.

Braff, D. L., & Geyer, M. A. (1990). Sensorimotor gating and schizophrenia. *Archives of General Psychiatry*, **47**, 181-188.

Brozoski, T. J., Brown, R. M., Rosvold, H. E., & Goldman, P. S. (1979). Deficit caused by regional depletion in prefrontal cortex of rhesus-monkey. *Science*, **205**, 929-932.

Carruthers, M., & Taggart, P. (1973). Vagotonicity of violence: Biochemical and cardiac responses to violent films and television programmes. *British Medical Journal*, **3**, 384-389.

Cazala, P. (1986). Self-stimulation behavior can be elicited from various aversive brain structures. *Neuroscience and Behavioral Reviews*, **2**, 115-122.

Colle, L. M., & Wise, R. A. (1988). Effects on nucleus accumbens amphetamine on lateral hypothalamus brain stimulation reward. *Brain Research*, **459**, 361-368.

Corbett, D. (1990). Differences in sensitivity to neuroleptic blockage: Medial forebrain bundle versus frontal cortex self-stimulation. *Behavioral Brain Research*, **36**, 91-6.

Crow, T. J. (1972). Catecholamine-containing neurons and electrical self-stimulation: 1. A review of some data. *Psychological Medicine*, **2**, 414-421.

Crow, T. J. (1973). Catecholamine-containing neurons and electrical self-stimulation: 2. A theoretical interpretation and some psychiatric implications. *Psychological Medicine*, **3**, 66-73.

Crow, T. J. (1980). Molecular pathology of schizophrenia: More than one disease process? *British Medical Journal*, **280**, 66-68.

Csernansky, C. A., Bonnet, K. A., & Hollister, L. G. (1985). Dopaminergic supersensitivity follows ferric chloride-induced limbic seizures. *Biological Psychiatry*, **20**, 723-733.

Early, T. S., Posner, M. I., Reiman, E. M., & Raichle, M. E. (1989). Hyperactivity of the left striato-pallidal projection, Part I: Lower level theory. Part II: Phenomenology and thought disorders. *Psychiatric Development*, **2**, 85-108, 109-121.

Einhorn, H. J., & Hogarth, R. M. (1980). Judging probable cause. *Psychological Bulletin*, **99**, 3-19.

Esposito, R. U., & Kornetsky, C. (1978). Opioids and rewarding brain stimulation. *Behavioral Brain Research*, **22**, 163-171.

Fantie, B. D., & Nakajima, S. (1987). Operant conditioning of hippocampal theta: dissociating reward from performance deficits. *Behavioral Neuroscience*, **191**, 626-633.

Fibiger, H. C., & Phillips, A. G. (1986). Reward, motivation, cognition: psychobiology of mesotelencephalic dopamine systems. In F. E. Bloom & S. R. Geiger (Eds.), *Handbook of Physiology. The Nervous System. Vol. 4: Intrinsic Regulatory Systems of the Brain* (Chapter 12). Bethesda, MD: American Physiological Society.

Freedman, R., Adler, L. E., Gerhardt, G. A., Waldo, M., Baker, N., Rose, G. M., Drebing, C., Nagomoto, H., Brickford-Wimer, P., & Franks, R. (1987). Neurobiological studies in sensory gating in schizophrenia. *Schizophrenia Bulletin*, **13**, 669-678.

Frith, C. D. (1987). Consciousness, information processing and schizophrenia. *British Journal of Psychiatry*, **134**, 225-235.

German, D. C., & Bowden, L. M. (1974). Catecholamine systems at the neural substrate for intracranial self-stimulation: A hypothesis. *Brain Research*, **73**, 381-419.

Glazunov, N. I., Rosenstein, G. S., & Jablonsky, A. J. (1971). Automata with "normal" and "disturbed" behavior in experiments with choice and reinforcement. *Reports of Academy of Science of USSR*, **197**(N6), 1280-1283.

Glazunov, N. I., Rosenstein, G. S., & Jablonsky, A. J. (1972). About the goal-oriented strategy of automata with "normal" and "disturbed" behavior in experiments with choice and reinforcement. In P. K. Anochin (Ed.), *Mechanisms and Principles of Goal-Oriented Behavior* (pp. 80-92). Moscow: Nauka (Russian).

Gratton, A., Hoffer, B. J., & Gerhardt, G. A. (1988). Effects of electrical stimulation of brain reward sites on release of dopamine in rat; an in vivo electrochemical study. *Brain Research Bulletin*, **21**, 319-24.

Gray, J. A. (1975). *Elements of a Two-process Theory of Learning*. New York: Academic Press.

Gray, J. A. (1982). *The Neuropsychology of Anxiety: An Inquiry Into the Functions of the Septo-hippocampal System*. Oxford and New York: Oxford University Press.

Gray, J. A., Feldon, J., Rawlins, J. N. P., Hemsley, D. R., & Smith, A. D. (1991) . The neuropsychology of schizophrenia. *Behavioral and Brain Sciences*, **14**, 1-84.

Gray, T., & Wise, R. A. (1980). Effects of pimozide on lever pressing behavior maintained on an intermittent reinforcement schedule. *Pharmacology, Biochemistry, and Behavior*, **2**, 931-935.

Hemsley, D. R. (1987). An experimental psychological model for schizophrenia. In H. Hafner, W. F. Gattar, & W. Janzavik (Eds.), *Search for the Causes of Schizophrenia*. Berlin: Springer Verlag.

Hoffman, R. E. (1986). Self-stimulation behavior can be elicited from various "aversive" brain structures. *Behavioral and Brain Sciences*, **9**, 503-548.

Huston, J. P., & Borbely, A. A. (1974). The thalamic rat: General behavior, operant learning with rewarding hypothalamic stimulation and effects of amphetamine. *Physiology of Behavior*, **12**, 433-448.

Iversen, S. D. (1977). Striatal function and stereotyped behavior. In A. R. Cools, A. H. M. Lohman, & J. H. L. Van Den Bercheen (Eds.), *Psychobiology of the Striatum*. Amsterdam: North-Holland.

Joyce, J. N. (1983). Multiple dopamine receptors and behavior. *Neuroscience and Behavioral Reviews*, **7**, 227-256.

Kamin, L. J. (1968). "Attention-like" processes in classical conditioning. In M. R. Jones (Ed.), *Miami Symposium on the Prediction of Behavior*. Miami, FL: University of Miami Press.

Katz, R. J. (1982). Dopamine and the limits of behavioral reduction — or why aren't all schizophrenics fat and happy? *Behavioral and Brain Sciences*, **5**, 60-61.

Klemm, W. R. (1982). Time for a new synthesis of hedonic mechanisms: Interaction of multiple and interdependent reinforcer system. *Behavioral and Brain Sciences*, **5**, 61-62.

Liddle, P. F. (1994). Volition and schizophrenia. In A. S. David & J. C. Cutting (Eds.), *The Neuropsychology of Schizophrenia*. Hillsdale, NJ: Lawrence Erlbaum Associates.

Ljungberg, T., & Enquist, M. (1987). Disruptive effects of low doses of d-amphetamine on the ability of rats to organize behavior into functional sequences. *Psychopharmacology*, **93**, 146-151.

Loscher, W., & Czuczwar, S. J. (1986). Studies on the involvement of dopamine D-1 and D-2 receptors in the anticonvulsant effect of dopamine agonists in various recent models of epilepsy. *European Journal of Pharmacology*, **128**, 55-65.

Lubow, R. E. (1973). Latent inhibition. *Psychological Bulletin*, **79**, 398-407.

Lubow, R. E. (1989). *Latent Inhibition and Conditioned Attention Theory*. Cambridge, UK: Cambridge University Press.

McKenna, P. J. (1987). Pathology, phenomenology and dopamine hypothesis of schizophrenia. *British Journal of Psychiatry*, **151**, 288-301.

Meschersky, R. M., & Rosenstein, G. S. (1969), Instability of the central nervous system. *Brain Research*, **13**, 367-375.

Miller, R. (1984). Major psychosis and dopamine: controversial features and some suggestions. *Psychological Medicine*, **14**, 779-789.

Mogenson, G. I., Jones, D. L., & Yim, C. V. (1980). From motivation to action: Functional interface between the limbic system and the motor system. *Progress in Neurobiology*, **14**, 69-97.

Mora, F., & Ferrer, J. M. (1986). Neurotransmitters, pathways and circuits as the neural substrates of self-stimulation of the prefrontal cortex: facts and speculations. *Behavioral Brain Research*, **22**, 127-240.

Nakahara, D., Ozaki, N., Niura, Y., Miura, H., & Nagatsu, T. (1989a). Increased dopamine and serotonin metabolism in rat nucleus accumbens produced by intracranial self-stimulation of medial forebrain bundle as measured by in vivo microdialysis. *Brain Research*, **495**, 178-181.

Nakahara, D., Ozaki, N., Kapoor, V., & Nagatsu, T. (1989b). The effect of uptake inhibition on dopamine release from the nucleus accumbens of rats during self- or forced stimulation of the medial forebrain bundle: a microdialysis study. *Neuroscience Letters*, 1989, Sept. 25, **104(1-2)**, 136-140.

Neill, D. (1982). Problems of concept and vocabulary in the anhedonia hypothesis. *The Behavioral and Brain Sciences*, **5**, 70.

Oades, R. D. (1985). The role of noradrenaline in tuning and dopamine in switching between signals in the CNS. *Neuroscience and Biobehavioral Reviews*, **9**, 261-282.

Oke, A. F., & Adams, R. N. (1987). Elevated thalamic dopamine: Possible line to sensory dysfunctions in schizophrenia. *Schizophrenia Bulletin*, **13(N4)**, 589-604.

Olds, M. E., & Yuwiler, A. (1972). Effect of brain stimulation in positive and negative reinforcing regions in the rat on content of catecholamines in hypothalamus and brain. *Brain Research*, **36**, 385-398.

Ono, T., Nakamura, K., Nishijo, H., & Fukuda, M. (1986). Hypothalamic neuron involvement in integration of reward, aversion and cue signals. *Journal of Neurophysiology*, **56**, 63-79.

Panksepp, J. (1981). Hypothalamic integration of behavior: Rewards, punishments and related psychological processes. In P. J. Morgan & J. Panksepp (Eds.), *Handbook of the Hypothalamus: Behavioral Studies of the Hypothalamus* (Vol. 3, part B, pp. 289-431). New York: Marcel Dekker.

Panksepp, J. (1982). The pleasure in brain substrates of foraging. *Brain and Behavioral Science*, **5**, 71-73.

Pay, R. G. (1979). Cognitive regulation of cortical activity by the reticular formation, hypothalamus and thalamus. *International Journal of Neuroscience*, **10**, 233-253.

Pay, R. G. (1981). Control of complex conation and emotion in the neocortex by the limbic entorhinal, subicular and cingulate cortices and the hypothalamus, mammillary body and thalamus. *International Journal of Neuroscience*, **15**, 1-30.

Poschel, B. P. H., & Ninteman, F. W. (1966). Hypothalamic self-stimulation: Its suppression by blockade of norepinephrine biosynthesis and reinstatement by methamphetamine. *Life Sciences*, **5**, 11-16.

Pribram, K. H. (1991). *Brain and Perception*. Hillsdale, NJ: Lawrence Erlbaum Associates.

Rado, S. (1969). *Adaptational Psychodynamics: Motivation and Control*. New York: Science House.

Reynolds, G. P. (1989). Beyond the dopamine hypothesis. The neurochemical pathology of schizophrenia. *British Journal of Psychiatry*, **155**, 305-316.

Ridley, R. M., Aitken, D. M., & Baker, H. F. (1989). Learning about rules not about reward is impaired following lesions of the cholinergic projections of the hippocampus. *Brain Research*, **502**, 306-318.

Robbins, T. W., & Koob, G. F. (1980). Selective disruption of displacement behavior by lesions of the mesolimbic dopamine system. *Nature*, **285**, 409-412.

Rosenstein, G. S. (1969). Some remarks on a qualitative model of epilepsy. *Biofizika*, **14**, 765-767.

Rosenstein, G. S. (1972). Income and choice in biological control systems. *Reports of the Sixth Symposium in Cybernetics (Part 4). Decision Making in Biological Systems of Control* (pp. 107-110). Tbilisi: Mezniereba. (Russian.)

Rosenstein, G.-Z. (1991). *Income and Choice in Biological Control Systems*. Hillsdale, NJ: Lawrence Erlbaum Associates.

Rosenstein, G.-Z. (1994). Cause of catecholamine distribution change in schizophrenic brain. *Journal of Basic and Clinical Physiology and Pharmacology*, **5**, 101-106.

Rotenberg, V. S. (1984). Search activity in the context of psychosomatic disturbances of brain monoamine, and REM sleep function. *Pavlovian Journal of Biology*, **19**, 1-15.

Rotenberg, V. S., & Arshavsky, V. V. (1979). Search activity and its impact on experimental and clinical pathology. *Activitas Nervosa Superior (Praha)*, **21**(2), 105-115.

Siegel, B. S. (1986). *Love, Medicine, and Miracles*. New York: Harper and Row.

Simon, H. B., Scatton, B., & LeMoal, M. (1980). Dopaminergic A10 neurons are involved in cognitive functions. *Nature*, **286**, 150-151.

Simonov, P. V. (1965). About the role of emotions on the adaptive behavior of living systems. *Problems of Psychology*, **4**, 75-84.

Sinnamon, H. M. (1982). The reward-effort model: An economic framework for examining the mechanism of neuroleptic action. *Behavioral and Brain Sciences*, **5**, 73-75.

Stein, L., & Wise, C. D. (1971). Possible etiology of schizophrenia: progressive damage to the noradrenergic reward system by 6-hydroxydopamine. *Science*, **171**, 1032-1036.

Stellar, J. R., & Corbett, D. (1989). Regional neuroleptic microinjections indicate a role for nucleus accumbens in lateral hypothalamic self-stimulation reward. *Brain Research*, **477(1-2)**, 126-143.

Stinus, L., LeMoal, M., & Cardo, B. (1972). Auto-stimulation et catecholamines. I. Intervention possible des deux "compartments" (compartment functional et compartiment de reserve). *Physiology of Behavior*, **9**, 175-185.

Swerdlow, N. R., & Koob, G. F. (1987). Dopamine, schizophrenia, mania and depression: Toward a unified hypothesis of cortico-striato-pallidothalamic function. *Behavioral and Brain Sciences*, **10**, 197-245.

Talbott, J. A., Hales, R. E., & Yudofsky, S.C. (Eds.) (1988). *Textbook of Psychiatry*. Washington, DC: American Psychiatric Press.

Unemoto, M., Takeichi, T., Kurumiya, S., & Olds, M. E. (1984). Selective neonatal depletion of dopamine has no effect on medial prefrontal cortex self-stimulation in the rat. *Neuroscience — Research*, 295-307.

Vaisbord, E. L., & Rosenstein, G. S. (1965). Life-time of non-stable stochastic automata. *Technical Izvestia Akademii Nauk SSSR, Seria Technicheskoja Kybernetika*, N4, 2-**59**.

Vaisbord, E. L., & Rosenstein, G. S. (1967). Unstable automata and models of active behaviour. In M. G. Goaze-Rapoport (Ed.), *Voprosi Bioniki* (Problems of Bionics) (pp. 292-294). Moscow: Nauka.

Vaisbord, E. L., & Rosenstein, G. S. (1968a). I. Qualitative models of disturbances in behavior. *Biofizika*, **13(3)**, 510-516.

Vaisbord, E. L., & Rosenstein, G. S. (1968b). II. Qualitative models of disturbances in behavior. *Biofizika*, **13(6)**, 1104-1110.

van Kammen, D. P. (1979). The dopamine hypothesis of schizophrenia revised. *Psychoneuroendocrinology*, **4**, 37-46.

Wise, R. A. (1978). Catecholamine theories of reward: a critical review. *Brain Research*, **152**, 215-247.

Wise, R. A. (1982). Neuroleptics and operant behavior: The anhedonia hypothesis. *Behavioral and Brain Sciences*, **5**, 39-87.

Wise, R. A., Spindler, J., de Wit, H., & Gerber, G. J. (1978). Neuroleptic-induced "anhedonia" in rats: Pimozide blocks reward quality of food. *Science*, **201**, 262-264.

20

Communication Cognition: Interactive Nets We Weave, When We Practice to Perceive

Sylvia Candelaria de Ram
Cognition and Communication, Las Cruces, New Mexico

   Sylvia Candelaria de Ram's chapter, **Communication Cognition: Interactive Nets We Weave, When We Practice to Perceive***, discusses the author's construct called* **Pragmase-mantic**[TM] **interactivity operations** *as optimizers. It deals with modeling the process of communication, and how it arises, functioning to satisfy hunger and other needs of individuals and societies. With a primary concern with human expression, verbal or gestural, the model differs from traditional linguists' work with surface forms and their syntax in that context is integrated at the very start of processing, explaining how words can be meaningful and provoke response. In a companion paper (Candelaria de Ram, 1994), the author applies the same principles both to human language and to context-dependent lower level processes in a variety of animals, such as shocking of potential predators by electric fish and motility of protozon.*
   Candelaria de Ram's modeling approach is somewhere on the boundary between connectio-nist neural network modeling (although she refers to connectionist models by others such as Helen Gigley and Donald Loritz) and production system modeling from mainstream artificial intelligence. She traces processes of sound production, both in infants and adults, through all the different neural levels of processing including interactions of auditory, visual, and motor systems, using the slogan that we are "cross-modal multi-layer and multi-domain action-percep-tion nets" (which is reminiscent of Michael Arbib's earlier statement that the human brain is an "action-oriented, layered, somatotopically organized computer"). The attempt to build a general theory of system interactions across different levels is quite similar in spirit to that of Prueitt's chapter in this volume. The emphasis on communication, involving more than one actor and depending heavily on context and affect, is also quite similar to the discussion of negotiation in Leven's chapter.

Candelaria de Ram's discussion of optimality seeks a unifying principle that applies both to lower level elements in the nets (and low-level motor processes) and to global processes like language use. She states that "Optimality is taken to be system functionality and continuance" — a definition reminiscent of the natural selection approach taken in Elsberry's chapter. This definition leads her away from search for a global, universal utility function: "... optimality is not everywhere and always the same, but is relative to temporal, local phenomena." This is not equivalent to satisficing; rather, it argues in favor of many linguistic and communication processes being optimal, but the nature of the optimization being flexible and acutely sensitive to context. She uses these ideas to explain not only changes in communication patterns but drift in language itself as cultural groups persist. Paradoxically, this does lead to discovery of an evolutionary motivational principle for sensor-based communication among cohorts: Discernment × entropy = "discerntropy" results as selective perception and reaction develop so that they reduce apparent entropy as evolution increases the complexity of our universe.

ABSTRACT

Communication, lending context sensitivity and then knowledge enhancement, becomes an ecological essential as cognition advances. Agents that can communicate have an extended range of knowing what is going on and of influencing it. From the viewpoint of interactive realism, real-world process rather than inventory of surface forms (words, syntax) becomes the focus. Communication is seen to be based on cognitive action to "simplify" in the face of obvious complexity, to reduce local apparent entropy in the face of ever-increasing system entropy. Communicating involves direct action supported by chemo-physical sensor-effectors that provide values along sensory parameters. The potential for communication goes way back in evolution, to molecular reactions like crystallizing and dissolving that are "sensitive" in "seeking" or "fleeing" certain things. Sensors, and cognition with them, develop through selective recombination. Then, in the face of ever-increasing complexity ("disorder," the ever-increasing entropy of the Second Law of Thermodynamics), cognition advances by selecting sensory parameters as relevant, and building up a "simpler" (or at least more helpful) cognitive meta-universe.

The key link between cohorts is for communicative gestures to be held in common gestures that cue cognitive parameter values and processing. Recent experimental evidence shows that in human infants, mimicry and discriminations matching sound to mouth position are instinctive (automatic behavior; Meltzoff & Moore, 1983). Automatic mimicry (along with other cognitive symmetries) appears to push individuals toward sharing code systems under dynamic conditions. Search for the cause of linguistic drift whenever languages are in use led Candelaria de Ram (1988) to posit the existence and critical function of automatic mimicry even in adults as providing ineluctable pull toward new variants.

Simple neural net models simulating phonemic alternation (Loritz) and multistage, multilayer nets for parsing (Gigley) point the way to adequate modeling. Pragmasemantics™ process descriptions can be cast within a special grounded, sortal logic (Candelaria de Ram 1992a) as high-level computer programs. They register real-parameter values and send out signals, so as to act as communicators. These artifices are, by virtue of their underlying operational logic, networks. The idea is to make them like natural communicators, exposing individual-agent communication reasoning and behavior-control detail. As shown in the companion work (Candelaria de Ram, 1994), the operators (functors) of Pragmasemantic logic apply for molecular and cellular level process components just as they do in processes at the speech act level. Rudimentary operations in a dynamic combinatory system eventually enable full-fledged communication.

1. LANGUAGE SCIENCE: CODE STATES VERSUS COMMUNICATOR STATES

Today's questions call for a viewpoint shift from language surface to intimate detail of communicative process.

1.1. Linguists' Language Vessels

Linguistics, even when examining language system change, has concentrated on verbal surface forms and patterns: "The evidence," wrote Samuels (1972, p. 4) in the introduction to *Linguistic Evolution with Special Reference to English* is "The substance of language, [that] from a descriptive point of view, exists in two equally valid and autonomous shapes — spoken and written. As a code, each exists in its own right ... for diachronic purposes our choice of evidence is far more limited: we must for the most part reconstruct from written records, ... [except for] comparisons of the speech of older and younger generations...[or other] present spoken forms of cognate dialects and languages." Limitation of evidence to grammar and philology of phonemically written language form has limited the methods of historical linguistics as well as other branches of language study, and the kinds of answers worked out.

"The shape of linguistics today was set in large part by Ferdinand de Saussure," wrote W. P. Lehmann (1968, p. 5), who drew linguistics into science with his "view of a linguist as an observer [that] leads to a concern for language as a state. Today there is a far greater concern for language in operation...at the morphological and semantic level, as well as at the phonological." But "recall how explicit Saussure was about the ineffective role of the speaker in initiating, and even in controlling, change in language. ... Kurylowicz in contrast views the speaker as deciding between alternate forms ...[so he] controls the effects of change in language. Others, as do Weinreich, Labov, and Herzog" in setting out *A Theory of Language Change* (1968, pp. 97-195) describe mechanisms and come upon "The Transition Problem," where viewpoint has to shift from surface artifacts of stimulus-response to language-user's mind. However, important as their work is, they do not make the radical shift from inventorying to modelling mental processing or allowing individual intent in selecting code items along with what to say as we do here. They vaguely embed formal analysis systems that include a marking of pronunciation forms "by the feature archaic/innovating" into theoretical members of a speech group. Any individual is seen as a passive "competence," simply a medium for passing code patterns around:

> This transition or transfer of features from one speaker to another appears to take place through the medium of bidialectal speakers, or more generally, speakers with heterogeneous systems characterized by orderly differentiation. Change takes place

(1) as a speaker learns an alternate form, (2) during the time that the two forms exist in contact within his competence, and (3) when one of the forms becomes obsolete. ... empirical evidence gathered to date indicates that children do not preserve the dialect characteristics of their parents, but rather those of the peer group which dominates their preadolescent years. Sometimes in the process ... New groups enter the speech community and reinterpret the on-going linguistic change ... The advancement of the linguistic change to completion may be accompanied by a rise in the level of social awareness of the change and the establishment of a social stereotype. Eventually, the completion of the change and the shift of the variable to the status of a constant is accompanied by the loss of whatever social significance the feature possessed. (Labov, Weinreich, & Herzog, 1968, pp. 184-187)

Our refocus on process and action might let us see instead change as stemming from persons whose own parameters for interpreting and producing bits of code changed. Communicating appears to be partly instinctive, partly volitional, with automatic and learned links in a dynamic mental complex. Solving the puzzle of how changing individuals can continue to communicate may link the two approaches, and permit us new practical uses of linguistic findings about "evidence."

1.2. Communication Shapers

The place where "we must search for the general principles of a science of language" is not in "the different arbitrary conventions of language codes" but rather in "common potentials for developing languages, the shared systems of symbolic representation, the universal mechanisms for metaphor and synesthesia — all formed in the interaction of human biology and psychology with a fundamentally common environment." So said psycholinguistic semanticist Charles Osgood (1963, p. 322). He continued:

Thus because the general laws of perceptual grouping and patterning apply to all humans, we ... have more discriminative labels for finger vs. hand and for hand vs. arm than for upper-arm vs. lower-arm or for chest vs. abdomen (i.e., independent movement is one criterion for perceptual organization). Similarly, ...[primate brain devotion to vision and audition encourages more synasthetic] visual and auditory metaphors for touch, taste, smell, and thermal experiences [than vice-versa]. On the other hand, since the mapping of nonlinguistic events into linguistic codes is essentially arbitrary[1] using the sensorimotor discrimination system but independent of its structure — we would expect to find psycholinguistic relativity.

---

[1] Rather than trying to say how it came about that different-shaped words are used for same-referents across languages, traditional linguistics begs the question by declaring the connection between form and meaning to be "arbitrary" (unknowable, rather than unknown).

Psycholinguistic relativity, within the present book's framework of analogous biological and artificial networks for reasoning, is where characteristic thinking in a member of a speech group using a certain code (language) uses its names as thinking terms, its social labels in interaction roles, its conventional syntax and genres' organization to order reasoning — either in a limiting way (strong linguistic relativity) or as tendency (weak linguistic relativity). Conversely, universality arises from cohorts' having copies of a certain kind of "sensorimotor discrimination system" with intrinsic affective response loading (Candelaria de Ram, 1990a, 1991, 1992a, and elsewhere). "In other words," wrote Osgood (1963, p. 309), "the dominant ways of qualifying experience, of describing aspects of objects and events, tend to be very similar, regardless of what language one uses or what culture one happens to have grown up in." More to the point, as I see it, cohorts that perceive and act "symmetrically" can learn to influence what one another will do, by attempting to induce the enabling cognitive states in each other. Therein lies the utility of communication. External states (and internal states) can be manipulated. How? Ask as Cognitive scientists.

2. OPTIMALITY AS WORKABILITY AND COMMUNICATION AS TOOL

Language abilities shown by communication-system makers and carriers like ourselves are intrinsic consequences of our structure as processors of energy-patterns.

2.1. Networks

Our communication stems from our being, structurally, cross-modal multilayer and multidomain action-perception nets. The nets characteristically have self-modulatory feedback as well as reflex response circuits. Our similarity and mimicry result in built-in transmission among agents of informational structures expressing perception of real-world states. Optimality is taken to be the same thing for this global phenomenon as for low-level mechanisms of molecular and cellular size. Optimality is taken to be system functionality and continuance. This is consistent because the larger scale phenomena are taken to be composites of low-level events in a re-combinant system. Thus, semantic interpretation stems from sensory input, and speech results from (myriad) efferent actions. Working models (computable nets) help in understanding communication, perhaps the most sophisticated "survival mechanism" evolved to date.

In being a self-adjusting tool for local, cooperative activity involving reward in the real world, communication may, if and when used to keep the world running well, perdure along with it. That is optimality, as concerns communication. This is to say that optimality is not everywhere and always the same, but is relative to temporal, local phenomena. The kind of "best operation" that appears to apply to true-to-life phenomena like communication is thus concisely described by saying that functionally, optimality is workability. Optimality is not function itself, but a meta-property that concerns how well a mechanism functions.

Our discussion now proceeds to show the relevance of networks given our definition of communicating as being a real-world phenomenon involving cognition (thinking-and-act-ing-and-remembering-and-feeling). Questions that cognitive science faces about workable systems, agency, mutuality, and will are duly taken note of. Then we get to our three cases of natural and artificial or computer-encodable nets for communication and language. Starting small, these are nets for real-world property measuring and integrating of parameters into concepts and name-systems or languages: First is a small neural net serving as a simple "meter" of the voice onset time (VOT) of speech sounds (like "voiced" /b/ vs. "voiceless" /p/ with long VOT), proposed by Loritz (1991). Second is a multilayer parser net; it is designed (Gigley, 1985b) for time-synchronized processes and proves its worth in reproducing mistiming malfunctions seen in aphasia. Third is a multilayer, multidomain system, Pragmasemantics™ (Candelaria de Ram, 1990b, 1990c, 1990d, 1992b). This is a grounded net of processing connectivities, connectivities that begin with energy patterns from peripheral sensors and proceed through interpretive and integrative operations to higher-level concept constructs and reasoning. In Pragmasemantic™ nets meaning and form are processed together from the outset. The illustration shows key communication functions, naming and speech acts. Detailed dependence on physiological mechanisms is indicated. The trio of networks culminates with one that involves an individual's being provoked to try to communicate.[2] Because Pragmasemantic™ operators are real processes, executing them takes time. Effort underlies the learning where instincts are combined into a higher perceptual option: So, for example, a child learns to mimic a grin and affect a hunger-appeasing agent.

2.2. What Communication Is: Interacting Extended to Cohorts

Possible Postulate: In order for you and me or anyone else to converse, our individual efforts must evoke like responses in each other.

That means a couple of things. For one, if no one makes the effort then no conversation will take place. There will be no discourse, no communication. For another, there has to be a cycle of responding between those making the effort to get through. Effort amounts to having and applying suitable operators even if that is not trivial. The interlocutors — a technical term for agents who talk to each other (loc as in loquacious) — must act, trying in turn to get the other guy(s) to jabber or gesture or whatever back. The need for interagent effort cycles may entail motivation mechanisms, such as molecular reward for sensing something expected (i.e., something whose subnet is primed).

Networks are quite suited to modeling intrinsic cycles and sequences of events. Further work to pinpoint cycle endpoints ought to look at exactly how communicators' "like responses" are alike; word-sequence identity is not sufficient; there may be many kinds of efforts that can be effective for communicating. Consider whole-body action, pheromones, speech, electric pulses, or even "brainwaves" as communication media. Nor is there any claim made at this juncture that if one interlocutor uses words the response will be words; response could just as well be a covert, cognitive adjustment, a change in what the interlocutor thinks due to what the conversational partner did. Communication is prior to verbal language.

---

[2] By speech acts we mean communication acts generally. A facial expression or a dance or a book chapter is as much a "speech act" as a telephone conversation; they carry an agent's intent-to-convey-or-express.

2.3. Interactive Realism and Optimality

As agent capabilities increase by aggregating molecules, cells, and organs, capacities for complex operations can develop. Cells palpate food versus poison molecules; cell aggregates like jellyfish share. Such is the nature of our universe that localized aggregates with special-function components can recur (in separated locations). Replication of complexes can occur, and, further, cohorts are reproduced with same-function components. A natural outgrowth of synthesizing operations into complexes is, then, that operations for monitoring of both internal state and external environment conditions may be shared among cohorts. Response to internal and external states extends naturally to actions pendant on perceived cohort states. Acting so that information is conveyed somehow to adjacent cohorts is next. As the agents become larger and range farther afield, perception of what is going on beyond immediate place and time becomes relevant.

The *interactive realism* approach we take is a fairly new one. It is related to ideas of affordances — the possibilities inherent in a configuration of objects and forces that frame an action. But interactive realism stems from basic principles of the way things in our chemo-physical real world necessarily react with each other (contrasts drawn in Candelaria de Ram, 1990b). As back-to-basics scientific philosophy, interactive realism lets us take a fresh look at prime questions like optimality.

A first question about optimality is whether having interactive realism's principle of necessary interaction is good or not when you are building a world. The answer is something we can read from the real world — our (best, worst and only) "interactive-real system." In reality, some optimality is seen under certain conditions:

Possible Postulate: The interactive-real system is Workably Optimal when local interaction necessarily occurs.

We have two new terms to deal with in this definition of optimality. First, what is a local interaction? It is a response to some action that happens only in "nearby" things; what is local for that action depends on what kind it is. Suppose an action is a magnetic change; nearness follows a "cylindrical-surround space" symmetry with an axis direction following that of the action, which generates a magnetic pole: Dendritic spines change their conformation in response this way. This localness effect is distinct from others that apply for gravity or radiowave or other light. Communication interaction is concrete. Thus if I try to speak to you but you can't hear me, you're not likely to answer unless you also see me writing or flapping my jaw or sending you some other signal to cue you to do so. Till this age of electronic communication, those who spoke the same language were invariably interlocutors living within speaking distance — not just in geolocal proximity but also within socially reachable roles that allowed them to make mutual efforts to communicate. Paper documents changed things so that, as now, some communication-artifact had to be brought into proximity. Artifact reception could be delayed. "Extended localness" admits of a lot more slippage. Still, there is some sort of contact that defines communication groups. We see special kinds of dialect boundaries these days in the lingos among stock traders, hackers, and TV-show watchers. Even for such complex phenomena our cognitive process perspective can show up multiple-stage linkings among communicators who, albeit separated in time and place, still influence one another via locally reproduced artifacts.

Second, what is workably optimal? There are standard quantitative definitions of optimality that set "optimal" as a statistical calculation result over numerical measures of a (relatively simple) completely and statically parametrized problem space, but that won't do for communication. To assume that kind of parametrizability for one of the most complex of known phenomena is at least premature, at worst wrong. An optimality concept more suitable for dynamic, complex systems is needed. "Workably Optimal" is a qualitative description of system functionality that entails the system's dynamic continuity. Necessary interaction according to the energy-matter conditions underlies stable dynamism. Our natural universe's properties of time monotonicity and ever-increasing total system entropy contribute the necessity of change, in balancing properties of conservation of energy-matter that underlie stability. Workably Optimal means that the thing works out as a whole over the long term — so we confidently expect our world to be chugging along merrily every sun-up, thank goodness. A system that continues to function shows itself to be workably optimal; further, dynamic system structure enhances optimality because it permits restructuring that enables nontrivial continuity, or workability.

For a complex system to be workable is very obviously a major achievement of what is optimal! Optimality as it is relevant to communication phenomena in the real world is, in some sense, a bigger, better version of Darwin's appeal that enduring is more powerful. Workably Optimal is "bigger and better" than the Law of Survival of the Fittest in the sense that workability applies not just to species, but is an absolute requirement on system structure as a whole. It is interesting that effort should be part of the scheme. That might be so because effort increases how workable the world is — trying makes it better. There wouldn't be communication without it, and maybe communication increases workability. We can see that communication is evolving; it is becoming ever more complex, with language varieties for science as well as poetry, with writing and electronic encoding joining speech, with graphical and verbal messages zooming across distances, both geolocal and social. If we ask, Why is communication evolving and becoming increasingly complex?, one part of the answer may be simply that everything is. The whole system is evolving and entropy is increasing, making communication complexity increasingly possible. Prior skill in communicating permits building fancier tools. Now maybe it isn't everything that is getting more complex, but there are pressures for selected parts of the system to do so. It could be that communication increases system workability, that the world works better for having communication in it. How?

On the Dynamics of Workable Optimality: There appears to be general growth and continual reoptimization of real-world communication systems.

Communication, consisting of processes that take place in real time, is part of the natural world and subject to its boundary conditions. Natural languages such as Malay and Spanish are specialized communication (sub)systems that have grown up in societies bounded geolocally. Environmental conditions such as visibility, noise, and social prestige affect the utility and longevity of each language variety. Gesture and writing have advantages for immediate conversing in a very noisy environment, for example. Some communication modes, such as music, have expressive potential that others lack. Arranged selections from systems of signs (e.g., letters, logos) for referents — "linguistic surface structures" like sentences, ads, diagrams, symphonies — can be used during social interaction, but residual cognitive and bodily changes may be the most significant results of communication — cognitive adjustment may be what makes communication enhance the workability of our dynamic world.

In summary, from the point of view of interactive realism, it isn't surprising that our communication (a) is systematic, (b) takes place within the boundary conditions of the real-world system, and so (c) is characterized by necessary local interactivity. With a new definition of optimality in hand we are now taking the first steps toward modeling the precise workings of what is very possibly the most sophisticated phenomenon yet evolved. We may not be able to ascertain yet whether communication makes our world more Workably Optimal, but the possibility should be kept in mind. Relevant information on the issue should be gathered.

2.4. To Know I Say: Bonus Questions of Agency, Will, and Self-Recognition

The present description of communicating as effortful, if mutual, response cycles avoids some sticky questions. One such question is whether planning is intending and means the responders have free will (rather than just will). Another one is the question of "I think therefore I am" — a philosophers' claim that awareness of one's own existence as a thinker is basic to having a mind or soul — or whether computers can be thinkers if and only if they know they are thinking (to think if they think they think they think ...! — an old tautology "they think because they think").

2.5. To Know We Answer: Essential Recognition of Mutuality

Nevertheless, describing communicating as a process like this leads straight into the question of how the interlocutors know there is response back and forth. That might look like an easy one — we observe the response, make the connection, and plan our own next effort. But how did we know that what is going on is response to each other in a mutual effort cycle? Aren't we starting to go around in circles saying we communicate because we communicate? No, because cohorts parametrize similarly. In this chapter we see, in the form of several networks, how an agent can begin to respond to systematic communication (discriminate distinctive features in the sound stream, coordinate sequencing expectations and syntax), how the connections between agents can be made, what context and instincts have to do with it. These understandings have immediate and valuable applications for cognitive interchange tasks ranging from socialization/teaching, to rehabilitation after brain injury or other cognitive trauma, to naturalistic human/computer interaction.

2.5.1. Voice Onset Time. Loritz accomplishes the decomposition of a phonemic or "distinctive sound feature" with a neural-network style meter that shows how switching could account for sensory discrimination on a universal parameter of voice onset time — a feature so universal that other mammals (chinchillas, even) can discriminate among human words on this basis. Thus a neural net for phoneme voicing accounts with a minor mechanism for what is customarily treated in standard linguistics as atomic or unanalyzable. Similarly, the theoretically categorical feature that linguists call [+ voice], noted Loritz (1991, p. 306), can be modeled as a process using a standard bit of neural net. The processor is a neural net junction construed as taking outputs from thousands of cilia in the inner ear. The junction is basically a cumulator, looking for fundamental frequency in a sound until a timer runs out so as to measure whether voice onset time (VOT) is over threshold. If the fundamental is on long enough, the speech sound is classified as voiced — [+ voice] for /b/, /d/, or /g/ as opposed to [– voice] for /p/, /t/, or /k/, for example (Lisker & Abramson, 1964). In practice, the harmonic structure of the whole sound would have to be recognized and first harmonic distinguished from fundamental, but the idea is simple. Loritz (1991; see Fig. 20.1) showed a standard neural net crossover gate but with a built-in timer measuring gaps in fundamental frequency in a speech stream for voiced sounds. (This is somewhat analogous to the use of gating to make decisions about the relative importance of stimulus attributes in Levine & Leven, 1992.) Gap threshold adjustment would allow reuse of such meters for different individuals, dialects, or languages; variation of VOT with language was a property that led to its discovery (Lisker & Abramson, 1964).

    2.5.2. Time-Taking Parsing Net. Integrating this phonetic-to-phonemic decision level with successive ones, Gigley (1985a, 1985b, 1988, 1992; see Fig. 20.2) uses a multilayer net with built-in timing. Gigley's HOPE takes speech sound analysis through "time-slice" diagrams representing linked-step parsing that yields the traditional sort of syntactic structures for sentences.

    As with Loritz' net, threshold settings and numerous timers are used to classify portions of the input stream into units used in traditional linguistic analysis. Using such things as activation duration and decay rates, phonemic, then word, then categories like Subject-Noun and Verb are associated with sound input from certain intervals. HOPE uses assumptions that Subject, Verb, and Object enter in fixed order, reasonable for some languages and for a first approximation. The abstracted words tagged with part-of-speech are strung together, forming standard-order SVO sentences with an associated traditional parse. This achievement is again using a net to turn morpho-syntactic and syntactic analysis into a series of processes that take time. (By virtue of this, messed-up timings cause it to produce various kinds of aphasic misfunctions, another achievement that shows the clinical applicability of cognitive modelling of communication.) Gigley's net is more complex than Loritz' in explicitly having several concurrently operating processing "layers," with the syntactic part-of-speech category conditional on what word was identified, the words having been built on the phonemes. Beyond showing how analytic rules adopted from linguistics can be expressed with nets, as Loritz' "feature meter" does with its durational threshold, Gigley's HOPE demonstrates that durational thresholds between stages more generally are so appropriate in modelling linguistic cognition that running the program with messed up settings replicates real-world cognitive misfunctions affecting communication — blocked access to verbs, for example, seen in certain kinds of aphasias from brain damage (cf. Vroman, 1987). These models make a start for constructing artificial nets that fairly directly model biological processes instrumental to use of language.

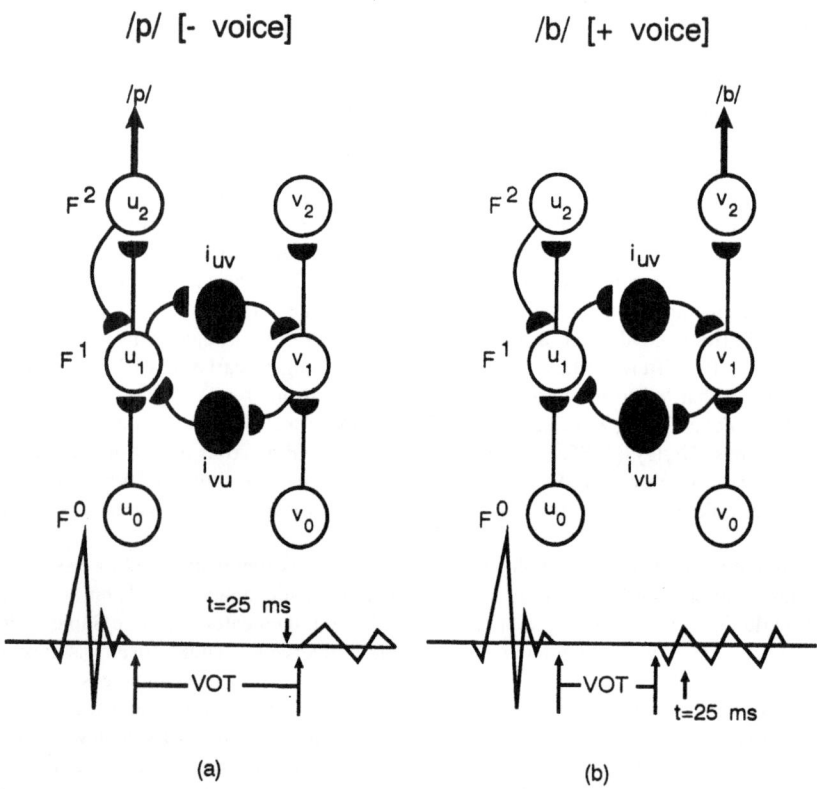

Fig. 20.1. Neural net model of energy flow measurement of a linguistic feature. The parameter is VOT (voice onset time); the measure is Δt, the ms duration of a consonant sans fundamental frequency vibrations.
In (a), /p/ stimulates the left pole of a voicing detector dipole. The right pole is inhibited via the inhibitory interneuron Iuv and a nonlinear feedback signal is established via u2. u1 persists in suppressing v1 even though Iuv subsequently stimulates the right pole. An unvoiced percept is emitted from the dipole. In (b), a /b/ also stimulates the left pole first, but before the left pole feedback loop can be established, it is inhibited by the more rapid onset voicing in the right pole. The dipole emits a voiced percept. (From Loritz, 1991. Adapted with permission.)

2.5.3. Communication Thinking-and-Acting. With an eye to modeling actual physiological events comprising communication reception, understanding, answering, and memory, Candelaria de Ram (1992b and elsewhere) developed a system for specifying even more cognitive process detail for high-level and low-level processes. Candelaria de Ram's multilayer, multidomain nets have tools for more specific description of process times; for input/output conversions and transformations (reinterpretations); for physiological structure conditions, current states, self-modifications in the system; and ofr parametrization at the level of social interaction as well. Emphasis in Candelaria de Ram's modelling is on cognition regardless of whether or not the exigencies of process are addressed by terms set out in linguistic analysis for contrasting language surface structures. The approach, Pragmasemantics[TM], seeks to discover how meaning is apprehended and expressed through cognition and communication. Primary vocal responses and concepts, for instance, may be more reflex than other vocalization: Bodily structures, connections, and motivations as well are provided in neonates, for example, for certain cries including pain, hunger, and lovely-eating. Absence of or difference in those cries or perceptions indicates abnormalities in the agent, much as do the aphasias where connections are not made at the right times.

Candelaria de Ram (1991a) illustrated how the simplest of concepts, such as loudness, may be built out of reflexive processing that enters into communication processing. (Decreased gain protecting the middle ear from loud sounds is measurable as muscle tension in the stapedius, so loudness is a graded qualitative concept with a direct basis in small physiological structures.) Cohorts can have the same processor structures and concept capabilities: critical among them are naming, case relations among referents, and speech acts, our next three examples.

2.6. Pragmasemantic[TM] Operations

The accompanying diagrams show some of the process operations and parametrizations of Pragmasemantics[TM], where the key linguistic relation, of naming, is an efficient and concrete operator or connective between energy patterns (Fig. 20.3). In this case a coreference connection is made between interpretations of two directly perceived events in the same modality. The referent is a brass bell ringing; the verbal token is another audible stimulus comprised of energy patterns proceeding through layers of the net where first phonetic and then phonemic identifications are made. Interpretation occurs as the stimuli flow through memory/skill processors with appropriate energy-passing and sorting capacities. Via neural/structural meters like the VOT subnet seen above, the stimulus is figured out to be first a voice- and dialect-specific phonetic image [bel] and then a more abstract phonemic form /bel/ suitable for the nets of morphophonemic tactics.

Pragmasemantic[TM] nets can be diagrammed. They can be set forth as high-level propositions in a grounded logic, and written in equation forms specifying appropriate parameters such as process locations, times, and energy pattern specifics as well as durations such as seen in Gigley's HOPE. These, illustrated above, can in turn be programmed readily in high-level programming languages like Prolog, as illustrated below.

A remarkable new feature of Candelaria de Ram's Pragmasemantic[TM] techniques is that interpreted stimuli and context are processed together from the beginning, so Pragmasemantics[TM] captures the cognitive fact that significance and sign are inextricably connected at all levels of processing. Babies always have the capacity for at least rudimentary "symbolic" thought in that sense. It further appears that from birth they have the capacity for making like vocal and gestural responses as well, as described next.

TIME INTERVALS

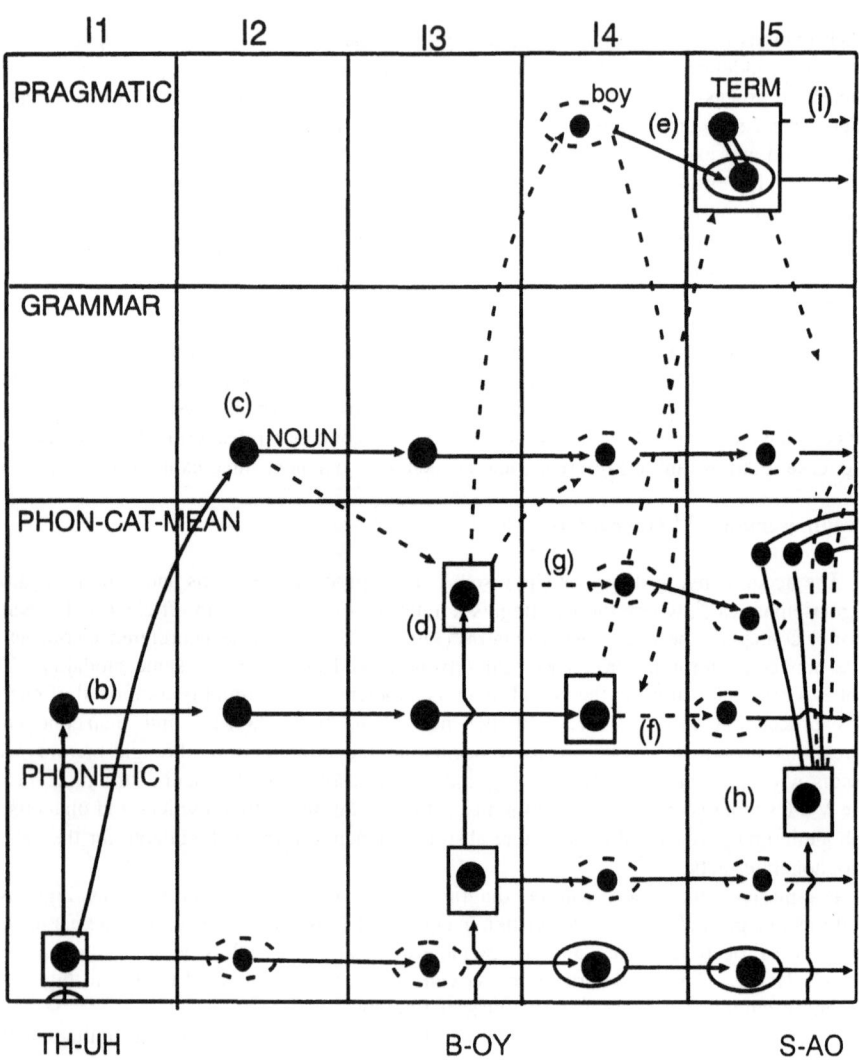

Fig. 20.2. Reinterpretive process stages: Expected sequencing in speech stream leads to a net with time steps for lookup of aspects of significance depending on what is found a bit earlier at a different level. Phonemicization is shown as the first level (cf. linguistics tradition and Loritz' feature detector nets). (From Gigley, 1985b. Adapted with permission.)

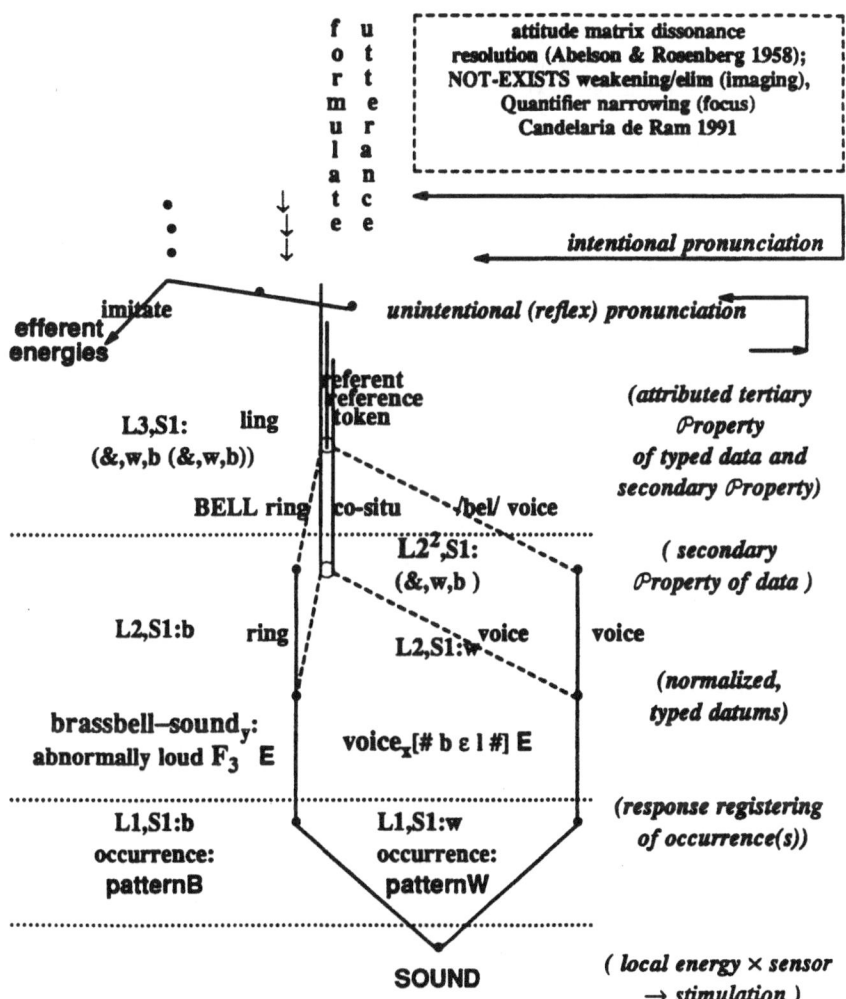

Fig. 20.3. Multistage discrimination of pattern types. Linguistic and referential appear parallel, contiguous (versus traditional separation into linguistic levels in Pragmasemantic™ nets.)

Consider again Fig. 20.3, showing a Pragmasemantic™ net transducing perceptual information, making various discriminations at various stages of processing by using the operators of that respective level, joining various streams, and coming up with a referential connection between a verbal name and a perceived thing. Naming, a key property of language, is shown in Pragmasemantic[TM] nets to be vested in multistage processing near the peripheral sensors. The process of discovery of a relation between two sound patterns where one refers to the other — a word token bel refers to the ringing sound of a brass bell — is a sequence of processes building on each other's input energy patterns. This form of Pragmasemantics[TM] notation sets out relations (e.g., &—coreferential) among energy-patterns occurring at specific times, durations and processor locations; the items appearing within angled brackets < > in the portion below of one formula rendered in high-level computing language propositions (Prolog) are just stand-ins for actual energy patterns which are repeatedly converted as energy is routed through the system for analysis and storage. In this example the two stimuli occurred at the same time and a simultaneity relation (e.g., &—cotemporal) serves as a prior condition for figuring out that one stimulus names the other.

++ Pragmasemantic[TM] INFORMATION STREAM LINKOR OPERATOR ++

Sensor S1 -> Language-domain S1', Level L2(conversion of outputs from +1,2)
Prolog rendition of part of intralevel processing where (1) data supports stimuli co-occurrence check (2), which fulfills condition for secondary recombination of data as referent and name that has been/can be used to point at it.
{Capitalization indicates variables for parameter values from sensors.}

```
[                              % high-level tagging in square brackets, then signal in <>
  [                            % referential relation pair
    [token, TOKEN],           % cross-binding of relational values TOKEN & REFT
    [referent, REFT]],
  ['&co-temp',                 % simultaneity condition (3rd branch of net node)
    [voice, TOKEN],           % "meter" finds this stimulus has voice structure
    [occurrence, REFT]]       % stimulus is left as phonetic image

  :: <TOKEN>, <REFT>,          % ACTUAL ENERGY VALUES, not tags here
    ['&co-temp',w,b]@[S1,L2],  % transduced/interpreted energy Sensor1,Level2
      [attime, T3, DUR6]],     % DURation within same-time window
  :: < C,V+>,                  % syllable, bell-sound patterned energies
    ['&co-ref', w,b]@[S1,L1,L2,L2,2],   % name relation node energies
      [attime, T4, DUR7]
```

A remarkable aspect of Candelaria de Ram's Pragmasemantics[TM] nets is that context and signal are processed together. (This is not true of many neural nets nor of classical semantics such as Weinreich, 1963/1966.) Its focus on cognition brings out this marvelous contrast with the long-traditional presumption in linguistics that semantics and/or pragmatics is not accessed until some late stage after syntax.

In Pragmasemantics[TM] communicative tokens have significance immediately because they occur in context; as discriminations accumulate during further processing, the significance accumulates. For example, a lexicon of parametric property-based condition definitions of basic concepts might contain reasoning network specifications that can be illustrated again in the propositional computing language Prolog. The variables (capitalized terms) here get values coming from "meters," not for VOT, but for things like locations of sensed objects in [loc,LA,of,A] and movements at some point like [state,S2,[move,A]], and amount of insertion like [degree,DIN1,[A,in,B]] or found at some level to fulfill some relation like further-in for [DIN1 > DIN2], or in-time-sequence for [tseq,S1,S2,S3], the last being rather in the manner of temporal logic except that here the arguments are not tags but energy patterns grounded in sensory perception.

++ Pragmasemantic™ CASE-RELATION OPERATORS (GROUNDED CONCEPTS) ++
Pragmasemantic™ reasoner Lexical Entries
(Prolog at left, explained at right after %)   (Candelaria de Ram 1992b, pp. 196-197)

```
interp([A,in,B],                        % concept "in"; used by "across,"
"into," etc.
  [[state, S1,                          % there is a state S1
      [loc,LA,of,A],[loc,LB,of,B],      % in which A has location LA, etc.
      [extent,EB,of,B], [extent,EA,of,A]],   % B is so-big, etc.
    [EA < EB], [LA,at,LB],              % B bigger than A, A & B same place
    [focus,A],[focus,LA]],             % linguistic use focuses on A &
where's A
  Evaln).                              % concrete energy processing
interp([A,into,B],                      % concept "into"; uses "in," is learned
later
  [[state,S1, [degree,DIN1,[A,in,B]] ],   % there's a state S1 when A is in B
    [state,S2,[move,A]],              % then A moves
    [state,S3,[degree,DIN2, [A,in,B]] ],  % so it's in B a different amount DIN2
    [DIN2 > DIN1],                     % namely farther in
    [tseq,S1,S2,S3], [focus,A],[focus,S2,S3],[ground,B] ],  % that's the point
  Evaln).                              % concrete energy processing
```

This structuring is designed to be consistent with processing phenomena indicated by psychological experimenting on semantic priming and other spreading activation phenomena. Pragmasemantics™ levels can be elaborated, so we can use and develop them as we learn more specifics about mechanism in the neurophysiologically differentiated, interconnected domains in the nervous system. Pragmasemantic™ nets have capacity for any number of processing levels and levels of various types.

2.7. Mimicry: Built-In Bootstrap Effects Self-Adjusting System Drift

Our most fundamental communication response seems to be mimicry. Reflexive mimicry of speech, hand, and body gestures from birth gives babies a no-delay start in social exchange. The physiological connectivities provide for chunking of information by cross-correlating stimuli from different modalities; equivalents in a network model (Pragmasemantics™) are complex measurement predicates or parameters. (So, for example, the measured degrees of "in"-ness in State1 and State3 above may come one from touch, one from vision.) Automatic mimicry also provides, by default, for constant system update across its speech community. Details follow.

Granted that certain vocalizations and concepts are ready to be used in the system, an agent has some tools to use in communicating. But how does the communication cycle start? It would be consistent with the observations made above that primary concepts and vocal responses are founded in reflex, automatically accessible, circuitry. In particular, there should be some reflexes in neonates that get them started performing communication actions as responses. There is in fact evidence from careful psychology experiments with 40 newborns averaging 32 hours old — less than 2 days — (Meltzoff, 1990; Meltzoff & Moore, 1983) that neonates imitate the position of oral articulators of someone in front of them. As shown in photos from the experiments (Fig. 20.4 from Meltzoff & Moore, 1977), newborn humans mimic pursing of the lips (lip-rounding, a universal phonetic feature), opening of the mouth and spreading of the lips (enabling capturing both /a/ and /i/ distinctions, as well as /u/ typing), and tongue protrusion (which gives them a handle on /l/ sounds and consonant frontness versus backness) as the voice is turned on and off in cooing. Imitation of head and finger movements are also found in newborns (Meltzoff, 1990, p. 35). Similar imitative exchanges are found at all ages in humans, indicating that we have identified the key agent feature that starts and restarts the communication cycle: built-in communication mimicry actions.

2.7.1. Sensor Integrator Subnets Connected to the Jawbone. By the age of 4 months infants do things like selectively match /i/ mouth position with /i/ sounds, although in a few months they take closer looks at a bad match. Capacities such as these require matching up visual cues and sounds. How is it possible for mimicry and cross-modal connections to be set up within the same system? Briefly, the deep portions of the superior colliculus at the top of the brainstem have multimodal image cells that get sensor image inputs and coordinate eye and sound, and so on. In fact, intracell response measurement shows that among these "meters" some are multimodal additive, whereas in some of the cells, characteristic firing rates for single-modality stimulus become inhibited by cross-modal co-occurrence (Stein & Meredith, 1990). Built-in lags in these multimodal image cells compensate for systematic differences in stimulus energy arrival times due to length of the nerves from eye (55-125 ms) and ear (10-30 ms).

Stein and Meredith (1990, p. 61) figure that the system provides for selective attention to multisignal events, probably from birth (Diamond in Stein & Meredith, 1990, p. 69). Such bundling of multimodal stimuli corresponds to multimodal parameters or higher-order predicates of Pragmasemantics[TM]; for instance, measurement of an object A's extent EA could be through multiple sensors (note specifications of sensor and level as well as time, in examples below shown as variables S2, L2; Ti, DURi; implemented through the evaluation map Evaln). Motor projections from the superior colliculus also provide for image-coordinated movement (e.g., Figure 2 of Stein & Meredith, 1990, p. 53, cf. ascending neural bundles). The brainstem, sensors, and cerebellum are tightly linked with reflex action circuits that control significant sequences of fine muscle movements such as vocal articulators, so that copying of communicative actions can be transferred quickly before conscious monitoring can interfere.

Copying is a low-load process, but apparently subject to emotion gating (to participate or not; Candelaria de Ram, 1989). Therefore, besides being instrumental in starting off a newborn's language skill exercises, imitation appears to be a source of linguistic drift, pushing speech communities into new language varieties, making communication adapt as time goes on, just because variants get automatically integrated into the system (Candelaria de Ram, 1990a). Hence related-language cognates such as *reign* in some communities and *rAjA* in their cousins'. Note that the intermodality of these automatic processes means that drift is to be expected for all modalities of communication at once, gesture, speech, and all.

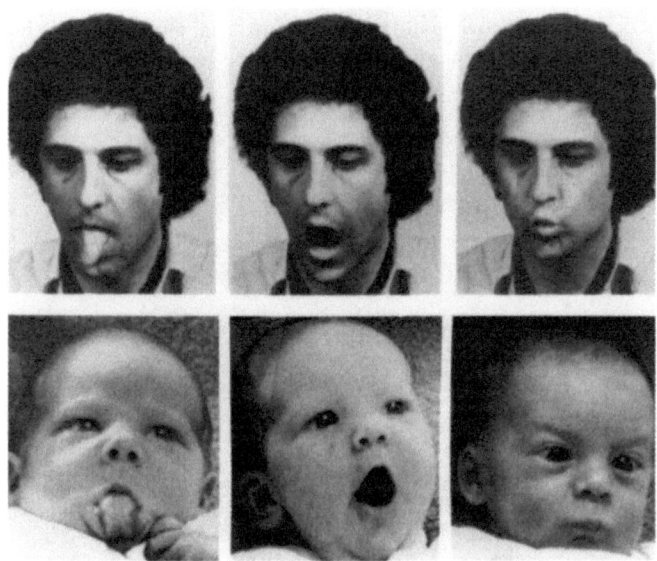

Fig. 20.4. Photographs of 2- to 3-week-old infants imitating facial gestures presented to them by an adult experimenter. Automatic mimicry and spontaneous vocalization at birth mean spoken communication can begin immediately (far before the age of 1 year as traditionally claimed). Quick learning of speech act cognition can appear as in the Pragmasemantic™ COMMUNICATION OPERATOR shown below. From A. N. Meltzoff & M. K. Moore, 1977, "Imitation of facial and manual gestures by human neonates." *Science*, 1977, *198*, 75-78. Reprinted by permission.

Although reflexive imitation may start us off as lifelong communicators, what motivates us to intentionally communicate? Reward. Simple principle, reward, and well documented by psychological experimentation at the macro level as action consolidator and motivator. Reward is beginning only now to be understood as an extraordinarily complicated molecular-level phenomenon of substance balancing (e.g., hunger and thirst satiation, opiates interacting with neurotransmitter manufacture). But, fortunately, simple high-level handling in Pragmasemantics™ works fine as we see in the example that follows. Note how the multilevel net is designed for being elaborated with details of component (or overall) process as they become available, so that, for instance, ACT can become a large relational complex of known and sought values, and GRIN can be specified in terms of muscle tightenings. Learning takes place as values for person P and EGO are interchanged so EGO observes and learns to use a grin.

++ Pragmasemantic™ COMMUNICATOR OPERATOR (NON-VERBAL) ++

See a Speech Act, Do a Speech Act: Pragmasemantic™ Cognispaces™ with Intent and Copy Originally, a believer Bx (Manuela) might not represent 'grin' as a speech act, but only as an environmental occurrence. How would grinning with social purpose come about? By mimicry, reasoning by analogy over intentionally caused event sequences: (For further discussion see Candelaria de Ram, 1992b, pp. 154-159 and prior work.)

```
Beliefs of Manuela, attime Tx, loc Lx
    ...
Domain: grin
 [EGO | MANUELA],
 [believe, Agent0 | Properties J],
 [PropertiesJ,
     [occur, [[attime | AnyCurrentT], EGO],
      A0, GRIN, A, P, S1, S2, X, ... ],
 [obj, X],
 [person, EGO],
 [person, P], [agent, P],
 [GRIN, [agent,P] | TEMPORARY-FACIAL-POSITION-OF-P],
 [tsequence, [attime,T02],              % a grin from P bridges states 1 and 2
 [state,S1], [act, P, GRIN], [state,S2]],
 [[state,S2], [do_act,G,A0],            % agent A0 does G during state 2
  [G,[give,[agent,A0],[patient,P],[obj,X]]],    % namely A0 gives X to P
 [[state,S1], [not, possessor_of,X,P]],
 [state,S2], [possessor_of,X,P],
    ...
 [want,X,P] ...
-------- Manuela applies reasoning/net-connection, this-level's combination-rule
  [sensor,Sx],[level,Lx],[combination_rule,Cx,    % rule execution defined
    [causal sequence,                             % rule constitutes recognition of cause-effect
        [tsequence, [attime | Ti],                % event-sequence
         [state | SAi], [act | ACTi], [state | SBi]],  % act(s) bridge state(s)
        [tsequence, [attime | Tj],
         [state | SAj], [act | ACTj], [state | SBj]],
        [verisimil, [SAi,SAj], [ACTi,ACTj], [SBi,SBj]]],
                                                  % pair of events similar
-------- Manuela wants to get like P did, mimics P's GRIN: rebinds agency to self
```

```
[want,X,P],                          % hmm, X wanted P while
[want,SB,P],                         % I want SB
[intend, GRIN,P],

                                     % and  P  grinned  (he  meant  to);  reflex
                                        recognition of P's agency
[intend,SB,P]                        % so I intend to too:
% WHEN INTENT IS BOUND TO AGENT EGO
% RUNNING COMPUTATION PUTS ENERGY INTO BEHAVIOR CIRCUITS.
```

This analogical reasoning is doubtless little different at base than babies' instinctive mimicry of communicative gesture. Grinning to get something and use of a name to point out something are the kinds of actions from which communication systems can be built up. Eventually any number of acts ACTj may be involved, leading to full-fledged discourse. Discourse structuring, with agent turn-taking, is a matter of chained time sequences of acts with agents (tsequence ...[act...agent as seen in the last example).

Note, however, that motivation, intent to act, effort, comes into the picture as pointed out earlier in this chapter. It seems to be a real-world fact that motivation involves reward as a necessary consequence in cyclical interaction sequence patterns. Further, effort to communicate with other agents results in recognizing their efforts and coordinating information about those events with other information.

3. LOCALIZED OPTIMIZING THROUGH DISCERNING REDUCTION OF APPARENT ENTROPY

In sum, specialized memory and perception organs develop. By measuring some domain of perception, say smell, taste, sound, or light, filtering organs let composite organisms impose order.

As the complexity of the agent-in-context system increases, then, percepts — which are made of critical measurable values or value ranges — reduce the apparent disorder. (Physics has a descriptive term for disorder — *entropy*. Disorder as it relates to kinetic processes refers to where a [sub]system's homogeneity lowers its capacity to fall into a lower-energy state and so to do work consisting of of moving something. Defined for the Second Law of Thermodynamics in terms of heat, entropy change from S later-state to S earlier-state is greater than or equal to the integral (from earlier to later) of the change in heat absorbed divided by the system temperature (Halliday & Resnick, 1960, pp. 546-552). For compound systems like agents, the trick is not to fall into an undifferentiated state.) Percepts, because they constitute added superstructure that is cognitive in nature, may not actually reduce the ever-increasing entropy of irreversible natural process as evolution spins forward in spacetime, but they counter apparent disorder for agents by selectively registering only certain real properties through real organs of perception. Sensors are therefore filters (see Candelaria de Ram, 1990b); they measure sequences of parameter values within select ranges; the cognitive processing they support reduces local entropy; the cognitive structures constitute an opposite quality to constrain entropy.

Possible Postulate: Reducing apparent complexity lets agents recognize things.

Recognition is an artifice to induce apparently conserved entropy so that certain parts of a later situation seem to depend "only on the state of the system and not on how it got there from a previous state" as with reversible processes where entropy does not increase (ibid.). When the description of a later state is reduced to the description of an earlier one with the same parameter value(s), the organism is able to isolate with what appears to be a reversible process, and can repeat a prior successful reaction to the state as perceived.

This idea about how communication comes about may be coalesced with a new term, *discerntropy*, defined as follows:

Discerntropy: The capacity of discernment to reduce apparent entropy locally (i.e., for agents), a capacity pendant on perception and developed naturally over the course of evolution to counter increasing complexity of the natural world which has concomitant ever-increasing entropy (following from general operation of the Second Law of Thermodynamics).[3]

Communication enhances discerntropy. According to Pragmasemantics™ (Candelaria de Ram's cognition theory), sensors introduce parameters, which can be combined in perceptual constructs, so that a "simplifying" meta-universe is built as cognition develops by successive recombination of parameterized percepts grounded in sensing of external (and internal) environs.

In sum, this is the scenario. As agent capabilities increase by aggregating molecules, cells, and organs, capacities for complex operations can develop. Cells palpate food versus poison molecules; cell aggregates like jellyfish share. Such is the nature of our universe that localized aggregates with special-function components can recur (in separated locations). Replication of complexes can occur, and, further, cohorts are reproduced with same-function components. A natural outgrowth of synthesizing operations into complexes is, then, that operations for monitoring of both internal state and external environment conditions may be shared among cohorts. Response to internal and external states extends naturally to actions pendant on perceived cohort states. Acting so that information is conveyed somehow to adjacent cohorts is next. As the agents become larger and range farther afield perception of what is going on beyond immediate place and time becomes relevant. So they set up to convey and recapture state information.

Discerning selected things in one's locale works with ability to react selectively. Eventually complex agents with similar perceptual structures can act so as to inform each other of extra-local conditions including their own internal states: En fin, communication arises ineluctably with evolution of social agents.

---

[3] Coined in multilingual form by Candelaria de Ram (in press): English discerntropy etymology is dis-cern (to separate by distinguishing, judging) + entropy (roughly, system inhomogeneity or "disorder"). Cognates of the English term discerntropy (already reduced from discernentropy, so as to incorporate sandhi) include Spanish discierntropía (DISCERNIR + ENTROPIA --> discierntropía) and hyphenated form for Hindustani (which normally borrows English scientific terms) DÉwy-entropy (dr;ST (discern) --> DÉwy - entropy). (See American Heritage, 1994; Apte, 1884/1974; Platts, 1884/1974; Random House, 1994; Smith, 1971.)

4. CONCLUSION

Communication is seen to arise as chemo-physical complexities of the world increase, with larger molecules, living entities, and eventually mobile agents. Communication appears to be a mechanism for meta-optimization: As real-world entropy increases, systematized communication among now-distinct agents allows some control over dangers from excessive disorganization. The more free-wheeling the agents, the more elaborate the communication system. Paradoxically, this leads to a new kind of complexity, namely, information complexity. Our decades have seen mighty increases in mobility of groups of agents sharing languages and cultures more generally. Ready transport of records of communications (e.g., book, phone, film, fax, computer file transfer) makes for delayed and extended local access. The deluge of communiqués is being channeled with tools for handing on and handling information which far exceed speech, traditionally held to be the root of language, while tools like electronic encoding are extending transcription techniques like phonetic and logographic writing (including music transcription). Communication evolution is an optimizing process now in full swing.

Although a great deal more study is essential for understanding how communication results in cognitive adjustment, we have now laid out some of the basic functional mechanisms of how it occurs. There are built-in similarities in how communicative agents perceive and reason, so they cognize each other (and their environments) similarly. They can (a) share communicative significance of acts, and (b) figure out what cognitive adjustment to expect in their interlocutors. They have group discerntropy. Communication is therefore a tool for shaping the world, dynamic by the very nature of our chemophysical universe, significant by virtue of the operations that constitute the very cognition that drives communication acts, both overt and covert. These are real world operations that are concrete and contextual — pragmatic — and semantic in that they interpret happenings as significant for a cognitive agents. Communicating communities of agents may hold operations in common. This leads to languages: changing communication conventions that form evolving usage systems. If communication is used to make the world continue on its merry way, it too will doubtless perdure as an optimality maker.

REFERENCES

American Heritage (1994). *American Heritage Dictionary III Deluxe*, electronic version 3.5 for Windows 3.1. Novato, CA: WordStar International.

Apte, V. S. (1974). *The Student's English-Sanskrit Dictionary*. Delhi: Motilal Banarsidass (sic). (discern, p. 110). (Original work published 1884.)

Candelaria de Ram, S. (1988, October). Neural feedback and causation of evolving speech styles. In *New Ways of Analyzing Language Variation* (NWAV-XVII), Centre de recherches mathematiques, Montreal.

Candelaria de Ram, S. (1989, October). Sociolinguistic style shift and recent evidence on "presemantic" loci of attention to fine acoustic difference. In *New Ways of Analyzing Language Variation joint with American Dialect Society* (NWAV-XVIII/ADS-C), Duke University.

Candelaria de Ram, S. (1990a). The sensory basis of mind: Feasibility and functionality of a phonetic sensory store. (Commentary on Risto Naatanen, The role of attention in auditory information processing as revealed by event-related potentials and other brain measures of cognitive function, *Behavioral and Brain Sciences*, 13, 201-233, 261-288), *Behavioral and Brain Sciences*, 13, 235-236.

Candelaria de Ram, S. (1990b, June). Real-world sensors, meaning, or mentalese. *Workshop on Artificial Intelligence: Emerging Science or Dying Art Form?* SUNY Binghamton (Tech. Rep. No. MCCS-90-189, Computing Research Lab, New Mexico State University).

Candelaria de Ram, S. (1990c, October). Belief/knowledge dependency graphs with sensory groundings. *Proceedings of the Third International Symposium on Artificial Intelligence Applications of Engineering Design and Manufacturing in Industrialized and Developing Countries* (pp. 103-110). Instituto Tecnologico de Estudios Superiores de Monterrey (ITESM), Monterrey, Mexico.

Candelaria de Ram, S. (1990d). *Sensors and concepts: Grounded cognition.* Preliminary notes of working session on algebraic approaches to problem solving and representation, June 27-29, Philips Laboratories, Briarcliff, NY.

Candelaria de Ram, S. (1991, April). From sensors to concepts: Pragmasemantic system constructivity. *International Conference on Knowledge Modeling and Expertise Transfer*, KMET'91, LISAN Centre for AI, University of Nice, Sophia-Antipolis, France. Reprinted in *Knowledge Modeling and Expertise Transfer*, pp. 433-448, Paris: IOS Press.

Candelaria de Ram, S. (1992a, December). Intelligent, context-bound discourse: Sensory sources for primitives for semantic relations from Pragmasemantics™. *International Symposium on Artificial Intelligence — The Artificial Intelligence Technology Transfer Conference (pp. 383-392). Sponsored by the Instituto Tecnologico de Estudios Superiores de Monterrey (ITESM), Centro de Inteligencia Artificial.* Cancun, Mexico. Menlo Park, CA: AAAI Press.

Candelaria de Ram, S. (1992b). *Pragmasemantics: Toward a computer-implementable model for linguistic cognition.* Unpublished doctoral dissertation, New Mexico State University.

Candelaria de Ram, S. (1994). *Pragmasemantic logic grounded operators: From molecular mechanisms to speech acts.* Unpublished book chapter.

Candelaria de Ram, S. (in press). Why to enter into dialogue is to come out with changed speech: Cross-linked modalities, emotion, and language shift — "He was surprised to hear himself repeating congenially ... ." In F. Neel & M. M. Taylor (Eds.), *Structure of Multimodal Dialogue.*

Diamond, A. (Ed.) (1990). *The Development and Neural Bases of Higher Cognitive Functions. Annals of the New York Academy of Sciences* (Vol. 608, pp. 1-751). New York: New York Academy of Sciences.

Gigley, H. (1985a, August). Computational neurolinguistics — What is it all about? *Proceedings of the Ninth International Joint Conference on Artificial Intelligence, 18-23 August, 1985, Los Angeles* (pp. 260-266).

Gigley, H. (1985b). Grammar viewed as a functioning part of a cognitive system. *Proceedings of the 23rd Annual Meeting of the Association for Computational Linguistics*, **ACL-85**, 324-332.

Gigley, H. (1988). Process synchronization, lexical ambiguity resolution, and aphasia. In S. L. Small, G. W. Cottrell, & M. K. Tanenhaus (Eds.), *Lexical Ambiguity Resolution: Perspectives from Psycholinguistics, Neuropsychology, and Artificial Intelligence* (pp. 260-267). San Mateo, CA: Morgan Kaufmann.

Gigley, H. (1992, May). *Neural processing constraints: Implications for natural language processing.* Talk presented at 2nd Behavioral and Computational Neuroscience Workshop, Georgetown University.

Halliday, D., & Resnick, R. (1960). *Physics for Students of Science and Engineering* (Combined Edition). New York: John Wiley & Sons.

Lehmann, W. P. (1968). Saussure's dichotomy between descriptive and historical linguistics. In W. P. Lehmann & Y. Malkiel (Eds.), *Directions for Historical Linguistics* (pp. 3-20). Austin: University of Texas Press.

Levine, D. S., & Leven, S. J. (1992). A theory of dynamic vigilance setting and preference reversal, based on the example of New Coke. In *Proceedings of the Fourteenth Annual Conference of the Cognitive Science Society* (pp. 933-938). Hillsdale, NJ: Lawrence Erlbaum Associates.

Lisker, L., & Abramson, A. S. (1964). A cross-language study of voicing in initial stops: Acoustical measurement. *Word, 20,* 384-422.

Loritz, D. (1991). Cerebral and cerebellar models of language learning. *Applied Linguistics, 12,* 299-318.

Meltzoff, A. N. (1990). Towards a developmental cognitive science: The implications of cross-modal matching and imitation for the development of representation and memory in infancy. In A. Diamond (Ed.), *The Development and Neural Bases of Higher Cognitive Functions* (pp. 1-37). New York: New York Academy of Sciences.

Meltzoff, A. M., & Moore, M. K. (1977). Imitation of facial and manual gestures by human neonates. *Science, 198,* 75-78.

Meltzoff, A. M., & Moore, M. K. (1983). Newborn infants imitate adult facial gestures. *Child Development, 54,* 702-709.

Osgood, C. (1963). Language Universals and Psycholinguistics. In J. H. Greenberg (Ed.), *Universals of Language* (2nd ed., pp. 299-322). Cambridge, MA: MIT Press.

Platts, J. T. (1974). *A Dictionary of Urdu, Classical Hindl, and English.* London: Oxford University Press (esp. p. 512). (Original work published 1884).

Random House (1994). *Random House Webster's Electronic Dictionary and Thesaurus,* College Edition 1.5 for Windows 3.1. Orem, UT: Word Perfect.

Samuels, M.L. (1972). *Linguistic Evolution with Special Reference to English.* Cambridge, UK: Cambridge University Press.

Smith, C. (with M. Bermejo Marcos and E. Chang-Rodrigues) (1971). *Collins Spanish-English/English-Spanish Dictionary — Diccionario Español-Ingles/Ingles-Español.* London: Collins.

Stein, B. E., & Meredith, M. A. (1990). Multisensory integration: Neural and behavioral solutions for dealing with stimuli from different sensory modalities. In A. Diamond (Ed.), *The Development and Neural Bases of Higher Cognitive Functions* (pp. 51-70). New York: New York Academy of Sciences.

Vroman, G. M. (1987). Aphasia as a communicative disorder: Application of the levels concept. In G. Greenberg & E. Tobach (Eds.), *Cognition, Language and Consciousness: Integrative Levels* (pp. 99-116). Hillsdale, NJ: Lawrence Erlbaum Associates.

Weinreich, U. (1966). On the semantic structure of language. In *Universals of Language* (pp. 142-216). Cambridge, MA: MIT Press. (Original work published 1963.)

Weinreich, U., Labov, W., & Herzog, M. I. (1968). Empirical foundations for a theory of language change. In W. P. Lehmann & Y. Malkiel (Eds.), *Directions for Historical Linguistics* (pp. 95-108). Austin, TX: University of Texas Press.

21

Communication and Optimality in Biosocial Collectives

Raymond Trevor Bradley
Institute for Whole Social Science

Karl H. Pribram
Radford University

*Raymond Trevor Bradley and Karl Pribram's chapter, **Communication and Optimality in Biosocial Collectives**, is unique among the chapters in this book in that it deals with optimality in social systems composed of people, rather than individual nervous systems composed of brain regions. It is related to this book's other chapters, however, because of the strong analogies between different levels within complex nonlinear dynamical systems. Just as typical neural networks, whose nodes are brain regions rather than individual neurons, include effects that can be described by the average of effects in single neurons, social networks include effects that can be described by the average of effects in individual brains.*

*Bradley and Pribram describe two key variables in the dynamics of urban communes that need to be in optimal ranges for a commune to have a good chance of surviving over periods of four years or more. These two variables they call **flux** and **control**. Flux is defined as the density of interactions between different members of the commune. Because flux is related in some sense to the amount of "excitement" or "stimulation," optimal flux appears to be related to the criterion, developed in Rosenstein's chapter of this book, of optimal "income" within an individual brain. Control is defined as the extent to which rules or hierarchies govern these interactions.*

The optimal patterns of flux and control can be related in many ways to theories of individual brains and neural networks. Pribram relates these variables, by analogy, to physical energy variables, and concludes that the stable states correspond to configurations that obey some kind of least action principle. Previously, in joint work with Diane McGuinness, Pribram had related such principles to optimal functioning of a brain system that includes the frontal lobes and hippocampus, among other areas. Also, the optimal states can be described as states that lie in a region, as Bradley and Pribram say, "between total randomness and total organization." These states are described by a thermodynamic analogy also used in neural networks developed, for example, by John Hopfield, Geoffrey Hinton, and Terrence Sejnowski.

Bradley and Pribram's chapter can be considered part of the same project as Leven's chapter in this book (and, to some degree, the chapters by Levine, Prueitt, and Werbos). That project aims to bring insights from neural network theory and cognitive science to bear on developing new theories in the social sciences (in the case of this chapter, sociology). This project should address the sore need in the social sciences for foundations that (a) are quantitative **and at the same time** *(b) integrate the dynamics of real human behavior and emotions.*

ABSTRACT

A theory of communication is developed to explain optimization in the social collective: to explain how energy expenditure interacts with control operations to form an efficient information processing system that results in a stable, effective collective. The theory shows how two orders of social relations, flux and control, act on the biosocial energy of the collective's members to create quantum-like, elementary units of information. Each unit of information contains a description of the collective's endogenous organization. Constructing and distributing such descriptions throughout the collective on a moment-by-moment basis, the interaction between the two orders operates as a communication system that *in*-forms (gives shape to) the expenditure of energy and results in stable, effective collective action. Results from a longitudinal study of 57 social collectives offer empirical support for the theory. Only those configurations of flux and control that produced a path of least action — one which entailed the smallest amount of turbulence — resulted in a stable and thus effective social collective.

One has the vague feeling that information and meaning may prove to be something like a pair of canonically conjugate variables in quantum theory, they being subject to some joint restriction that condemns a person to the sacrifice of the one as he insists on having much of the other. (Shannon & Weaver, 1949, p. 117).

1. INTRODUCTION

The picture of reality that science portrays reflects the way science is organized. Broadly speaking, this organization divides science into distinct disciplines (e.g., sociology, psychology, biology, chemistry, and physics), each perceiving the natural world by way of its own perspective and techniques. Reflecting this organization, most behavioral science is a single-level enterprise generating bodies of data and theory specific to the phenomena of concern. This is especially true of contemporary sociological research which, since Emile Durkheim's *Les Regles de la Methode Sociologique* (1895/1938), has treated (human) social interaction as a separate order best understood by studying the ways social organization constrains the behavior of the individual and that of collectives of individuals (e.g., Burt, 1992; Coleman, 1990; White, 1992; a notable exception is Collins, 1975).

Essential as single-level enterprises are, some of the most exciting moments in the history of science have come when data collected from adjoining levels provide insights that point to the possibility that scientific knowledge can be woven into a single coherent tapestry. Such insights are the product of multilevel investigations that focus on the interlinkages between systems of organized behavior at adjacent levels.

The broader aim of this chapter, therefore, is to relate social phenomena to basic concepts that have developed in physics and control engineering. Beyond the urgency and importance of the development of a common scientific language (Bishop, 1995), we take this approach because single-level research in social science has not been successful in predicting the behavior of social collectives, and because we believe that the development of this understanding can be informed by concepts that have proven useful in these fields.

Two steps are usually required to obtain insights when using a multi-level approach. The first step is to discern commonalities in the behavior of collectives operating at different levels and to describe these in a common terminology. For instance, an assembly of neurons in the brain is conceived to obey the same laws of communication as an assembly of people in a social group. A formal approach to this step was taken by General Systems Theory (Miller, 1968; von Bertalanffy, 1969). The second step seeks an understanding of the intimate relations that connect two adjacent levels of inquiry. Ideally, the operation of these relations, formally described as transfer functions (transformations), must account for the results obtained in the first step (see Nicolis & Prigogine, 1977; Pribram, 1991). As an explanation of the periodic table of chemical elements, atomic number theory is a prime example of this second step (Bohr, 1921a, 1921b). In this report we take only the first of these steps to explain optimization in the social collective: to explain how energy expenditure interacts with control operations to form an efficient communication system that results in a stable, effective collective.

We draw our insights and formalisms from thermodynamics and information measurement theory to help understand the communicative structure of small social collectives. In a subsequent work, we plan to describe the transfer functions (rules) by which the processes these formalisms embody are translated into psychological and sociological mechanisms operative in the collective. Thus the multilevel strategy is *not* being employed in the service of reductionism; indeed, as essential as single-level science is, we believe it must be complemented with multilevel work if a general scientific account of the behavior of collectives is to be achieved.

Scientific exploration is often dependent on the invention and application of new technology. One of our premises is that ideas derived from the formalisms of mathematics provide a technology that can be applied to data acquisition and analysis, especially as implemented in computer programs. The formalisms provide ways of expressing, in precise form, problem-solving algorithms, that is, ways of thinking about data sets. Mathematics in this sense is the technology of thinking.[1]

---

[1] This view of mathematics was expressed explicitly by Grassmann (Lewis, 1977, p. 104) who founded an algebra to represent the thought process. On the basis of Grassmann's work, Clifford developed an eight-valued (two-quaternion) algebra used by David Bohm and Basil Hiley (1993) in formulating issues in quantum mechanics.

Our report can be conceived, therefore, as an experiment in which we are tentatively applying computational devices, algorithms, found useful in stating and solving problems in other scientific endeavors. Computers are to the behavioral sciences what test tubes are to biochemistry. Both are "in vitro," silicon-based technologies, means by which energy relationships, control processes, and information transmission can be studied.

In what follows, we present a theory of global communication in social collectives. By *global communication* we mean a process by which information about the collective's internal organization is gathered, processed, and distributed throughout the system as a whole. A social collective is defined as a durable arrangement of individuals distinguished by shared membership and interaction in relation to a common purpose or goal. The theory shows how two orders of social relations act on the biosocial energy of the collective's members to create elementary units of information. Each of these units of information contains a description of the collective's endogenous organization. Constructing and distributing such descriptions throughout the collective on a moment-by-moment basis, the interaction between the two orders operates as an efficient communication system that informs the expenditure of energy to effect stable, effective collective action. We ground our understanding on the empirical results of a study of 57 social collectives.

2. PURPOSE AND APPROACH

The approach we develop differs in a number of important respects from the kind of understanding generally offered by social science in accounting for social behavior. A basic difference is our focus on an elemental order of communication, an order that is more inclusive than the restricted concept of (human) communication generally employed in social science. Emphasizing the cultural basis of *human* sociation, the term is generally used to denote interaction that involves the exchange of normatively defined meanings and understandings among purposeful social actors (Cherry, 1966). Irrespective of whether it occurs in an interpersonal or a collective context, communication is viewed as centered on the individual — transpiring between or among self-conscious actors, either in the pursuit of their own goals or in the roles they play as agents for collectives (Rogers & Kincaid, 1981; Jablin, Putnam, Roberts, & Porter, 1987).

The broader concept of communication that we develop here is similar to the notion of communication that underlies the "connectionist" computational models of "brain-style processing" (Rumelhart, 1992, p. 69). In these models synchronous parallel distributed processing among densely connected artificial "neural networks" is shown capable of encoding and "learning" quite complex patterns of "knowledge" and behavior (see Rumelhart, McClelland, & the PDP Research Group, 1986, and McClelland, Rumelhart, & the PDP Research Group, 1986, for examples). Here, information processing (computation) occurs *in the pattern of excitatory and inhibitory relations that interconnect all of the "neuron-like" nodes of the "neural network;"* it does *not* occur in the nodes themselves. This is the same core idea in our concept: a field of relations in which it is the interaction among different orders of social connection that processes and transmits information throughout the collective. Rather than being centered on the individual social actor, as is the case in the "block model" analyses (Freeman, White, & Romney,

1989; White, Boorman, & Breiger, 1976) and the "system dynamics" models of social systems (Forrester, 1968; Legasto, Forrester, & Lyneis, 1980), the locus of communication in our concept is the interaction among networks of social relationships connecting all members.

Another difference concerns our use of formalisms in place of the metaphorical analogies often used to portray aspects of the social collective that appear to endow it with the qualities of a sentient entity. Emile Durkheim used the notion of "collective consciousness" to portray what he believed to be the collective's psychic capabilities:

> The collective consciousness is the highest form of the psychic life, since it is the consciousness of the consciousnesses. Being placed outside of and above individual and local contingencies, it sees things only in their permanent and essential aspects, which it crystallizes into communicable ideas. At the same moment of time that it sees from above, it sees farther; at every moment of time it embraces all known reality; that is why it alone can furnish the mind with the molds which are applicable to the totality of things and which make it possible to think of them. (Durkheim, 1915/1965, p. 492)

More recently, largely in response to the emergence of so-called "cognitive science," a growing number of social scientists have drawn parallels between the organization of information processing in the brain and communication and behavior in the social collective (Bougon, 1983; Bradley, 1987; El Sawy, 1985; Garud & Kotha, 1994; Glazer, 1986; Hutchins, 1991; MacKenzie, 1991; Morgan & Ramirez, 1984; Sandelands & Stablein, 1987; Weick & Roberts, 1993). For instance, Hutchins (1991) drew on the distributed properties of neural processing to describe how redundancy in "overlapping" cognitive knowledge among individuals within a collective forms a system of mutual constraints to coordinate actions at the collective level. Sandelands and Stablein (1987) extended the analogy further and argued that in the same way that connections among neurons encode concepts and ideas in the brain, connections among social activities encode concepts and ideas in the collective. And although Weick and Roberts (1993) acknowledged that such metaphorical reasoning is a "shaky basis" for a theory of "organizational mind," they nonetheless contended that connectionism's sociological utility lies in the "insight" it offers, namely, that "relatively simple [social] actors may be able to apprehend complex inputs if they are organized in ways that resemble neural networks" (Weick and Roberts, 1993, p. 359; our addition).

However, although these analogies may offer descriptive imagery with which to characterize these poorly understood features of collective organization, there is always the risk of false attribution, which can yield obfuscation instead of explanation. Thus, we use the more neutral term *communication* instead. Because the formalisms we employ provide explicit principles that appear to account for such properties of collective organization, they offer a rational basis — one based on reason and logic — for building scientific understanding.

Social science has long recognized the importance of two basic patterns of social organization. Although these have been expressed in a variety of terms — formal versus informal organization (Roethlisberger & Dickson, 1939), rational versus natural systems (Selznick, 1948), sociotechnical versus socioemotional systems (Trist & Bamforth, 1951), mechanistic versus organic organization (Burns & Stalker, 1961), instrumental versus expressive leadership (Bales, 1958) among others — underlying these conceptualizations is a deeper (often implicit) dimensionality: a distinction between *hierarchical* and *heterarchical* forms of organization, between a pattern of organization based primarily on explicit relations of social rank and social control and a pattern of social connection that is more fluid and transitory involving an equivalence among individuals.

Previous analyses of the groups in this study (Bradley, 1987; Bradley & Roberts, 1989a, 1989b), have shown that these two patterns of organization form the communicative structure of the social collective — a heterarchical field of energy expenditure (that we refer to here as flux) and a constraint system of hierarchical controls (see Fig. 21.1). The heterarchical field, a distributed, massively parallel network of symmetrical relations in which members of the collective are essentially interchangeable, activates and unifies the biosocial energy of individuals. The hierarchical controls, a densely interconnected stratified order of asymmetrical relations in which the position of each individual is unique, operate on this field to produce an information processing network. By constraining the paths of energy expenditure, the controls render the potential for an informed pattern of collective action.

Following up on these earlier findings, the question posed here is whether insights and formalisms derived from thermodynamics can illuminate the biosocial interactions that compose the heterarchical order, and whether insights and formalisms derived from control engineering can illuminate the functions of the hierarchical order. Finally, our purpose is to enquire whether the interaction between heterarchy and hierarchy can be best understood as an instance of information processing in which data about flux (unfolding sequences of relations) and position (spatial-temporal location) are combined to create elementary units of information that provide optimal, moment-by-moment descriptions of the collective's endogenous organization.

3. THEORY

3.1. Assumptions

We begin our experiment in theory by limiting our task in four ways. First, our interest is restricted to collectives that have an explicit boundary distinguishing members from nonmembers; our account does not include partially bounded structures such as cliques or open-ended entities such as social networks.[1] Second, we leave aside any influence that normative elements, such as cultural values, norms, and roles, may have on the organization and action of social collectives, and on the behavior of their members. Third, apart from their biosocial potential — their capacity for physical and social activity — we ignore effects that the characteristics of the

---

[1] It is important to note that *all* members of the collective are included; this follows from our concept of communication, the interaction among networks of relations connecting all individuals in a collective. As mentioned, it is the same notion that underlies connectionist models of "neural networks." This is a different approach than that employed by most social networks researchers and system dynamics modelers in which the criterion of "mutual relevance" (Laumann, Marsden, & Prensky, 1982) is used to include only those actors who are (contextually) relevant to each other in the system.

collective's members, as individuals (e.g., gender, age, personality etc.), may have on system behavior.

Our fourth restriction is to limit our focus to the endogenous operations that characterize the collectives under study. Here we make the simplifying assumption that, to be exogenously effective, the collective must be stable. Our interest lies in exploring the efficiency of the endogenous processes by which stability is achieved, of developing an understanding of which patterns of endogenous organization are *optimal* for the collective's actions that result in stability (Coleman, 1990, p. 42). We will leave for a later discussion the question of the collective's effectiveness in its environment.

3.2. Energy and Least Action

The distinction between effectiveness and efficiency is derived from a rigorous definition of energy. Because *energy* is a rarely used term in social science — one that when it is used, is used as a metaphor (e.g., Collins', 1990, notion of "emotional energy") instead of as a scientific concept, we turn to the natural sciences for our use of the concept.

In the physical and biological sciences, energy is a measure of an amount of (physical) work that can be accomplished (McFarland, 1971). Two types of energy can be distinguished, kinetic and potential. When work is actually being done in producing change, it is defined as *kinetic energy*; the measure is directly proportional to the amount of kinesis, that is, to the amount of physical activity required to produce change. *Potential energy* is inferred from an estimate of the amount of possible work that a situation provides. It is an inference based on similarity to conditions that have been previously observed to transform potential into work.

In most physical and biological systems, there is a tendency to minimize work in order to conserve energy. This is known as the *least action principle* or the system's Hamiltonian function. In its general formulation, the principle holds that a system is maximally stable (i.e., at equilibrium) under conditions that maintain potential energy at a minimum (Considine, 1976, p. 1454). This means that any departure from equilibrium — any disequilibrating change in the system's structure — creates potential energy. For example, a pendulum at rest is at equilibrium: any change in conditions that disequilibrates the pendulum pushes or pulls it into positions in which the potential energy for returning to equilibrium becomes greater than that at equilibrium. In order to return to equilibrium, the system must expend the potential energy by performing work to use it up.

The initial measure of efficiency came from building steam engines. The aim was to convert the action of steam into useful work by minimizing its dissipation into friction and other useless generators of heat. Much experimentation with different engines was required to achieve this objective; it took effort to develop an efficient steam engine. *Effort*, in this sense, is directly related to *internally* attaining efficiency; whereas *effectiveness* deals with the total amount of work necessary to accomplish an *external* goal, irrespective of how much effort is expended (Pribram & McGuinness, 1975; Pribram, 1991, Chapter 9).

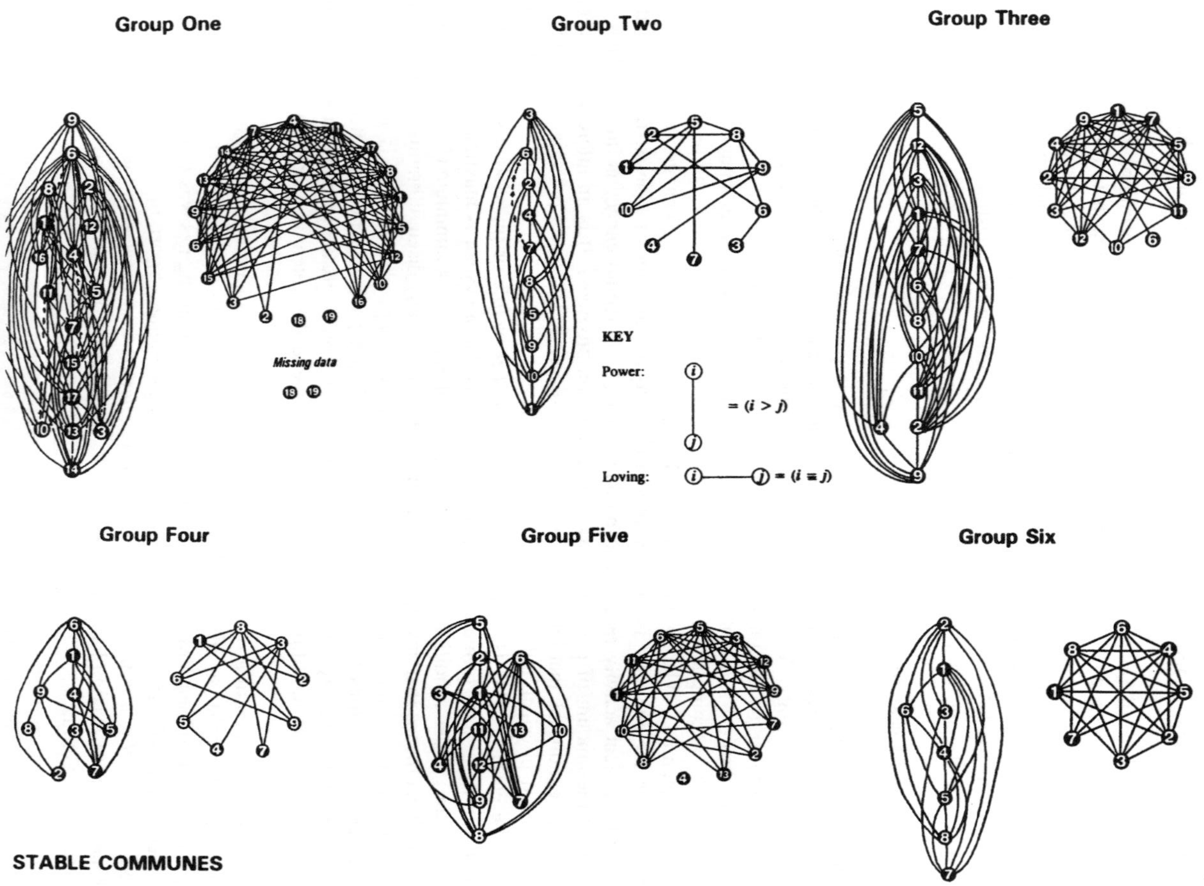

Group One

Group Two

Group Three

Group Four

Group Five

Group Six

KEY

Power: i —— j = ($i > j$)

Loving: i —— j = ($i = j$)

Missing data

STABLE COMMUNES

(a)

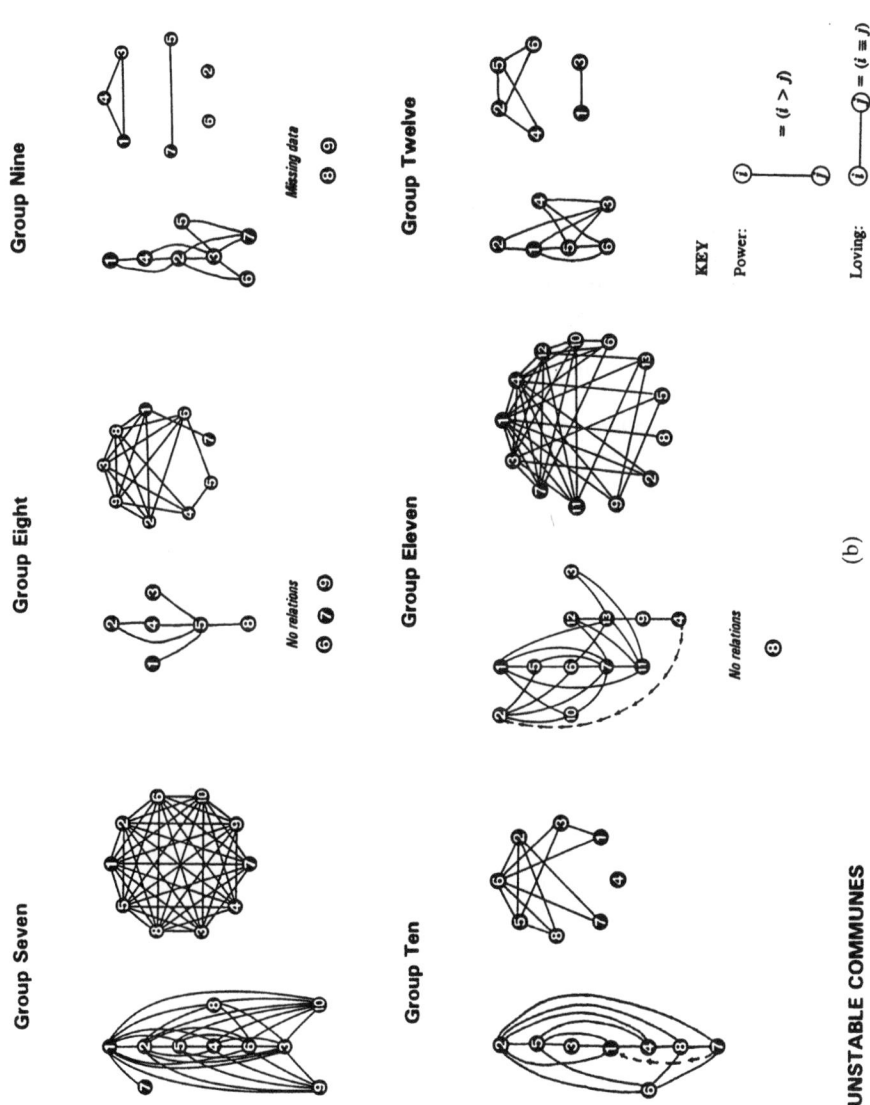

UNSTABLE COMMUNES

(b)

Fig. 21.1. (a) Sociometric structure of "power" (hierarchy) and "loving" (heterarchy) relations — selected stable communes. (b) Sociometric structure of "power" (hierarchy) and "loving" (heterarchy) relations — selected unstable communes.

To apply these concepts of energy, we assume that the members of the social collective are biologically capable of work — of engaging in physical behavior and activity, and that this capability is measurable as potential energy. When activated by the collective, the members' potential energy is converted into biosocial potential, the capability for engaging in social behavior or interaction. Actualizing this biosocial potential entails work; work is measured as kinetic energy. Because the dynamic operation of a collective requires an almost constant transformation of energy back and forth between potential and kinesis, as the collective continuously adjusts to internal and external changes, we use the term *flux* to characterize the medium of this continuous transfer of energy.

In its striving towards an efficient use of energy, the collective transforms potential to kinetic energy. The tendency to energy conservation requires effort on the part of the collective to explore and identify alternative paths of action to devise those that allow work to proceed efficiently, that is, with the least amount of dissipation.

How is a course of least action implemented? In the physical sciences a least action path (one that is *optimal* for the system) is determined by piecewise subtraction of potential by kinetic energy. Potential energy is reduced — through a series of successive fluctuations between potential and work — until its minimum level is reached.

One may conceive the path in this process as being determined by a landscape of constraints that channel the pattern of the actualization of potential into work. An example is when a river emerges from a mountain lake to course its way, by virtue of gravity, down the hillsides to the sea. Although at each point of the flow the river's potential energy is determined by the point's elevation above sea level, the flow path of the river — the pattern of its actualization of potential into work — is influenced by constraints obtaining to the terrain such as climate, vegetation, topography, geology, and so forth. A region of hard rock will interpose turbulence and require more effort than that of soft rock for the river to carve a direct, and thus an energy-efficient, path to the sea.

In an analogous fashion, a social collective constructs a landscape of social constraints to channel the actualization of the potential energy of its members into useful collective work. To the extent that least action holds, the effort demanded is that of seeking and then implementing an optimal landscape of endogenous relations that promote efficient group action. For instance, Henry Ford experimented with different ways of joining together the energy of his factory workers to find the maximally efficient organization for manufacturing cars (Lacey, 1986). To do this, he implemented a set of constraints, based on his invention of the production line and its associated techniques of mass production, that directed and thus optimized the action paths among the collective of workers; he produced automobiles at minimum cost, which, in turn, proved effective in the market place.

3.3. Conjunction and Control

Within this theoretical framework, two processes can be identified that act to generate descriptions of the collective's internal organization. The first is *conjunction*, which joins the biosocial potentials of the individuals composing the collective. The second is *control*, the construction of a landscape of social constraints that efficiently directs the transformation of potential into action. As detailed shortly, the landscape determines a communication processing network that generates patterns of actualization of the potential of the collective. The intricacy (complexity) of each successive configuration of these patterns is processed by the collective as a measure of information — as the means for describing its endogenous organization. Thus, the measure of information provides moment-by-moment descriptions of the endogenous order that are communicated by a holographic-like process throughout the collective (Bradley, 1987, Chapter 9).

Conjunction is achieved within a field of reciprocally equivalent *(equi-valent*, of equal value) relations among individuals. Within such a field a heterarchical order operates in which there is an absence of differentiation in terms of social status or rank, so that all individuals share in common a connection of social equality. As a result, the individuals are mutually open to each other and, by extension, open to the collective as a whole. Thus, the field of heterarchical relations describes the potential biosocial energy — the potential for work — of the collective.

In the absence of other factors, initial conditions (such as negative feelings like fear, hatred, or jealousy) will block the efficient conversion of potential to kinetic energy; in non-linear dynamics such systems are characterized by negative Lyapunov exponents leading to stasis, ossification (complete [physical] equilibrium), or to fluctuations described by relaxation oscillators (Abraham, 1991). On the other hand, as elaborated below, initial conditions such as admiration, awe, or love create a kind of harmonic resonance in the relations among members that will enhance the conversion of potential to kinetic energy. The danger here, if this enhanced kinetic energy is unconstrained, is that undue dissipation of energy will ensue: in the language of nonlinear dynamics, chaos will result.

The second process is control, a landscape of social constraints that influences the conversion of potential energy to kinetic energy, that is, the patterning of flux. Control is achieved by a transitively ordered structure of social relations among members that prevents the dissipation of kinetic energy. By precluding undue dissipation, the controls shape the paths of flux, thereby *in-forming* — giving shape to — the relations among individuals.[1]

3.4. Information and Efficiency of Communication

Surprisingly, given the rich, dense flow of verbal and nonverbal signals that comprise human interaction, information is rarely used as a rigorous concept in social research; in three recent influential works (Coleman, 1990; Burt, 1992; White, 1992) it is employed as an undefined term. Irrespective of whether the term is explicitly defined (e.g., Rogers & Kincaid, 1981, pp. 48-51) or not, its use in social science corresponds to Shannon's (1949) concept of information, that is *as a reduction of uncertainty through choice among alternatives*. In this conception the smallest unit of information is the *bit*, the binary digit — nowadays corresponding to the smallest standard unit of information in computational information systems. Shannon's concept of information applies to symbol-based communication systems, like human language. In such systems each unit of information in a sequence contributes to resolution of the signal's message by reducing the probability of alternative meanings.

---

[1] This conception is similar to Bohm and Hiley's notion of "active information" (see Bohm & Hiley, 1993, pp. 35-42, 59-71).

There is, however, a second concept of information used in the physical and biological sciences, virtually unknown in the social sciences. Because we employ this second concept to show how the interaction between heterarchy and hierarchy operates as a communication system to construct and distribute information, we turn to the natural sciences for our use of this concept.

Our discussion is informed by work on signal processing in telecommunications (see Cherry, 1966, for an excellent review). Two distinct properties of the signal have been utilized for transmission in telecommunication. One is the signal as a sequence of discrete pulses encoded in time, as in the use of Morse (or similar) code for telegraphic communication. The other encodes the signal as a pattern of energy oscillations across a waveband of frequencies, as in the encoding and transmission of vocal utterances for telephonic communication. Although the frequency aspects of a signal's oscillation are irrelevant for telegraphy, for telephone communication they are critical as fidelity is dependent on the spectral components of the signal (frequency, amplitude, and phase). It took some time, though, in the early part of this century, to realize that there was a relation between the rate of transmission of a given quantity of information and frequency bandwidth (Nyquist, 1924; Kupfmueller, 1924). This relation was generalized by Hartley (1928, p. 525): "the total amount of information which may be transmitted ... is proportional to the product of frequency range which is transmitted and the time which is available for the transmission."

Gabor (1946) formalized this relationship in his "Theory of Communication." He noted that there is a restriction to the efficiency with which a set of telephone signals can be processed and communicated. The restriction is due to the limit on the precision to which concurrent measurements of spectral components and the (space)time epoch of the signal can be made. This restriction is illustrated in the top of Fig. 21.2, in which time and frequency are treated as orthogonal coordinates. Although the frequency of a harmonic oscillation, represented by a vertical line, is exactly defined, its duration in time is totally undefined. Conversely, a sudden surge (a "unit impulse function") or change in the signal, the horizontal line, is sharply defined in time, but its energy is distributed evenly throughout the whole frequency spectrum. Thus, although accurate measurement can be made in time or in frequency, *it cannot be simultaneously made in both beyond a certain limit* (Gabor, 1946, pp. 431-432).

Gabor was able to show, mathematically, that this limit could be given formal expression by Heisenberg's uncertainty principle (Heisenberg had developed his mathematical formulation of uncertainty to define the discrete units of energy, *quanta*, emitted by subatomic radiation). In its rigorous form the uncertainty relation is given as $\Delta t\, \Delta f \geq \frac{1}{2}$, which states that t (time) and f (frequency) cannot be simultaneously defined in exact terms, but only with a latitude of greater than or equal to one-half in the product of their uncertainties. Since certainty can be obtained only by minimizing uncertainty on both coordinates, the *minimum measurement* of the signal in time *and* frequency is $\Delta t\, \Delta f = \frac{1}{2}$, which defines an *elementary unit of information* (Gabor, 1946, pp. 431-437).

This unit of information both minimizes uncertainty and provides the maximally efficient description of communication (the minimum space or time of transmission occupied by the signal that still maintained the fidelity of telephonic communication). Gabor called his unit of optimal efficiency a *logon*, or a *quantum of information* (illustrated in the middle of Fig. 21.2), and showed that the signal that occupies this minimum area "is the modulation product of a harmonic oscillation of any frequency with a pulse in the form of a probability function" (Gabor, 1946, p.

435). This fundamental unit of information is a sinusoid variably constrained by space-time coordinates; it differs from Shannon's unit of information, the binary digit (BIT), which is a Boolean choice between alternatives (Pribram, 1991, p. 28).

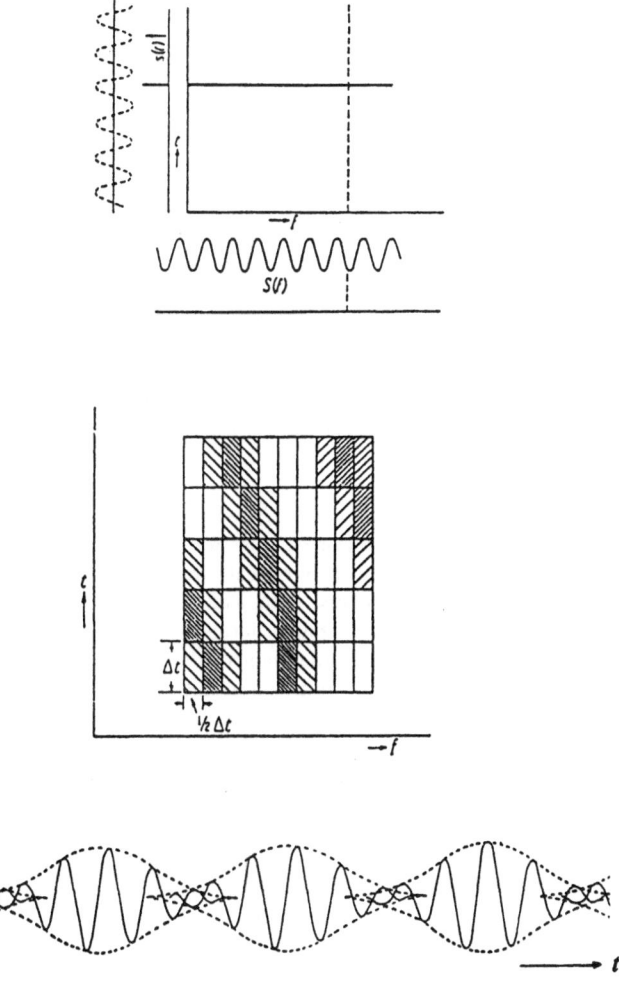

Fig. 21.2. (Top) Limits of concurrent measurement of time (t) and frequency (f) of a signal. (Middle) Representation of a signal by logons. (Bottom) Representation of the overlap of logons. (From Gabor, 1946, Figs. 1.3 and 1.7. Adapted with permission.)

A final point concerns an important implication of the use of these mathematics for the role of causality in communication involving these elementary units of information. Because logons are not discrete units but occur as overlapping sinusoids, each wrapped in a Gaussian probability envelope (illustrated in the bottom of Fig. 21.2), each logon contains, in Gabor's words, an "overlap (with) the future." This is a result of using time as one of the (measurement) dimensions because "the principle of causality requires that any quantity at an epoch t can depend *only* on data belonging to epochs earlier than t. ... In fact, *strict causality exists only in the 'time language'*" (Gabor, 1946, p. 437; our emphasis). What is of special interest here is the extent to which this overlap (interference) among logons yields a communicative system in which the data in succeeding logons is contained, in a nontrivial way, in the logons that preceded them: in other words, that information about the "future" order is enfolded into the elementary units of information being processed in the "present" (see Bradley, 1996, for more on this).

The Gabor elementary function, as it is often referred to, has been found to characterize perceptual processing in the cerebral cortex (see Pribram, 1991, for a review). It is, therefore, an alternative unit for biological information processing to Shannon's (1949) unit of information, the BIT. Moreover, two previous findings from the social collectives examined in this study document an order of communication that does not seem describable within the terms of Shannon's concept but appears more readily understood within Gabor's terms. One is a holographic-like order in which information about the organization of the collective as a whole was found to be distributed as a nonlocalized order to all individuals, and the second is that this order was found to be constrained by a system of hierarchical relations (see Bradley, 1987, Chs. 8 and 9, respectively).

To summarize (Fig. 21.3), two very different modes of organization characterize the hierarchical and heterarchical operations of relations within the collective. Because individuals are asymmetrically connected in the hierarchical order, the system of controls operates differentially on the collective's members, both with respect to their particular socio-spatial location as well as with respect to actualization during particular frames of time. By contrast, the symmetric bonds of the heterarchical order indicate that individuals are essentially equivalent in terms of the pattern of distribution of flux within this endogenous field. As this field is an energy field, it lies within the spectral domain (energy is measured in terms of frequency times Planck's constant) and is related to space and time by way of a transformation (the Fourier transform).

The operation of hierarchical controls on the heterarchical distribution of flux generates a moment-by-moment — *quantized* — description of the collective in terms of both structure (spatial-temporal position) and flux (frequencies of oscillation of unfolding relations). By providing, thus, a succession of descriptions within space-time and spectral coordinates, quantum-like Gabor units of information are constructed and communicated, via a holographic-like process, throughout the collective. These units of information characterize the endogenous order as it evolves in a continuing series of interactions. Because each unit of information overlaps with the unit that succeeds it, each unit contains information about the future (potential) order of the collective. However, whenever there is an imbalance between the amount of distribution of flux and the amount of control, the efficient operation of the collective becomes impaired and the probability of instability is increased. This impairment is due to what Ashby (1956) characterizes as the necessity for "requisite variety."

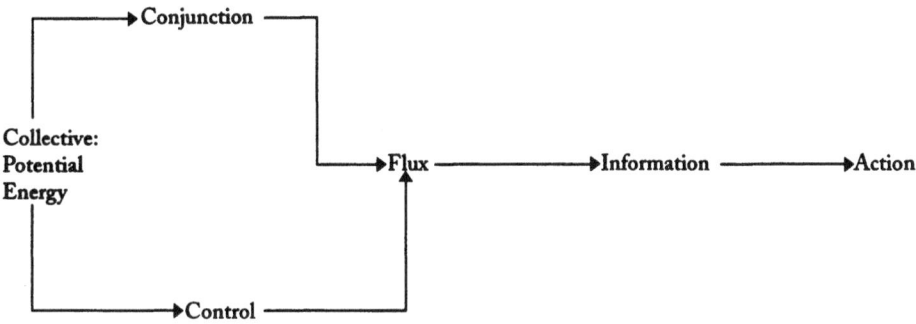

Fig. 21.3. Logic of theoretical model.

4. METHOD AND DATA

The data were gathered over a decade ago as part of a nationwide longitudinal field study of 60 urban communes (Zablocki, 1980); a commune was operationally defined as a minimum of three families, or five non-blood-related adults (persons aged 15 years or older), who shared, to some degree, common geographical location, voluntary membership, economic interdependence, and some program of common enterprise, usually spiritual, social-psychological, political, cultural, or some combination of these (Bradley, 1987, p. 14). Stratified on a number of basic social characteristics, and sampled in equal numbers from six Standard Metropolitan Statistical Areas (Atlanta, Boston, Houston, Los Angeles, Minneapolis-Saint Paul, and New York), various formal and informal methods were used to study the communes. Data from 57 communes are used in this report; three communes from the original sample were not included as membership in these groups was not completely voluntary (for more detail on the methods and sample of the original study, see Zablocki, 1980, & Bradley, 1987).

In terms of the sample's social characteristics (Table 21.1) at the time of the first wave of data collection (the summer of 1974), the communes ranged in size from 5 to 35 permanent adult members (mean size = 10 members) and had been in existence from 3 months to 9 years (mean commune age = 3 years). A total of 566 adults (15 years and older; mean age = 25 years), with slightly more men than women, were residing in the communes; most had never been married. Being a generally well-educated population, most reported working at a full-time white collar or professional job. In terms of social organization, the communes covered a wide spectrum of cultural values and included Christian religious, Eastern religious, personal growth, family, countercultural, and political ideologies. Most communes had special requirements for membership, and most also had incorporated elements of formal organization into their social structure (e.g., chore rotation, mandatory rules, positions of leadership and office, decision-making procedures, group rituals etc.).

Characteristics of Adult Population (15 years and older; N = 545)

Median age	25 years
Percentage male	54%
Percentage single, never married	72%
Percentage with college diploma	50%
Percentage with white collar or professional occupation	63%
Percentage with full time or part time job	67%

Characteristics of Communes (N = 57)

Mean size (adult members)	9.9
Percentage existed two or more years	42%
Percentage with "many" rules	21%
Percentage assign or rotate chores	51%
Percentage have communal business or jobs	16%
Percentage requiring novitiate or trial membership	33%
Mean percentage members holding formal positions or office	41%[1]
Percentage ideology "important" to group	79%
Percentage without leaders	30%

Ideological Type:

Religious	40%
Political or counter-cultural	26%
Personal growth, household, or family	34%
	100%

Table 21.1. Urban Communes Sample: Social Characteristics of Adult Population and Communes.

Formal and informal methods were used to collect two panels of data, 12 months apart, during the summers of 1974 and 1975. Data on commune survival status were also gathered for an additional 2 years. A number of structured interviews and questionnaires were administered to all permanent adult members to gather information on social background, communal involvement, self-concept, and attitudes. Data on the organization and activities of each commune were collected by field worker observations and taped interviews during the summer. A sociometric instrument (Table 21.2), the primary source of the data presented in this report, was administered to map the structure of social relationships in each commune. Each adult member was asked about the content of his or her relationship to each other member, thus providing an exhaustive mapping of the $N(N-1)$ possible pairwise (dyadic) relations in the group (where N = the number of permanent adult members). The instrument was administered under strict field worker supervision to ensure that there was no collusion among members in answering the questions.

Although all except 1 of the 57 communes cooperated with the administration of the sociometric instrument, the quality of responses in 11 groups was unsatisfactory in that missing

---

[1] N = 273, respondents to the "Long Form" interview.

data ("no answer," "incomplete," or an "uncodeable" response) were greater than 25% of the total of possible relations in these groups, and contained, therefore, an unacceptable level of potential structural bias (see Bradley, 1987, p. 24, note 19; p. 98, note 3). As in the original study, these 11 communes were excluded from the structural analysis. This means that the sociometric results presented in this report are based on data from 46 communes.

4.1. Operationalization

To test the theory, the following procedures were used to operationalize the primary concepts, flux, control, and stability, with the data from the communes. Sociometric procedures, following Bradley and Roberts' (1989b) measurement guidelines, were used to operationalize the concepts of flux and control. Administered to every adult member in each commune, the sociometric instrument, mentioned above, generated an enumeration of all possible dyadic relations in which the relation between each pair of individuals, i and j, was measured from both sides of the dyad, that is, from i to j, and from j to i.

Flux, the activation of biosocial potential, was indicated by a positively reciprocated response (an answer of "yes") by both individuals to either the "loving," "improving," or "exciting" questions (see Question 5g, Table 21.2). This operationalization follows both from the theory regarding the expected role that positive affect plays in enhancing the distribution of flux (see above under Conjunction and Control), and also from the original study in which it was found that measures of negative affect had little descriptive or explanatory utility (see Bradley, 1987, pp. 83-94). To translate these responses into a group-level dyadic measure of flux, the mean proportion of loving, improving, or exciting relations in each commune was calculated.

Control, the operation of constraints on the activation of potential energy, was measured by the "power" question (Question 5e, Table 21.2). Only those responses that indicated the asymmetric ordering of the relationship — that is, which of the two individuals (the respondent, i, or the other member, j) held the "greater amount of power" — were used.

The subsets of relations that met these two operational definitions were then translated into a symmetric and an asymmetric sociomatrix of relations of flux and control, respectively, to encode the disposition of these dyadic relations among all members in each group. A binary coding was used in which, for flux, a value of 1 indicated the presence of a reciprocated relation, and for control, a value of 1 indicated the presence of an ordered relationship (i.e., $i \rightarrow j = 1$, control flows from i to j; $j \rightarrow i = 1$, control flows from j to i); any other condition, for either flux or control, was indicated by a value of 0. The mean results for all communes on these dyadic definitions of flux and control are provided in the Appendix.

The final step entailed the use of triadic analysis (Holland & Leinhardt, 1976) — a technique for analyzing the structural organization of social networks — as the means to build structural indices of flux and control. This technique first subdivides the sociomatrix into triads, and then, through a census of all possible triadic configurations, classifies the array of triads for the group into 16 isomorphic triad types (see Fig. 21.4).

The following set of items is from "page three" of the "Relationship Questionnaire" (see Bradley, 1980, for the complete instrument) and is the source of most of the relational data Bradley employed in his study. Each respondent received a questionnaire with multiple copies of "page three" inserted in it — one page for each other adult resident. A respondent in a commune with a population of nine, for example, would receive a questionnaire with eight page threes. Each of these page threes had one of the commune members' names typed in at the top (e.g., "This sheet is about _____"). By completing this questionnaire, each respondent supplied information systematically describing his/her realtionship with each of the other members of their commune.

5. This sheet is about _____
    a. How long have you known the above named person?
       Years _____ Months _____
    b. In your own words briefly characterize the changes which have occurred in your unique relationship with this person as a fellow commune member over the last twelve months or, if less, for the time you have known each other.
    _____
    c. How many hours in a typical week do you spend just by yourselves?
    _____
    d. If you happen to know it, state what kind of work (his/her) father did while the person named above was growing up.
    e. Even the most equal of relationships sometimes has a power element involved. However insignificant it may be in your relationship with this person, which of you do you think holds the greater amount of power in your relationship?
    _____
    f. If this commune did not exist, would you want to have a close relationship with this person?
    _____
    g. For the list of descriptions below, indicate if the following are involved in your relationship with the person named above by checking the appropriate answer. *Please answer each of the following*:

Work together	Yes ____	No ____	Sometimes ____
Spend free time together	Yes ____	No ____	Sometimes ____
Mind children together	Yes ____	No ____	Sometimes ____
Sleep together	Yes ____	No ____	Sometimes ____
Confide in each other	Yes ____	No ____	Sometimes ____
Loving	Yes ____	No ____	Sometimes ____
Exciting	Yes ____	No ____	Sometimes ____
Awkward	Yes ____	No ____	Sometimes ____
Feel close to each other	Yes ____	No ____	Sometimes ____
Tense	Yes ____	No ____	Sometimes ____
Jealous	Yes ____	No ____	Sometimes ____
Agree on communal policy	Yes ____	No ____	Sometimes ____
Feel estranged from each other	Yes ____	No ____	Sometimes ____
Exploitive	Yes ____	No ____	Sometimes ____
Hateful	Yes ____	No ____	Sometimes ____
Improving	Yes ____	No ____	Sometimes ____
Sexual	Yes ____	No ____	Sometimes ____

    h. Do you feel that the overall relationship between the two of you is more important to you, or do you feel it is more important to the above named person?
       ____ More important to you   ____ More important to him/her
    i. In your relationship with this person, does he/she ever act to you as a father or mother, sister or brother, son or daughter, or none of these?
    _____

Table 21.2. Sociometric Instrument.

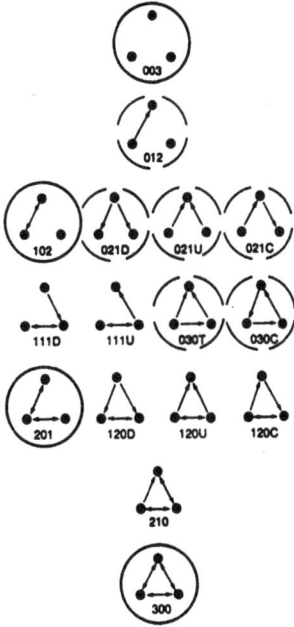

Fig. 21.4. Holland and Leinhardt's 16 isomorphic triad types: the 16 isomorphism classes for digraphs with $g = 3$ (that is, the triad types). Triad labeling convention: the first digit is the number of mutual dyads; the second digit is the number of asymmetric dyads; the third digit is the number of null dyads; trailing letters further differentiate among the triad types. Four symmetric triad types (unbroken circle) were used in the structural analysis of flux relations and seven asymmetric triad types (broken circle) were used in the analysis of control relations; the "vacuous" 003 triad was used in both analyses. From Holland and Leinhardt (1976, p. 6, Fig. 2. Adapted with permission.)

The triad types are distinguished from one another structurally by their composition in terms of three kinds of dyads: *mutual* dyads (M), in which a symmetric relation connects the two individuals; *asymmetric* dyads (A), involving an ordered or directed relation between the two; and *null* dyads (N), in which there is no relation between the two. Hence each triad type can be uniquely identified and labeled in terms of its dyadic composition. For example, the 012 triad (see Fig. 21.4) has no *mutual* relations, one *asymmetric* relation, and two *null* relations.

Of the 16 triad types, 3 are symmetric in form in that they are composed exclusively of positively reciprocated dyads (see Fig. 21.4: the 102, 201, and 300 triad types, enclosed by a solid circle). Aggregated across the "loving," "improving," and "exciting" relations, the mean sum of these three triads as a proportion of all possible triads in a commune was used to measure the amount of flux. A bar graph in Fig. 21.5 plots the results of this procedure and shows a positively skewed distribution for the measure of flux (mean sum = .629, standard deviation [SD] = .196).

Seven other triad types (enclosed by a broken circle in Fig. 21.4) are composed exclusively of asymmetric dyads. Based on their successful use in previous analyses (Bradley, 1987; Bradley & Roberts, 1989a), three of these triad types (the 021C, the 021D, and the 030T, summed and expressed as the proportion of all possible triads) were used to measure the amount of control. Aggregated for "power" relations, the three triad types constituted just over half (.509), on average, of all possible triads of control in the communes. A bar graph of the result for all communes is shown in Fig. 21.5 in which a flatter distribution is evident (mean sum = .510, SD = .218). The mean results of the triadic census for all communes for symmetric relations of flux and the asymmetric relations of control are provided in the Appendix.

Stability, the degree to which the collective is able to maintain itself as an enduring, self-sustaining entity, was measured by a commune's survival status at a specific moment in time. Classified into one of two categories, *survivor* or *nonsurvivor*, each commune's stability was determined at each of the four successive 12 month intervals that observations were collected; Time 0 is the point in time when a commune was founded and Time 1 is moment of the first wave of data collection (August 1974). Starting with Time 1, measurement of each commune's stability (survival status) was made at 12-month intervals for the succeeding 4 years, that is, through Time 5. Twenty-two (48%) of the 46 communes survived the 48-month observation period. A pattern of declining instability over time was observed, from 24% by the end of the first 12 months, at Time 2, to 8% after 60 months, at Time 5 (see Appendix).[1]

5. EMPIRICAL ANALYSES

5.1. Verification of the Theory

Testing the theory entailed an analysis to determine the degree to which the observed patterns of commune behavior with respect to our measures of flux, control, and stability were consistent with the expected patterns. Among other techniques, a spatial representation of the data was derived as the theory rests on a field-theoretic concept of energy — a conceptualization of the collective's potential for action as an endogenous *field* of biosocial energy that operates along two dimensions: an unordered dimension of *equi*-valent, symmetrical relations (flux); and an ordered dimension of (transitive) asymmetric relations differentiated by spatiotemporal position (control).

---

[1]Although the communes ranged in group age from 3 months to 9 years at Time 1, there is little evidence that "period effects" (differences in group age at the time data collection commenced) explain the variability in survival status. Dividing the sample into "young" (2 or less years; $N = 23$) and "old" (more than 2 years; $N = 23$) categories of group age at Time 1, and cross-tabulating these classifications by survival status grouped in three categories (dissolved by Time 2 or Time 3: $N = 17$; dissolved in Time 4 or Time 5: $N = 7$; survived beyond Time 5: $N = 22$) shows nonexistent (0%) to modest (12%) nonstatistically significant differences between the "young" and "old" categories of communes (*chi*-square coefficient with two degrees of freedom = 1.260, pr. = .533).

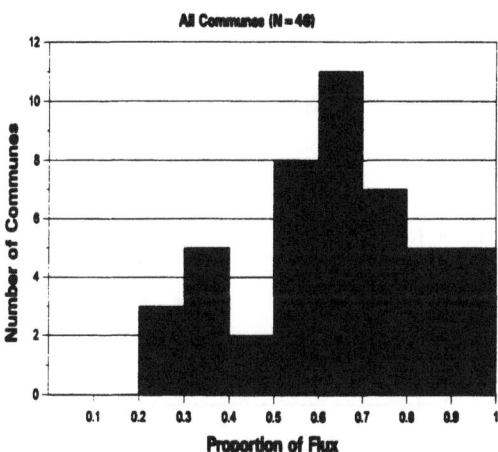

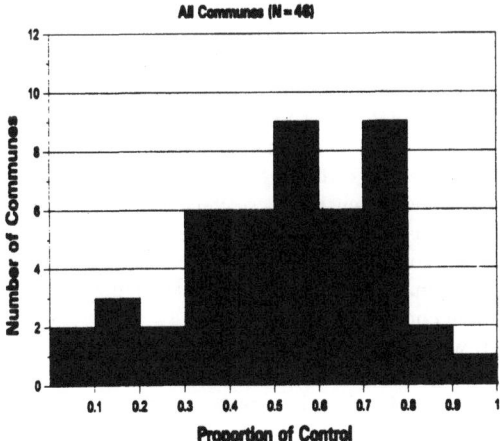

Fig. 21.5. Bar charts showing commune distribution for flux and control at Time 1.

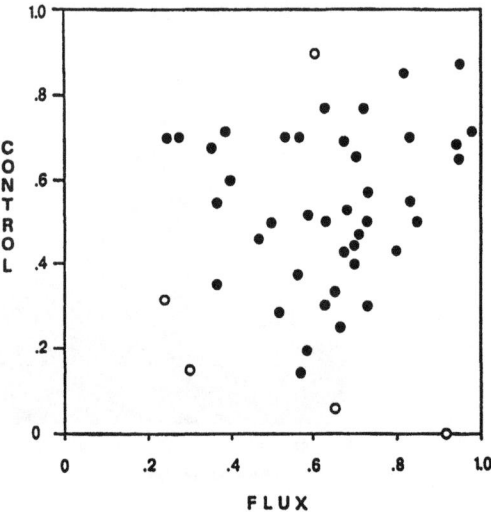

Fig. 21.6. Scatterplot of all communes ($N = 46$) by flux and control at Time 1 (outlying cases are hollow dots).

Theoretically, it was expected that an efficient communication processing network required a certain amount of flux in interaction with a certain amount of control, and that the tendency toward least action would result in an increasingly closer correspondence in their respective values at higher magnitudes of flux. This is because there would be an increased risk of turbulence, resulting from undue dissipation of kinetic energy, when the conversion of potential to kinetic energy is enhanced at higher levels of flux.   Data bearing on the relationship between flux and control are shown in the scatterplot in Fig. 21.6. In accordance with the dimensionality of our concepts of flux and control, the measure of control is the vertical ordinate and that for flux is the horizontal ordinate. It can be seen that the null hypothesis — an equal or random distribution of groups over all locations in the endogenous field — does not hold, and that the low nonsignificant correlation (Pearson's $r = .12$; pr. $> r$, .43) actually masks a nonlinear association. Moreover, with the exception of five outlying cases (hollow dots in Fig. 21.6), the observed pattern — a triangular distribution with a wide base and its apex in the high-flux/high-control region (upper right quadrant) — is consistent with the theory. Thus, there is an absence (with one exception) of communes in the high-flux/low-control region (lower right quadrant), an absence in the low-flux/high-control region (upper left quadrant), and (with two exceptions) an absence of communes in the low- flux/low-control region (lower left quadrant).

In addition to the relative amounts of flux and control involved, theoretically it was expected that there also was a limit on the *total amount* of information processed by a collective in a given unit of space or time. Thus, when the total amount of information exceeds the processing capacity of the communication system, it is processed as "noise" and results in unsynchronized action with a high likelihood of turbulence and instability; on the other hand, when the total amount of information is insufficient to convey current descriptions of the ever-evolving endogenous order, collective action will be largely uninformed and ineffective, with dissolution a likely result.

A measure of the total amount of information processed by a collective at a given moment in time was computed by summing and averaging, for each commune, the total for flux and control at Time 1 (mean for all communes = .569, median = .552, and SD = .155). The values for all communes were grouped into .10 intervals and, holding these values on this measure constant at Time 1, the sample was partitioned by survival status and the distribution of survivors and nonsurvivors was plotted on a time series of bar charts at 12-month intervals, that is, from Time 2 through Time 5 (see Fig. 21.7).

Examining the pattern of results in Fig. 21.7, two things stand out. First, the distribution for the total amount of information for all communes at Time 1 is bell-shaped with 67% falling within one standard deviation of either side of the mean. Second, this bell-shaped distribution gradually devolves over time into *two contrasting patterns* that are virtually the *inverse* of each other by Time 5: a single-peaked distribution for the 22 survivors with its mode (9 cases, 41%) in the .500—.599 interval; a bi-modal distribution for the twenty-four nonsurvivors with its trough (2 cases, 8%) in the .500—.599 interval and its twin peaks (6 cases (24%) each) in the two adjacent intervals of .400—.499 and .600—.699. This difference in survival rates between the groups in .500-.599 interval and the other groups outside this range is statistically significant (chi-square = 6.695, pr. = .010).

Taken together, these two patterns appear to mark the bounds of a region where the probability of stability is maximized, that is in the .500—.599 interval. So that although, in this interval, the rate of instability for all communes is lowest (18%, 2 of 11 groups), it rises sharply in the adjoining intervals: 60%, 6 of 10 groups in the each of the .400—.499 and .600—.699 intervals; 75%, three of four groups in each of the .300—.399, .700—.799, and .800—.899 intervals. When computed for the communes in these adjoining intervals at Time 1, the rate of instability by Time 5 for these two sets of intervals is 63% (twelve of nineteen groups at .600 and above, and ten of sixteen groups at .499 and below), which is significantly different than the 18% for the eleven groups in the .500-.599 interval (chi-square = 6.966, pr. = .035). Thus it would appear that the total amount of information in the intervals above .599 was excessive in terms of information processing capacity, whereas the amount of information in the intervals *below* .500 was insufficient to sustain a viable collective.

Two further interrelated theoretical expectations were investigated. The first of these was that there would be restrictions on the *relative* amounts of flux and control involved in communication — that there would be both a *lower* and an *upper* limit on the amounts of each of these processed by an effective collective. Combinations of flux and control that fall outside the limits were expected to result in collective dysfunction. The second expectation, for collectives operating within these limits, was that the disposition of the collective at a future moment would be enfolded in the information processed by the communicative structure in the present. This follows from the overlap among these logon-like elementary units of information by means of which the present order is *in*-formed (given shape to) by the order implicit in the

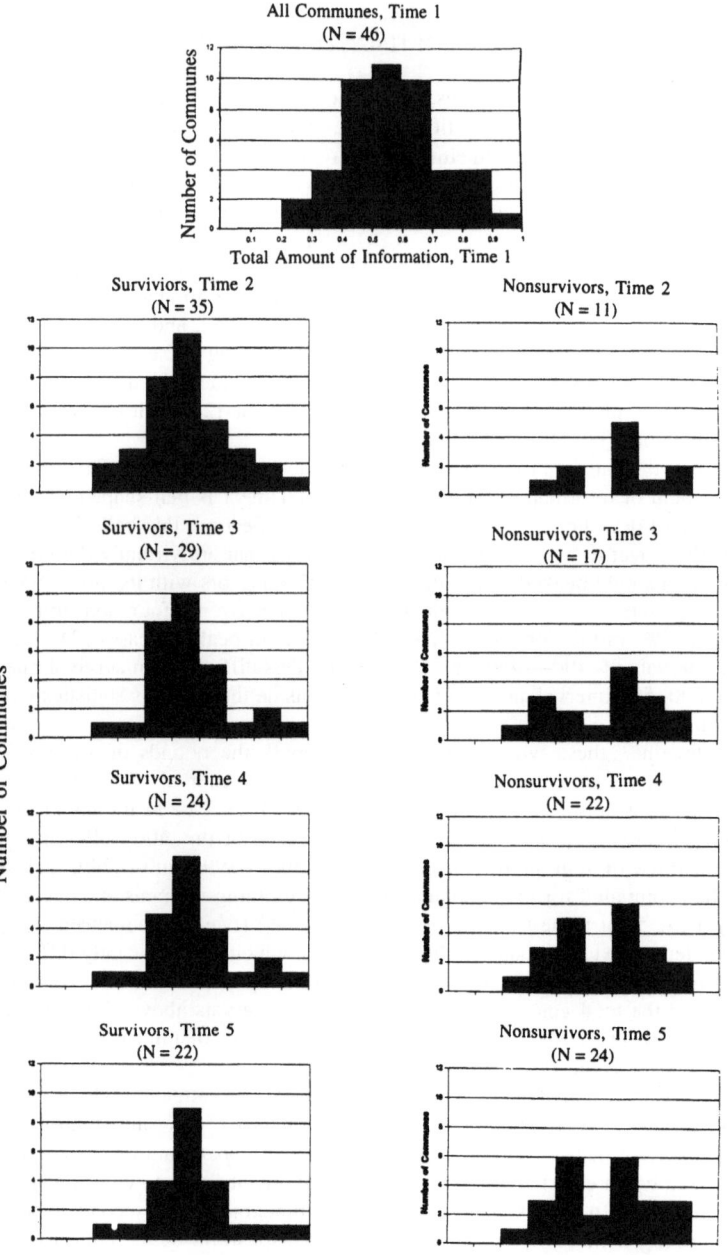

Fig. 21.7. Bar charts showing distribution of communes for total amount of information at Time 1 by survival status at Time 2 through Time 5.

series of succeeding moments (see Bradley, 1996). Thus, combinations of flux and control within the limits at a given moment were expected to yield an increased potential for effective action in succeeding moments.

Figure 21.8 presents a time-series of scatterplots showing the relationship between flux and control at Time 1 to stability at Time 2 and at Time 3--in other words, the relationship between the composition (in terms of flux and control) of the information provided by a collective's communication system at a given point in time, and the stability of the collective at two successive moments in the future. The scatterplot on the far left-hand side is for all communes plotted by their values for flux (horizontal ordinate) and control (vertical ordinate) at Time 1, the first point of measurement. Holding the values for each commune on flux and control constant at Time 1, the scatterplots for Time 2 and Time 3 are divided into a plot for survivors (top row of scatterplots in Fig. 21.8) and a plot for nonsurvivors (bottom row). This provides a view of the relationship between information on the endogenous order at a given moment in time and collective stability at twelve and at twenty-four months later.[1]

Starting with the baseline pattern at Time 1 for all communes, three related patterns become increasingly evident as survival status is plotted at Time 2 and Time 3. First, instability tends to be highest for groups in the peripheral regions of the field — that is, for groups with the greatest imbalance between flux and control. Second, with the exception of three stable groups in the high-flux/high-control region, survivors tend to form a triangular pattern with most groups clustered together in the mid-region. And third, that location in this mid-region at Time 1 is strongly related to survival at Time 3, 24 months into the future. What is most striking about the results is that the pattern for survivors is virtually the complement of that for nonsurvivors: *there is a complete absence of nonsurvivors in the mid-region where the greatest concentration of survivors is observed.*

Looking more closely at the pattern for the 17 nonsurvivors, two bands of instability become clearly apparent by Time 3: the upper band of 12 (71%) nonsurvivors, marks a region of high instability; the lower band formed by the other 5 nonsurvivors, appears to define a lower bound to the region of stability. In short, the two bands of instability seem to distinguish functional from dysfunctional combinations of flux and control.

To test the veracity of this interpretation, we divided the full sample of communes into stable and unstable sets such that the probability of survival was maximized for the former while being minimized for the latter. Operationally, this entailed establishing partitions that would mark the upper and lower bounds to the region where stability is optimized.

The boundary of the lower bound was established by the four communes (see the scatterplot for nonsurvivors, Time 3, Fig. 21.8) on a line in the lower band of instability orthogonal to the low-flux-low control/high-flux-high control axis. A total of six communes was observed in this region, of which five (83%) had become nonsurvivors by Time 3. For

---

[1]This time series of scatterplots on stability was run out across the full 48 months (i.e., Time 1 through Time 5) for which observations were collected on the communes. The results for the first 24 months (i.e., through Time 3 as shown in Fig. 21.8) suggest this is a reasonable period over which to aggregate survival status to accumulate enough nonsurviving cases (nonsurvivors at Time 3 = 17 cases) for the analysis; the scatterplots for Time 4 and Time 5 (not displayed) show evidence of a deterioration in the "predictive power" of the information provided by flux and control at Time 1 for stability beyond 24 months.

comparison, the baseline rate of instability over all communes was 37%, 17 nonsurvivors out of 46 groups.

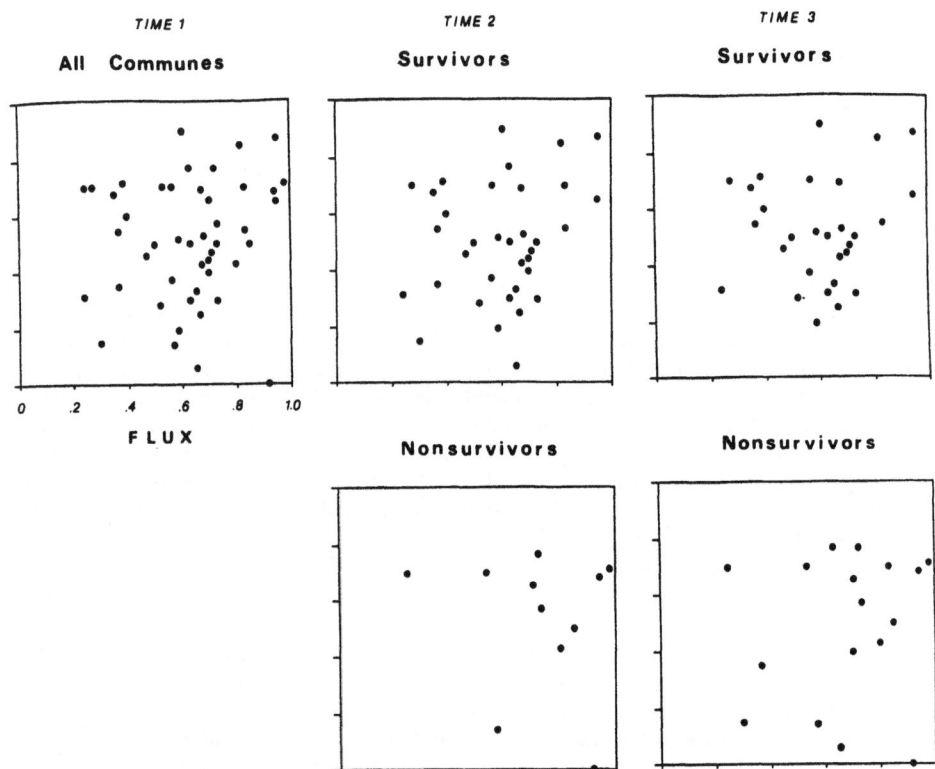

Fig. 21.8. Scatterplots of communes on flux and control at Time 1 by stability (survival status) at Times 2 and 3.

For a boundary marking the upper bound to the region of maximal stability, there were two possibilities. The first is the line (orthogonal to the axis just mentioned) established by the three communes at the bottom of the upper band of instability; this is *not* an optimal partition because although the probability of survival is maximized (100%; there are no nonsurvivors) for the 15 groups in the area defined by this line and the lower bound, the probability of instability is not maximized for the 25 groups classified by this line as belonging to the upper band region of instability (nonsurvivors = 12 communes, 48%). The second possibility is the line (orthogonal to the same axis) established by the four nonsurvivors immediately above the three communes. This second line meets our two criteria for an optimal partition. First, between the lower bound and this line marking the upper bound, 25 communes were observed, 22 (88%) of which survived through Time 3 — some 24 months beyond the initial measures of flux and control at Time 1. And second, on this line and above, 15 communes were observed, 9 (60%) of which had become nonsurvivors by Time 3.

The results of this procedure are shown in the scatterplot for all communes in Fig. 21.9. This scatterplot is identical to the scatterplot at Time 1 in Fig. 21.8 with the following additions: first, the two lines marking the thresholds of lower and upper regions of instability, as just established, are indicated; and second, the survival status for each commune is shown at Time 3 (nonsurvivors are shown as hollow dots in Fig. 21.9). It is clearly evident that the two partitions separate an area of stability in the mid-region from two adjoining regions characterized by a high probability of collective instability; the differences in the rates of instability, by Time 3, between the groups in the three regions are statistically significant (chi-square = 15.641, pr. = .0004). In addition to its extraordinarily high stability over the twenty-four month period from the point of initial measurement, the mid-region also is distinguished by the lack of dispersion of communes along the low control-high flux/high control-low flux axis. Instead, there is a strong tendency to locate between these extremes of rapid (high) flux and rigid (high) control in the area expected to define efficient information processing.

Finally, also shown in Fig. 21.9 are four communes that had a charismatic leader living in residence with the group (circled in Fig. 21.9). Of all communes in the sample, these were the collectives most intent on achieving a radical restructuring of social order. Although there are too few cases for a (statistically) reliable result, all four of these transformation-oriented (charismatic) communes — three of which were still in existence by Time 3 — are concentrated exclusively in the apex of the high flux/high control region; the fifth group (a nonsurvivor) is a noncharismatic commune whose members expressed a strong desire for charismatic leadership as the means to facilitate their efforts at social change. As established elsewhere (Bradley, 1987, pp. 167-193; 264-268), charismatic leadership is not only correlated with enormous increases in flux and control, but when these two are linked in a balanced coupling, charismatic leadership also is associated with an increase in the probability of group survival. However, for other (noncharismatic) collectives, not only were such high levels of flux and control rare, but when these conditions were observed they were found to be highly associated with instability.

5.2. A Multivariate Model of Optimality

To this point, our analysis has employed simple largely bivariate statistical techniques, which, given the small number of cases available, has been both necessary and appropriate. But because it was possible that the optimality we identified was the result of a more complex relationship between flux and control and a number of other sociological factors measured in the original study, a *discriminant function analysis* was conducted to ensure that this was not the case and, therefore, to confirm the veracity of our results.

Two features of discriminant analysis made it especially appropriate: first, the procedure aims to construct a multivariate linear (discriminant) function that maximizes the separation between two or more mutually exclusive categorical groupings of data; second, it offers a test of predictive power by comparing the a priori group classifications against those made by the discriminant function/s. As a measure, thus, of *statistical* optimality, discriminant analysis provides a rigorous means of verifying the finding that, in relation to the other factors examined here, our measures of flux and control provide the best means of predicting optimal collective action.

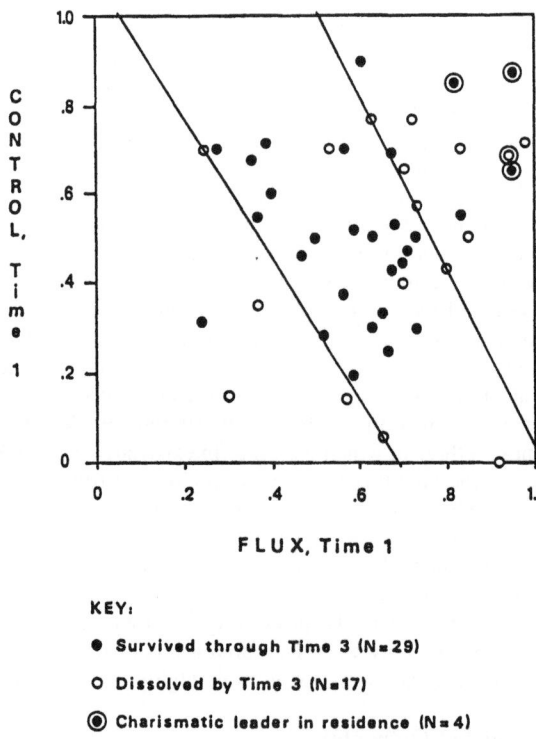

FLUX, Time 1

KEY:

● Survived through Time 3 (N = 29)

O Dissolved by Time 3 (N = 17)

◉ Charismatic leader in residence (N = 4)

Fig. 21.9. Scatterplots of communes on flux and control at Time 1 by stability (survival status) at Time 3, and showing transformational communes (charismatic leader in residence).

To perform the discriminant analysis, the communes were classified into one of the three categories of stability at Time 3 established above, as shown in Fig. 21.9: namely, location in the upper region of instability ($N = 15$; survivors = 6 [40%] communes); location in the stable (mid) region ($N = 25$; survivors = 22 [88%] communes); or location in the lower region of instability ($N = 6$; survivors = 1 [17%] commune). For the purposes of this analysis, we will refer to these three groupings of the communes as *nonoptimal — upper region, optimal — mid region,* and *nonoptimal — lower region,* respectively. Along with our measures of flux and control, eight sociological variables with limited missing data were used as independent variables for the stepwise multivariate analysis. The univariate statistics (means, SDs, Wilks' lambda, and univariate F ratio) are given in Table 21.3a.

Table 21.3a
Discriminant Function: Univariate Statistics

Variable	Optimality Groupings								Wilks' Lambda	Univar- iate F Ratio	Pr.
	Optimal — Mid Region		Nonoptimal — Upper Region		Nonoptimal — Lower Region		Total				
	Mean	SD	Mean	SD	Mean	SD	Mean	SD			
(N)	(25)		(15)		(6)		(46)				
Admission requirements[a]	1.96	.93	1.87	.99	2.00	.89	1.93	.93	.997	.062	.940
Extent of authority[b]	1.52	.51	1.47	.52	1.33	.52	1.48	.51	.985	.326	.723
Affiliated to a larger organization[c]	.48	.51	.33	.49	.33	.52	.41	.50	.978	.484	.620
Control	.457	.178	.688	.136	.286	.231	.510	.218	.603	14.156	.000
Flux	.581	.149	.802	.122	.396	.174	.629	.196	.517	20.055	.000
Mean group age, years	3.36	1.91	2.20	1.08	2.67	1.51	2.89	1.69	.899	2.413	.102
Degree of ideological consensus[d]	1.40	.50	1.47	.52	1.00	.00	1.37	.49	.908	2.174	.126
Mean propn. old members[e]	.46	.29	.48	.34	.41	.35	.46	.31	.995	.117	.890
Formal rules[f]	1.40	.50	1.40	.51	1.50	.55	1.41	.50	.995	.101	.904
Group size[g]	9.20	4.44	8.20	2.08	8.67	2.73	8.80	3.59	.984	.358	.702

Note: SD, standard deviation; Wilks' lambda, U-statistic; pr., statistical significance with 2 and 43 degrees of freedom.
[a]Admission requirements: 1 = if room/see individual; 2 = trait required/group ready; 3 = trial membership/novitiate required/group closed
[b]Extent of authority: 1 = none/a little; 2 = some/a lot
[c]Affiliated to a larger organization: 0 = not affiliated; 1 = affiliated
[d]Degree of ideological consensus: 1 = a little/some; 2 = a lot/unity
[e]Mean propn. old members = proportion of adult members who joined commune before 1973
[f]Formal rules: 1 = none/few; 2 = some/many
[g]Group size = number of adult members (≥ 15 years old)

Table 21.3b. Discriminant function analysis of optimality classification of communes by selected characteristics: Stepwise results and canonical analyses

Variable	Step	Wilks' Lambda	Pr.	Minimum D^2	Pr.	Equivalent F	Pr.
Flux	1	.517	.0000	1.656	.0070	8.011	.0070
Control	2	.214	.0000	6.028	.0000	14.245	.0000

*Maximum significance of F-statistic to enter = .050; minimum significance of F-statistic to remove = .100.

Summary of Stepwise Analysis*
Test of Differences Between Pairs of Groupings After Step 2

Pairs of Groupings	F-statistic	Significance*
Optimal/Nonoptimal — Upper	43.848	.0000
Optimal/Nonoptimal — Lower	14.246	.0000
Nonoptimal — Upper/ Nonoptimal — Lower	64.451	.0000

*With 2 and 42 degrees of freedom.

Canonical Discriminant Functions

	Function 1	Function 2
Canonical Correlation	.887	.002
Squared Canonical Correlation	.786	.0005
Percent of Variance	99.99	.01
Eigenvalue	3.674	.001

Unstandardized Canonical Discriminant Function Coefficients

	Function 1	Function 2
Control	6.080	3.546
Flux	7.652	−3.574
(Constant)	−7.911	.440

Table 21.3c. Discriminant Function ... Classification Results

Actual Group	Predicted Group Optimal — Mid Region N	%	Nonoptimal — Upper Region N	%	Nonoptimal — Lower Region N	%	Total N	%
Optimal — Mid Region	25	100.0	0	0.0	0	0.0	25	100.0
Nonoptimal — Upper Region	1	6.7	14	93.3	0	0.0	15	100.0
Nonoptimal — Lower Region	1	16.7	0	0.0	5	83.3	6	100.0
Total	27	58.7	14	30.4	5	10.9	46	100.0
Prior Probability		.54		.33		.13		100.0

Table 21.3. Discriminant Function Analysis of Optimality Classification Of Communes by Selected Characteristics. (a) Univariate statistics. (b) Stepwise results and canonical analyses. (c) Classification results.

Maximizing the minimum Mahalanobis distance (D^2 — a measure of separation) between the three groupings of communes was the selection rule used for the stepwise multivariate analysis; the statistical significance of the F statistic was used as the criterion to enter (pr. ≤ .050) and remove (pr. ≥ .100) the independent variables. A summary of the stepwise results is presented in the first section of Table 21.3b.

As is clearly evident from the results (see Table 21.3b), the only two variables selected in the stepwise procedure are our two measures of flux and control; all of the other variables fail the selection criteria. Flux, the variable with the strongest discriminating power, was entered into the stepwise analysis at the first step (D^2 = 1.656, pr. = .0070; Wilks' lambda, a multivariate measure of association between groups = .517; pr. = .0000). At Step 2, control was entered as the next most powerful discriminating variable (D^2 = 6.028: pr. = .0000.) Wilks' lambda decreased substantially (to .214; pr. =.000), indicating that only a low association between the three groupings of communes remained. The (F-statistic) test, after Step 2, of the differences between each pair of groupings shows that there are moderate (14.246) to strong (64.451) differences between each pair of groupings that cannot be explained by chance.

The rest of Table 21.3b provides information on the nature and statistical power of the two canonical (multivariate) discriminant functions formed by flux and control; a minimum of two discriminant functions is required to discriminate among three groupings of data. (The canonical discriminant functions are statistically independent of each other; each is a linear combination of the variables entered in the stepwise analysis — similar to a multiple regression equation, and should be thought of as a latent variable (not measured directly), a statistical artifact comparable to a factor constructed by factor analysis.) Comparing the statistical information on the two discriminant functions shows that first function possesses much greater discriminating power than the second function. The canonical correlations (a measure of the association between the discriminant scores and the groupings) are .887 and .022, respectively, and indicate that first function possesses most of the discriminating power — approximately 79% compared to .05% (squared canonical correlation = .786 and .0005, respectively). This is confirmed by the huge difference in the eigenvalues for the two functions, 3.674 versus 0.001.

Table 21.3b also presents the unstandardized canonical discriminant function coefficients, which were used to compute the discriminant scores (one for each discriminant function) for each case. The two discriminant scores were then used to classify individual cases into one of the three optimality groupings of communes established prior to the discriminant analysis. Comparing the a priori grouping to the posterior classification provides a means of determining the predictive power of the two discriminant functions in correctly assigning cases.

The results in Table 21.3c show that the two discriminant functions are able to correctly predict the optimality grouping for each commune in 44 of 46 cases, an overall success rate of 96%. Thus, all 25 (100%) of the communes belonging to the optimal — mid region category were correctly classified, 14 (93%) of the 15 in the nonoptimal — high region grouping were correctly classified, and 5 (83%) of the 6 communes in the nonoptimal — low region grouping were correctly classified. Moreover, these prediction rates are substantially higher than the prior probabilities of commune membership in these groupings (.54, .33, and .13, respectively).

The results of the discriminant function analysis confirm our conclusion based on more simple statistical procedures: namely, that flux and control are predictive of optimal collective action. The fact that the multivariate results show no evidence of any statistically significant latent relationships between our measures of flux, control, optimality, and the other sociological variables we examined is particularly noteworthy, for it flies in the face of conventional sociological theory (see Turner, 1986). This would suggest that location in the region of optimal stability may have its basis in a different logic and dynamics than that embodied by current sociological thinking (e.g., Burt, 1992; Coleman, 1990; White, 1992). It is toward an understanding of these dynamics and their implications that the following discussion is directed.

6. DISCUSSION

6.1. Model of Global Communication

Drawing on the theory and empirical results presented above, a model of the communicative structure of the collective was constructed (see Fig. 21.10). In the terms of this model, the communicative structure is formed by the interaction of networks of endogenous relations organized along two dimensions in which the values allocated in each dimension define points within a relational field (Bradley & Roberts, 1989a). The values ascribed to the horizontal dimension represent flux, the amount of activation of potential energy in a social collective. The values ascribed to the vertical dimension represent the amount of control (the degree to which individuals are interconnected by a transitive network of relations) exercised at that location. The coordinates representing the dimensions bound a phase space within which each value represents an amount of information (in Gabor's terms) characteristic of the communicative structure of the collective.

Two regions of stability can be distinguished within the phase space. These are regions associated with viable patterns of global communication. They are located within a larger region in which the minimum values for global communication are not met so that various forms of collective dysfunction result.

All regions are separated from each other, marked, in the terms of nonlinear dynamics, by a phase transition from psychosocial instabilities to [far-from-(physical)-equilibrium] psychosocial stabilities in collective organization (Prigogine & Stengers, 1984). The region of optimal function represents, therefore, a qualitative change in psychosocial organization. The phase transition from dysfunctional to stable collective forms (which includes the area between the two stable regions) is described by fluctuations in potential and control that end in a point (the bifurcation point) where the patterns of energy activation and expenditure no longer dissipate into the environment — no longer average out to equal the energy levels of the surrounding context, but coalesce to crystallize as an emergent stable collective order. To defy the tendency toward entropy (disorder), and sustain the stable collective order, requires minimizing the fluctuations by linking the activation of potential to the control operations so that the energy expenditure of all members is *in*-formed in relation to the collective's action.   In the dysfunctional region, the patterns of potential and control are therefore either unable to establish or unable to sustain stable forms of collective organization. Values of low potential and low control (the area labeled as *insufficiency* in Fig. 21.10) fail to provide stability because, in addition to requiring a certain minimum of kinetic energy, stability also requires at least a minimum of direction given to that energy. As shown above, this direction comes from the interaction between flux and control that in-forms the paths by which kinetic energy is expended in action. Thus, in terms of the data presented in Fig. 21.1, stable organization requires *both* a minimum of flux *and* a minimum of control: a network of reciprocal *equi*-valent connections linking every individual to at least one other person; this connection must be coupled to a transitive ordering of asymmetric relations linking the action of each individual to that of at least one other person. Failing to meet these minima, a collective would devolve into a loose aggregation of disjointed cliques and isolated individuals unable to communicate and, consequently, function as a social collective.

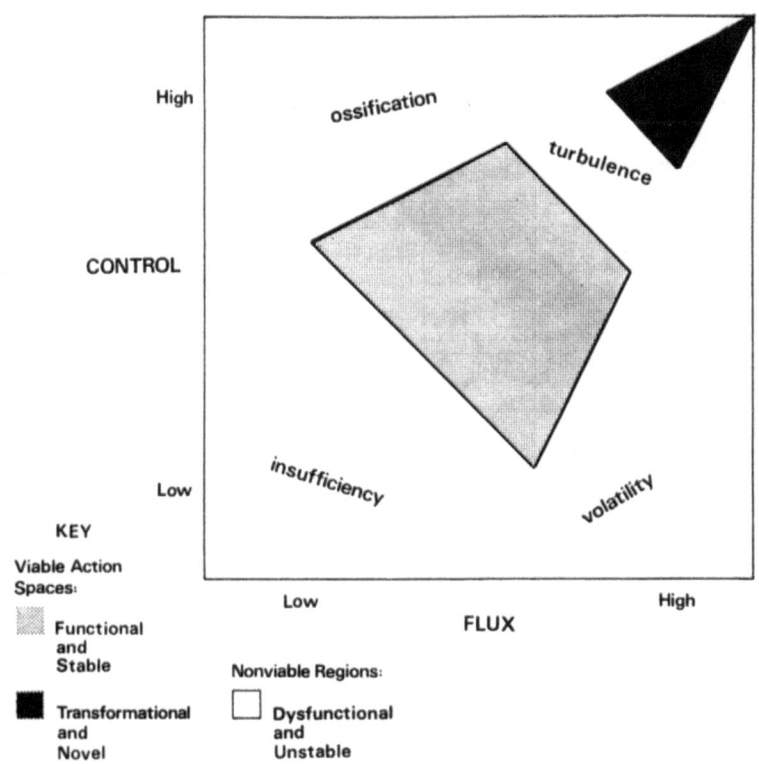

Fig. 21.10. Model of dynamics of communication and collective action.

Two other combinations are also shown to produce instability. Coordinate values representing high control and low flux (labelled *ossification* in Fig. 21.10) delineate a rigid organization in which insufficient flux is available for global communication. The lack of communication means that the paths to action are fixed, not adequately informed by current circumstances, and are therefore unable to adapt and evolve as the situation changes.

At the other extreme, combinations of high flux and low control (labelled *volatility*) delineate a turbulent situation in which little of the enormous flux is guided by hierarchic controls.   Communication is inadequate as insufficient information about the ever-changing situation is distributed.

The region of dysfunction surrounds the region of optimal collective function which is centered along a main diagonal of the phase space, and which, as noted, embodies a qualitative change in psychosocial organization. The lower and upper boundaries of this region define the values representing efficient information processing; this region is consistent with thermodynamically inspired connectionist models of neural networks (e.g., Hinton & Sejnowski, 1986; Hopfield, 1982). In such models efficient pattern matching is found to occur in a region between total randomness and total organization: in our terms, between rapid flux and rigid organization. The relationship between flux and control narrows progressively from many degrees of freedom at the low end of the phase space, to an almost one-to-one correspondence at the high end. Thus the shape of the space of optimal function is triangular.   Figure 21.10 shows that this space can be subdivided into two distinct kinds of global communication: functional and transformational. Transitions from one subregion to another are not gradual but involve qualitative change; distinct types of communication can be defined. In between subregions is a phase transition characterized by turbulence and instability. Each subregion is composed of different combinations of values of flux and control so that a social collective can only have one of these communication patterns at any given time.

Furthermore, there is considerable difference in vulnerability to collective dysfunction between the patterns constituting the subregions. At the low end of the functional subregion, the range of combinations of flux and control is great and there are thus many different viable patterns of communication possible. As a result of this loose articulation between flux and control, communication tends to be effortless but minimally efficient. At the high end of this subregion, there is a close articulation between flux and control so that the patterns of information processing here tend to be optimal — maximally efficient and highly stable.

Beyond this, at the apex of the viable region, is a small subregion, labeled *transformational* (separated from the region of stability by a turbulent gap), defined by an almost one-to-one ratio between flux and control. To assure stability this ratio must be maintained, a not-so-easy task: the greater the flux the more control must be exercised and vice versa, taking much effort. Often, when such an effortful course is in operation, a sudden organizational spasm occurs. The spasm has two possible outcomes. One is a structural transformation in the pattern of information processing, resulting in total reorganization to form a novel, qualitatively different collective. The other is structural devolution, the complete breakdown and collapse of the collective as a viable social entity.

7. SUMMARY AND IMPLICATIONS

7.1. Efficiency of Communication and Optimal Collective Action

Our model concerns the internal structure of the collective. This internal structure is conceived to be based on the biological potential of the individuals composing the collective to engage in work, measured as energy. This biological potential appears to be heterarchically organized and, when activated by the collective, is made available for social interaction and behavior as a pool or field of latent biosocial energy. We have labeled this dimension of the endogenous order *flux*.

In the other dimension, individuals are connected hierarchically. We have labeled this dimension *control* because it appears to direct and regulate the activation of the biosocial energy of the collective. Controls over the activation and distribution of flux result in global communication by way of quantum-like units of information (logons) — moment-by-moment descriptions, in terms of space-time and spectral coordinates, of the collective's endogenous organization. Because these elementary units of information overlap as a series, the collective's order at a given moment is informed by the order implicit in the units of the succeeding moments.

A simplifying assumption was that stability can be identified with survival. Unless the collective remains a stable, durable social entity, there is little to enquire about. Thus in order to understand how stability is accomplished, we have restricted our concern to the structure and internal dynamics of the collective, and have left aside its behavioral effectiveness as an entity operating on its environment. It may well be that less stable collectives could be more effective under certain conditions than hyperstable ones (Roberts & Bradley, 1988).

The efficiency of the internal dynamics, and its relationship to the collective's action, was found to display an optimal (energy conserving) combination of flux and control that is associated with stable collective action. Our results thus show that for the group to survive as an effective working unit, an efficient communicative structure was required. Only those configurations of heterarchy and hierarchy that produced a path of least action — one that entailed the smallest amount of turbulence — resulted in a stable, effective collective. The findings indicate, in the terms of Shannon and Weaver (1949) at the opening of this chapter, that heterarchy and hierarchy are related as conjugate orders: that within the limits of its biosocial energy, the social collective must achieve an optimal combination of flux and control to produce efficient communication.

A final point concerns the implications of this multilevel investigation for the single-level approach generally pursued by social science. In contrast to such discipline-specific accounts of collectives, our approach has been to seek a common scientific language — to explore the degree to which insights and principles from thermodynamics and information measurement theory could be used to build a rigorous and testable understanding of the communicative structure of social collectives. Although this account has drawn little from (and found little empirical support for) the normative sociological concepts usually employed to explain the behavior of social collectives (viz., ideology, values, formal organization, role structure, leadership, member commitment etc.; see the review by Turner, 1986, or Jablin et al., 1987, for examples), we believe that our results demonstrate the utility of an approach that grounds explanation in the commonalities and dynamics of collectives more generally. This is *not* to say that the factors identified by single-level descriptions are unimportant, but rather that a general understanding of collectives will only arise as a result of complementary work on the interlinkages between systems of organized behavior at different levels (e.g., Csányi, 1989). In this way the relationship between general principles of system behavior and the specific conditions that obtain at different structural levels will be addressed and a general science of social behavior will be born.

ACKNOWLEDGMENTS

We would like to thank Vadim Glezer, Daniel Levine, Peter Reynolds, Nancy Roberts, G. Rosenstein, and an anonymous reviewer for their helpful comments on an earlier draft of this manuscript; Winnie Y. Young assisted with the data analysis. A previous version was presented to the Fourth Holonomic Processes in Social Systems Conference, 1989; the conference was held at, and kindly sponsored by, the Esalen Institute, Big Sur, CA.

REFERENCES

Abraham, F. D. (1991). *A Visual Introduction to Dynamical Systems Theory for Psychology.* Santa Cruz, CA: Aerial Press.
Ashby, W. R. (1956). *An Introduction to Cybernetics.* London: Chapman & Hall.
Bales, R. F. (1958). Task roles and social roles in problem-solving groups. In E. E. Maccoby, R. M. Newcomb, & E. L. Hartley (Eds.), *Readings in Social Psychology.* New York: Holt.
Bishop, M. (1995). Paradoxical strife: Science and society. *Bulletin of the American Academy of Arts and Sciences,* Vol. XL, No. 4: 10-30.
Bohm, D., & Hiley, B. J. (1993). *The Undivided Universe.* London: Routledge.
Bohr, N. (1921a). "Letter" in *Nature,* **107,** 104.
Bohr, N. (1921b). "Letter" in *Nature,* **108,** 208.
Bougon, M. (1983). Uncovering cognitive maps: the self-Q technique. In G. Morgan (ed.), *Beyond Method: Strategies for Social Research* (pp. 173-188). New York: Sage.
Bradley, R. T. (1987). *Charisma and Social Structure: A Study of Love and Power, Wholeness and Transformation.* New York: Paragon House.
Bradley, R. T. (1996). The anticipation of order in biosocial collectives. *World Futures: The Journal of General Evolution,* in press.
Bradley, R. T., & Roberts, N. C. (1989a). Relational dynamics of charismatic organization: The complementarity of love and power. *World Futures: The Journal of General Evolution,* **27,** 87-123.
Bradley, R. T., & Roberts, N. C. (1989b). Network structure from relational data: Measurement and inference in four operational models. *Social Networks: An International Journal of Structural Analysis,* **11,** 89-134.
Burns, T. & G. M. Stalker (1961). *The Management of Innovation.* London: Tavistock.
Burt, R. S. (1992). *Structural Holes: The Social Structure of Competition.* Cambridge, MA: Harvard University Press.
Cherry, C. (1966). *On Human Communication: A Review, a Survey, and a Criticism.* Cambridge, MA: MIT Press.
Coleman, J. S. (1990). *Foundations of Social Theory.* Cambridge, MA: Harvard University Press.
Collins, R. (1975). *Conflict Sociology: Toward an Explanatory Science.* New York: Academic Press.

Collins, R. (1990). Stratification, emotional energy, and the transient emotions. In T. D. Kemper (Ed.), *Research Agendas in the Sociology of Emotions* (pp. 27-57). Albany: State University of New York Press.

Considine, D. M. (Ed.) (1976). *Van Nostrand's Scientific Encyclopedia, Fifth Edition.* New York: Van Nostrand Reinhold.

Csányi, V. (1989). *Evolutionary Systems and Society: A General Theory of Life, Mind, and Culture.* Durham, NC, and London: Duke University Press.

Durkheim, E. (1938). *The Rules of Sociological Method.* Eighth Edition (S. A. Solovay & J. H. Mueller, Trans.; G. E. G. Catlin, Ed.). New York: Free Press. (Original work published 1895.)

Durkheim, E. (1965). *The Elementary Forms of the Religious Life.* New York: The Free Press. (Original work published 1915.)

El Sawy, O. A. (1985). *From separatism to holographic enfolding: the evolution of the technostructure of organizations.* Unpublished paper presented to the TIM/ORSA Conference, Boston.

Forrester, J. W. (1968). *Principles of Systems,* 2nd ed. Cambridge, MA: MIT Press.

Freeman, L. C., White, D. R., & Romney, A. K. (Eds.) (1989). *Research Methods in Social Network Analysis.* Fairfax, VA: George Mason University Press.

Gabor, D. (1946). Theory of communication. *Journal of the Institute of Electrical Engineers,* **93,** 429-457.

Garud, R., & Kotha, S. (1994). Using the brain as a metaphor to model flexible production systems. *Academy of Management Review,* **19,** 671-698.

Glazer, R. (1986). *A holographic theory of decision making.* Unpublished paper, Graduate School of Business, Columbia University, New York.

Hartley, R. V. L. (1928). Transmission of information. *Bell System Technical Journal,* **7,** 535.

Hinton, G. E., & Sejnowski, T. J. (1986). Learning and relearning in Boltzmann machines. In D. E. Rumelhart & J. L. McClelland (Eds.), *Parallel Distributed Processing: Explorations in the Microstructure of Cognition* (Vol. 1, pp. 282-317). Cambridge, MA: MIT Press.

Holland, P. W., & Leinhardt, S. (1976). Local structure in social networks. In D. R. Heise (Ed.), *Sociological Methodology 1976* (pp. 1-45). San Francisco: Jossey-Bass.

Hopfield, J. J. (1982). Neural networks and physical systems with emergent collective computational abilities. *Proceedings of the National Academy of Sciences,* **79,** 2554-2558.

Hutchins, E. (1991). The social organization of distributed cognition. In L. B. Resnick, J. M. Levine, & S. D. Teasley (Eds.), *Perspectives on Socially Shared Cognition* (pp. 283-307). Washington, DC: American Psychological Association.

Jablin, F. M., Putnam, L. L., Roberts, K. H., & Porter, L. W. (Eds.) (1987). *Handbook of Organizational Communication.* Newbury Park, CA: Sage.

Kupfmueller, K. (1924). Über Einschwingvorgange in Wellenfiltern. *Elektronische Nachrichten-Tech,* **1,** 141.

Lacey, R. (1986). *Ford: The Man and the Machine.* Boston & Toronto: Little, Brown.

Laumann, E. O., Marsden, P. V., & Prensky, D. (1982). The boundary specification problem in network analysis. In R. S. Burt & M. J. Minor (Eds.), *Applied Network Analysis: A Methodological Introduction* (pp. 18-34). Beverly Hills, CA: Sage.

Legasto, A. A., Forrester, J. W., & Lyneis, J. M. (Eds.) (1980). *Studies in the Management Sciences: System Dynamics.* New York: North-Holland.

Lewis, A. C. (1977). Mathematics to represent thought: A review of Grassmann's algebra. *Annals of Science,* **34**, 104.

MacKenzie, K. D. (1991). *The Organizational Hologram: The Effective Management of Organizational Change.* Norwell, MA: Kluwer.

McClelland, J. L., Rumelhart, D. E., & the PDP Research Group (1986). *Parallel Distributed Processing: Explorations in the Microstructure of Cognition, Vol. 2: Psychological Models.* Cambridge, MA: MIT Press.

McFarland, D. J. (1971). *Feedback Mechanisms in Animal Behavior.* New York: Academic Press.

Miller, J. (1968). *The Nature of Living Systems: An Exposition of the Basic Concepts in General System Theory.* Washington, DC: Academy for Educational Development.

Morgan, G., & Ramirez, R. (1984). Action learning: A holographic metaphor for guiding social change. *Human Relations,* **37**, 101-106.

Nicolis, G., & Prigogine, I. (1977). *Self-Organization in Nonequilibrium Systems: From Dissipative Structures to Order Through Fluctuation.* New York: Wiley-Interscience.

Nyquist, H. (1924). Certain factors affecting telegraph speed. *Bell System Technical Journal,* **3**, 324.

Pribram, K. H. (1991). *Brain and Perception: Holonomy and Structure in Figural Processing.* Hillsdale, NJ: Lawrence Erlbaum Associates.

Pribram, K. H., & McGuinness, D. (1975). Arousal, activation and effort in the control of attention. *Psychological Review,* **82**, 116-149.

Prigogine, I., & Stengers, I. (1984). *Order out of Chaos: Man's New Dialogue with Nature.* New York: Bantam Books.

Roberts, N. C., & Bradley, R. T. (1988). Limits to charisma. In J. Conger & R. Kanungo (Eds.), *Charismatic Leadership: The Elusive Factor in Organizational Effectiveness* (pp. 253-275). San Francisco: Jossey-Bass.

Roethlisberger, F. J., & Dickson, W. J. (1939). *Management and the Worker.* Cambridge, MA: Harvard University Press.

Rogers, E. M., & Kincaid, D. L. (1981). *Communication Networks: Towards a New Paradigm for Research.* New York: Free Press.

Rumelhart, D. E. (1992). Towards a microstructural account of human reasoning. In S. Davis (Ed.), *Connectionism: Theory and Practice* (pp. 69-83). New York: Oxford University Press.

Rumelhart, D. E., McClelland, J. L., & the PDP Research Group (1986). *Parallel Distributed Processing: Explorations in the Microstructure of Cognition, Vol. 1: Foundations.* Cambridge MA: MIT Press.

Sandelands, L. E., & Stablein, R. E. (1987). The concept of organization mind. In S. Bacharach & N. DiTomaso (eds.), *Research in the Sociology of Organizations* (Vol. 5, pp. 135-161.) Greenwich, CT: JAI.

Selznick, P. (1948). Foundations of the theory of organization. *American Sociological Review,* **13**, 25-35.

Shannon, C. E. (1949). The mathematical theory of communication. In C. E. Shannon & W. Weaver, *The Mathematical Theory of Communication* (pp. 3-91). Urbana: University of Illinois Press.

Shannon, C., & Weaver, W. (1949). *The Mathematical Theory of Communication*. Urbana: University of Illinois Press.

Trist, E. L., & Bamforth, K. W. (1951). Some social and psychological consequences of the Longwal method of goal setting. *Human Relations*, **4**, 3-38.

Turner, J. H. (1986). *The Structure of Sociological Theory*. Chicago: Dorsey Press.

von Bertalanffy, L. von (1969). *General System Theory: Foundations, Development, Applications*. New York: G. Braziller.

Weick, K. E., & Roberts, K. M. (1993). Collective mind in organizations: Heedful interrelating on flight decks. *Administrative Science Quarterly*, **38**, 357-381.

White, H. C. (1992). *Identity and Control: A Structural Theory of Social Action*. Princeton, NJ: Princeton University Press.

White, H. C., Boorman, S. A., & Breiger, R. (1976). Social structure from multiple networks. I. Block models of roles and positions. *American Journal of Sociology*, **81**, 730-780.

Zablocki, B. D. (1980). *Alienation and Charisma: A Study of Contemporary Communes*. New York: Free Press.

APPENDIX

Summary Statistics for the Operational Procedures Used to Measure Flux, Control, and Stability
(N = 46 communes)

	Mean Dyadic Density[1]	Triadic Structure (Mean Proportions) Symmetric Triad Types				
Flux		003	102	201	300	Total
Loving (L)	.44	.260	.341	.208	.192	1.001
Improving (I)	.46	.232	.348	.224	.196	1.000
Exciting (E)	.17	.622	.285	.067	.027	1.001
Mean (L+I+E/3)	.36	.371	.325	.166	.138	1.000

	Mean Dyadic Density	Asymmetric Triad Types							
Control		003	012	021D	021U	021C	030T	030C	Total
Power	.30	.097	.261	.137	.113	.129	.243	.020	1.000

Stability: Survival status, Time 1 - Time 5 (12 month intervals)

	T1 1974		T2 1975		T3 1976		T4 1977		T5 1978		Total	
	N	%	N	%	N	%	N	%	N	%	N	%
Survived	46	100	35	76	29	83	24	83	22	92	22	48
Disintegrated	0	0	11	24	6	17	5	17	2	8	24	52
Total	46	100	46	100	35	100	29	100	24	100	46	100

---

[1] Number of relations of a selected dyad type / all possible relations. For the three indicators of flux (*loving*, *improving*, and *exciting*) the numerator was the number of relations formed as a dyad of positively reciprocated relations (i.e., both *i* and *j* answered "yes"); for the indicator of control (*power*) the numerator was the number of dyads for which an asymmetric ordering was evident in the relationship between *i* and *j* (i.e., either *i* had greater power in the relationship than *j*, or *j* had greater power than *i*).

Author Index

Subject Index

A

Adaptation, 80, 81, 85, 231, 232, 235, 236, 239, 257, 366
Adaptive critic, 28
Adaptive recognition, 106
Adaptive resonance theory (ART), 14, 50, 51, 110, 232, 288, 291, 294, 295, 297, 309, 363, 374, 375, 380
Adaptive threshold rules, 288, 293, 301, 308, 313
Adjunctive behavior, 398, 415
Advanced computing, 34
Affect (see *Emotion*)
Affordance, 112, 113, 117
Amphetamine, 403, 404, 406
Amygdala, 9, 12, 114, 118, 408
Analogies, 7, 8
Anhedonia, 398, 409
Anthropic principle, 119
Anticipation, 115
Aplysia, 265, 266, 274
Appropriateness, 9, 15
    (see also *Proprieties*)
Arithmetic, learning by humans, 29
Artificial intelligence (AI), 162, 206, 364, 365, 424
    attempt to reduce it to classical physics, 106
    strong, 127, 134
    strong versus weak, 126
    underground in, 45
Artificial life, 60, 73
Artificial neural networks
    (see *Neural networks*)
ARTMAP, 291
Associative memory, 212, 318, 320, 321
Atrophy due to disuse, 288, 289, 291-293, 296, 298, 299, 302, 308, 309

Attentive scanning, 369, 371, 374
Attractors, 5, 13, 156
Autoassociator, 317-327
    classical, 318-321, 327, 331-334
    generalized, 317-320, 322-327, 331-334
Automatic control, 46
Automaticity, 44, 46, 50, 53
Avoidance learning, 387

B

Back propagation, 5, 28, 31, 32, 35, 38, 40, 50, 51, 82, 147, 153-155, 187, 232, 263, 285, 289, 293, 331
    at neuronal level, via nitric oxide, 232
Basal ganglia, 118
Behaviorism, 23, 365
Beliefs, 442
Belief structure, 149, 151
Bias, 364, 383, 410
    among features, 317, 319, 323, 325
Bias-variance dilemma, 364
Blocking, 230, 235, 237, 239, 247-252, 259, 274, 410
Body resistance, 415, 416
Boltzmann machine, 10, 139, 155, 218
Boolean functions, 273, 278-280
Boolean satisfiability, 203, 208, 209, 217, 221
Boundary completion, 339
BP-SAM, 52
Brain
    ability to assemble disparate features, 339
    analogous to social collective, 453
    computational power of, 131, 132, 135
    computer simulation of, 130-132
    reward system of, 399-410, 413-415, 417
Brain-state-in-a-box
    (BSB), 153, 155, 156